LANDOLT-BÖRNSTEIN

Numerical Data and Functional Relationships
in Science and Technology

New Series
Editor in Chief: K.-H. Hellwege

Group V: Geophysics and Space Research

Volume 1

Physical Properties of Rocks

Subvolume a

V. Čermák · H.-G. Huckenholz · L. Rybach · R. Schmid
J. R. Schopper · M. Schuch · D. Stöffler · J. Wohlenberg

Editor: G. Angenheister

Springer-Verlag Berlin · Heidelberg · New York 1982

LANDOLT-BÖRNSTEIN

Zahlenwerte und Funktionen aus Naturwissenschaften und Technik

Neue Serie
Gesamtherausgabe: K.-H. Hellwege

Gruppe V: Geophysik und Weltraumforschung

Band 1

Physikalische Eigenschaften der Gesteine

Teilband a

V. Čermák · H.-G. Huckenholz · L. Rybach · R. Schmid
J. R. Schopper · M. Schuch · D. Stöffler · J. Wohlenberg

Herausgeber: G. Angenheister

Springer-Verlag Berlin · Heidelberg · New York 1982

CIP-Kurztitelaufnahme der Deutschen Bibliothek

Zahlenwerte und Funktionen aus Naturwissenschaften und Technik/Landolt-Börnstein. - Berlin, Heidelberg; New York: Springer.
Parallelt.: Numerical data and functional relationships in science and technology

NE: Landolt, Hans [Begr.]; PT. N.S./Gesamthrsg.: K.-H. Hellwege. N.S., Gruppe 5, Geophysik und Weltraumforschung. N.S., Gruppe 5, Bd. 1. Physikalische Eigenschaften der Gesteine/Hrsg.: G. Angenheister. N.S., Gruppe 5, Bd. 1, Teilbd. a./V. Čermák ... – 1982.

ISBN 3-540-10333-3 (Berlin, Heidelberg, New York);
ISBN 0-387-10333-3 (New York, Heidelberg, Berlin)

NE: Angenheister, Gustav [Hrsg.]; Čermák, Vladimir [Mitverf.]; Hellwege, Karl-Heinz [Hrsg.]

This work is subject to copyright. All rights are reserved, whether the whole or part of the material is concerned specifically those of translation, reprinting, reuse of illustrations, broadcasting, reproduction by photocopying machine or similar means, and storage in data banks.

Under § 54 of the German Copyright Law where copies are made for other than private use a fee is payable to 'Verwertungsgesellschaft Wort' Munich.

© by Springer-Verlag Berlin-Heidelberg 1982

Printed in Germany

The use of registered names, trademarks, etc. in this publication does not imply, even in the absence of a specific statement, that such names are exempt from the relevant protective laws and regulations and therefore free for general use.

Typesetting, printing and bookbinding: Universitätsdruckerei H. Stürtz AG Würzburg

2163/3020 – 543210

Editor

G. Angenheister
Institut für Allgemeine und Angewandte Geophysik und Geophysikalisches Observatorium, Ludwig-Maximilians-Universität, München, FRG

Contributors

V. Čermák
Českoslovenká Akademie VED, Geofyzikální Ústav, Praha – Spořilov, ČSSR

H.-G. Huckenholz
Mineralogisch-Petrographisches Institut, Ludwig-Maximilians-Universität, München, FRG

L. Rybach
Institut für Geophysik, Eidgenössische Technische Hochschule, Hönggerberg, Zürich, Switzerland

R. Schmid
Institut für Kristallographie und Petrographie, Eidgenössische Technische Hochschule, Zürich, Switzerland

J. R. Schopper
Institut für Geophysik, Technische Universität Clausthal, Clausthal-Zellerfeld, FRG

M. Schuch
Bayerische Landesanstalt für Bodenkultur u. Pflanzenbau, München, FRG

D. Stöffler
Institut für Mineralogie der Westfälischen Wilhelms-Universität, Münster, FRG

J. Wohlenberg
Lehrgebiet für Angewandte Geophysik, Rheinisch-Westfälische Technische Hochschule, Aachen, FRG

Vorwort

Im Tabellenwerk „Landolt-Börnstein" erschien die Geophysik zum ersten Mal im Jahre 1952 zusammen mit der Astronomie in einem von P. ten Bruggencate (Astronomie) und J. Bartels (Geophysik) gemeinsam herausgegebenen Band der 6. Auflage (Band III). In der „Neuen Serie" des Landolt-Börnstein sind dagegen von vornherein zwei getrennte Buch-Gruppen, eine für Astronomie und Astrophysik (Gruppe VI) und eine für Geophysik und Weltraumforschung (Gruppe V) vorgesehen.

Die beiden ersten Bände der Gruppe V enthalten die geophysikalischen Daten der festen Erde. (Die Hydro- und Atmosphäre werden später in anschließenden Bänden dargestellt.) Im Band V/1 werden die physikalischen Eigenschaften der Gesteine behandelt, im Band V/2 die der Erde als Ganzes. Der Band V/1 enthält also die im Labor an Gesteins-Proben gemessenen Daten. Im Band V/2 sind dagegen vorwiegend die Werte der physikalischen Größen des Erdinnern, also des Materials in situ aufgeschrieben, die durch Messungen an oder dicht an der Erdoberfläche und mit Beachtung weiterer Beobachtungen, z.B. der Messungen im Labor, abgeleitet worden sind. Man spricht in diesem Fall von in-situ-Bestimmungen, in-situ-Messungen etc. Zum Beispiel findet man im Band V/1 die Dichte der Gesteine, gemessen an Proben, die man im Gelände, gegebenenfalls in Steinbrüchen oder im Bergwerk oder durch Bohrungen, erhalten hat. Im Band V/2 findet der Leser dagegen die Dichte im Inneren der Erde als Funktion der Tiefe, also als Ergebnis von in-situ-Messungen.

Im Band V/1 sind die Daten nach den physikalischen Größen wie Dichte, Porosität, elektrische Leitfähigkeit etc. geordnet. Von dieser Einteilung sind nur das Eis und die Gesteine des Mondes ausgenommen, für die die Werte aller physikalischen Größen in zwei getrennten Kapiteln zusammengestellt sind. Dieses wird den Lesern, die sich nur für diese Materialien interessieren, das Finden der gesuchten Werte erleichtern.

Die physikalischen Eigenschaften der Minerale sind nur soweit behandelt, wie es für das Verständnis der physikalischen Eigenschaften der Gesteine erforderlich ist. Auch sind nicht alle physikalischen Eigenschaften der Gesteine behandelt worden. Bevorzugt sind diejenigen Größen, die auch durch in-situ-Messungen bestimmt werden können.

Für die vollständige Beschreibung der physikalischen Eigenschaften der Gesteine benötigt man auch die Terminologie der Petrographie und der Mineralogie. Am Anfang des Bandes V/1 sind daher die Gesteins-Einheiten und deren Namen beschrieben, die man an oder dicht an der Erdoberfläche findet und die man im tieferen Erdinneren vermutet. Es hat sich gezeigt, daß die für die Gesteine der Erde entwickelte Terminologie nur um wenige Begriffe erweitert werden mußte, um die Gesteine der Mondoberfläche beschreiben zu können. Es darf vermutet werden, daß auch bei fortschreitenden Kenntnissen von den Oberflächen und von den inneren Strukturen der festen Planeten und der festen Satelliten die gegenwärtige Terminologie der Petrographie und Mineralogie zum größten Teil verwendet werden kann.

Die Gesteine der Erdoberfläche bestehen aus Körnern verschiedener Minerale. Die Durchmesser der Körner sind meist kleiner als 1 cm. Sie können aber auch kleiner als 10 μm sein. Sie sind durch schmale Korngrenzen voneinander getrennt. Die innere Struktur der intergranularen Grenzen ist wenig bekannt. Typisch für die Gesteine ist auch, daß sie kleine Hohlräume in Form von Poren z.B. zwischen den Körnern oder in Form von feinen Rissen enthalten. Oft haben die Körner eine interne, nur mit dem Mikroskop erkennbare Struktur. Kann man mit einem stark vergrößernden Mikroskop keine Korngrenzen und keine internen Strukturen erkennen, so wird dieses Material oft zunächst als

glasige Grundmasse bezeichnet. Ob eine solche glasige Grundmasse, die die Körner umschließt, eine submikroskopische Ordnung, z.B. in Nestern, enthält, oder kryptokristallin ist, kann nur durch weitere Untersuchungen entschieden werden.

Man nimmt allgemein an, daß die Porosität in einer Tiefe von 20 bis 30 km sehr klein ist. Vermutlich haben die Gesteine aber in dieser Tiefe noch eine Kornstruktur. Dies gilt auch, jedoch mit einigen Modifikationen, für partiell geschmolzene Gesteine. Es ist aber unbekannt, bis in welche Tiefe die Kornstruktur existiert, oder – falls sie existiert – welcher Art die Kornstruktur ist. Es ist also noch fraglich, ob das Material des Erdmantels, insbesondere des unteren Erdmantels, als Gestein im oben definierten engeren Sinn bezeichnet werden kann. Es können daher viele der in diesem Band aufgeschriebenen Daten physikalischer Größen nicht ohne ausführliche Diskussion und nur mit Beachtung der Hochdruck- und Hochtemperatur-Experimente auf das tiefere Erdinnere übertragen werden. Für einige Größen findet man diese Diskussion in diesem Band.

Bei der Erforschung der festen Planeten und der festen Satelliten (z.B. des Mondes) wird man die gleichen physikalischen Größen zu messen haben wie bei der Erforschung der festen Erde. Man wird auch zum großen Teil die gleichen Meßprinzipien anwenden können, jedoch in anderer technischer Ausführung. Will man z.B. Seismographen auf der Oberfläche des Mondes aufstellen, so hat man die große Temperaturvariation, die geringe Schwere, den lockeren Boden etc. zu beachten. Obwohl die Meßprinzipien die gleichen sind, so ist doch zu erwarten, daß die Zahlenwerte der physikalischen Größen der festen Planeten und der festen Satelliten sehr verschieden sind von denen der Erde. Für den Mond ist dies bereits bekannt, da sowohl Messungen an Mondproben als auch in-situ-Messungen auf dem Mond durchgeführt worden sind.

Aus Umfangs- und Zeitgründen ist der Band V/1 in zwei Teilbände V/1a und V/1b aufgeteilt worden. Teilband V/1a wird hiermit vorgelegt, Teilband V/1b erscheint Anfang 1982. Es muß als ein Gewinn für die Physik und für die Geowissenschaften gewertet werden, daß im Landolt-Börnstein ein so umfangreicher Band für das Spezialgebiet „physikalische Eigenschaften der Gesteine" zur Verfügung gestellt werden konnte.

Zu danken ist zunächst den Autoren, insbesondere für die Sammlung der Daten, dann dem Springer-Verlag und der Landolt-Börnstein Redaktion. In großzügiger Weise und mit viel Geduld wurden die Wünsche der Autoren erfüllt. Mit besonderer Sorgfalt hat Frau H. Weise die Vorbereitung der Manuskripte, die Korrekturen und Umbrüche besorgt.

Wie alle Bände des Landolt-Börnstein ist auch dieser ohne finanzielle Unterstützung von anderen Stellen veröffentlicht worden.

München, September 1981 **Der Herausgeber**

Preface

The first publication concerning geophysical data within the Landolt-Börnstein Tables appeared 1952 in a volume (vol. III of the 6th Edition) concerning Geophysics as well as Astronomy; it was edited by J. Bartels (Geophysics) and P. ten Bruggencate (Astronomy). The amount of data accumulated since then makes it necessary to separate these fields into two groups in the "New Series" of Landolt-Börnstein, namely group V for Geophysics and Space Research and group VI for Astronomy and Astrophysics.

The geophysical data of the solid earth are compiled in the first two volumes of group V, i.e. V/1 and V/2 (further volumes will present data relating to the hydrosphere and the atmosphere). Volume V/1 contains a compilation of the physical properties of rocks while volume V/2 treats physics of the earth as a whole. Thus, volume V/1 contains data of rock samples measured in laboratories whereas volume V/2 will contain predominatly the values of the physical quantities of the earth's interior, i.e. of the material in-situ. These values have been derived from measurements at the surface of the earth or very close to it under consideration of further observations such as laboratory measurements. The results are the so-called in-situ determinations or in-situ measurements. In volume V/1, for instance, rock densities which have been measured on samples from ground, quarries or even mines or bore-holes are to be found. In volume V/2, the reader can find the density of the interior of the earth as function of the depth as a result of in-situ measurements.

In volume V/1, the data are arranged in tables according to physical quantities such as density, porosity, electric conductivity etc. Only data on ice and the moon are treated in two separate chapters. This will be helpful for the reader interested only in these materials.

The physical properties of minerals are treated in so far as they are necessary for the understanding of the physical properties of rocks. Nevertheless these properties have not been treated exhaustively. Results which can be determined also by in-situ measurements have been preferred.

The terminology used in petrography and mineralogy is required to describe the physical properties of rocks completely. Therefore, volume V/1 begins with a description of the different rock types found at the surface of the earth or very close to it and of those thought to be situated deep in the earth's interior. It has been shown that the terminology needed for the description of the earth's rocks can be used to describe the rocks on the moon's surface if one adds some terms. It may be presumed that it will be possible to use the actual terminology of petrography and mineralogy as the knowledge of the surfaces and internal structure of the solid planets and solid satellites advances.

The rocks of the earth's surface are composed of the grains of different minerals. The grain diameters are generally less than 1 cm. They can also be less than 10 μm. Thin boundaries separate the grains from one another. The internal structure of the intergranular boundaries is not well known. The rocks are characterized by minute cavities e.g. pores between the grains or fine cracks. In many cases, the internal structure of a grain can only be observed under a microscope. If the internal structure or grain boundaries cannot be detected even under a highly magnifying microscope, for the present the material is called a glassy base material. Further investigations are needed in order to decide if a submicroscopic order (perhaps in clusters) of the glassy base material surrounding the grains or a kryptorcrystalline state exist.

It is generally supposed that at a depth of 20 to 30 km the porosity is very low. However, it is probable that rocks still consist of an aggregate of individual mineral grains at such a depth. This is also true – still with some modifications – for partially melted rocks. Never-

theless, it is not known up to what depth a grain structure can exist or – if it exists – of what kind it is. Therefore it is still an open question whether the material of the earth's mantle and especially of the lower mantle can be considered as a rock according to the above restrictive definition. That is the reason why many data on physical quantities given in this volume cannot be applied to the interior of the earth without a detailed discussion and without considering the experiments at high pressure and temperature. Such a discussion concerning some quantities is reported in this volume.

The exploration of the solid planets and of the solid satellites (e.g. the moon) will require measurements of the same physical quantities as those for the exploration of the solid earth. It will be possible to apply the same measuring principles in the most cases even though with other technical means. For instance, seismographs should not be set up on the moon's surface without considering the large temperature variations, the low gravity and the loose soil. Although the numerical values of the physical quantities both on the earth's surface and on the solid planets and solid satellites are obtained by the same measuring principles, they differ considerably. This is already known for the moon since both measurements on samples from the moon and in-situ measurements on the moon have been performed.

For technical reasons, volume V/1 has been divided into two subvolumes V/1a and V/1b. Subvolume V/1a is presented herewith, subvolume V/1b will appear at the beginning of 1982. It should be regarded as an advantage for the physics and geosciences that Landolt-Börnstein publishes such a comprehensive compilation on a specialized field like "physical properties of rocks".

The authors deserve our thanks, especially for compiling and evaluating the data, the Springer Verlag and the editorial office of Landolt-Börnstein for having respected and fulfilled the authors' wishes in a patient and liberal way; Frau H. Weise has cared for the preparation and correction of the manuscripts, galleys and final pages.

Like all other volumes of Landolt-Börnstein, this volume is published without outside financial support.

München, September 1981 **The Editor**

Übersicht Band V/1

		Teilband
0	**Die Gesteine der Erde**	**a**
1	**Dichte**	**a**
2	**Porosität und Permeabilität**	**a**
3	Elastizität und Inelastizität	b
4	**Thermische Eigenschaften**	**a**
5	Elektrische Eigenschaften	b
6	Magnetische Eigenschaften	b
7	Radioaktivität der Gesteine	b
8	Physikalische Eigenschaften des Eises	b
9	Physikalische Eigenschaften der Gesteine des Mondes	b
	Zwei-dimensionale Inhaltsübersicht für beide Teilbände	**a, b**

Survey of Volume V/1

		Subvolume
0	**The rocks of the earth**	**a**
1	**Density**	**a**
2	**Porosity and permeability**	**a**
3	Elasticity and inelasticity	b
4	**Thermal properties**	**a**
5	Electrical properties	b
6	Magnetic properties	b
7	Radioactivity of rocks	b
8	Physical properties of ice	b
9	Physical properties of lunar rocks	b
	Two-dimensional survey of contents for both subvolumes	**a, b**

Inhaltsverzeichnis

0 Die Gesteine der Erde (H.-G. Huckenholz) ... 1
0.1 Einleitung ... 1
0.2 Die magmatischen Gesteine ... 3
    0.2.1 Nomenklatur ... 3
    0.2.2 Quantitativer Mineralbestand der magmatischen Gesteinstypen ... 9
    0.2.3 Chemische Zusammensetzung der magmatischen Gesteine ... 17
    0.2.4 Das Gefüge der magmatischen Gesteine ... 18
    0.2.5 Die Herkunft der magmatischen Gesteine ... 21
0.3 Die metamorphen Gesteine ... 26
    0.3.1 Nomenklatur ... 26
    0.3.2 Die mineralische Zusammensetzung der Metamorphite ... 29
    0.3.3 Die chemische Zusammensetzung der Metamorphite ... 32
    0.3.4 Das Gefüge der metamorphen Gesteine ... 34
    0.3.5 Die Entstehung der metamorphen Gesteine ... 37
0.4 Die sedimentären Gesteine ... 44
    0.4.1 Nomenklatur ... 44
    0.4.2 Die mineralische Zusammensetzung der sedimentären Gesteine ... 45
    0.4.3 Das Gefüge der sedimentären Gesteine ... 46
    0.4.4 Klastische oder Trümmer-Sedimente ... 50
    0.4.5 Chemische und biochemische Sedimente ... 56
    0.4.6 Die chemische Zusammensetzung der sedimentären Gesteine ... 59
    0.4.7 Die Herkunft und Entstehung der sedimentären Gesteine ... 63
0.5 Literatur zu 0.1 ··· 0.4 ... 65

1 Dichte ... 66
1.1 Dichte der Minerale (J. Wohlenberg) ... 66
    1.1.0 Einleitung ... 66
    1.1.1 Dichte und Röntgen-Dichte von Mineralen ... 67
    1.1.2 Literatur zu 1.1 ... 113
1.2 Dichte der Gesteine (J. Wohlenberg) ... 113
    1.2.0 Einleitung ... 113
    1.2.1 Dichte von Tiefengestein ... 114
    1.2.2 Dichte von Ergußgestein ... 115
    1.2.3 Dichte von vulkanischen Gläsern ... 116
    1.2.4 Dichte von metamorphen Gesteinen ... 116
    1.2.5 Dichte von Sedimentgesteinen und -Böden ... 118
    1.2.6 Dichte von einigen metallischen Erzen, mineralischen Erzen und Substanzen aus organischen Lagerstätten ... 119
    1.2.7 Literatur zu 1.2 ... 119
1.3 Dichte von Mineralen und Gesteinen bei Stoßwellenkompression (D. Stöffler) ... 120
    1.3.1 Theorie der Stoßwellen in Festkörpern ... 120
        1.3.1.1 Grundlegende thermodynamische Beziehungen ... 120
        1.3.1.2 Berechnung der Zustandsgleichung aus Stoßwellendaten ... 122
        1.3.1.3 Der Einfluß der Materialfestigkeit und der Phasenumwandlungen auf die Struktur der Stoßwelle ... 124
    1.3.2 Experimentelle Methoden ... 126
    1.3.3 Tabellen ... 127
        1.3.3.1 Standard-Metalle ... 129
        1.3.3.2 Minerale ... 131

Contents

0 The rocks of the earth (H.-G. HUCKENHOLZ) 1
0.1 Introduction 1
0.2 The igneous rocks 3
    0.2.1 Nomenclature 3
    0.2.2 Quantitative mineral composition of igneous rock types 9
    0.2.3 The chemical composition of igneous rocks 17
    0.2.4 The texture and structure of igneous rocks 18
    0.2.5 The origin and genesis of igneous rocks 21
0.3 The metamorphic rocks 26
    0.3.1 Nomenclature 26
    0.3.2 The mineral composition of metamorphic rocks 29
    0.3.3 The chemical composition of metamorphic rocks 32
    0.3.4 The texture and structure of metamorphic rocks 34
    0.3.5 The origin and genesis of metamorphic rocks 37
0.4 The sedimentary rocks 44
    0.4.1 Nomenclature 44
    0.4.2 Mineral composition of sedimentary rocks 45
    0.4.3 The texture and structure of sedimentary rocks 46
    0.4.4 Clastic or fragmental sedimentary rocks 50
    0.4.5 Chemical and biochemical sedimentary rocks 56
    0.4.6 The chemical composition of sedimentary rocks 59
    0.4.7 The origin and genesis of sedimentary rocks 63
0.5 References for 0.1 ··· 0.4 65

1 Density 66
1.1 Densities of minerals (J. WOHLENBERG) 66
    1.1.0 Introduction 66
    1.1.1 Density and X-ray density of minerals 67
    1.1.2 References for 1.1 113
1.2 Densities of rocks (J. WOHLENBERG) 113
    1.2.0 Introduction 113
    1.2.1 Density of intrusive rocks 114
    1.2.2 Density of extrusive rocks 115
    1.2.3 Density of volcanic glasses 116
    1.2.4 Density of metamorphic rocks 116
    1.2.5 Density of sedimentary rocks and soils 118
    1.2.6 Density of some common metallic ores, mineral ores, and substances of organic deposits 119
    1.2.7 References for 1.2 119
1.3 Density of minerals and rocks under shock compression (D. STÖFFLER) 120
    1.3.1 Theory of shock waves in solids 120
        1.3.1.1 Basic thermodynamics 120
        1.3.1.2 Calculation of equation of state data from shock wave data 122
        1.3.1.3 The influence of material strength and phase transitions on the shock wave structure . . . 124
    1.3.2 Experimental techniques 126
    1.3.3 Tables 127
        1.3.3.1 Standard metals 129
        1.3.3.2 Minerals 131

                1.3.3.2.1 Elemente 131
                1.3.3.2.2 Oxide 135
                1.3.3.2.3 Carbonate 143
                1.3.3.2.4 Halogenide 145
                1.3.3.2.5 Sulfide 148
                1.3.3.2.6 Silikate 149
        1.3.3.3 Gesteine 156
                1.3.3.3.1 Magmatische Gesteine 157
                1.3.3.3.2 Metamorphe Gesteine 169
                1.3.3.3.3 Sedimente und sedimentäre Gesteine 174
        1.3.3.4 Gläser 180
    1.3.4 Literatur zu 1.3.1⋯1.3.3 181

2 Porosität und Permeabilität 184

2.1 Porosität der Gesteine (J.R. Schopper) 184
    2.1.0 Einleitung 184
    2.1.1 Definitionen 184
    2.1.2 Entstehungsursache und unterschiedliche Art der Porosität 185
        2.1.2.1 Intergranulare Porosität 185
        2.1.2.2 Intragranulare Porosität 187
        2.1.2.3 Riß- und Kluftporosität 187
        2.1.2.4 Kavernöse Porosität 187
    2.1.3 Direkter Einfluß der Temperatur auf die Porosität 188
    2.1.4 Einfluß des Druckes auf die Porosität 188
    2.1.5 Verknüpfung mit der Dichte 189
    2.1.6 Messung 190
        2.1.6.1 Die Archimedische Methode 190
        2.1.6.2 Die Methode nach dem Boyle-Mariotteschen Gesetz 191
    2.1.7 Symbolliste 192
    2.1.8 Tabellen und Diagramme 192
    2.1.9 Literatur zu 2.1 262
2.2 Spezifische innere Oberfläche und Kapillarität der Gesteine (J.R. Schopper) . . 267
    2.2.1 Einleitung 267
        2.2.1.1 Allgemeine Bemerkungen 267
        2.2.1.2 Definitionen 267
        2.2.1.3 Petrographische Aspekte 269
        2.2.1.4 Meßmethoden für die innere Oberfläche 269
                2.2.1.4.1 Die BET-Methode 270
                2.2.1.4.2 Stereologische Methoden 270
        2.2.1.5 Meßmethoden für die Kapillarität 271
                2.2.1.5.1 Die „Restored-State"-Methode 271
                2.2.1.5.2 Die Quecksilber-Injektions-Methode 272
                2.2.1.5.3 Die Zentrifugen-Methode 273
        2.2.1.6 Verknüpfungen mit anderen petrophysikalischen Größen . . 273
        2.2.1.7 Symbolliste 273
    2.2.2 Tabellen und Abbildungen 274
        2.2.2.1 Spezifische Oberfläche 274
        2.2.2.2 Kapillaritätskurven 275
    2.2.3 Literatur zu 2.2.1 und 2.2.2 277
2.3 Permeabilität der Gesteine (J.R. Schopper) 278
    2.3.1 Definitionen 278
    2.3.2 Entstehungsursache und unterschiedliche Arten der Permeabilität 282
    2.3.3 Verknüpfungen mit anderen petrophysikalischen Größen 282
    2.3.4 Messung 284
        2.3.4.1 Messung mit Flüssigkeiten 284
        2.3.4.2 Messung mit Gasen 284

	1.3.3.2.1 Elements	131
	1.3.3.2.2 Oxides	135
	1.3.3.2.3 Carbonates	143
	1.3.3.2.4 Halides	145
	1.3.3.2.5 Sulfides	148
	1.3.3.2.6 Silicates	149
1.3.3.3	Rocks	156
	1.3.3.3.1 Igneous rocks	157
	1.3.3.3.2 Metamorphic rocks	169
	1.3.3.3.3 Sediments and sedimentary rocks	174
1.3.3.4	Glasses	180
1.3.4 References for 1.3.1···1.3.3		181

2 Porosity and permeability . . . 184

2.1 Porosity of rocks (J. R. SCHOPPER) . . . 184
- 2.1.0 Introduction . . . 184
- 2.1.1 Definitions . . . 184
- 2.1.2 Origin and different types of porosity . . . 185
  - 2.1.2.1 Intergranular porosity . . . 185
  - 2.1.2.2 Intragranular porosity . . . 187
  - 2.1.2.3 Fissure and fracture porosity . . . 187
  - 2.1.2.4 Vugular porosity . . . 187
- 2.1.3 Direct effect of temperature on porosity . . . 188
- 2.1.4 Effect of pressure on porosity . . . 188
- 2.1.5 Relation to density . . . 189
- 2.1.6 Measurement . . . 190
  - 2.1.6.1 The Archimedian method . . . 190
  - 2.1.6.2 The Boyle's law method . . . 191
- 2.1.7 List of symbols . . . 192
- 2.1.8 Tables and diagrams . . . 192
- 2.1.9 References for 2.1 . . . 262

2.2 Specific internal surface and capillarity of rocks (J. R. SCHOPPER) . . . 267
- 2.2.1 Introduction . . . 267
  - 2.2.1.1 General remarks . . . 267
  - 2.2.1.2 Definitions . . . 267
  - 2.2.1.3 Petrographical aspects . . . 269
  - 2.2.1.4 Methods of measuring internal surface area . . . 269
    - 2.2.1.4.1 The BET-method . . . 270
    - 2.2.1.4.2 Stereological methods . . . 270
  - 2.2.1.5 Methods of measuring capillarity . . . 271
    - 2.2.1.5.1 The restored-state method . . . 271
    - 2.2.1.5.2 The mercury-injection method . . . 272
    - 2.2.1.5.3 The Centrifuge method . . . 273
  - 2.2.1.6 Relations to other petrophysical quantities . . . 273
  - 2.2.1.7 List of symbols . . . 273
- 2.2.2 Tables and figures . . . 274
  - 2.2.2.1 Specific surface . . . 274
  - 2.2.2.2 Capillarity curves . . . 275
- 2.2.3 References for 2.2.1 and 2.2.2 . . . 277

2.3 Permeability of rocks (J. R. SCHOPPER) . . . 278
- 2.3.1 Definitions . . . 278
- 2.3.2 Origin and different types of permeability . . . 282
- 2.3.3 Relations to other petrophysical quantities . . . 282
- 2.3.4 Measurement . . . 284
  - 2.3.4.1 Liquid flow measurement . . . 284
  - 2.3.4.2 Gas flow measurement . . . 284

        2.3.5 Symbolliste 285
        2.3.6 Tabellen und Abbildungen 286
        2.3.7 Literatur zu 2.3.1···2.3.6 303

3 Elastizität und Inelastizität . . . . siehe Teilband b, S.1 ff.
    3.1 Geschwindigkeiten elastischer Wellen und Elastizitäts-Konstanten von Gesteinen und gesteinsbildenden Mineralen
        3.1.1 Einleitung (H. GEBRANDE)
        3.1.2 Geschwindigkeiten elastischer Wellen und Elastizitäts-Konstanten bei Normalbedingungen (H. GEBRANDE)
        3.1.3 Geschwindigkeiten elastischer Wellen und Elastizitäts-Konstanten von Gesteinen bei Zimmertemperatur und Drucken bis 1 GPa (H. GEBRANDE)
        3.1.4 Geschwindigkeiten elastischer Wellen und Elastizitäts-Konstanten bei erhöhten Drucken und Temperaturen (H. KERN)
    3.2 Bruch und Inelastizität von Gesteinen und Mineralen (F. RUMMEL)

4 Thermische Eigenschaften 305
    4.1 Wärmeleitfähigkeit und Wärmekapazität der Minerale und Gesteine (V. ČERMÁK, L. RYBACH) 305
        4.1.1 Einführung 305
            4.1.1.1 Definitionen 305
            4.1.1.2 Einheiten 307
            4.1.1.3 Anisotropie 307
            4.1.1.4 Temperatur- und Druckabhängigkeit 307
            4.1.1.5 Einfluß von Dichte/Porosität und Wassergehalt 309
        4.1.2 Tabellen 310
            4.1.2.1 Minerale 310
            4.1.2.2 Gesteine 315
        4.1.3 Literatur zu 4.1 341
    4.2 Wärmeleitfähigkeit des Bodens (M. SCHUCH) 344
        4.2.1 Einleitung 344
        4.2.2 Daten 344
        4.2.3 Literatur zu 4.2.1 und 4.2.2 344
    4.3 Schmelztemperaturen der Gesteine (R. SCHMID) 345
        4.3.1 Einführung 345
        4.3.2 Schmelztemperaturen einzelner Gesteinsgruppen 346
            4.3.2.1 Allgemeine Bemerkungen zu Fig. 5 und 6 346
            4.3.2.2 Bemerkungen zur Solidusdarstellung (Fig. 5) 350
            4.3.2.3 Bemerkungen zur Liquidusdarstellung (Fig. 6) 350
        4.3.3 Literatur zu 4.3.1 und 4.3.2 352
    4.4 Radioaktive Wärmeproduktion in Gesteinen (L. RYBACH, V. ČERMÁK) 353
        4.4.1 Einleitung 353
            4.4.1.1 Allgemeine Bemerkungen 353
            4.4.1.2 Wärmeproduktion durch natürliche Radioaktivität 353
            4.4.1.3 Geochemische Gesetzmäßigkeiten 354
                4.4.1.3.1 Magmatische Gesteine 354
                4.4.1.3.2 Sedimentgesteine 354
                4.4.1.3.3 Metamorphe Gesteine 355
            4.4.1.4 Anordnung der Wärmeproduktionsdaten 355
        4.4.2 Daten 356
            4.4.2.1 Magmatische Gesteine 356
            4.4.2.2 Sedimentgesteine 364
            4.4.2.3 Metamorphe Gesteine 365
        4.4.3 Literatur zu 4.4.1 und 4.4.2 371

Zwei-dimensionale Inhaltsübersicht 373

Contents

2.3.5 List of symbols	285
2.3.6 Tables and figures	286
2.3.7 References for 2.3.1···2.3.6	303

3 Elasticity and inelasticity . . . . see Subvolume b, p. 1 ff.
3.1 Elastic wave velocities and constants of elasticity of rocks and rock-forming minerals

    3.1.1 Introduction (H. GEBRANDE)
    3.1.2 Elastic wave velocities anc constants of elasticity at normal conditions (H. GEBRANDE)

    3.1.3 Elastic wave velocities and constants of elasticity of rocks at normal temperature and pressures up to 1 GPa (H. GEBRANDE)
    3.1.4 Elastic wave velocities and constants of elasticities of rocks at elevated pressures and temperatures (F. KERN)
3.2 Fracture and flow of rocks and minerals (F. RUMMEL)

4 Thermal properties 305
4.1 Thermal conductivity and specific heat of minerals and rocks (V. ČERMÁK, L. RYBACH) 305
    4.1.1 Introductory remarks 305
        4.1.1.1 Definitions 305
        4.1.1.2 Units 307
        4.1.1.3 Anisotropy 307
        4.1.1.4 Temperature and pressure dependence 307
        4.1.1.5 Effects of density/porosity and water content 309
    4.1.2 Tables 310
        4.1.2.1 Minerals 310
        4.1.2.2 Rocks 315
    4.1.3 References for 4.1 341
4.2 Thermal conductivity of soil (M. SCHUCH) 344
    4.2.1 Introduction 344
    4.2.2 Data 344
    4.2.3 References for 4.2.1 and 4.2.2 344
4.3 Melting temperatures of rocks (R. SCHMID) 345
    4.3.1 Introduction 345
    4.3.2 Melting temperatures of individual rock groups 346
        4.3.2.1 General comments to Figs. 5 and 6 346
        4.3.2.2 Comments on the solidi (Fig. 5) 350
        4.3.2.3 Comments on the liquidi (Fig. 6) 350
    4.3.3 References for 4.3.1 and 4.3.2 352
4.4 Radioactive heat generation in rocks (L. RYBACH, V. ČERMÁK) 353
    4.4.1 Introduction 353
        4.4.1.1 General remarks 353
        4.4.1.2 Heat generation by radioactive decay 353
        4.4.1.3 Geochemical control of heat generation 354
            4.4.1.3.1 Igneous rocks 354
            4.4.1.3.2 Sedimentary rocks 354
            4.4.1.3.3 Metamorphic rocks 355
        4.4.1.4 Arrangement of heat generation data 355
    4.4.2 Data 356
        4.4.2.1 Igneous rocks 356
        4.4.2.2 Sedimentary rocks 364
        4.4.2.3 Metamorphic rocks 365
    4.4.3 References for 4.4.1 and 4.4.2 371

Two-dimensional survey of contents 372

0 The rocks of the earth — Die Gesteine der Erde

0.1 Introduction — Einleitung

Rocks are natural mineral assemblages crystallized by rockforming processes. Rocks form the earth's crust and mantle. Their nomenclature is based on origin and composition. This leads to the distinction of **igneous rocks** (magmatites; magmatic rocks), **metamorphic rocks** (metamorphites), and **sedimentary rocks** (sediments).

The formation of the planet earth and its evolution in space and time caused rocks different in chemical composition and physical state within the crust and mantle. Crustal rocks are well exposed and accessible over a vertical section of at least 20 km from the earth's surface. Rocks of the (upper) mantle and the deeper crust cannot be observed *in situ*. Knowledge of their chemical composition and physical state comes from fragmental inclusions in igneous rocks. These rock fragments are torn off the upper mantle and deeper crust and are brought to the surface by rising magma from depths as deep as 150 km or more. The chemical composition and physical state of zones deeper than the upper mantle can be extrapolated from the geophysical behavior of the earth. Laboratory experiments are able to simulate the pressure-temperature regime of regions as low as the mantle/core boundary.

The crust of the earth – approximately 1 % of the planet's volume (equal to about 0.5 % of its mass) – is about 30 km thick. Its thickness decreases under the oceans to less than 10 km but increases within the orogenic belts to 60 km and more. In general, the crust consists of a thin cover of sediments which is about 3 to 5 km thick. Toward greater depths, the sediments change gradually into metamorphic rocks forming the basement. Igneous rocks penetrate the sedimentary cover and the basement as dikes, extrude from the surface and build volcanic edifices, and intersect the sedimentary cover and the basement as large scale intrusions.

The thin crust is underlain by a (very) thick mantle which extends to depths of about 2900 km. It constitutes about 80% of the planet's volume (equal to about 68% of its mass). The rocks within the upper mantle from the Mohorovičić-discontinuity (the crust/mantle boundary) to about 200 km depth have metamorphic textures caused by flow and partial melting-processes associated with physical and chemical re-equilibration of the mineral phases. Seismic data obtained for depths between 100 and 150 km (low velocity zone, Gutenberg zone) interpret the material of the mantle to be solid (crystalline) with small amounts of

Gesteine sind natürliche Mineralgemenge, die durch gesteinsbildende Prozesse entstehen. Gesteine bauen die Erdkruste und den Erdmantel auf. Ihre Einteilung erfolgt nach Herkunft und Zusammensetzung, dabei werden **magmatische Gesteine** oder Magmatite, **metamorphe Gesteine** oder Metamorphite und **sedimentäre Gesteine** oder Sedimente unterschieden.

Die Entstehung des Planeten Erde und seine Entwicklung in Raum und Zeit bedingte hinsichtlich chemischer Zusammensetzung und physikalischer Zustände unterschiedliche Gesteine in Erdkruste und Erdmantel. Die Gesteine der Erdkruste sind gut erschlossen und vertikal über mindestens 20 km von der Erdoberfläche zugänglich. Die Gesteine der tieferen Erdkruste und der oberen Zone des Erdmantels (Oberer Mantel) sind nicht direkt beobachtbar. Die Kenntnis ihrer chemischen Zusammensetzung und ihrer physikalischen Zustände ergibt sich aus Einschlüssen in magmatischen Gesteinen, die aus Teufen bis hinunter zu 150 km und mehr stammen. Beim Aufstieg wurden Fragmente der tieferen Kruste und des Oberen Mantels mitgerissen und an die Oberfläche gebracht. Stoffzusammensetzungen und Phasenzustände tieferer Erdzonen können aus dem geophysikalischen Verhalten der Erde abgeleitet werden. Laborexperimente simulieren den Druck-Temperatur-Bereich bis zur Grenze des Unteren Mantels und des Erdkerns.

Die Erdkruste, die rund 1 % des Volumens (das sind etwa 0.5 % der Gesamtmasse) unseres Planeten ausmacht, ist etwa 30 km mächtig. Sie dünnt unter den Ozeanen auf unter 10 km aus, schwillt aber unter (geologisch jungen) Faltengebirgen auf über 60 km an. Zum größten Teil besteht die Erdkruste aus einer (unterschiedlich mächtigen) Decke von Sedimenten. Diese bilden das sogenannte Deckgebirge (im Mittel 3···5 km mächtig). Zu größerer Teufe gehen die Sedimente in Metamorphite über, die das Grundgebirge bilden. Magmatite durchziehen als Gänge (Ganggesteine) das Grund- und Deckgebirge, erscheinen als überlagernde Vulkanbauten (Oberflächengesteine) und durchsetzen Grund- und Deckgebirge als (großräumige) Intrusionen (Tiefengesteine).

Die dünne Erdkruste wird von einem sehr dicken Mantel unterlagert, der bis etwa 2900 km Teufe reicht und etwa 80 % des Volumens (das sind rund 68 % der Masse) unseres Planeten ausmacht. Im Prinzip sind die Gesteine im oberen Teil des Mantels von der Mohorovičić-Diskontinuität (der Grenze zwischen Kruste und Mantel) bis etwa 200 km durch Fließvorgänge und Teilschmelzprozesse verbunden mit chemischer und physikalischer Reequilibration der Phasen in ihrem Gefüge metamorph. In Teufen zwischen 100 und 150 km (Gutenberg-Zone; low velocity layer) können die seismischen Daten so interpretiert werden, daß

a pore melt which may be responsible for the (continuous) basalt generation. The dynamic turnover within the crust and mantle – visible in drifting plates on the earth's surface (plate tectonics) – recycles the material of the upper mantle during geologic times. Juvenile ocean floor is formed by outpouring basalt magma from vents along fault-systems (rift systems) within the drifting plates (sea floor spreading). At the margin of the plate, old ocean floor is overthrusted by adjacent plates and dives along subduction zones (Benioff zones) into the upper mantle where it is altered and incorporated.

The spherical shell structure of the earth represents the chemical and physical separation of the planet's material since its formation and during following evolution within the last 4.5 billion years ($=10^9$ a). The chemical composition of crust, mantle, and core as calculated, extrapolated, and estimated is depicted in Table 1. Oxygen (O) and silicon (Si) are enriched in the crust and mantle (crust > mantle). The core preponderously consists of iron (Fe), nickel (Ni), and sulfur (S). The crust contains silicates of sodium (Na), potassium (K), calcium (Ca), and aluminum (Al). The crystal structures of these minerals are loosely packed and their densities are low. The upper mantle consists of denser silicates of magnesium (Mg), iron (Fe), and, to a lesser extent, of calcium (Ca) and aluminum (Al). The chemical composition of the crust is *granitoid* (granite

neben kristallinen Anteilen eine Porenlösung (Teilschmelze) in der Größenordnung weniger Prozente vorliegt, die für die (kontinuierliche) Basalt-Magmenproduktion in Frage kommt. Die Dynamik der Erdkruste und des Erdmantels, die ihren Ausdruck in der Drift von Platten an der Oberfläche des Planeten findet (Plattentektonik), trägt dazu bei, daß das Material des Mantels im Laufe geologischer Zeiten stofflich ständig einer Erneuerung unterworfen ist. Entlang großer Störungs-Systeme (Rift-Systeme) wird fortwährend neuer Ozeanboden durch austretendes Basalt-Magma gebildet. Älterer Ozeanboden wird an Plattengrenzen von benachbarten Platten überschoben und taucht entlang von Subduktionszonen (Benioff-Zonen) in den Mantel unter, wo er umgewandelt und aufgenommen wird.

Der Schalenbau der Erde repräsentiert die chemische und physikalische Sonderung des Planetenmaterials seit Entstehung und Fortentwicklung innerhalb der letzten 4,5 Milliarden Jahre. Die berechneten, extrapolierten und geschätzten chemischen Zusammensetzungen von Erdkruste, Erdmantel und Erdkern gibt die Tab. 1 wieder. Sauerstoff (O) und Silicium (Si) überwiegen in Kruste und Mantel, während der Kern vorwiegend aus Eisen (Fe), Nickel (Ni) und Schwefel (S) besteht. Die Kruste enthält vorwiegend Silikate des Natriums (Na), Kaliums (K), Kalziums (Ca) und Aluminiums (Al). Ihre Kristallstrukturen sind (relativ) locker gepackt und damit ihre spezifischen Gewichte niedrig. Der (Obere) Mantel besteht dagegen aus relativ dichtgepackten Silikaten des Magnesiums (Mg), Eisens (Fe) und Kalziums (Ca). Die chemische Zu-

Table 1. Estimated and calculated chemical composition of the earth's crust, mantle, and core [Cor68].

	Crust wt-%	Mantle wt-%	Core wt-%
O	47.25	43.7	
Si	30.54	22.5	
Al	7.83	1.6	
Fe	3.54	9.88	86.3
Ca	2.87	1.67	
K	2.82	0.11	
Na	2.45	0.84	
Mg	1.39	18.8	
Ti	0.47	0.08	
P	0.08	0.14	
Mn	0.07	0.33	
Ni	0.0044		7.36
Co	0.0012		0.40
S	0.031		5.94
	99.3466	99.65	100.00

like) and more or less resembles rocks with the composition of granites, granodiorites, and quartz diorites. These are mineral assemblages which contain (large amounts of) feldspars and (smaller amounts of) quartz. The composition of the upper mantle appears to be *peridotitic*, i.e. mineral assemblages containing large amounts of olivine and orthopyroxene but smaller amounts of clinopyroxene, garnet, and spinel.

sammensetzung der Kruste ist in erster Annäherung *granitähnlich*, d.h. sie besteht mehr oder weniger aus Graniten, Granodioriten und Quarzdioriten, also Mineralgemengen, die sich durch (hohe) Feldspatanteile neben Quarz-Führung auszeichnen. Die Zusammensetzung des (Oberen) Mantels ist in erster Annäherung *peridotitisch*, d.h. sie ist gekennzeichnet durch Mineralgemenge, die neben Olivin und Orthopyroxen Granat, Klinopyroxen und Spinell enthalten.

0.2 The igneous rocks — Die magmatischen Gesteine
0.2.1 Nomenclature — Nomenklatur

Igneous rocks (magmatites; magmatic rocks; eruptive rocks; eruptiva) are mineral assemblages which have crystallized from a melt **(magma)**. According to their natural occurrences three types of igneous rocks can be distinguished. Igneous rocks which have reached the surface are called **volcanic rocks** (volcanics; lava; effusive rocks; effusiva; extrusive rocks; extrusiva). Igneous rocks which have intruded into the sedimentary cover or the basement and crystallized in greater depths are called **plutonic rocks** (plutonites; intrusive rocks; abyssal rocks). Igneous rocks protruding from the sedimentary cover and basement as (small) dikes and sills are called **dike rocks** (dikes; hypabyssal rocks).

Effusive, plutonic, and hypabyssal igneous rocks can also be distinguished by their texture. In general, effusive, and hypabyssal rocks are (very) fine-grained and/or glassy, plutonic rocks are all-crystalline and coarse-grained. Further classification of igneous rocks is achieved according to their qualitative and quantitative mineral composition (modal composition; mode) and chemical composition (chemistry). The mode and texture are obtained by microscopic examination (petrographic microscopy) of thin sections. X-ray and chemical methods complete and/or refine the optical analysis.

Feldspars are the most abundant silicate minerals in assemblages of igneous rocks. The average mineral composition of igneous rocks is shown in Table 2. Over 80% of the total igneous rocks are made up by feldspars with about 62% and quartz with about 21%. Thus, feldspar- und quartz-bearing intrusive types such as granites, granodiorites, and quartz diorites make up approximately 85% of all intrusive rocks (Tables 3 and 4). Feldspar-rich tholeiites and andesites are the most abundant volcanic rocks (Table 5). Feldspar-free igneous rocks are scarce in the crust and only a very small group within the magmatic rock family. On the basis of feldspar, quartz,

Magmatische Gesteine (Magmatite, Magmagesteine, Eruptivgesteine, Eruptiva) sind Mineralgemenge, die aus einer Schmelze **(Magma)** auskristallisieren. Aus dem natürlichen Auftreten der magmatischen Gesteine lassen sich generell drei Gruppen unterscheiden für die die Definitionskriterien zutreffen. Magmatite, die bis zur Oberfläche durchgebrochen sind, bezeichnet man als **Vulkanite** (vulkanische Gesteine; Lava; effusive Gesteine, Effusiva, Effusiv-Gesteine; Extrusiva, Extrusiv-Gesteine; Oberflächengesteine; Ergußgesteine). Magmatite, die in der Tiefe steckengeblieben sind, bezeichnet man als **Plutonite** (plutonische Gesteine; intrusive Gesteine; abyssische Gesteine; Tiefengesteine). Magmatite, die Grund- und Deckgebirge in schmalen Gängen durchsetzen, nennt man **Ganggesteine** (hypabyssische Gesteine; subvulkanische Gesteine).

Effusive, plutonische und hypabyssische Magmatite unterscheiden sich außerdem in den Eigenschaften des Gesteinsgefüges. Im allgemeinen sind effusive und hypabyssische Gesteine feinkörnig und/oder glasig, plutonische Gesteine vollkristallin und grobkörnig. Die weitere Einteilung der Magmatite geschieht nach dem qualitativen und quantitativen Mineralbestand **(Modus)** und nach der chemischen Zusammensetzung. Der qualitative Mineralbestand und das Gesteinsgefüge eines magmatischen Gesteins werden mikroskopisch am Gesteinsdünnschliff bestimmt (Polarisationsmikroskopie). Röntgenographische und chemische Methoden ergänzen und/oder verfeinern den optisch qualitativ ermittelten Mineralbestand.

In Mineralzusammensetzungen der magmatischen Gesteine herrschen die Feldspäte mengenmäßig bei weitem vor. Die mittlere Zusammensetzung der Magmatite in Tab. 2 zeigt 62% Feldspäte. Das zweitwichtigste Mineral ist Quarz, sodaß Feldspäte+Quarz bereits $>80\%$ des gesamten Mineralbestandes der Magmatite ausmachen. Feldspat-+Quarz-führende Intrusiv-Gesteine wie Granite, Granodiorite und Quarzdiorite sind demnach bei den plutonischen Magmatiten vorherrschend. Sie machen volumenmäßig in der Erdkruste etwa 85% aller Magmatite aus (Tab. 3 und 4). Bei den effusiven Gesteinen herrschen die feldspatreichen Tholeiite und Andesite vor (Tab. 5). Feldspat-

and feldspathoids (minerals which replace feldspars in certain rocks) a first division of rocks into:

1. Quartz-rocks,
2. Quartz+feldspar-rocks,
3. Feldspar-rocks,
4. Feldspar+feldsparthoid-rocks,
5. Feldsparthoid-rocks, and
6. Quartz-, feldspar-, and feldspathoid-free rocks

is obtained (Table 6). This division is almost identical with the SiO_2/metal-oxides ratios and expresses as a first approximation the state of Si-saturation: free SiO_2 – as quartz, tridymite, or cristobalite – reveals oversaturation: only feldspars saturation and feldspathoids with or without feldspars undersaturation.

freie Magmatittypen sind selten und treten volumenmäßig stark zurück (<1%). Qualitativ lassen sich die Magmatite deshalb in Tab. 6 einteilen in

1. Quarz-Gesteine,
2. Quarz-Feldspat-Gesteine,
3. Feldspat-Gesteine,
4. Feldspat-Foid (=Feldspatvertreter)-Gesteine,
5. Foid-Gesteine,
6. Quarz-, Feldspat- und Foid-freie Gesteine.

Diese Einteilung ist weiterhin identisch mit dem Verhältnis von SiO_2/Metalloxide und drückt in erster Näherung den Si-Sättigungsgrad aus: Freies SiO_2 als Quarz (oder Tridymit oder Cristobalit) zeigt Übersättigung neben Feldspäten; Feldspäte allein Sättigung und Feldspäte+Foide oder Foid allein Untersättigung an.

Table 2. Approximate average mineral composition of the earth's (upper) crust and of major intrusive rocks [Wed69].

	Crust vol-%	Granite vol-%	Granodiorite vol-%	Quartz diorite vol-%	Diorite vol-%	Gabbro vol-%
Plagioclase	41	30	46	53	63	56
Alkalifeldspar	21	35	15	6	3	
Quartz	21	27	21	22	2	
Amphibole	6	1	13	12	12	1
Biotite	4	5	3	5	5	
Orthopyroxene	2				3	16
Clinopyroxene	2				8	16
Olivine	0.6					5
Magnetite, Ilmenite	2	2	2	2	3	4
Apatite	0.5	0.5	0.5	0.5	0.8	0.6

Table 3. The composition of the earth's crust expressed in proportions of intrusive rocks [Wed69].

Intrusive rock	vol-%
Granites, quartz monzonites	44
Granodiorites	34
Quartz diorites	8
Diorites	1
Gabbros	13
Peridotites	<0.5
Syenites, alkalic rocks	<0.5
Anorthosites	<0.5

Table 4: see p. 5.

Table 5. Relative abundance of volcanic rocks in percentages of the area covered [Wed69].

	Volcanic rocks on Earth	Volcanic rocks Cordillera U.S.A.	Tertiary and pleistocene volcanics, Japan 175000 km^2
Tholeiitic basalts	82 (>80)	32	45
Alkaliolivine basalts	(<20)		
Andesites	16	44	
Trachytes, alkalic rocks		0.5	
Dacites		0.9	55
Rhyolites	2	23	

Table 4. Portions of intrusive igneous rocks of specific areas [Wed69].

	Finland	Cordillera U.S.A.	Cordillera U.S.A.	Appalachians U.S.A.	Canadian Shield	Basement Midwestern U.S.A.
Area [km²]	191400	14100	44800	3950		
	vol-%	vol-%	vol-%	vol-%	vol-%	vol-%
Granites (+ quartz monzonites)		46.7	34.6		(<50)	
Granodiorites	86.8	37.3	19.1	88	87	87
Quartz diorites		1.0	33.7		(>25)	
Diorites		4.1	1.9	0.8		
Gabbros	13.2	6.9	10.7	10.8	13	13
Peridotites		1.4				
Syenites, alcalic rocks		2.0	0.1	0.3		
Anorthosites		0.9				

Table 5: see p. 4.

Table 6. Major intrusive igneous rocks classified on the basis of quartz—feldspar—feldspathoid (mainly after [Trö35/38]).

Intrusive igneous rocks	with feldspars			without feldspars
	alkalifeldspar	alkalifeldspar + plagioclase	plagioclase	
with quartz	alkali granites	granites grano diorites	quartz diorites	peracidites
without quartz without feldspathoids (Foide)	alkali syenites	monzonites monzodiorites monzogabbros syenites	diorites gabbros anorthosites	ultramafitites
with feldspathoids (Foiden)	nepheline syenites shonkinites	theralites	essexites	feldspathoidites (Foiditite)

Table 7. Estimated amounts of minor and trace elements in common rockforming minerals [Cor68], after Wedepohl.

	x %	0.x %	0.0x %	0.00x %	0.000x %
plagioclase	K	Sr	Ba, Rb, Ti, Mn	Ga, V, Zn, Ni	Pb, Cu, Li, Cr, Co, B
potash feldspar	Na	Ca, Ba, Sr	Rb, Ti	Pb, Li, Ga, Mn	B, Zn, V, Cr, Ni, Co
quartz				Al, Ti, Fe, Mg, Ca	Na, Ga, Li, Ni, B, Zn, Ge, Mn
amphibole		Ti, F, K, Mn, Cl, Rb	Zn, Cr, V, Sr, Ni	Ba, Cu, Co, Ga, Pb	Li, B
pyroxene	Al	Ti, Na, Mn, K	Cr, V, Ni, Cl, Sr	Cu, Co, Zn, Li, Rb	Ba, Pb, Ga, B
biotite	Ti, F	Ca, Na, Ba, Mn, Rb	Cl, V, Cr, Li, Ni	Cu, Sr, Co, Pb, Ga	B
magnetite	Ti, Al	Mg, Mn, V	Cr, Zn, Cu	Ni, Co	Pb, Mo
olivine		Ni, Cr, Ti, Ca	Mn, Co	Zn, V, Cu, Sc	Rb, B, Ge, Sr, As, Ga, Pb

An igneous rock nomenclature on the basis of the mode (quantitative mineral composition in % by volume) requires the detailed knowledge of the chemical composition of the rockforming minerals and of their physical state. The average composition of igneous rocks includes feldspars, quartz, amphiboles, pyroxenes, micas, ores, and olivines as major constituents (see Table 2). There are, however, other minerals present in small amounts. The most important of these minerals are the feldspathoids and melilites. Feldspathoids occur in quartz-free rocks with or without a feldspar-content; melilites in feldspar-free but feldspathoid-bearing igneous rocks. The accessory mineral apatite is responsible for the phosphorus-content of the rocks. Zircon and sphene contain the rare earth elements, uranium, and thorium. The radioactive decay of the latter two elements can be used to determine the (absolute) age of the crystallization of the mineral. Minor- and trace-element concentration in the most abundant (magmatic) rockforming minerals is given in Table 7. The most abundant rockforming minerals for magmatic rocks as well as for metamorphic and sedimentary rocks are given in Table 8.

Die Einteilung der magmatischen Gesteine nach dem Modus (quantitativer Mineralbestand in Volumen-%) setzt die detaillierte Kenntnis der chemischen Zusammensetzung der gesteinsbildenden Minerale und ihrer physikalischen Zustände voraus. Neben den Feldspäten und Quarz zeigt Tab. 2 Amphibole, Pyroxene, Glimmer, Erze und Olivin in abnehmender Reihenfolge als gesteinsbildende Minerale. Bei der mittleren Magmatit-Zusammensetzung nicht berücksichtigt sind die Minerale der Feldspatvertreter und die Minerale der Melilithgruppe. Sie sind in magmatischen Gesteinen enthalten, die mengenmäßig hinter den Quarz+Feldspat- und Feldspat-führenden Gesteinen zurücktreten. Der Übergemengteil (Akzessorie) Apatit ist verantwortlich für den Phosphorgehalt, daneben findet man meist noch Zirkon und Titanit, die vor allem neben ihren Gehalten an Seltenen Erden, durch ihre Uran- und Thorium-Führung für eine Altersbestimmung der Mineralkristallisation wichtig sind. Neben- und Spurenelementkonzentrationen in den wichtigsten gesteinsbildenden Mineralen sind in Tab. 7 zusammengestellt. Die wichtigsten gesteinsbildenden Minerale der magmatischen, der metamorphen und der sedimentären Gesteine zeigt Tab. 8.

Table 8. Rockforming minerals of magmatic, metamorphic, and sedimentary origin. × × × : only in a specific assemblage; × × : mainly in a specific assemblage; × : in other assemblages as well.

No.	Mineral name	Composition	Occurrence in assemblages of		
			magmatic origin	metamorphic origin	sedimentary origin
1	acmite (aegirine) (Ägirin)	$NaFe^{3+}Si_2O_6$	× ×	× ×	
2	actinolite	$Ca_2(Mg, Fe^{2+})_5Si_8O_{22}(OH, F)_2$		× × ×	
3	akermanite	$Ca_2MgSi_2O_7$	× ×	× ×	
4	albite	$NaAlSi_3O_8$	× ×	× ×	×
5	alkalifeldspar	$(Na, K)AlSi_3O_8$, solid solutions of Nos. 4, 81, 91, and 107.	× ×	× ×	
6	almandine	$Fe_3^{2+}Al_2Si_3O_{12}$		× × ×	
7	amphibole	$(Ca, Na, K)_2(Mg, Fe^{2+}, Fe^{3+}, Al, Ti^{4+}, Mn)_5$ $\cdot (Si, Al)_8O_{22}(OH, F)_2$; see Nos. 2, 14, 19, 40, 49, 52, 121	× ×	× ×	
8	andesine	plagioclase of anorthite 30–50	× ×	× ×	
9	analcite (Analcim)	$NaAlSi_2O_6 \cdot H_2O$	×	× ×	×
10	andalusite	Al_2SiO_5 (orthorhombic)		× × ×	
11	andradite	$Ca_3Fe_2^{3+}Si_3O_{12}$		× × ×	
12	ankerite	$CaFe^{2+}(CO_3)_2$			× ×
13	anorthite	$CaAl_2Si_2O_8$	× ×	× ×	
14	anthophyllite	$Mg_7Si_8O_{22}(OH, F)_2$		× × ×	
15	antigorite	$Mg_3Si_2O_5(OH)_4$		× × ×	
16	apatite	$Ca_5(F, Cl, OH, CO_3)(PO_4)_3$	× ×	× ×	× ×
17	aragonite	$CaCO_3$ (orthorhombic)		× × ×	
18	augite	$(Ca, Na)(Mg, Fe^{2+}, Fe^{3+}, Al, Ti^{4+}, Mn)$ $\cdot (Si, Al, Fe^{3+})_2O_6$	× ×	× ×	

continued

Table 8 (continued)

No.	Mineral name	Composition	magmatic origin	metamorphic origin	sedimentary origin
			\multicolumn{3}{c}{Occurrence in assemblages of}

No.	Mineral name	Composition	magmatic origin	metamorphic origin	sedimentary origin
19	barkevikite	$Ca_2(Na, K)(Fe^{2+}, Mg, Fe^{3+}, Mn)_5 \cdot Si_{6.5}Al_{1.5}O_{22}(OH, F)_2$	× ×	× ×	
20	beidellite	$Al_4(Ca_{0.5}, Na)_{0.66}Si_{7.3}Al_{0.66}O_{20}(OH)_6 \cdot nH_2O$			× × ×
21	biotite	$K(Mg, Fe^{2+})_3AlSi_3O_{10}(OH, F)_2$ and see No. 95	× ×	× ×	×
22	bytownite	plagioclase of anorthite 70–90	× ×	× ×	
23	calcite	$CaCO_3$ (trigonal)		× ×	× ×
24	cancrinite	$(Na_4Ca)_4Al_6Si_6O_{24}CO_3 \cdot 0\cdots3H_2O$	× ×	× ×	
25	chlorites	$(Mg, Al, Fe^{2+})_{12}(Si, Al)_8O_{20}(OH)_{16}$		× ×	× ×
26	chloritoid	$(Fe^{2+}, Mg)Al_2SiO_5(OH)_2$		× × ×	
27	chromite	$Fe^{2+}Cr_2O_4$	× ×	× ×	
28	chrysotile	$Mg_3Si_2O_5(OH)_4$		× × ×	
29	clay minerals (Tonminerale)	see Nos. 20, 37, 51, 56, 58, 65, 70, 85, 108		× ×	× × ×
30	clinopyroxenes	solid solutions of Nos. 1, 18, 38, 43, 59, 67, 89, 96	× ×	× ×	
31	clinozoisite	$Ca_2Al_3Si_3O_{12}(OH)$ (monoclinic)		× × ×	
32	cordierite	$(Mg, Fe)_2Al_4Si_5O_{18} \cdot 0\cdots xH_2O$	×	× ×	
33	corundum	Al_2O_3	×	× ×	
34	cristobalite	SiO_2	× ×		
35	diamond	C	× × ×		
36	diaspore	α-AlO(OH)			× × ×
37	dickite	$Al_4Si_4O_{10}(OH)_8$		× ×	× ×
38	diopside	$CaMgSi_2O_6$	× ×	× ×	
39	dolomite	$CaMg(CO_3)_2$		× ×	× ×
40	edenite	$NaCa_2(Mg, Fe^{2+})_5Si_7AlO_{22}(OH, F)_2$	× ×	× ×	
41	enstatite	$MgSiO_3$	× ×	× ×	
42	epidote	$Ca_2(Al, Fe^{3+})_3Si_3O_{12}(OH)$		× × ×	
43	fassaite	$Ca(Mg, Fe^{2+}, Al, Fe^{3+})(Si, Al, Fe^{3+})_2O_6$	×	× ×	
44	fayalite	$Fe_2^{2+}SiO_4$	× ×	× ×	
45	feldspars	$(K, Na, Ca)(Si, Al)_4O_8$, solid solutions of Nos. 5 and 97	× ×	× ×	× ×
46	feldspathoids (Foide)	see Nos. 69, 75 and 84	× × ×		
47	forsterite	Mg_2SiO_4	× ×	× ×	
48	garnets	$(Ca, Mg, Fe^{2+}, Mn)_3(Al, Fe^{3+}, Cr, Ti^{4+})_2 \cdot (Si, Al, Fe^{3+})_3O_{12}$, solid solutions of Nos. 6, 11, 53, 79, 101, 115, and 125	×	× × ×	
49	gedrite	$(Mg, Fe^{2+})_{6\ldots5}Al_{1\ldots2}Si_6(Si, Al)_2O_{22}(OH, F)_2$		× × ×	
50	gehlenite	$Ca_2Al_2SiO_7$		× × ×	
51	glauconite	$(K, Na, Ca)_{1.2\ldots2}(Fe^{3+}, Al, Fe^{2+}, Mg)_4Si_{7\ldots7.6} \cdot Al_{1\ldots0.4}O_{20}(OH)_4 \cdot nH_2O$			× × ×
52	glaucophane	$Na_2Mg_3Al_2Si_8O_{22}(OH)_2$		× × ×	
53	grossular	$Ca_3Al_2Si_3O_{12}$		× × ×	
54	gypsum	$CaSO_4 \cdot 2H_2O$			× × ×
55	halite (Steinsalz)	NaCl			× × ×
56	halloysite	$Al_4Si_4O_{10}(OH)_8$			× × ×
57	hauynite (Hauyn)	$(Na, Ca)_{4\ldots8}Al_6Si_6O_{24}(SO_4)_{1\ldots2}$	× × ×		
58	hectorite	$Mg_{5.3}Li_{0.66}(Ca_{0.5}, Na)_{0.66}Si_8O_{20}(OH)_4 \cdot nH_2O$			× × ×
59	hedenbergite	$CaFe^{2+}Si_2O_6$	× ×	× ×	
60	hematite	Fe_2O_3	× ×	× ×	

continued

Table 8 (continued)

No.	Mineral name	Composition	Occurrence in assemblages of		
			magmatic origin	metamorphic origin	sedimentary origin
61	hercynite	$Fe^{2+}Al_2O_4$		××	
62	heulandite	$CaAl_2Si_7O_{18} \cdot 6H_2O$	(××)		×××
63	hornblende, common	solid solution of Nos. 2, 19, 40, 94 and 121	××	××	
64	hydrargillite	$Al(OH)_3$			×××
65	illite	$K_{1...1.5}(Si_{7...6.5}Al_{1...1.5})O_{20}(OH)_4$			×××
66	ilmenite	$FeTiO_3$	××	××	
67	jadeite	$NaAlSi_2O_6$		×××	
68	kaersutite	$Ca_2(Na, K)(Mg, Fe^{2+}, Fe^{3+})_4Ti^{4+}$ $\cdot (Si_6Al_2)O_{22}(O, OH, F)_2$	×××		
69	kalsilite	$KAlSiO_4$	×××		
70	kaolinite	$Al_2Si_2O_5(OH)_4$		×	××
71	kyanite (Disthen)	Al_2SiO_5 (triclinic)		×××	
72	labradorite	plagioclase of anorthite 50–70	××	××	
73	laumontite	$CaAl_2Si_4O_{12} \cdot 4H_2O$		×××	
74	lawsonite	$CaAl_2Si_2O_7(OH)_2 \cdot H_2O$		×××	
75	leucite	$KAlSi_2O_6$	×××		
76	magnesite	$MgCO_3$		×	×
77	magnetite	$Fe^{2+}Fe_2^{3+}O_4$	××	××	
78	margarite	$CaAl_4Si_2O_{10}(OH)_2$		×××	
79	melanite	$Ca_3(Fe^{2+}, Fe^{3+}, Ti^{4+}, Al)_2(Si, Fe^{3+}, Ti^{4+})_3O_{12}$	××	××	
80	melilites	$(Ca, Na)_2(Al, Mg)(Al, Si)_2O_7$, solid solutions of Nos. 3, 50, and 114	××	××	
81	microcline	$KAlSi_3O_8$ (triclinic)	××	××	×
82	muscovite	$KAl_3Si_3O_{10}(OH)_2$	××	××	×
83	nakrite	$Al_4Si_4O_{10}(OH)_8$			×××
84	nepheline	$NaAlSiO_4$	××	×	
85	nontronite	$Fe^{3+}(Ca_{0.5}, Na)_{0.66}Si_{6.7}Al_{1.3}O_{20}(OH)_4 \cdot nH_2O$			×××
86	noseane	$Na_8Al_6Si_6O_{24}SO_4$	×××		
87	oligoclase	plagioclase of anorthite 10–30	××	××	
88	olivine	$(Mg, Fe^{2+})_2SiO_4$, see Nos. 44 and 47	××	××	
89	omphacite	$(Ca, Na)(Mg, Fe^{2+}, Al)(Si, Al)_2O_6$		×××	
90	opal	$SiO_2 \cdot H_2O$			××
91	orthoclase	$KAlSi_3O_8$ (monoclinic)	××	××	
92	orthopyroxenes	modifications of Nos. 41 and 99	××	××	
93	paragonite	$NaAl_3Si_3O_{10}(OH)_2$		×××	
94	pargasite	$NaCa_2(Mg, Fe^{2+})_4(Al, Fe^{3+})Si_6Al_2O_{22}(OH, F)_2$	××	××	
95	phlogopite	$KMg_3AlSi_3O_{10}(OH)_2$ see No. 21	××	××	
96	pigeonite	$(Mg, Fe^{2+}, Ca)_2Si_2O_6$	××	××	
97	plagioclase	$(Ca, Na)(Al, Si)_4O_8$, solid solutions of Nos. 4 and 13. Members are Nos. 4, 8, 13, 22, 72 and 87	××	××	×
98	prehnite	$Ca_2Al_2Si_3O_{10}(OH)_2$		×××	
99	protoenstatite	$MgSiO_3$ (high-temperature form) see No. 92	×××		
100	pumpellyite	$Ca_2Al_3Si_3O_{11}(OH)_3$		×××	
101	pyrope	$Mg_3Al_2Si_3O_{12}$	××	××	
102	pyrophyllite	$Mg_3Si_4O_{10}(OH)_2$		×××	
103	pyroxenes	$(Ca, Na)(Mg, Fe^{2+}, Al, Fe^{3+}, Ti^{4+}, Mn)$ $\cdot (Si, Al, Fe^{3+})_2O_6$	××	××	
104	quartz	SiO_2	×	×	×
105	rhodochrosite	$MnCO_3$		××	××

continued

Table 8 (continued)

No.	Mineral name	Composition	Occurrence in assemblages of magmatic origin	Occurrence in assemblages of metamorphic origin	Occurrence in assemblages of sedimentary origin
106	rutile	TiO_2	×	×	×
107	sanidine	$KAlSi_3O_8$ (monoclinic)	× × ×		
108	saponite	$Mg_6(Ca_{0.51}Na)_{0.66}Si_{7.34}Al_{0.66}O_{20}(OH)_4 \cdot nH_2O$			× × ×
109	serpentine	see Nos. 15 and 28		× × ×	
110	siderite	$Fe^{2+}CO_3$		× ×	× ×
111	sillimanite	Al_2SiO_5 (orthorhombic)		× × ×	
112	scapolite	$(Na, Ca)_8(Cl_2, SO_4, CO_3)_{1...2}Al_6Si_{18}O_{48}$	× ×	× ×	
113	sodalite	$Na_8Al_6Si_6O_{24}Cl_2$, see Nos. 57 and 86	× × ×		
114	soda-melilite (Natrium-Melilith)	$CaNaAlSi_2O_7$	× × ×		
115	spessartine	$Mn_3Al_2Si_3O_{12}$	× ×	× ×	
116	sphene (Titanit)	$CaTiSiO_5$	× ×	× ×	
117	spinel	$(Mg, Fe^{2+})(Al, Cr, Fe^{3+})_2O_4$, see Nos. 27, 61, 77, and 124	× ×	× ×	
118	staurolite	$(Fe^{2+}, Mg)Al_{18}Si_8O_{44}(OH)_4$		× × ×	
119	stilpnomelane	$(K, Na, Ca)_{0...1.4}(Fe^{3+}, Fe^{2+}, Mg, Al, Mn)_{5.9...8.2} \cdot Si_8O_{20}(OH)_4(O, OH, H_2O)_{3.6...8.5}$		× × ×	
120	talc	$Mg_3Si_4O_{10}(OH)_2$		× ×	×
121	tremolite	$Ca_2Mg_5Si_8O_{22}(OH, F)_2$		× × ×	
122	tridymite	SiO_2	× × ×		
123	tourmaline	$(Na, Ca)(Al, Mg, Fe^{2+}, Mn, Li)_3(Al, Mg)_6 \cdot (BO_3)_3Si_6O_{18}(OH, F)_4$	× × ×		
124	ulvite	$Ti^{4+}Fe_2^{2+}O_4$	× ×	× ×	
125	uvarovite	$Ca_3Cr_2Si_3O_{12}$		× × ×	
126	vesuvianite (idocrase)	$Ca_{38}Al_8Fe_2^{2+}(Al, Mg, Fe^{2+})_{16}Si_{36}O_{140}(OH)_{16}$		× × ×	
127	wairakite	$CaAl_2Si_4O_{12} \cdot 2H_2O$	×		× ×
128	wollastonite	$CaSiO_3$	×	× ×	
129	xenotime	YPO_4	× × ×		
130	zeolites	see Nos. 62 and 73 (and philippsite $(Ca_{0.5}, Na, K)_3 \cdot Al_3Si_5O_{16} \cdot 6H_2O$ as a further example)	×	×	×
131	zircon	$ZrSiO_4$	× ×	× ×	×
132	zoisite	$Ca_2Al_3Si_3O_{12}(OH)$ (orthorhombic)		× × ×	

0.2.2 Quantitative mineral composition of igneous rock types — Quantitativer Mineralbestand der magmatischen Gesteinstypen

The classification of igneous rocks requires knowledge of the amounts (volume) of the rockforming minerals (mode, modal composition) making up a rock. Determination of composition is achieved mainly optically by (a statistical) measurement of mineral grains in thin sections of the rock, by an optical counting of the grains in grain-mounts, and/or by X-raying the rock powder. The amounts of feldspars, quartz, and feldspathoids as light (or salic) constituents contrast with the dark (or mafic) constituents of a rock. Mafic constituents or mafites are ore, olivine, pyroxene, amphibole, mica, apatite, melilite, and calcite. The amount (volume) of these minerals is responsible for the color of the rock and defines the **color index**. It

Für die quantitative Einteilung der magmatischen Gesteine werden die Mengengehalte der gesteinsbildenden Minerale in Volum-Prozent (Modus, modale Zusammensetzung) benötigt. Die Bestimmung der quantitativen Zusammensetzung geschieht durch (statistisches) Ausmessen oder Auszählen der Mineralkörner im Gesteinsdünnschliff unter dem Polarisationsmikroskop, durch mikroskopisches Auszählen des Körnerpräparates und durch röntgenographische (quantitative) Bestimmung des Gesteinspulvers. Die Anteile an Quarz, Feldspäten und Foiden werden als helle Gemengteile den dunklen Gemengteilen (Mafiten) gegenübergestellt. Mafite ist der Sammelbegriff für Erz, Olivin, Pyroxen, Melilith, Amphibol, Glimmer, Apatit

indicates whether a rock specimen is dark (high color index) or light (low color index) in color. Accessory minerals (<1%) are not considered in the nomenclature of the rocks. Subdivisions of igneous rocks (groups or families) can be achieved according to the kind of feldspar (alkalifeldspar vs. plagioclase), the anorthite content of the plagioclase (plagioclase with less than 10 mol-% anorthite is considered as alkalifeldspar), the kind of feldspathoidal minerals (nepheline, leucite, analcite, cancrinite, sodalite), and the kind of mafic minerals. The classification of (magmatic) glass as a component is problematic and a detailed chemical analysis is required to reveal its composition.

A nomenclature scheme on the basis of modal quartz-feldspar-feldspathoid-composition in association with the color index is shown in Fig. 1. Detailed information about intrusive and effusive rocks is given on the basis of the internationally accepted double triangle of quartz-alkalifeldspar-plagioclase-feldspathoids introduced by [Str67] and modified by [Wed69] and for this study (modified and extended in Tables 9, 10).

und Calcit. Der Anteil an Mafiten ist gleichzeitig der **Farbindex** (=Farbzahl), der approximativ angibt, ob ein Gestein im Handstück hell oder dunkel gefärbt ist. Akzessorien (Gemengteile <1%) werden nicht berücksichtigt. Die Unterteilung der Magmatite in Gruppen oder Familien geschieht nach Art der Feldspäte (Alkalifeldspäte und Plagioklase), nach dem Anorthitgehalt der Plagioklase (Plagioklas mit <10 Mol-% Anorthitkomponente zählt als Alkalifeldspat), nach Art der Mafite und Foide (Nephelin, Leucit, Analcim, Cancrinit, Sodalith). Die Zuordnung von Glas zu bestimmten Komponenten ist schwierig. Sie ist ohne Kenntnis der chemischen Zusammensetzung des Glases nur schwer möglich.

In Fig. 1 sind die wichtigsten magmatischen Gesteinstypen auf der Grundlage der Nomenklatur von [Str67], modifiziert nach [Wed69] dargestellt und in Tab. 9, 10 (modifiziert und erweitert) zusammengestellt.

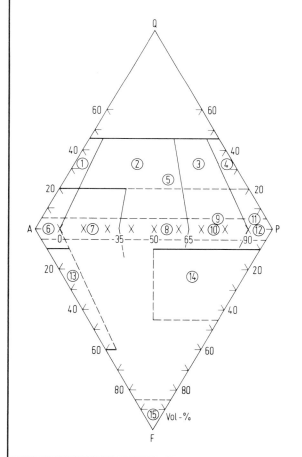

Fig. 1. Modal composition of igneous rocks in terms of alkalifeldspar (A), plagioclase (P), and quartz (Q) or feldspathoids (F) projected on the Q+A+P- and A+P+F-triangles [Str67]. Only igneous rocks having a color index <85 are considered. Field boundaries slightly modified by [Wed69]. Numbers refer to rock names as listed in Tables 9 and 10. Not shown are rocks from group No. 16 with a color index of >85.

Table 9. The major intrusive and volcanic rocks (after [Wed69]; slightly modified); quartz + plagioclase + alkalifeldspar – and plagioclase + alkalifeldspar + feldspathoid – proportions are displayed in Fig. 1. (i = intrusive; v = volcanic; numbers refer to Fig. 1 and Table 10).

No.	Igneous rock name	Common synonym (including altered igneous rocks)	Color index	Remarks
1 i	alkali granite		0– 20	
1 v	alkali rhyolite	pantellerite	0– 12	alkali pyroxene
2 i	granit		5– 20	
2 v	rhyolite	liparite rhyolitic obsidian (quartz porphyry, pitchstone (Pechstein))	0– 15	mainly glass altered rhyolitic glass
3 i	granodiorite		5– 25	biotite, amphibole
3 v	rhyodacite	(quartz porphyrite)	0– 20 (0– 30)	
4 i	quartz diorite	tonalite (dark) trondhjemite (light)	15– 40	
4 v	dacite	(quartz porphyrite)	5– 25 (0– 30)	
5 i	quartz monzonite	adamellite	5– 25	
5 v	quartz latite	dellenite	0– 20	
6 i	alkali syenite		0– 25	
6 v	alkali trachyte	(quartz keratophyre)	0– 20 (0– 20)	
7 i	syenite		10– 35	
7 v	trachyte	(quartz keratophyre, keratophyre)	5– 25 (0– 20; 10– 25)	
8 i	monzonite		15– 45	
8 v	latite	lamprophyre: minette, vogesite	5– 35 30– 50	pyroxene + biotite amphibole
9 i	monzodiorite		20– 50	
9 v	latite andesite	doreite mugearite	15– 40	K-rich alkalifeldspar olivine
10 i	monzogabbro	mangerite	25– 60	
10 v	latite basalt	trachybasalt	40– 60	
11 i	diorite		25– 50	andesine, biotite, amphibole
11 v	andesite	(porphyrite) lamprophyre: kersanite	20– 40 20– 40	biotite > pyroxene
12 i	gabbro		35– 65	labradorite-bytownite; pyroxene
		norite	35– 65	orthopyroxene > clinopyroxene
		anorthosite	0– 10	mostly basic plagioclase
12 v	tholeiitic basalt	tholeiite	40– 75	hypersthene in norm
	alkali (olivine) basalt		40– 75	nepheline in norm
		diabase (spilite)		euhedral plagioclase in pyroxene matrix (veränderte Basalte)
	picrite basalt	picrite oceanite ankaramite	40– 75	> 20% olivine; pyroxene < 20% olivine; pyroxene
		lamprophyre: camptonite	40– 75	pyroxene + amphibole

continued

Table 9 (continued)

No.	Igneous rock name	Commonsynonym (including altered igneous rocks)	Color index	Remarks
13 i	nepheline syenite		0– 85	
		foyaite	0– 30	
		malignite	30– 60	
		shonkinite	60– 85	
13 v	phonolite		0– 25	
14 i	essexite		30– 60	nepheline as feldspathoid (Foid) plagioclase > alkalifeldspar
	theralite			nepheline as feldspathoid (Foid) plagioclase < alkalifeldspar
	teschenite			analcite as feldspathoid (Foid)
14 v	nepheline basanite		30– 70	> 5% olivine
	nepheline tephrite			< 5% olivine
	leucite basanite			> 5% olivine
	leucite tephrite			< 5% olivine
	analcite basanite			> 5% olivine
	analcite tephrite			< 5% olivine
15 i	ijolite		30– 60	nepheline
	melteigite		60– 85	nepheline
	urtite		0– 30	nepheline
	fergusite		40– 90	leucite
	tawite		40– 90	sodalite
	turjaite		40– 90	feldspathoids (Foide) + melilite
15 v	nephelinite		40– 75	
	ankaratrite	olivine nephelinite	60– 85	> 50% pyroxene, ± melilite
	leucitite		40– 75	
	melilitite		40– 75	feldspathoids (Foide) + melilite
		lamprophyre: alnöite	40– 75	melilite; biotite > pyroxene
16 i	peridotite		85–100	olivine 30–90
		lherzolite	85–100	olivine + pyroxenes (spinel, garnet)
	dunite		100	olivine ≫ pyroxenes
	griquaite	garnet-pyroxenite	100	garnet + pyroxenes + olivine
	pyroxenite		100	pyroxenes > olivine
		websterite	100	ortho- + clinopyroxene
		wherlite	100	clinopyroxene + olivine
	hornblendite		100	amphibole
16 v	komatiite		100	(with devitrified glass)
	kimberlite		100	phlogopite-bearing

Table 10. Average chemical and normative*) compositions (in wt-%) of common igneous rocks (after [Noc54]). Numbers refer to Table 9 and Fig. 1; i are intrusive and v are volcanic igneous rocks.

	Alkali granites	Alkali rhyolites	Granites	Rhyolites	Quartz monzonites	Quartz latites	Granodiorites	Rhyodacites	Quartz diorites
	1 i	1 v	2 i	2 v	5 i	5 v	3 i	3 v	4 i
chemical composition									
SiO_2	73.86	74.57	72.08	73.66	69.15	70.15	66.88	66.27	66.15
TiO_2	0.20	0.17	0.37	0.22	0.56	0.42	0.57	0.66	0.62
Al_2O_3	13.75	12.58	13.86	13.45	14.63	14.41	15.66	15.39	15.56
Fe_2O_3	0.78	1.30	0.86	1.25	1.22	1.68	1.33	2.14	1.36
FeO	1.13	1.02	1.67	0.75	2.27	1.55	2.59	2.23	3.42
MnO	0.05	0.05	0.06	0.03	0.06	0.06	0.07	0.07	0.08
MgO	0.26	0.11	0.52	0.32	0.99	0.63	1.57	1.57	1.94
CaO	0.72	0.61	1.33	1.13	2.45	2.15	3.56	3.68	4.65
Na_2O	3.51	4.13	3.08	2.99	3.35	3.65	3.84	4.13	3.90
K_2O	5.13	4.73	5.46	5.35	4.58	4.50	3.07	3.01	1.42
H_2O^+	0.47	0.66	0.53	0.78	0.54	0.68	0.65	0.68	0.69
P_2O_5	0.14	0.07	0.18	0.07	0.20	0.12	0.21	0.17	0.21
CO_2									
Cl									
SO_3									
normative composition *)									
qz (quartz)	32.2	31.1	29.2	33.2	24.8	26.1	21.9	20.8	24.1
or (K-feldspar)	30.0	27.8	32.2	31.7	27.2	26.7	18.3	17.8	8.3
ab (albite)	29.3	35.1	26.2	25.1	28.3	30.9	32.5	35.1	33.0
an (anorthite)	2.8	2.0	5.6	5.0	11.1	9.5	16.4	14.5	20.8
cor (corundum)	1.4		0.8	0.9					
ne (nepheline)									
di (diopside)		0.2				0.4		2.4	0.6
hy (hypersthene)	1.7	0.8	3.0	0.8	4.5	2.2	6.8	4.1	8.7
fo (forsterite)									
mt (magnetite)	1.2	1.9	1.4	1.9	1.9	2.5	1.9	3.0	2.1
ilm (ilmenite)	0.5	0.3	0.8	0.5	1.1	0.8	1.1	1.4	1.2
ap (apatite)	0.3	0.2	0.4	0.2	0.5	0.3	0.5	0.3	0.5
cc (calcite)									
lc (leucite)									
hm (hematite)									
la (larnite)									

*) For the CIPW-Norm, see p. 17.

continued

Table 10 (continued)

	Dacites	Alkali syenites	Alkali trachytes	Syenites	Trachytes	Monzonites	Latites	Monzodiorites	Latite andesites
	4 v	6 i	6 v	7 i	7 v	8 i	8 v	9 i	9 v
chemical composition									
SiO_2	63.58	61.86	61.95	59.41	58.31	55.36	54.02	54.66	56.00
TiO_2	0.64	0.58	0.73	0.83	0.66	1.12	1.18	1.09	1.29
Al_2O_3	16.67	16.91	18.03	17.12	18.05	16.58	17.22	16.98	16.81
Fe_2O_3	2.24	2.32	2.33	2.19	2.54	2.57	3.83	3.26	3.74
FeO	3.00	2.63	1.51	2.83	2.02	4.58	3.98	5.38	4.36
MnO	0.11	0.11	0.13	0.08	0.14	0.13	0.12	0.14	0.13
MgO	2.12	0.96	0.63	2.02	2.07	3.67	3.87	3.95	3.39
CaO	5.53	2.54	1.89	4.06	4.25	6.76	6.76	6.99	6.87
Na_2O	3.98	5.46	6.55	3.92	3.85	3.51	3.32	3.76	3.56
K_2O	1.40	5.91	5.53	6.53	7.38	4.68	4.43	2.76	2.60
H_2O^+	0.56	0.53	0.54	0.63	0.53	0.60	0.78	0.60	0.92
P_2O_5	0.17	0.19	0.18	0.38	0.20	0.44	0.49	0.43	0.33
CO_2									
Cl									
SO_3									
normative composition*)									
qz (quartz)	19.6	1.7		2.0			0.5	2.0	7.2
or (K-feldspar)	8.3	35.0	32.8	38.4	43.9	27.8	26.1	16.7	15.6
ab (albite)	34.1	46.1	54.0	33.0	28.8	29.3	27.8	31.9	29.9
an (anorthite)	23.3	4.2	3.3	10.0	9.7	15.8	19.2	21.1	22.2
cor (corundum)									
ne (nepheline)			0.6		2.0				
di (diopside)	2.4	5.5	3.7	5.6	7.9	11.7	8.4	8.4	7.6
hy (hypersthene)	7.0	2.0		4.5		6.7	8.2	11.4	8.0
fo (forsterite)					1.6	1.2			
mt (magnetite)	3.3	3.3	3.3	3.3	3.7	3.7	5.6	4.9	5.3
ilm (ilmenite)	1.2	1.2	1.4	1.5	1.2	2.1	2.3	2.1	2.4
ap (apatite)	0.3	0.5	0.4	1.0	0.5	1.0	1.2	1.0	0.8
cc (calcite)									
lc (leucite)									
hm (hematite)									
la (larnite)									

*) For the CiPW-Norm, see p. 17.

continued

Table 10 (continued)

	Diorites	Andesites	Gabbros	Tholeiitic basalts	Alkali olivine basalts	Anorthosite	Nepheline syenites	Phonolites	Essexites
	11 i	11 v	12 i	12 v	12 v	12 i	13 i	13 v	14 i
chemical composition									
SiO_2	51.86	54.20	48.36	50.83	45.78	54.54	55.38	56.90	46.88
TiO_2	1.50	1.31	1.32	2.03	2.63	0.52	0.66	0.59	2.81
Al_2O_3	16.40	17.17	16.81	14.07	14.64	25.72	21.30	20.17	17.07
Fe_2O_3	2.73	3.48	2.55	2.88	3.16	0.83	2.42	2.26	3.62
FeO	6.97	5.49	7.92	9.00	8.73	1.46	2.00	1.85	5.94
MnO	0.18	0.15	0.18	0.18	0.20	0.02	0.19	0.19	0.16
MgO	6.12	4.36	8.06	6.34	9.39	0.83	0.57	0.58	4.85
CaO	8.40	7.92	11.07	10.42	10.74	9.62	1.98	1.88	9.49
Na_2O	3.36	3.67	2.26	2.23	2.63	4.66	8.84	8.72	5.09
K_2O	1.33	1.11	0.56	0.82	0.95	1.06	5.34	5.42	2.64
H_2O^+	0.80	0.86	0.64	0.91	0.76	0.63	0.96	0.96	0.97
P_2O_5	0.35	0.28	0.24	0.23	0.39	0.11	0.19	0.17	0.48
CO_2							0.17		
Cl								0.23	
SO_3								0.13	
normative composition *)									
qz (quartz)	0.3	5.7		3.5		1.4			
or (K-feldspar)	7.8	6.7	3.3	5.0	6.1	6.7	31.1	31.7	15.6
ab (albite)	28.3	30.9	18.9	18.9	18.3	39.3	32.0	36.2	14.7
an (anorthite)	25.8	27.2	34.2	25.9	24.7	45.9	2.8	1.7	16.1
cor (corundum)									
ne (nepheline)					2.3		23.3	18.7	15.3
di (diopside)	10.4	7.8	14.9	19.2	20.8	0.6	4.1	5.2	22.2
hy (hypersthene)	19.0	12.6	14.5	18.1		3.6			
fo (forsterite)			6.8		16.5		0.2		3.6
mt (magnetite)	3.9	5.1	3.7	4.2	4.6	1.2	3.5	3.3	5.3
ilm (ilmenite)	2.9	2.4	2.4	3.8	5.0	0.9	1.4	1.2	5.3
ap (apatite)	0.8	0.7	0.6	0.5	1.0	0.3	0.4	0.3	1.2
cc (calcite)							0.4		
lc (leucite)									
hm (hematite)									
la (larnite)									

continued

Table 10 (continued)

	Nepheline tephrites	Leucite tephrites	Ijolites	Olivine nephelinites	Olivine leucitites	Olivine melilitites	Peridotites	Kimberlite[1]	Komatiite[2]
	14v	14v	15i	15v	15v	15v	16i	16v	16v
chemical composition									
SiO_2	44.82	47.05	42.58	40.29	43.64	37.08	43.54	27.93	45.58
TiO_2	2.65	1.54	1.41	2.90	2.54	3.31	0.81	2.73	0.40
Al_2O_3	15.42	16.05	18.46	11.32	10.82	8.08	3.99	4.47	8.91
Fe_2O_3	4.28	3.49	4.01	4.87	5.11	5.12	2.51	7.04	
FeO	6.61	5.78	4.19	7.69	5.89	7.23	9.84	5.12	11.29 (as total Fe)
MnO	0.16	0.17	0.20	0.22	0.15	0.18	0.21	0.23	0.23
MgO	7.27	6.20	3.22	13.28	13.86	16.19	34.02	25.42	22.57
CaO	10.32	10.80	11.38	12.99	10.66	16.30	3.46	10.01	9.46
Na_2O	5.30	2.35	9.55	3.14	2.16	2.30	0.56	0.21	0.64
K_2O	1.26	5.38	2.55	1.44	4.09	1.36	0.25	1.18	0.09
H_2O^+	1.56	0.60	0.55	1.08	0.72	1.89	0.76	7.89	0.02
P_2O_5	0.35	0.59	1.52	0.78	0.63	0.96	0.05	1.07	
CO_2			0.38					5.61	
Cl									
SO_3									
normative composition *)									
qz (quartz)									
or (K-feldspar)	7.8	22.2	10.0		6.9		1.7	7.0	0.5
ab (albite)	12.6						4.7	1.8	5.5
an (anorthite)	14.5	17.5		12.8	6.1	7.5	7.5	7.3	21.3
cor (corundum)								0.2	
ne (nepheline)	17.3	10.8	43.7	14.2	9.9	10.5			
di (diopside)	27.2	26.1	28.9	32.5	33.4	20.1	7.3		20.9
hy (hypersthene)							14.0	0.5	9.1
fo (forsterite)	7.0	6.0		16.8	15.3	24.8	58.7	44.5	41.9
mt (magnetite)	6.3	5.1	5.8	7.2	7.4	7.4	3.7	9.4	
ilm (ilmenite)	5.0	2.9	2.7	5.5	4.9	6.2	1.5	5.2	0.8
ap (apatite)	0.8	1.3	3.6	1.8	1.5	2.3	0.1	2.6	
cc (calcite)			0.9					12.9	
lc (leucite)		7.4	3.6	6.5	13.8	6.5			
hm (hematite)								0.6	
la (larnite)				1.6		12.8			

[1] Chemical composition from [Daw62].
[2] Chemical composition from [Nes76].

0.2.3 The chemical composition of igneous rocks — Chemische Zusammensetzung der magmatischen Gesteine

The chemical composition of igneous rock is obtained with the methods of analytical chemistry. It is given (conventionally) in wt-% of the oxides of the elements. The average chemical composition of igneous rocks is granitoid (granite like) as granites, granodiorites, and quartz diorites represent about 85% of the total (continental) crust (see Table 3). Specific chemical compositions of magmatic rocks of intrusive and volcanic origin are listed together in Table 10. It can be demonstrated that in igneous rocks large amounts of SiO_2 correspond with low amounts of MgO, FeO, and CaO and with relatively large amounts of alkalies. On the other hand, rocks with low SiO_2 contents have low alkali and high MgO, FeO, and CaO contents. Thus, the silica (SiO_2) content of a rock can be used for a first (and rough) classification of the igneous rocks on the basis of their acidity. Magmatites with more than 66% SiO_2 are considered as *acid*, with 66 to 52% SiO_2 as *intermediate*, with 52 to 45% SiO_2 as *basic*, and with less than 45% SiO_2 as *ultra-basic*. Any further and more specific classification of igneous rocks on the basis of the rock chemistry with about 8 major and up to 10 secondary oxides (not counting the trace elements) is problematical. Certain elements, however, substitute each other (e.g. $Mg \rightleftharpoons Fe$ or $Ca \rightleftharpoons Na$) in the rockforming minerals (because of their similar ionic sizes) and reduce the number of components in the natural rock system.

Thus, based on the rock chemistry a calculation of a virtual mineral composition can be conducted which represents the norm or normative composition of an igneous rock (after **C**ross, **I**ddings, **P**irson, and **W**ashington, 1902, CIPW-Norm). Normative mineral compositions of the specific igneous rocks listed in Table 10 have been calculated in terms of the CIPW-procedure.

Normative and modal compositions of igneous rocks are not identical in most cases. The normative composition or the norm can be considered as a translation of the chemistry of the rocks into a virtual or fictive mineral assemblage. It offers a possibility to compare chemically different igneous rocks and is used for a (chemical) nomenclature scheme. Igneous rocks of identical chemistry must not have identical modes. According to the pressure-temperature-regime in the crust (and upper mantle) chemically identical magmas may crystallize in very distinct mineral assemblages. For example, a granitic magma will crystallize as a rhyolite (the volcanic equivalent of granite) containing quartz and alkalifeldspar with (almost) no biotite at the earth's surface, or as a real

Die chemische Zusammensetzung eines magmatischen Gesteins wird mit Hilfe analytischer Methoden der Chemie bestimmt und (konventionell) in Gewichtsprozenten der Element-Oxide gegeben. Die mittlere chemische Zusammensetzung der Magmatite ist granitoid; d.h. Granite, Granodiorite und Quarzdiorite, die etwa 85% des Volumenanteils der Magmatite ausmachen, sind repräsentativ für die Erdkruste (siehe Tab. 3). Chemische Zusammensetzungen der (wichtigsten) magmatischen Gesteinstypen sind in Tab. 10 zusammengestellt, in der neben der Zusammensetzung des Tiefengesteins auch die des entsprechenden Ergußgesteins gegeben wird. Es zeigt sich deutlich, daß in Gesteinen mit hohen SiO_2-Werten die MgO-, FeO- und CaO-Werte (bei hohen Alkali-Werten) im allgemeinen klein sind, während solche mit niedrigen SiO_2-Werten im allgemeinen hohe MgO-, FeO- und CaO-Werte (bei niedrigen Alkali-Werten) aufweisen. Die Gehalte an Kieselsäure können zu einer groben Einteilung benutzt werden: Gesteine mit >66% SiO_2 werden als *sauer*, mit 66···52% SiO_2 als *intermediär*, mit 51···45% SiO_2 als *basisch* und mit <45% SiO_2 als *ultrabasisch* bezeichnet. Eine (spezielle) Einteilung auf der Grundlage der Oxidprozente ist bei wenigstens 10 Haupt- und Nebenelementen schwierig.

Da sich aber bestimmte Elemente in den gesteinsbildenden Mineralen der magmatischen Gesteine (wie auch in den Sedimenten und Metamorphiten) wegen ihrer gleichartigen Kationengröße ersetzen, ist die Berechnung eines (virtuellen) Mineralbestandes aus der Analyse (der sogenannten CIPW-Norm, nach **C**ross, **I**ddings, **P**irson & **W**ashington, 1902) möglich. Die nach der CIPW-Norm aus den chemischen Analysen berechneten normativen Mineralbestände sind für die wichtigsten magmatischen Gesteine in Tab. 10 gegeben.

Die Norm ist nicht mit dem Modus identisch. Sie ist vielmehr die Übersetzung der chemischen Analyse in einen (nach bestimmten Regeln und theoretisch-mineralogischen Gesichtspunkten) normativen und damit virtuellen Mineralbestand. Dieser dient dazu, chemisch unterschiedliche Gesteine (auf der Grundlage ihrer normativen Mineralzusammensetzung) vergleichen und einteilen zu können. Demnach können chemisch identische Gesteine einen unterschiedlichen Modus (aber gleiche Norm!) aufweisen. Entsprechend dem Druck-Temperatur-Regime in der Kruste (und oberem Mantel) kristallisieren chemisch identische Magmen in unterschiedlichen Mineralassoziationen aus. Zum Beispiel kristallisiert granitisches Magma an der Oberfläche als Rhyolith mit nur einem Alkali-

granite (the plutonic equivalent) containing quartz, biotite as well as alkalifeldspar and acidic plagioclase within the crust. Thus, chemically identical magmas can occur in different mineral assemblages, a feature called **heteromorphism.**

Chemical analyses of igneous rocks cannot measure the real amounts of former volatile components (or volatiles) which were present in the magma. Only fractions of H_2O, CO_2, F, Cl, S, and P_2O_5 are incorporated in the structures of the crystallizing minerals because the larger amount of the volatiles which had been dissolved in the magma will have escaped during solidification

Feldspat und keinem Glimmer; aber in der Tiefe als Granit mit Alkali-Feldspat und saurem Plagioklas sowie Glimmer. Die Eigenschaft, daß chemisch identische Magmatite in verschiedenen Mineralgemengen (oder - Kombinationen) existieren können, bezeichnet man als **Heteromorphismus.**

Mit der chemischen Analyse eines Gesteins sind die ehemals im Magma gelösten, leichtflüchtigen oder volatilen Stoffe nicht mehr (vollständig) zu finden. Nur ein Teil des H_2O, CO_2, F, S, Cl oder P_2O_5 wird in Minerale der plutonischen oder hypabyssischen Assoziationen eingebaut. Der überwiegende Teil der volatilen Bestandteile muß bei der Erstarrung aus dem Magma entwichen sein.

0.2.4 The texture and structure of igneous rocks — Das Gefüge der magmatischen Gesteine

The texture and structure of igneous rocks are determined microscopically on oriented thin sections of the rocks and or by X-ray methods. The most obvious feature of an igneous rock is its **texture**. It is the result of the consolidation of the magma and reflects the physico-chemical conditions of the melt during crystallization. The **structure** is the result of the hydro-mechanical behaviour of the crystal-bearing melt when moving relatively (crystal settling, flow-direction of magma etc.). The texture is established by the order of the minerals precipitated from the melt. The structure represents the spatial orientation of the minerals composing an igneous rock.

Whether a crystalline rock or a glass is formed from a magma depends mostly on its viscosity and the speed of cooling. Igneous rocks containing crystalline phases only are called *holocrystalline*, rocks consisting wholly of glass are called *holohyaline*. There are, however, *hypocrystalline* rocks containing both the crystals and the glass.

The grain size (or granularity) of mineral phases composing an igneous rock can only be obtained by statistical examination of thin- or polished-sections of the rock. Rockforming minerals are, however, scarcely ideal spheres or isometric polyhedra. They are prisms, needles, plates rather than (ideal) cubes. The average grain size of the mineral phases can be deduced approximately from their length/width ratios or from an arbitrary intersection along the profile line across the thin- or polished-section of the rock. A classification of igneous rocks on the basis of their grain size distribution can be achieved by the DIN 4022 classification scheme. This scheme was made for clastic (sedimentary) rocks in Germany in particular but is also applicable for grains of any other rocks. Grain size fractions are derived by a division of the grain radii on the basis of logarithmic scales. Igneous rocks whose mineral grains are so

Das **Gefüge** der magmatischen Gesteine wird mikroskopisch im Gesteinsdünnschliff und/oder röntgenographisch mit dem Texturgoniometer bestimmt. Beim Gefüge der magmatischen Gesteine sind zwei Eigenschaften zu unterscheiden, die **Struktur (= texture)** und die **Textur (= structure)**. Die **Struktur** ist das Abbild der Verfestigungsvorgänge des Magmas und gibt die chemisch-physikalischen Bedingungen bei der Erstarrung des Magmas bzw. Gesteinsschmelze wider. Die **Textur** erlaubt Hinweise auf das (hydro-) mechanische Verhalten der Schmelze bei ihrer Erstarrung. Prinzipiell ist das **Gefüge** der magmatischen Gesteine gekennzeichnet durch die Kristallisations-Abfolge der gesteinsbildenden Mineralphasen.

Von der Abkühlungsgeschwindigkeit und Viskosität des Magmas hängt es ab, ob ein kristallines Gestein oder ein Gesteinsglas entsteht. Gesteine, in denen alle Phasen als Kristalle vorliegen bezeichnet man als *holokristallin*, Gesteinsgläser als *hyalin*, Gesteine mit glasigen und kristallinen Anteilen als *hypokristallin*.

Die absolute (Korn-) Größe der Mineralphasen eines magmatischen Gesteins ist für das kompakte Gestein nur im Dünn- oder Anschliff mit statistischen Näherungsmethoden zu ermitteln. Die wenigsten gesteinsbildenden Minerale sind Kugeln oder isometrische Körper, sie sind prismatisch, nadelig oder blättchenförmig. Ihr mittlerer Korndurchmesser oder ihre mittlere Korngröße ergibt sich nur annähernd aus den Längen/Breitenverhältnissen der Mineralphasen oder aus willkürlichen Anschnitten einer Meßtraverse im Schliff. Eine Einteilung läßt sich am besten nach DIN 4022 mit Hilfe einer Korngrößenverteilungs- oder Korngrößensummenkurve vornehmen. Diese Einteilung ist speziell für die klastischen Sedimente in Deutschland entwickelt worden. Die Korngrößen-Fraktionen sind logarithmisch eingeteilt und erlauben auch eine Klassifizierung der magmatischen Gesteine

minute that a microscopic identification is difficult are called *cryptocrystalline* (grains less than 0.002 mm in diameter). Rocks containing mineral grains with diameter between 0.002 and 0.200 mm are called *fine-grained*, between 0.200 and 2.000 mm *medium-grained*, between 2.000 and 20.000 mm *coarse-grained*, and > 20.000 mm *gigantic-grained*.

The grain-size-distribution indicates whether the rock has grains of identical (= equigranular) or different (= inequigranular) sizes. In the first case, the grain-size-distribution of the rock contains one or a few size fractions only. Grain sizes are distributed over a larger range in the second case. Their grain-size distribution-curve (size vs. wt-% of grains) may contain more than one maximum and minimum. The latter grain-size-distribution is typical for the porphyritic texture, i.e. a (relatively) fine-grained groundmass embedding (relatively) large crystals as phenocrysts.

nach der (absoluten) Korngröße. Gesteine deren Minerale mikroskopisch nicht mehr auflösbar sind, bezeichnet man als dicht oder *kryptokristallin* (Korndurchmesser < 0,002 mm). Gesteine mit Mineralphasen, deren Durchmesser zwischen 0,002 bis 0,20 mm liegt, nennt man *feinkörnig*, zwischen 0,2 bis 2000 mm, *mittelkörnig*, zwischen 2000 bis 20000 mm *grobkörnig* und > 20000 mm *riesenkörnig*.

Die Korngrößen-Verteilung zeigt auch an, ob die Gesteine gleich- oder ungleichkörnig (= wechselkörnig) sind. Gleichkörnige Gesteine verteilen ihre Korngrößen auf eine oder nur wenige Korngrößenklassen. Die Korngrößen-Verteilungs-Kurve der ungleichkörnigen Gesteine kann Maxima und Minima aufweisen. Solche Verteilungen sind typisch für porphyrische Strukturen: Diese zeigen eine (relative) feinkörnige Grundmasse, in die (relativ) grobe Kristalle als Einsprenglinge (Phänokryste) eingebettet sind.

Fig. 2a⋯c. Thin section drawings for illustrating rock textures; from [Cor68].

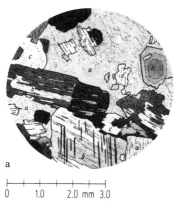

a

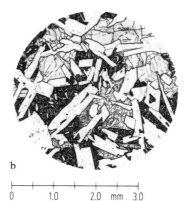

b

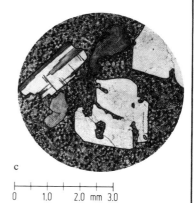

c

a) Granit, La Ginnea, Genova, Italy. Holocrystalline and equigranular; hypidiomorphic crystals of orthoclase (a), quartz (c), and mica (d). Plagioclase (b) is more or less idiomorphic.

b) Tholeiitic basalt, Sababurg, Lowe Saxony. Almost holocrystalline, idiomorphic phenocrysts of plagioclase (light) with interstices of (xenomorphic) pyroxenes (gray) and of matrix (dark) consisting of fine-grained crystals and minute patches of (devitrified) glass.

c) Rhyolite, Colmitz, Saxony. Holocrystalline, porphyritic, more or less idiomorphic phenocrysts of quartz (light and corroded) and plagioclase (with cleavage and twin lamella) in a fine-grained matrix of quartz, feldspar, and mica.

In the nomenclature of igneous rock textures the shape and mutual relationship between the grains is considered as the fabric. Mineral phases which articulate their crystal phases while the magma is on cooling are called *euhedral* or *idiomorphic* (automorphic). Fine examples are the euhedral phenocrysts of the porphyritic textures. Crystals precipitated early disable those which crystallize from the magma during a later stage. They may partly be bounded by crystal faces and their fabric is called *subhedral* or *hypidiomorphic*. If mineral grains are devoid of crystal boundaries, the term *anhedral* or *xenomorphic* (allotriomorphic) must be applied.

Textural features such as fine- or coarse-grain, hypohyaline or holocrystalline, porphyritic or granular are not only an indication for volcanic or plutonic consolidation of magmas. The speed of cooling, the viscosity, and the amount of volatiles dissolved in the magma determine the textural features in the first place. The (geologic) position within the crust and mantle at which solidification takes place is of secondary importance. Contacts of plutonic rocks with the (cold) country rocks usually causes chilled margins which are mostly fine-grained or aphanitic. Large-scale dikes reveal chilled margins which are hypocrystalline or holohyaline because of rapid cooling. Thick dikes or sills contain coarse-grained textures in their center portion because of slow cooling. (Undersaturated) alkalibasalts can produce mineral assemblages containing nepheline + albite in the chilled margins but analcite + albite in the center parts because rapid crystallization of their outer parts prevented diffusion of H_2O to the outside.

Preferred orientation of mineral grains within the three-dimensional space of the rock can cause banding and lineation. Such large-scale features are called structure. An igneous rock can crystallize without any preferred orientation of its rockforming minerals (isotropic structure). This structure type is hardly realized in nature. Most igneous rocks, however, do reveal – obscure or strongly visible – preferred orientation of their mineral grains (anisotropic structure). Gravitational descent (or ascent) of non-isometric crystals (as micas, feldspars, pyroxene, etc.) in a solidifying magma or floating of elongated crystals in a flowing magma causes banding and lineation. The descent of mineral grains arranges them with the greatest surface (almost) perpendicular to the direction of deposition (i.e. sub-parallel to the floor of the magma reservoir). The results are magmatic sediments (cumulates) formed by gravity from a magma. Magmatic flow orientates crystals with their longest axis sub-parallel to the flow direction (a feature well displayed in the so-called trachytic texture of thin-section scale). Preferred orientation can also be produced by tensional forces while the magma is under consolidation.

Bei den magmatischen Strukturen spielt die Ausbildung der Gemengteile, d.h. ihre Korn- bzw. Kristall-Form eine Rolle. Mineralphasen, die im abkühlenden Magma ihre Eigengestalt ausbilden können, nennt man **idiomorph** (automorph). Beispiele sind die idiomorphen Einsprenglinge der porphyrischen (ungleichkörnigen) Strukturen. Diese früh ausgeschiedenen Kristalle behindern solche, die in der Reihenfolge zeitlich nach ihnen kristallisieren. Diese weisen Fremdgestalt auf und sind nur zum Teil idiomorph oder **hypidiomorph.** Ist die Eigengestalt vollständig unterdrückt, spricht man von **xenomorph** (allotriomorph).

Struktureigenschaften wie grob- oder feinkörnig, glasig oder kristallin, ungleichkörnig oder gleichkörnig sind nicht immer ein Hinweis auf effusive oder plutonische Erstarrung eines Magmas. Die Abkühlungsgeschwindigkeit, die Viskosität und Gehalte an leichtflüchtigen Bestandteilen bestimmen die Struktureigenschaften des Magmatits und erst in zweiter Linie die geologische Position. So sind die Kontaktzonen von Tiefengesteinen zu (relativ kälteren) Nebengesteinen feinkörnig und dicht (Salband). Bei mächtigen (Basalt-) Gängen ist der Kontaktbereich durch die Temperaturunterschiede Gang/Nebengestein durch Abkühlung (oder Abschreckung) sogar glasig ausgebildet. Im Inneren des Ganges verlief die Abkühlung langsam, die Strukturen sind grobkörnig (=doleritisch). Schnelle Erstarrung läßt in Si-untersättigten Basalten im Salband die Mineralparagenesen Nephelin + Albit, im Inneren (durch Verbleib des H_2O im System) Analcim + Albit kristallisieren.

Unter der Textur oder den texturellen Eigenschaften eines magmatischen Gesteins versteht man die räumliche Anordnung der Mineralphasen im Gesteinsverband. Ein magmatisches Gestein kann richtungslos (isotrop) erstarrt sein. Die Kristalle zeigen entsprechend ihrer Gestalt oder Kornform keine Regelung oder Ausrichtung. Dieser Fall ist jedoch selten verwirklicht. Die meisten Gesteine zeigen verborgene oder offen sichtbare anisotrope Texturen. Schon beim (gravitativen) Absinken (oder Aufsteigen) von anisometrischen Kristallen (wie Glimmer oder tafelige Feldspäte) wird sich eine Regelung der Kristalle einstellen. Ebenso können magmatische Kristalle beim Strömen des Magmas eingeregelt werden. Beim Absinken (oder Aufsteigen) setzen sich die Kristalle mit ihrer größten Oberfläche senkrecht zur Absatzrichtung ab, beim Strömen ordnen sich die Kristalle mit ihrer längsten Achse in die Fließrichtung ein. Das Ergebnis sind magmatische Sedimente (Kumulate), deren Absatz nach der Schwerkraft in einer Gesteinsschmelze erfolgte. Steht das Magma beim Erstarren unter Spannungen (Stress) so kann ebenfalls eine Einregelung der Kristalle erfolgen.

0.2.5 The origin and genesis of igneous rocks — Die Herkunft der magmatischen Gesteine

In the previous sections the mode and the norm (chemical composition) have been used to classify igneous rocks. Igneous rocks, however, are the product of molten silicate material, the magma. This implies the question where these magmas originate, i.e. where and how are they formed. Generally speaking, the deeper crust and the upper mantle are able to produce all the specific magmas exposed as igneous rocks on the earth's surface. Granit-like magmas are (mainly) produced by partial melting of (deeper) crustal material, basalt-like magmas by partial melting of (upper) mantle material. The trigger mechanism for partial melting is the temperature distribution as a function of pressure (depths) and (mineral) composition in the crust and upper mantle. Partial melting takes place when the temperature intersects the solidus of the specific mineral assemblages in the crust or the upper mantle. Due to the presence of a (multicomponent) vapor phase (which is predominantly H_2O-rich) in the crustal environment, the solidus of the silicate material is lowered with increasing pressure (depth). Increasing pressure also increases the incorporation of the volatiles in the melt. Previous sediments and igneous rocks such as shales, graywackes, arkoses, and granitic volcanics which have undergone high grade metamorphism in the deeper crust will produce melt-fractions with the composition of (ideal) granite in the presence of a H_2O-vapor phase. For example, at about 4 kbar (about 12 km depth) metasediments containing alkalifeldspars + quartz + H_2O have a solidus which is about 650 °C and partial melting will occur at that particular pressure-temperature-regime in the deeper crust. Separation of granitic liquid from the solid remnants and the rising of the liquid into the upper crust or to the surface causes intrusive bodies, dike rocks, or volcanics respectively.

From a physico-chemical point of view similar partial melting processes take place within the upper mantle. Compared with the deeper crust, the pressure-temperature-regime of the upper mantle is > 10 kbar (> 30 km depth) and > 900 °C. The composition of the upper mantle is predominantly peridotitic with CO_2 + H_2O as major components in the vapor phase. The mol fraction of CO_2 in the vapor phase is (mainly) responsible for the nature of the melts generated at a given temperature and pressure within the upper mantle. A low molar fraction of CO_2 in the (mainly) binary vapor phase decreases the garnet peridotite solidus considerably and produces quartz-normative magmas (andesites, quartz-tholeiites, tholeiites, etc.). Increasing molar fractions of CO_2 require higher solidus temperatures at a given pressure and the partial melt will

Neben den allgemeinen (deskriptiven) Einteilungskriterien wie Modus, chemische Analyse und Gefüge steht die Frage nach der Herkunft der magmatischen Gesteine und den Prozessen, die Magmen auf oder in unserer Erde entstehen lassen. Prinzipiell sind die (tiefere) Erdkruste und der (Obere) Mantel in der Lage, spezifische Magmen hervorzubringen. Granitoide Magmen entstehen durch partielle Aufschmelzungsprozesse von Gesteinen in der tieferen Erdkruste, basaltoide Magmen durch Aufschmelzungsprozesse im Oberen Erdmantel. Der auslösende Mechanismus der Aufschmelzung ist die Druck-Temperatur-Verteilung in Erdkruste und Erdmantel. Durch Überschreiten der Gesteinssolidi werden (in unterschiedlicher Tiefenlage) Krusten- und Mantelgesteine partiell aufgeschmolzen oder flüssig. Von entscheidender Bedeutung ist die Tatsache, daß sowohl Kruste als auch Oberer Mantel mit leichtflüchtigen Anteilen (wesentlich H_2O in der Kruste, H_2O + CO_2 im Oberen Mantel) gesättigt sind. Die Anwesenheit der (multikompositionellen) Gasphase bewirkt bei den silikatischen Krusten- und Mantelgesteinen (als Funktion des Druckes) eine (drastische) Erniedrigung der Solidustemperatur. Mit Zunahme des Druckes (also der Tiefenlage) werden zunehmende Anteile der leichtflüchtigen Komponenten im Magma gelöst und die Solidustemperatur (bei Zunahme des Druckes) erniedrigt. Aus ehemaligen (jetzt metamorphen) Sedimenten wie Tonschiefer, Grauwacken, Arkosen und granitischen Vulkaniten kann so bei H_2O-Sättigung des Gesteinssystem eine (ideal-) granitische Fraktion ausgeschmolzen werden, die vom (kristallinen unlöslichem) Rest oder Restbestand getrennt als Granit in höhere Krustenteile intrudieren kann. Bei einem Druck von ≈ 4,0 kbar H_2O betragen die Solidustemperaturen etwa 650 °C, eine Temperatur bei der die granitische Schmelze gebildet werden kann. Trennung der Schmelze vom kristallinen Bestand, der Aufstieg der Schmelze in die Kruste oder zur Oberfläche verursacht intrusive Körper, Ganggesteine oder Vulkanite.

Ähnlich sind die Aufschmelzungsvorgänge im Oberen Mantel; nur ist die Tiefenlage gegenüber der Schmelzbildung in der Kruste viel größer und beträgt in der Größenordnung etwa 100 km. Das entspricht einem Belastungsdruck von > 10 kbar (> 30 km) und > 900 °C. Im Oberen Mantel ist CO_2 neben H_2O wesentlicher Bestandteil der Gasphase. Als zusätzlicher Freiheitsgrad spielt deshalb für die Bildung spezifischer, basaltischer Schmelzen die spezifische Zusammensetzung der Gasphase eine Rolle. Ist die Gasphase im Oberen Mantel CO_2-arm, aber H_2O-reich, so wird (als Funktion des Druckes) der Granat-Peridotit-Solidus stark erniedrigt, das Ergebnis sind andesitische, quartholeiitische und tholeiitische Schmelzen, die ausgeschmolzen werden können. Wird die Gasphase zunehmend CO_2-reicher, so nimmt die Si-Untersättigung

change from quartz- to olivine- or even larnite-normative composition (olivine basalts, nepheline basanites, olivine nephelinites, melilitites, etc.).

The decrease of temperature (in a magma reservoir) causes consolidation of a magma. If the composition is basaltic (gabbric), the first phases which crystallize on the liquidus are Mg-rich olivine (forsterite) and Ca-rich plagioclase (anorthite, bytownite). Early crystallization of olivine and plagioclase, however, increases

des Basaltmagmas zu, die Solidustemperaturen von Granat-Peridotit steigen an und mit steigendem CO_2-Gehalt der Gasphase werden nacheinander Olivinbasalte, Nephelin-Basanite, Olivinnephelinite, Melilitholithe usw. gebildet.

Die Erstarrung einer Gesteinsschmelze erfolgt mit fallender Temperatur. Hat die Schmelze basaltoiden (gabbroiden) Chemismus, so scheiden sich als silikatische Phasen Mg-reicher Olivin zuerst mit Anorthitreichem Plagioklas aus. Die Schmelze wird dadurch SiO_2-reicher und mit fortschreitender Abkühlung rea-

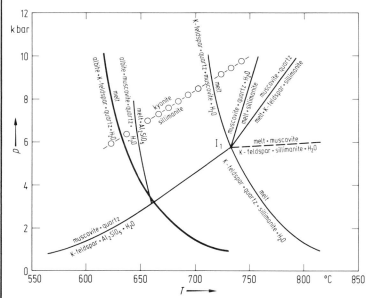

Fig. 3. Melting relations of alkalifeldspar-, quartz-, and muscovite-bearing crustal rocks in the presence of H_2O [Win76]. I_1, invariant point with liquid + vapor + muscovite + quartz + K-feldspar + one Al_2SiO_5 phase in equilibrium. Pressure p, temperature T.

Fig. 4. Melting relations of upper mantle peridotite (spinel lherzolite) from [Yod76]. Curve A is the solidus under anhydrous conditions; curve B is the estimated solidus in the presence of CO_2; curves C, D, and E, respectively, are the solidi with H_2O/H_2O+CO_2-mol fractions of 0.25, 0.50, and 0.75; curve F is the solidus with only excess H_2O.

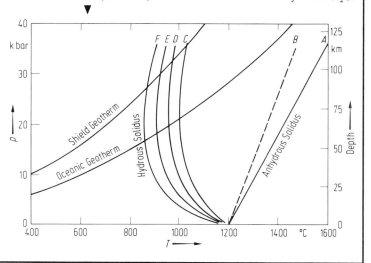

the SiO$_2$-content of the remaining melt fraction. During further cooling olivine reacts with the SiO$_2$-rich melt to form pyroxene: plagioclase changes its composition toward a more albitic plagioclase which is joined by a potash-feldspar if the magma is K$_2$O-rich. Quartz is the last phase to be precipitated if the (remaining) liquid is oversaturated in SiO$_2$. The crystallization behavior of a cooling basaltic liquid is illustrated from a physicochemical point of view by the two binary temperature vs. composition diagrams which are modelling a simple (and iron-free) basalt system (Fig. 5).

giert der frühausgeschiedene Olivin mit der SiO$_2$-reichen Schmelze unter Bildung von Pyroxenen; der Plagioklas wandelt sich sukzessive in Albit-reichere Plagioklase um und zum Schluß kristallisiert (bei Si-Sättigung des Magmas) eine SiO$_2$-Phase, meist Quarz. Ist das Magma K$_2$O-reich, so kann sich neben der Plagioklasreihe auch eine entsprechende Alkalifeldspatreihe mit Sanidin als thermisch höchstem und Albit als thermisch tiefstem Glied ausscheiden. Physikalisch-chemisch kann das Kristallisationsverhalten einer sich abkühlenden basaltischen Schmelze an zwei (isobar) binären Systemen erklärt werden (Fig. 5), die zusammen ein einfaches (eisenfreies) Basaltsystem ausmachen.

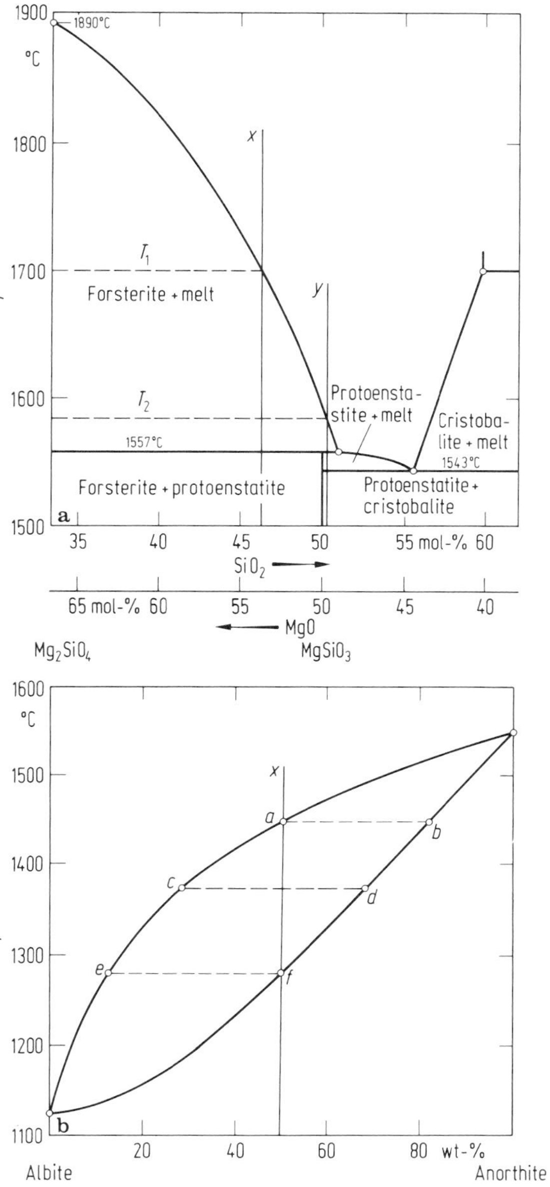

Fig. 5a, b. a) Forsterite-SiO$_2$: portion of the binary system MgO−SiO$_2$ from [Bar52]. Clinoenstatite (which is formed by rapid quenching) is changed to protoenstatite (the stable modification of MgSiO$_3$) at high temperature.
Crystallization at composition x yields forsterite+liquid at $T=T_1$, forsterite+protoenstatite+liquid at $T=1557$ °C, and forsterite+protoenstatite for $T<1557$ °C after liquid is consumed completely.
At composition y: forsterite+liquid at $T=T_2$, forsterite+protoenstatite+liquid at $T=1557$ °C, for $T=1557$ °C forsterite reacts with liquid under complete consumption and protoenstatite+liquid become stable, protoenstatite+cristobalite+liquid are in coexistence at $T=1543$ °C, liquid reacts out and protoenstatite+cristobalite become stable below 1543 °C. Cristobalite will transform into tridymite and quartz, and protoenstatite into enstatite, respectively, on further cooling.
b) Albite-anorthite system from [Bar52]. Crystallization at composition x yields plagioclase of b and liquid of a at a temperature of $a-b$, the composition of plagioclase has changed to d and the liquid to c at a temperature of $c-d$, liquid of e will be consumed completely and the final plagioclase has changed to f at a temperature of $e-f$.

Basaltic magmas which have lost the major portion of their volatiles during the ascent to the surface (in connection with a drastic pressure release) generally crystallize in a volatile-free environment very similar to that outlined in Fig. 5, but as a multicomponent system at lower temperatures. Basaltic magmas which do not reach the surface consolidate within the crust as basaltic dike rocks or gabbric intrusions. Essential portions of the volatiles in the magma are incorporated into the rockforming minerals to the prevailing partial pressures of the specific gas species. In these cases pyroxenes are usually accompanied by amphiboles and mica that is biotite in most cases.

Gravity affects the crystallization of the mineral phases in a cooling magma reservoir. Because of the density difference between the melt and the (size of the) mineral phases, gravity is mainly responsible for ascent or descent of the crystals in the reservoir. Mafic minerals and anorthite-rich plagioclases are generally more dense than the magma. They settle down and form magmatic sediments (or cumulates) on the bottom. This process substracts crystalline material from the magma fractionation and changes its chemical composition. The remaining magma fraction is now enriched with SiO_2 and alkalies. In this way the magma changes its composition continuously from basic to acid. The resultant rock unit consists of a sequence with the denser mineral assemblages at the bottom grading into less dense assemblages toward the top.

Thermal convection operating in the reservoir, however, will stir up the minerals deposited on the bottom or still suspended in the melt. Crystals may float to the top of the convective roller and become partly or wholly consumed (the latter will occur if the liquidus is intersected by the convective roller) but (re-)crystallize on the way downward. As long as the temperature of the magma system is on the hypersolidus (between solidus and liquidus), mineral phases can recycle from bottom via top to bottom forming rhythmic layered (magmatic) sediments.

The process of magma consolidation which forms sequences of ultrabasic, basic, intermediate, and acid rock units is called **differentiation.** It is the result of the fractionation of the magma during crystallization under the influence of gravity and thermal convection. Differentiation is mainly responsible for the production of the major igneous rock types exposed on the earth's surface. The main feature of differentiation is displayed in Fig. 6.

In addition to differentiation there are still a few other possibilities to explain specific rock types. Thus, a magma still in the liquid state may split into two immiscible fractions. At liquidus temperatures oxide- or sulfide-rich liquid will unmix from a silicate melt. Another process is filter-press action, whereby the last residual liquid in a liquid-crystal mush is squeezed from one place to another in the reservoir or migrates

Beim Aufstieg an die Erdoberfläche verlassen (fast) alle leichtflüchtigen Anteile durch die Druckentlastung das Gesteinssystem. Phasen, die H_2O oder CO_2 enthalten, können an der Erdoberfläche nicht mehr gebildet werden. Erstarrt das (basaltoide) Magma abyssich als gabbrisches Gestein, so verbleiben wesentliche Anteile der leichtflüchtigen Bestandteile durch den höheren Druck im Gesteinssystem. Frühausgeschiedene Pyroxene reagieren mit der H_2O-reichen Schmelze unter Bildung von Amphibolen und Glimmer, das ist im wesentlichen Biotit.

Während der Erstarrung bewirkt die Schwerkraft, daß die Kristalle je nach spezifischem Gewicht in der Schmelze absinken (oder aufsteigen). Durch diesen Prozeß werden dem Magma kontinuierlich Stoffe entzogen und die chemische Zusammensetzung der Schmelze verändert. Im allgemeinen sind die Mafite und Anorthit-reichen Plagioklase spezifisch schwerer als die Schmelze, diese sinken nach unten und bilden ein magmatisches Sediment oder Kumulat. Dadurch wird die Schmelze immer SiO_2- und alkalireicher, d.h. sie hat sich von basisch nach sauer verändert. Es entsteht eine geschichtete Magmamasse mit den spezifisch schweren Mineralassoziationen also basischen bzw. ultrabasischen Gesteinen im Liegenden, während die spezifisch leichten Mineralassoziationen bzw. sauren Gesteine sukzessive zum Hangenden folgen.

Thermische Konvektion im Magmareservoir kann bereits sedimentierte oder noch suspendierte Kristalle aufwirbeln. Kristalle können mit der Konvektionsrolle aufsteigen, auf- oder angelöst werden und auf dem Wege nach unten wieder kristallisieren. Solange die Temperatur des magmatischen Systems auf dem Hypersolidus (zwischen Solidus und Liquidus) ist, können die Mineralphasen in Zyklen auf- und absteigen wodurch rhythmisch geschichtete (magmatische) Sedimente entstehen.

Die komplexen Vorgänge, die sich bei der Erstarrung eines Magmakörpers abspielen und bei dem kontinuierlich ultrabasische, basische, intermediäre und saure Gesteine gebildet werden, bezeichnet man als gravitative **Kristallisationsdifferentiation.** Es ist ein (Magma)-Fraktionierungsprozeß (physikalisch-chemisch gesehen ist es ein fortwährendes Ungleichgewicht) bei dem aus einem einheitlichen (Mutter-) Magma eine Vielzahl von Gesteinstypen gebildet werden kann (Fig. 6).

Auch wenn die verschiedenen Typen der magmatischen Gesteine sich durch Kristallisationsdifferentiation erklären lassen, gibt es noch weitere Möglichkeiten aus einem einheitlichen (Mutter-) Magma bestimmte Magmenderivate abzuleiten. Ein Magma kann sich in flüssigen Zustand in zwei chemisch unterschiedliche Schmelzen trennen. Dieser Prozeß ist in der Petrologie als liquid-magmatische Entmischung bekannt. So ent-

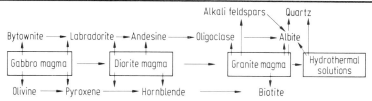

Fig. 6. Illustration of the differentiation process of a gabbric (basaltic) magma in a reservoir within the earth's crust (from [Bar52]). Reaction of forsterite to pyroxene according to Fig. 5a, crystallization of plagioclase according to Fig. 5b. Remaining hydrothermal solutions after precipitation of the silicates very often lead to formation of mineral- or ore-deposits due to ions which cannot be incorporated in common silicate structures.

into cracks in igneous material already solidified. Mingling of magmas or assimilation of any solid material by a magma may also generate new (hybrid) igneous rock types when crystallized.

The following sketch (Fig. 7) generally outlines the differentiation process of natural magmas. Magmas primarily generated in crust and mantle are *primary* or *parental magmas*. They may split into *derivative magmas* by differentiation, or less frequently by filter-press action, assimilation and unmixing in the liquid state. It should be noted, however, that derivative rocks of the basalt kindreds can also be generated from garnet peridotite within the upper mantle itself.

mischt sich bereits bei hohen (Liquidus-) Temperaturen eine Oxid- oder Sulfid-Schmelze von der Silikat-Schmelze. Weiterhin kann die Aufnahme von Fremdmaterial (Festkörper oder Schmelze) ein Magma verändern und ein Abtrennen durch Ausquetschen von Magmenanteilen während der Kristallisationsdifferentiation kann ebenfalls chemisch unterschiedlich Magmen-Zusammensetzungen hervorrufen.

Das nachfolgende Schema (Fig. 7) gibt einen Überblick welche Magmatittypen auf unserem Planeten im Prinzip durch Differentiation entstehen können. Magmen, die primär in Kruste und Mantel hervorgebracht worden sind, heißen **primäre Magmen**. Solche Magmen können durch Differentiation, Abquetschung, Assimilation und durch liquidmagmatische Entmischung Magmen-Derivate bilden. Es muß jedoch betont werden, daß die derivativen Produkte der Basalte auch direkt durch partielles Aufschmelzen von Granatperidotit im Oberen Mantel entstehen können.

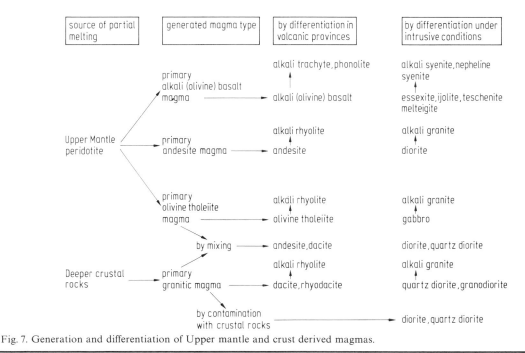

Fig. 7. Generation and differentiation of Upper mantle and crust derived magmas.

0.3 The metamorphic rocks — Die metamorphen Gesteine
0.3.1 Nomenclature — Nomenklatur

Metamorphic rocks (metamorphites, metamorphica, schists) are the results of **metamorphism.** Metamorphism is the solid-state-conversion of igneous and sedimentary rocks under the pressure-temperature-regime of the deeper earth's crust. A separation between sedimentary and metamorphic rocks is obtained by (critical) mineral assemblages which are conventionally considered as being of sedimentary (diagenetic) and of (very-low grade) metamorphic origin. The distinction between metamorphic and igneous rocks is blurred because metamorphic rocks may contain magmatic portions formed by partial melting (anatexis) that have not left the (metamorphic) rock system. These rocks are called **migmatites.**

Metamorphic rocks are named according to their mineral assemblages and their textures. In particular, a metamorphic rock can be a gneiss, an amphibolite, or a hornfels, for example. A summary of common metamorphic rock names is given in Table 11. The mineral composition of metamorphites as crustal rocks is feldspar- and quartz-rich. Because of a (possible) sedimentary origin phyllosilicates and carbonate minerals may also be present. Like the sediments, metamorphic rocks can be classified on the basis of quartz + feldspars + phyllosilicates + carbonates. Such a classification scheme is depicted in Fig. 8a and b. For the listed metamorphic rock types the state of metamorphism can also be expressed by critical minerals or mineral combinations. For example, a gneiss may contain the critical mineral combination *muscovite + quartz* or *kyanite/sillimanite + K-feldspar*. In the first case, the metamorphic rock is a *quartz-muscovite gneiss*, in the second, a *kyanite/sillimanite + K-feldspar gneiss*.

Metamorphe Gesteine oder Metamorphite (Umwandlungsgesteine; Metamorphica; Kristalline Schiefer) sind die Produkte der **Metamorphose**. Metamorphose ist die Umwandlung von magmatischen und sedimentären Gesteinen im festen Zustand unter den Bedingungen der tieferen Erdkruste. Die Abgrenzung Sediment/Metamorphit erfolgt durch Mineralparagenesen, die konventionell als (diagenetisch) sedimentär und (sehr schwach-gradig) metamorph definiert werden. Die Abgrenzung Metamorphit/Magmatit ist unscharf. Oft enthalten Metamorphite bereits (ehemals) magmatische Anteile, die durch partielle Aufschmelzung (Anatexis) entstanden sind und sich nicht von ihrem Alt- oder Restbestand getrennt haben. Solche Gesteine nennt man **Migmatite**.

Metamorphe Gesteine werden nach ihrem Gefüge und Mineralbestand benannt. Im besonderen kann ein Metamorphit z.B. ein Gneis, Amphibolit oder Hornfels sein. Eine Übersicht der wichtigsten Metamorphitbezeichnungen gibt Tab. 11. Der Mineralbestand der Metamorphite als Krustengesteine ist Quarz + Feldspat-betont, kann auf Grund einer sedimentären Herkunft aber auch wesentliche Karbonat- und Phyllosilikat-Anteile enthalten. Ebenso wie die sedimentären Gesteine lassen sich deshalb die metamorphen Gesteine auf der Basis von Quarz + Feldspäten + Phyllosilikaten + Karbonaten klassifizieren. Eine solche Einteilung für metamorphe Gesteine ist in Fig. 8a und b veranschaulicht. Zusätzlich zu den in Fig. 8 aufgeführten Metamorphittypen kann der Metamorphosegrad des Gesteins jeweils durch kritische Mineralkombinationen gekennzeichnet werden. So kann ein Gneis durch die kritische Mineralkombination *Muskovit + Quarz* oder durch *Disthen/Sillimanit + Kalifeldspat* bestimmt sein. Im ersten Falle würde man von einer *Quarz-Muskovit-Gneis*, im zweiten von einem *Disthen-* oder *Sillimanit-Kalifeldspat-Gneis* sprechen.

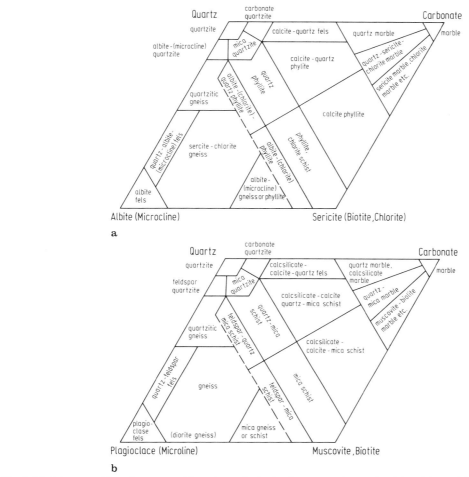

Fig. 8. Modal composition of metamorphic rocks on the basis of quartz, carbonate (mainly calcite and dolomite), alkali feldspar (microcline, albite) + plagioclase, and phyllosilicates (muscovite/sericite, biotite, chlorite) from [Win76].
a) modal composition-triangles for low grade metamorphic rocks;
b) modal composition-triangles for medium and high grade metamorphic rocks.

Table 11. Common metamorphic rock names listed and desribed in an order of their approximate abundance.

Phyllites (Phyllite)	are fine-grained and schistose rocks. More than 50% of the modal composition are phyllosilicates. The major mineral is muscovite (sericite) giving the schistosity planes the silky lustre in most cases. Phyllites with >50% chlorite are *chlorite phyllites*, with >50% quartz are *quartz phyllites*. Albite-rich plagioclase appears to be present mostly in moderate amounts.	sind feinkörnige, schiefrige Gesteine. Mehr als 50% des modalen Mineralbestandes sind Phyllosilikate. Hauptmineral ist Muskovit (Serizit), der dem Gestein auf den Schieferungsflächen einen seidigen Glanz verleiht. Phyllite mit mehr als 50% Chlorit heißen *Chlorit-Phyllite*, mit mehr als 50% Quarz *Quarz-Phyllite*. Albitreicher Plagioklas ist fast immer in mäßigen Gehalten vertreten.
Schists (Schiefer)	are mainly medium-grained rocks exhibiting excellent schistosity and are fissile in plates and stalks of less than 1 cm in thickness. Varieties are *(muscovite-) mica schists*, *chlorite schists*, *talc schists*, and *tremolite schists*. Epidote-chlorite-albite-sericite is the major mineral assemblage of the *greenschists*.	sind meist mittelkörnige Gesteine von ausgezeichneter Schiefrigkeit, die in Platten und Stengeln von weniger als 1 cm spalten. Varietäten sind *(Muskovit-)Glimmerschiefer*, *Chloritschiefer*, *Talkschiefer* und *Tremolitschiefer*. *Grünschiefer* enthalten hauptsächlich Epidot-Chlorit-Albit-Serizit.

continued

Table 11 (continued)		
Gneisses (Gneise)	are coarse-grained, schistose, and quartz+ feldspar-rich metamorphic rocks. They are fissile in plates, blocks, and stalks of more than 1 cm in thickness. Varieties may contain biotite, muscovite, amphibole, etc. in greater amounts. Gneisses can be of *orthogenic* origin (that are metamorphosed igneous rocks such as granites, granodiorites, quartz diorites, etc.) and are called *orthogneisses*. Gneisses can be of *paragenic* origin (that are sediments such as metamorphosed graywackes, arkoses, and shales) and are called *paragneisses* (or *kinzigites*).	sind grobkörnige, schiefrige, Quarz-Feldspatbetonte, metamorphe Gesteine. Sie spalten in Platten, Stengeln und Quadern von mehr als 1 cm Dicke. Gneisvarietäten können Biotit, Muskovit, Amphibol und Chlorit in größeren Mengen enthalten. Bei den Gneisen werden solche *orthogener* Herkunft (aus Magmatiten wie Graniten, Granodioriten, Quarzdioriten etc.) als *Orthogneise* von solchen *paragener* Herkunft (aus Sedimenten wie Grauwacken, Arkosen, Tonschiefer etc.) als *Paragneise* (oder *Kinzigite*) unterschieden.
Skarns (Skarne)	are metamorphic (impure) limestones dolomites, and marls which are rich in iron and/or manganese. The name is derived from an (old) skandinavian miner's expression meaning a shooting star due to bright porphyroblasts of (mainly) garnet. Skarn mineral assemblages bear garnet, hedenbergite, vesuvianite, epidote, actinolite, and ores such as magnetite, hematite and sulfides, for example.	(skandinavisch „die Lichtschnuppe") sind metamorphe (unreine) Kalksteine und Dolomite, die hohe Gehalte an Eisen und/oder Mangan aufweisen. Skarnmineralparagenesen enthalten Granat, Hedenbergit, Vesuvian, Epidot, Aktinolith und (verschiedene) Erze wie zum Beispiel Magnetit, Hämatit und Sulfide.
Serpentinites (Serpentinite)	consist mainly of serpentine minerals (chrysotile, antigorite, lizzardite). They are hydrothermally (retrograde) converted peridotites. Huge serpentinite bodies can be found in orogenic belts. They are intercalated with sediments and are mostly associated with basaltic igneous rocks.	sind metamorphe Gesteine, die (fast) vollkommen aus Mineralen der Serpentin-Gruppe (Antigorit, Chrysotil, Lizzardit) bestehen und durch hydrothermale (retrograde) Umwandlung von Peridotiten entstanden sind. Serpentinit-Körper großen Ausmaßes sind in den orogenen Gürteln zu finden. Sie sind die Geosynklinalsedimente eingeschaltet und meist mit basaltischen Gesteinen assoziiert.
Felses (Felse)	are massive and solid metamorphic rocks which mostly lack any visible parallel structure. *Hornfelses* are (thermo-) metamorphic pelites of fine-grained but granoblastic texture. The name hornfels is derived from the appearance of their edges which are translucent like horn material.	sind massive und kompakte Metamorphite, die meistens eine Parallel-Textur vermissen lassen. Feinkörnige, splittrig-brechende und an den Kanten hornartig-durchscheinende (thermometamorphe) Gesteine werden als *Hornfelse* bezeichnet. Es sind ehemals pelitische Gesteine, die während der Metamorphose granoblastisch umkristallisierten und Mosaikgefüge aufweisen.
Quartzites (Quarzite)	or *quartz felses* are metamorphic sandstones being rich in quartz. *Quartzites*, however, can also be formed diagenetically under sedimentary conditions.	oder *Quarzfelse* sind metamorphe Sandsteine, die hohe Gehalte an Quarz aufweisen. *Quarzite* können aber auch diagenetisch durch Lösungs- und Ausfällungsreaktionen während der sedimentären Gesteinsbildung entstehen.
Granulites (Granulite)	are fine- to coarse-grained feldspar-rich rocks with or without quartz. Mafic minerals are garnet, pyroxene, kyanite, and sillimanite but cordierite, amphibole, and biotite can also be present in (very) small amounts. Their texture is granoblastic. Lamination and foliation as well as disk-like quartz grains are very common. Granulites are the products of high-grade metamorphism in a H_2O-deficient or poor environment.	sind fein- bis grobkörnige feldspatbetonte Gesteine mit oder ohne Quarz. Mafite sind Granat, Pyroxen, Disthen, Sillimanit aber auch – in kleinen Mengen – Cordierit, Amphibol und Biotit. Das Gefüge ist (vorwiegend) granoblastisch; plattige und gneisige Texturen sind typisch. Granulite können linsenförmige Quarze oder Quarz-Aggregate (disk quartz) enthalten. Granulite werden bei der hochgradigen (H_2O-freien bis -armen) Metamorphose gebildet.

continued

Table 11 (continued)

Eclogite (Eklogit)	are orthogenic rocks of basaltic or gabbric chemistry containing clinopyroxene and garnet as essential mineral phases. The clinopyroxene is omphacite (that is mainly a diopside-jadeite solid solution), the garnet is pyrope-rich. Eclogites are the product of anhydrous metamorphism of basalts or gabbros. Further subdivision of eclogites can be achieved by the jadeite content of the clinopyroxenes, by the pyrope content of the garnet, and by the occurrence of glaucophane and/or zoisite.	sind Orthogesteine von basaltischem oder gabbrischem Chemismus, die im wesentlichen aus zwei Mineralphasen, Klinopyroxen und Granat, bestehen. Klinopyroxen ist Omphacit (das ist ein vorwiegend Diopsid-Jadeit-Mischkristall), die Granate sind Pyrop-betont. Eklogite sind die (anhydrischen) Metamorphose-Produkte von Basalt oder Gabbro. Eklogite können nach dem Jadeit-Gehalt der Klinopyroxene, nach dem Pyrop-Gehalt der Granate sowie nach der Glaukophan- und Zoisit-Führung weiter unterschieden werden.
Amphibolites (Amphibolite)	are metamorphic rocks which consist predominantly of amphibol + plagioclase. Amphiboles are mostly arranged parallel or subparallel with their prismatic shapes to the schistosity planes if they are develloped.	sind metamorphe Gesteine, die vorwiegend aus Amphibol + Plagioklas bestehen. Die Amphibole sind bei Vorhandensein von Schieferung meist parallel oder subparallel zur Schieferungsfläche eingeregelt.
Marbles (Marmore)	are coarse-grained recrystallized limestones and dolomites. There are pure *calcite-* or pure *dolomite-marbles* as well as transitional types. The sedimentary (and clastic) silicate components of impure limestones and dolomites react with the carbonates under the formation of talc, tremolite, diopside, wollastonite, forsterite, grossular, anorthite, melilite, and others during (contact) metamorphism. Such rocks are called *calcsilicate marbles*. Special names are *forsterite* and *diopside marbles* for example. *Calcsilicate-felses* contain more than 50% of silicate phases.	sind grobkörnige, rekristallisierte Kalksteine und Dolomite. Es gibt reine *Kalzit-* und *Dolomitmarmore* aber auch alle Übergänge zwischen beiden. Die sedimentären (und klastischen) Silikatkomponenten verunreinigter Kalksteine und Dolomite (Mergel) reagieren bei der (Kontakt-)Metamorphose mit den Karbonaten unter Bildung von Talk, Tremolit, Diopsid, Wollastonit, Forsterit, Grossular, Anorthit, Melilith und anderen (Ca-Al-)Silikaten. Solche Gesteine werden *Kalksilikat-Marmore* genannt, *Forsterit-* und *Diopsid-Marmore* sind spezielle Varietäten. *Kalksilikat-Felse* enthalten mehr als 50% an Silikat-Mineralen.

0.3.2 The mineral composition of metamorphic rocks — Die mineralische Zusammensetzung der Metamorphite

Specific (critical) minerals of mineral assemblages will occur after metamorphism according to the initial bulk composition of the non-metamorphic rock. There are minerals stable only under metamorphic (pressure-temperature) conditions, there are others which are abundant in metamorphic as well as in igneous and sedimentary rocks. These are quartz, the feldspars, the micas and certain amphiboles and pyroxenes. Critical, however, is the combination of minerals which appears to exist under (near to) metamorphic equilibrium conditions. These combinations can be called **parageneses** (or assemblages). Typical metamorphic minerals are listed in Table 8 and natural metamorphic mineral combinations are shown in Table 12.

Bei der Metamorphose bilden sich je nach Ausgangsgestein spezifische (kritische oder typomorphe) Mineralgemenge. Manche der Mineralphasen in diesen Gemengen sind nur unter metamorphen Bedingungen stabil, andere sind Durchläufer, die wie Quarz, die Feldspäte, Amphibole, Glimmer oder Pyroxene auch in magmatischen oder sedimentären Gesteinen anzutreffen sind. Kritisch ist jedoch die Kombination von bestimmten Mineralphasen, die einen annähernden Gleichgewichtszustand widergibt und als **Paragenese** bezeichnet wird. Typisch metamorphe Mineralphasen sind in Tab. 8 zusammengestellt. Sie treten je nach Ausgangszusammensetzung und Metamorphosegrad in spezifischen Kombinationen in einer (metamorphen) Mineralparagenese auf (Tab. 12).

Table 12. Selected modal compositions of common metamorphic rocks. Crosses indicate the presence as a major component.

	Phyllite, Furuland, Norway [Cor68]	Phyllite, Grand Paradise France [Cor68]	Pelitic schist, Vermont [Mue77]	Pelitic schist, Koto-Bizan, Japan [Mue77]	Quartz-rich pelitic schist, Koto-Bizan, Japan [Mue77]	Biotite schist, Quebec [Mue77]	Magnetite-hematite-bearing pelitic gneiss, Glen Clova, Scotland [Mue77]	Plagioclase-quartz-biotite gneiss, West Balmat, New York [Mue77]	Pyroxene gneiss, Lützow Holm Bay, Antarctica [Mue77]	Granite gneiss, Quebec [Mue77]	Staurolite-garnet-plagioclase gneiss, Spessart, Germany [Cor68]	Kyanite-andalusite-sillimanite gneiss, Idaho [Mue77]	Amphibolite, Adirondack, New York [Mue77]
	vol-%	wt-%	vol-%	wt-%	vol-%	vol-%	vol-%	vol-%	vol-%	vol-%	vol-%	vol-%	vol-%
Quartz	1.1	5.5	5	30.8	79.0	30	2.6	32.3		25	30	3.2	8.6
Plagioclase				30.6		30	13.8	44.4	+	30	26.4	15.6	18.5
Alkalifeldspar	39.9	35						1.7	+	35			
Muscovite			34		20.2		0.6	5	53.7	0.9	3	13.2	0.2
Paragonite			19									17.9	
Andalusite/Kyanite/Sillimanite								2.7				17.9	
Stilpnomelane													
Biotite						30	13.3	20.0		3	21.3	25.1	1.3
Garnet				0.2	6.1		2.4		+		1.3		
Melilite													
Vesuvianite													
Cordierite												37.1	
Staurolite							4)	0.5			9.1		
Chloritoid			9										
Chlorite	29.4	25	31	7.1	0.1						<2		
Epidote/Clinozoisite/Zoisite	23.0	18		2.3		<5							
Pumpellyite													
Lawsonite													
Amphiboles	3.5	9		5.4					+				68.6
Pyroxenes				4.0					+				0.9
Calcite/Aragonite	2.6			2.3				0.2				0.5	
Sphene				0.9									
Rutil		1											
Opaques				4.7			11.2	0.1	+				2.1
Others (Graphite, Apatite, Pyrite)		7.5		5.6	0.1	<1		0.1	+	<1	2.1		1.1

*) Unpublished data.   1) + mica.   2) = 10 apatite, 25 fluorite, 4 scapolite.   4) Trace.

Table 12 (continued)

Amphibolite, Adirondack, New York [Mue77]	Epidote amphibolite, Sulitelma, Norway [Cor68]	Granulitic paragneiss, New York [Cor68]	Hypersthene granulite, Lappland [Cor68]	Kyanite eclogite, Silberbach, Fichtelgebirge, Germany [Cor68]	Eclogite, Glenelg, Scotland [Cor68]	Metabasalt of very-low grade, Northern California [Cor68]	Metachert of very-low grade, Northern California [Cor68]	Glaucophane-bearing schist of very-low grade, Northern California [Cor68]	Aragonite marble, California [Mue77]	Calcsilicate rock, Monzoni Alpe, Italy [Huc]*)	Andalusite-cordierite hornfels, Oslo region [Cor68]	Anorthite-hypersthene hornfels, Oslo region [Cor68]	Anorthite-diopside hornfels, Oslo region [Cor68]	Fluorite-bearing skarn, Hudderfield township, Quebec [Cor68]	Hedenbergite-andradite-magnetite skarn, Hartenstein, Austria [Huc71]
vol-%	vol-%	vol-%	vol-%	vol-%	vol-%	vol-%	vol-%	vol-%	vol-%	vol-%	vol-%	vol-%	vol-%	vol-%	vol-%
2.2		22.6	2.5	4	8.1		57.9	26.6	+		21.0	22	2.4	4)	+
33.5		46.8	49.5							+	0.4	24.9	4.3		
	42.8	3.1	7.1								55.1	21.9	53.1		
			0.5		0.3			23.9			5.0				
				18							6.9				
							7.6		+						
4)		15.7									1.0	25	4.0		
		9.7		18	30.2		5.8		+	+					+
										+					
										+	13.8	21			
	2.9	0.8				0.5									
	12.3														
						1.2									
						34.0			+						
35.3	42.2				8.1	56.3	17.6	48.4						4)	
23.6			34.9	58.5	53.6					+			32	10 1)	+
						2.5		> 50					0.8	45···50	
						5.2	1.1						1.2	4)	
	0.5										1.0				
3.2		0.1	4.9	0.9		1.3					1.3	1.0			+
2.2		0.8		0.6					+		3.6	1.5	2.2	39 2)	

*) Unpublished data.   1) + mica.   2) =10 apatite, 25 fluorite, 4 scapolite.   4) Trace.

0.3.3 The chemical composition of metamorphic rocks — Die chemische Zusammensetzung der Metamorphite

Text: see p. 34

Table 13. Selected chemical compositions of common metamorphic rocks. For comparison with igneous and sedimentary rocks the CIPW-norm or normative composition is given.

	Phyllite, Furulund, Norway [Cor68]	Phyllite, Grand Paradise, France [Cor68]	Pelitic schist, Vermont [Mue77]	Actinolite-albite-epidote, Agnew Lake, Ontario [Mue77]	Pelitic schist, Agnew Lake, Ontario [Mue77]	Magnetic-hematite-bearing pelitic gneiss, Glen Clova, Scotland [Mue77]	Plagioclase-quartz-biotite gneiss, West Balmat, New York [Mue77]	Pyroxene gneiss, Lützow Holm Bay, Antarctica [Mue77]	Staurolite-garnet-plagioclase gneiss, Spessart, Germany [Cor68]	Kyanite-andalusite-sillimanite gneiss, Idaho [Mue77]	Amphibolite, Adirondack, New York [Mue77]
	wt-%										
SiO_2	49.22	47.50	39.41	55.07	48.60	44.09	67.92	56.81	58.71	48.20	48.20
TiO_2	0.18	2.25	0.66	0.98	0.32	1.69	0.70	1.01	0.83	0.14	1.89
Al_2O_3	18.56	18.79	31.21	21.50	9.16	23.64	15.53	17.33	20.78	32.54	14.45
Fe_2O_3	2.22	4.65	2.10	2.72	4.65	12.01	0.77	1.87	4.24	0.23	3.50
FeO	5.35	6.30	10.36	6.04	6.48	3.66	3.51	5.55	3.46	2.24	10.53
MnO	0.12	0.10	0.16	0.05	0.25	0.37	0.05	0.13	0.18	0.05	0.25
MgO	8.15	5.92	3.41	3.00	18.41	2.61	2.04	3.46	2.56	9.30	6.62
CaO	7.17	7.68	0.14	0.98	9.57	0.85	2.22	6.55	1.15	0.84	10.25
Na_2O	4.65	3.76	1.55	1.43	0.53	2.03	3.90	3.54	1.65	1.66	1.94
K_2O	0.10	0.30	4.01	3.01	0.20	6.01	2.67	2.24	4.05	2.32	0.96
H_2O^+	3.15	1.12	6.54	3.53	1.29	3.09	0.72	0.85	1.70	2.26	1.31
H_2O^-				0.02		0.12		0.14		0.12	0.01
P_2O_5	n.d.	0.46		0.11	0.03	0.15	0.11	0.26	0.24	0.01	0.18
CO_2	0.43	1.50		0.12	0.07					0.22	
F								0.07			
S		0.02									

normative composition				wt-%							
qz (quartz)		2.7	2.2	26.3		4.0	25.5	7.2	27.3	12.9	0.3
cor (corundum)		2.9	24.2	14.9		12.6	2.6		12.2	26.3	
or (K-feldspar)	0.6	1.8	23.8	18.0	1.2	35.4	15.7	13.3	24.0	13.7	5.7
ab (albite)	39.6	31.7	13.2	12.3	4.5	17.1	32.9	30.0	14.0	14.0	16.4
an (anorthite)	29.7	25.5	0.7	3.4	22.2	3.3	9.8	24.8	4.2	2.7	27.9
di (diopside)	2.7				19.7			5.0			17.9
hy (hypersthene)	0.3	18.9	2.5	15.0	40.6	6.5	9.8	13.4	8.2	26.9	21.4
ol (olivine)	19.3				3.0						
mt (magnetite)	3.2	6.7	3.1	4.0	6.8	8.1	1.1	2.7	6.2	0.3	5.1
hm (hematite)						6.4					
ilm (ilmenite)	0.3	4.3	1.3	1.9	0.61	3.2	1.3	1.9	1.6	0.2	3.6
ap (apatite)		1.1		0.3	0.1	0.4	0.3	0.6	0.6		0.4
ce (calcite)	1.0	3.4		0.3	0.2					0.5	
ne (nepheline)											

Table 13 (continued)

Amphibolite, Adirondack, New York [Mue77]	Epidote amphibolite, Sulitelma, Norway [Cor68]	Granulitic paragneiss, New York [Cor68]	Hypersthene granulite, Lappland [Cor68]	Kyanite eclogite, Silberbach, Fichtelgebirge, Germany [Cor68]	Eclogite, Glenelg, Scotland [Cor68]	Metabasalt of very-low grade, Northern California [Cor68]	Glaucophane-bearing schist of very-low grade, Northern California [Cor68]	Andalusite-cordierite hornfels, Oslo region [Cor68]	Anorthite-hypersthene hornfels, Oslo region [Cor68]	Anorthite-diopside hornfels, Oslo region [Cor68]	Hedenbergite-andradite-magnetite skarn, Hartenstein, Austria [Huc71]
47.69	52.45	61.27	52.03	50.24	50.05	46.5	69.3	62.80	56.59	57.24	49.48
1.76	0.38	0.78	2.27	0.26	1.55	1.2	0.41	1.36	0.29	0.65	
14.52	17.23	17.94	16.39	19.98	13.37	13.8	12.0	19.74	18.15	12.30	0.90
2.06	4.36	0.67	0.82	1.44	3.71	1.3	1.1		4.23	1.77	20.45
11.31	4.96	6.81	9.13	3.12	10.39	7.6	3.9	1.98	5.21	2.95	3.20
0.23	0.08	0.05	0.17	0.10	0.25	0.20	0.06	0.02	0.21	0.09	2.57
7.26	6.71	3.01	7.04	9.84	6.49	7.4	4.2	1.34	5.01	4.80	0.43
11.16	8.55	3.29	8.78	12.95	11.00	12.1	0.67	0.87	5.14	10.31	23.37
2.13	4.94	3.51	2.14	1.93	2.38	3.1	2.2	1.22	1.41	2.78	0.09
0.67	0.39	1.93	1.21	0.09	0.36	0.18	3.3	6.56	3.64	5.41	trace
0.81	0.69			0.31	0.39	3.6	2.1	0.86 ⎫	0.64	0.18	0.25
0.02			0.35		0.06	0.16	0.09	0.27 ⎭		0.06	
0.18	trace	0.56	0.06	0.02	0.12	0.16	0.14	0.60	0.10	0.90	
						0.14	2.6	0.05		0.35	trace
		0.04									
			0.04	0.35	0.08	0.09	0.05	0.52			
							Cl 0.58				
		18.6	2.6		2.1		34.3	28.8	13.5	1.4	22.7
							4.1	10.7	2.8		
4.0	2.3	11.4	7.11	0.5	2.1	1.1	19.6	39.5	21.4	32.0	
18.1	41.0	29.7	18.0	16.2	20.1	26.2	18.7	10.5	11.9	23.6	0.8
28.1	23.5	12.7	31.4	45.3	24.7	23.2	2.1	0.4	24.7	5.1	2.0
21.6	14.9			14.5	23.1	15.74				30.3	2.3
11.2		18.3		16.53	18.4	12.7	16.1	3.8	18.3	0.6	
9.5	10.4			3.2		6.7					
3.0	6.3	1.0	1.2	2.1	5.4	1.9	1.6		6.1	2.6	18.6
											7.5
3.3	0.7	1.5	4.3	0.3	2.9	2.3	0.8	2.6	0.5	1.2	
0.4		1.3	0.1	0.05	0.3	0.4	0.3	1.4	0.2	2.1	
					0.3	5.9	0.1			0.8	
	0.3										

Metamorphic rocks are converted from sediments and magmatites. It can be assumed that the bulk chemistry of the initial rock is preserved in most cases during metamorphism. By convention, removal or addition of volatile components – these are mainly H_2O and CO_2 – is allowed. Granitic igneous rocks are metamorphosed to schists and gneisses while the bulk chemistry is kept constant albeit the modal composition is changed according to the metamorphic grade. Basalts change to amphibolites or eclogites yielding the critical mineral combinations of *amphibole + plagioclase* and of *clinopyroxene + garnet*, respectively. Examples of the chemistry of common metamorphic rock types are given in Tables 13 and 14.

Metamorphite sind umgewandelte Magmatite und Sedimente. Unter der Voraussetzung einer isochemischen Metamorphose – was in der Mehrzahl der Fälle zutrifft – wird der primäre Chemismus des prämetamorphen Ausgangsgesteins beibehalten, jedoch ist definitionsgemäß Zufuhr und Abfuhr von leichtflüchtigen Anteilen, vor allem H_2O und CO_2, statthaft. Werden granitoide Gesteine zu Glimmerschiefer oder Gneis metamorphosiert, so bleibt der Chemismus (isochemisch) granitoid (plus H_2O), obwohl der Mineralbestand (qualitativ und quantitativ) unter den bestimmten Metamorphosebedingungen entsprechend umgewandelt wurde. Aus Basalten werden Amphibolite oder Eklogite mit den kritischen Mineralkombinationen *Amphibol + Plagioklas* bzw. *Klinopyroxen + Granat*. Die chemischen Zusammensetzungen der wichtigsten metamorphen Gesteine sind in Tab. 13 und 14 zusammengestellt.

Table 14. Average trace element composition (in ppm) of low, medium, and high grade metamorphic pelites in New Hampshire (from [Meh69]).

	Low grade metamorphic rock	Medium grade metamorphic rock	High grade metamorphic rock
	ppm	ppm	ppm
Ga	20.8	15.9	19.8
Cr	116	113	109
V	109	125	120
Li	54.7	108	127
Ni	80.5	63.7	57.4
Co	16.8	19.4	18.0
Cu	23.1	23.8	12.5
Sc	11.3	11.9	15.6
Zr	191	213	203
Y	38.8	37.9	51.7
Sr	524	731	760
Pb	16.1	23.3	27.3

0.3.4 The texture and structure of metamorphic rocks — Das Gefüge der metamorphen Gesteine

Sedimentary and igneous rocks recrystallize during metamorphism. This process is called **crystalloblastesis.** Unlike igneous rocks, the mineral phases of metamorphic assemblages crystallize simultaneously, thus providing new textural features. The preservation of previous mineral assemblages with a tendency to coarser grain-size is called **isophase recrystallization** (or **crystalloblastesis**). A typical example of that sort is the conversion of a fine-grained (sedimentary)

Bei der Metamorphose kristallisieren sedimentäre und magmatische Gesteine um. Dieser Prozeß wird **Kristalloblastese** genannt. Im Gegensatz zu den magmatischen Gesteinen kristallisieren die Mineralphasen einer metamorphen Paragenese zeitlich nebeneinander unter Ausbildung eines typischen metamorphen Reaktionsgefüges. Bewirkt die metamorphe Umkristallisation nur eine Kornvergröberung unter Beibehaltung der sedimentären oder magmatischen Mineralphasen,

limestone into a coarse-grained marble due to metamorphic recrystallization. Metamorphic recrystallization, however, produces new mineral assemblages in most cases. This so-called **allophase recrystallization** provides new and crystalloblastic textures. The articulation of a new texture requires initial homogeneity of the pre-metamorphic rock. All the phases participating in the recrystallization process should be in contact with one another and the new mineral phases will form at the junctions of critical phases. Fine-, medium-, or coarse-grained metamorphic rocks may have precipitated minerals which exhibit granoblastic (granular), nematoblastic (stalk-like), fibroblastic (fibrous) and lepidoblastic (flaky) shapes. Porphyroblastic textures consist of large crystals – the **porphyroblasts** – in a fine-grained matrix. The porphyroblasts can exhibit their crystal shape, a feature very often seen in **idioblasts** of sphene, magnetite, kyanite, staurolite, and garnet, unlike feldspars and quartz which mostly form **xenoblasts**. Porphyroblasts riddled by other minerals are called **poikiloblasts** and the texture **poikiloblastic**. Metamorphism can frequently, but by no means always, overprint previous textures. Sedimentary and magmatic (and previous metamorphic) textures are, however, very often preserved as relics. Initial inhomogeneities such as sedimentary layering or (magmatic) porphyritic textures cannot be exterminated completely by metamorphic recrystallization. Minerals of non- or previous metamorphic stages can be preserved when encompassed by the newly formed (metamorphic) phases, which prohibit their participation in the metamorphic mineral reactions. These remnants are called **armored relics**.

Most of the metamorphic rocks are distinguished by a parallel or subparallel structure of their minerals which includes a more or less planar fissility of the rock (gneisses, schists, slates, etc.). Felses and hornfelses, however, (usually) lack parallel structures. Structural features of metamorphic rocks can be inherited from their sedimentary and magmatic ancestors. **Schistosity** or **foliation**, however, is mostly produced by directed pressure (shear stress) which deforms the

so spricht man von **isophaser Umkristallisation**. Das typische Beispiel ist die Umkristallisation von diagenetisch-sedimentären Kalksteinen in grobkörnige Marmore. In den meisten Fällen erfolgt die Umkristallisation aber unter Ausbildung eines neuen (metamorph stabilen) Mineralgemenges. Diese **allophase Umkristallisation** (die ebenso wie die isophase Umkristallisation eine **Blastese** darstellt) schafft neue (metamorphe) Gefüge. Im Gegensatz zu magmatischen Gesteinen kristallisieren die Mineralphasen einer metamorphen Paragenese zeitlich nebeneinander unter Ausbildung eines neuen Gefüges. Die vollständige Ausbildung eines neuen Gefüges setzt die primäre Homogenität des prämetamorphen Mineralbestandes voraus. Alle an der metamorphen Reaktion beteiligten Mineralphasen müssen (sollten) miteinander in Berührung stehen. An den Berührungspunkten der Reaktionspartner können dann die Produkte der metamorphen Reaktion (durch Kristalloblastese) entstehen. Bei unterschiedlicher Korngröße (feinkörnig, mittelkörnig, grobkörnig etc.) lassen sich je nach Kornform granoblastische (körnige), nematoblastische (stengelige), fibroblastische (faserige), lepidoblastische (schuppige) oder porphyroblastische (ungleichkörnige) Strukturen unterscheiden. Bei den porphyroblastischen Strukturen unterscheidet man **Idioblasten** (mit Eigengestalt wie Titanit, Magnetit, Disthen, Staurolith und Granat) und **Xenoblasten** (mit Fremdgestalt). Sind Porphyroblasten siebartig durchwachsen, so spricht man von **poikiloblastischen** Strukturen. In der Mehrzahl der Fälle ist die metamorphe Umkristallisation nicht in der Lage, prämetamorphe Gefüge auszulöschen. Sedimentäre und magmatische Gefüge (aber auch ältere metamorphe Gefüge) bleiben meist reliktisch erhalten. Primäre Inhomogenitäten, wie sedimentäre Schichtung oder ungleichkörnige Gefüge von Magmatiten können während der Umkristallisation nur sehr schwer (vollkommen) verwischt und durch (vollständig) neue metamorphe Gefüge ersetzt werden. Ebenso können Mineralphasen eines prämetamorphen (d.h. eines zeitlich älteren, metamorphen) Stadiums reliktisch erhalten bleiben, wobei neue Phasen die alten von außen abschließen, sodaß sie nicht mehr an den Reaktionen teilnehmen können. Solche Überreste tragen die Bezeichnung **gepanzerte Relikte**.

Die meisten metamorphen Gesteine zeichnen sich durch plattige und schiefrige Anordnung der gesteinsbildenden Anteile aus. Felse und Hornfelse sind dagegen kompakt und massig. Solche Eigenschaften können durch sedimentäre aber auch durch ältere Metamorphitgefüge bedingt sein. In den meisten Fällen wird die „Schiefrigkeit" eines metamorphen Gesteins durch mechanische Verformung hervorgerufen, die die Gesteine im festen Zustand verformen. Die Beziehun-

metamorphic mineral assemblages in the solid state. The relationship between recrystallization and deformation produces specific rock structures. Recrystallization before deformation (postcrystalline deformation) produces broken and ground (**cataclastic** and **mylonitic**) structures. Simultaneous recrystallization and deformation (paracrystalline deformation) arrange the precipitating minerals in accordance with the directed pressure. The products of paracrystalline deformation are strictly orientated mineral assemblages as exhibited in schists, gneisses, and granulites. Deformation followed by recrystallization (precrystalline deformation) overprints the effect of deformation by a (renewed) recrystallization of the (former) cataclastic and mylonitic products (blastocataclasites, blastomylonites).

gen von Umkristallisation und Deformation bewirkt je nach Zeitlichkeit der beiden Prozesse typisch texturierte Gesteine. Umkristallisation vor der Deformation (postkristalline Deformation) schafft mechanisch zerbrochene (**kataklastische**) und zermahlene (**mylonitische**) Texturen. Gehen Deformation und Umkristallisation Hand in Hand, (parakristalline Deformation), so werden die sich neubildenden Mineralphasen der metamorphen Paragenese entsprechend dem Verhalten ihrer Kristallstruktur gegenüber gerichtetem Druck (Streß) eingeregelt. Das Produkt der parakristallinen Deformation (paratektonische Kristallisation) sind streng geregelte, das heißt schiefrige, plattige oder lagige Metamorphite (wie Phyllite, Schiefer, Gneise, Granulite etc.). Umkristallisation nach der Deformation (präkristalline Deformation) überprägt die Spuren der mechanischen Beanspruchung und rekristallisiert die kataklastischen und mylonitischen Produkte (Blastokataklasite, Blastomylonite).

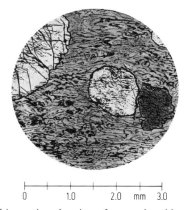

Fig. 9. Thin section drawing of a prophyroblastic texture of a metamorphic rock. Porphyroblasts of garnet (light; hypidioblastic to xenoblastic) and mica (dark; xenoblastic) in a fine-grained matrix of quartz+sericite (very fine-crystalline muscovite)+carbonaceous material. Garnet-sericite schist, boulder, near Bolzano, Italy; from [Cor68].

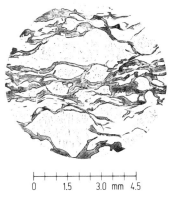

Fig. 10. Thin section drawing of a gneissic structure displaying large crystals of feldspars (light) smoothly surrounded by biotite. Augengneis, Niedere Tauern, Austria; from [Cor49].

The analysis of rock deformation is achieved by optical measurement of the spatial orientation of mineral grains with the universal stage or by X-ray with a structure goniometer. Display and interpretation of structural features (structural analysis) is done by means of stereographic projection. According to the symmetry of deformation, isotropic or anisotropic structures can be distinguished. The latter type is divided into axial, orthorhombic, monoclinic, and triclinic symmetry. Examples of anisotropic structures of metamorphic rocks are given in Fig. 11.

Die Analyse der Gesteinsverformung geschieht durch räumliches Einmessen bestimmter Minerale mit dem Universaldrehtisch unter dem Polarisationsmikroskop und/oder röntgenographisch mit dem Texturgoniometer. Darstellung und Interpretation der Gefüge erfolgt stereographisch (statistische Gefügeanalyse). Nach dem Deformationsplan lassen sich isotrope und anisotrope Gefüge unterscheiden, wobei die anisotropen nach ihrer Symmetrie axial oder wirtelig, rhombisch, monoklin und triklin sein können. Verschiedene Beispiele sind in Fig. 11 gegeben.

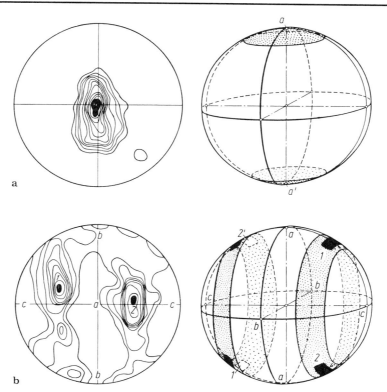

Fig. 11a, b. Examples of anisotropic structures in metamorphic rocks displaying axial (a) and orthorhombic (b) symmetry; from [Cor68].
a) spatial orientation of 138 crystallographic c-axes of quartz. Their piercing points are projected on the equatorial surface of the sphere (left figure). The complete sphere is shown by the right figure. Contours in the left figure refer to point densities (in per cent of the area as taken) arranged outward from the center: >18–16–14–12–10–8–6–4–2–1–0.5–0. Deformed granite, Odenwald, Germany;
b) spatial orientation of 380 crystallographic c-axes of quartz, projection as in (a). Contours outward from the maxima of 1 and 2 are >10–8–6–5–4–3–2–1–0.5–0. Granulite, Saxony.

0.3.5 The origin and genesis of metamorphic rocks — Die Entstehung der metamorphen Gesteine

Metamorphism is a transformation of solids in the presence and/or participation of a (polynary) vapor phase. Metamorphic transformations (or reactions) are initiated by a change in temperature and pressure. Metamorphism converts pre-existent mineral assemblages into new and stable assemblages. Volatiles are (mainly) H_2O and CO_2. The valency change of iron releases or consumes oxygen which participates in metamorphic reactions together with H_2O and CO_2. The pressure of the volatiles equals the total pressure in most cases. The composition of the vapor phase (and the fugacities of the gas species), the temperature and the (total) pressure determine the products of metamorphism in a specific rock system.

Jede Metamorphose ist eine Umwandlung von Festkörpern in Gegenwart oder unter Beteiligung einer (polynären) Gasphase. Metamorphe Umwandlungen erfolgen durch Änderung der Temperatur und des Druckes. Metamorphose wandelt präexistente Mineralparagenesen in neue, stabile Mineralparagenesen um. Die leichtflüchtigen Anteile sind hauptsächlich H_2O und CO_2. Beim Wertigkeitswechsel des Eisens wird Sauerstoff freigesetzt oder verbraucht, der zusammen mit H_2O und CO_2 an metamorphen Reaktionen teilnimmt. Der Druck der Gasphase ist in den meisten Fällen mit dem Auflastdruck identisch. Die Zusammensetzung der Gasphase (und die Fugazitäten der Gasspezies) spielen für die Produkte der metamorphen Reaktionen eine wesentliche Rolle.

A temperature and pressure change within the earth's crust can be caused by different types of metamorphism. Occurence and origin distinguish

i) contact or thermal metamorphism,
ii) dislocation or cataclastic metamorphism, and
iii) regional metamorphism.

Contact or **thermal metamorphism** is the (static) conversion of cold crustal rocks by a hot (magmatic) body. This type of metamorphism is locally restricted. According to the volume and heat capacity of the magmatic body, the country rocks become heated. The (metamorphic) heat effect decreases from the contact outwards and zones of specific mineral assemblages may develop. Contact metamorphic heating (usually) produces fels-like and granoblastic textures (e.g. hornfels, calcsilicate-fels, etc.). The pressure regime during contact or thermal metamorphism is a function of the depth. Most magmatic (intrusive) bodies are shallow seated and occur only a few kilometers deep in the crust. Thus, total pressures of one kilobar or less appear to be normal, higher pressures are exceptions.

Dislocation or **cataclastic metamorphism** is restricted to fault and thrust systems. Solid rocks (of any kind) become strongly deformed in narrow zones. The results are cataclastic and mylonitic products. Friction can be converted into heat causing partial melting on a very fine scale. Pseudo-tachylite fissures (mixtures of glasses and crystals) document that the solidus of the rock was exceeded. The impact of extraterrestial bodies can create textural features very similar to those of dislocation or cataclastic metamorphism (see impact or shock wave metamorphism or transformation).

Regional metamorphism produces rocks which cover huge areas. Occurrence and metamorphic deformation distinguish two types of regional metamorphic rocks. These are rocks crystallized under the conditions of

i) burial metamorphism and
ii) regional dynamothermal metamorphism.

Diagenesis of sedimentary rocks is followed by **burial metamorphism.** Sediments with intercalations of igneous rocks (mainly volcanics or minor intrusives) are buried continuously by detrital material in quickly sinking troughs (synclines). Cold rock material is transported quickly to great depths. The rock pile (usually) lacks deformation and sedimentary textures are well preserved. The temperature in these depths (15 to 30 km, corresponding to about 5 to 10 kbar) does not exceed 400 °C. Specific (exotic) mineral

Temperatur- und Druckänderungen können in der Erdkruste durch verschiedene Metamorphosearten verursacht werden. Nach Auftreten und Eigenart werden unterschieden:

i) Kontakt- oder Thermometamorphose,
ii) Dislokations- oder kataklastische Metamorphose und
iii) Regionalmetamorphose

Kontakt- oder **Thermometamorphose** ist eine (statische) Veränderung von kalten Krustengesteinen durch heiße (magmatische) Körper. Kontakt- oder Thermometamorphosen haben lokale Ausmaße und die Wärmequelle ist erkennbar. Je nach Volumen und Wärmekapazität des Magmakörpers wird das Nebengestein aufgeheizt, und es entstehen Zonen von spezifischen Mineralparagenesen, die vom Kontakt nach außen die metamorphe Umwandlung erkennen lassen. Durch die Aufheizung entstehen dichte und massige Felse, die granoblastische Gefüge aufweisen (wie Hornfels, Kalksilikat-Fels etc.). Der Belastungsdruck hängt von der Teufe der Intrusion ab; im allgemeinen ist diese seicht und liegt in der Größenordnung von nur wenigen Kilometern. Drucke um 1 kbar sind die Regel, 3 kbar und mehr werden nur selten erreicht.

Dislokations- oder **kataklastische** Metamorphosen sind an Verwerfungen oder Überschiebungen gebunden. Es kommt lokal zum Zerbrechen (Kataklase), Zerreiben (Mylonitisierung) oder durch Umsatz der Reibungsenergie in thermische Energie auch zu einem partiellen oder vollkommenen Aufschmelzen der betreffenden Gesteinspartien. Die Gesteine der Dislokations- oder kataklastischen Metamorphose sind Kataklasite und Mylonite, sowie Pseudotachylite. Posttektonische Umkristallisation der mechanischen Deformationsgefüge ergibt Blastokataklasite und Blastomylonite. Der Aufschlag von extraterrestrischen Körpern auf die Erdoberfläche schafft ähnliche Strukturen. (Siehe Stoßwellenmetamorphose oder Impaktmetamorphose).

Die **Regionalmetamorphose** schafft metamorphe Gesteine, die weite Areale bedecken. Nach Auftreten und metamorpher Beanspruchung lassen sich bei der Regionalmetamorphose prinzipiell zwei Typen

i) die Regionale Versenkungsmetamorphose und
ii) die Regionale Dynamo-Thermometamorphose, unterscheiden.

Die **Regionale Versenkungsmetamorphose** schließt unmittelbar an die Diagenese an. Sedimente mit zwischengelagerten Vulkaniten werden in Ablagerungströgen (Synklinalen) kontinuierlich versenkt und mit Detritus bedeckt. Kaltes, sedimentäres Material kommt schnell in große Teufe, tektonische Durchbewegung fehlt (meist) und das sedimentäre Gefüge bleibt erhalten. Die Temperaturen in so großer Teufe (15 bis 30 km, das entspricht 5 bis 10 kbar) gehen nicht über 400 °C hinaus. Metamorphe Umkristallisation wird

assemblages crystallize at high to very-high confining pressures and at low to moderate temperatures (200° to 400 °C).

Regional dynamothermal metamorphism (or strictly speaking, regional metamorphism) is restricted to orogenic zones. Unlike thermal metamorphism, the heat source is unknown. Thermal energy is produced in the crust itself (e.g. by radioactive decay of uranium, thorium, and potassium) or originates in the upper mantle. Metamorphic zones covering huge areas reflect the pressure-temperature regime of regional metamorphism. The (geo-)thermal gradient was abnormally high and local temperatures up to 800 °C must have been reached during metamorphism. The pressure was in the order of >2 kbar but may have reached 8 kbar and more locally. Deformation due to directed pressure in the (shallow) crust produced orientated and crystalloblastic rock structures.

Mineral assemblages of contact metamorphic, burial metamorphic, and regional metamorphic rocks reflect the stage of metamorphism in terms of temperature and pressure according to the bulk chemistry. The pressure-temperature stability fields of specific mineral assemblages are bounded by univariant curves (containing one degree of freedom according to the phase rule) which are mostly temperature dependent. According to [Win76] four stages or grades of metamorphic conditions can be discerned on the basis of critical mineral assemblages:

durch hohe hydrostatische Drucke und mäßigen Temperaturen (200° bis 400 °C) erreicht, die spezifische Mineralparagenesen schafft.

Die **Regionale Dynamo-Thermometamorphose** (d.h. die Regionalmetamorphose im engeren Sinn) ist ursächlich an Orogenzonen gebunden. Im Gegensatz zur Thermometamorphose ist die Wärmequelle nicht bekannt. Thermische Energie wird zum Teil in der Erdkruste selbst produziert (radioaktiver Zerfall von U, Th und K) und zum Teil aus dem Erdmantel zugeführt. Der geothermische Gradient zur Zeit der Metamorphose ist anomal hoch und lokale Temperaturen bis zu 800 °C müssen erreicht worden sein. Metamorphe Zonen großer (regionaler) Ausdehnung entstehen entsprechend der Temperatur- und Druck-Beeinflussung. Regionale Dynamo-Thermometamorphose findet bei Drucken von über 2 kbar statt, örtlich auch bis 8 kbar und größer. Tektonische Durchbewegung (im oberen Teil) der Kruste schafft die typischen kristalloblastischen (und gerichteten) Metamorphitgefüge (z.B. Kristalline Schiefer).

Thermometamorphose, Regionale Versenkungsmetamorphose und Regionale Dynamo-Thermometamorphose schaffen Gesteine, die nach ihren thermischen Beeinflussungen unterschiedliche Metamorphosegrade in ihren spezifischen Mineralparagenesen erkennen lassen. Die Druck-Temperatur-Stabilitätsfelder von natürlichen Mineralparagenesen sind durch univariante Kurven (mit einem Freiheitsgrad nach dem Gibbs'schen Phasengesetz) abgegrenzt, die vorwiegend von der Temperatur (und kaum oder nur mäßig vom Druck) abhängig sind. Nach [Win76] lassen sich auf Grund kritischer Mineralparagenesen vier Metamorphosegrade unterscheiden (Fig. 12):

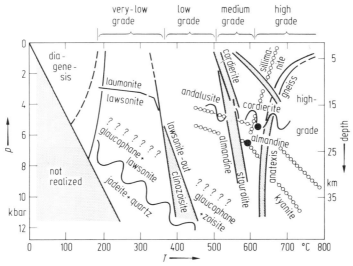

Fig. 12. The four metamorphic grades according to [Win76]. Their pressure and temperature ranges are indicated by typical mineral reactions.

1. Very-low grade metamorphism
   (about 190° to 360 °C)
2. Low grade metamorphism
   (about 360° to 510 °C)
3. Medium grade metamorphism
   (about 510° to 600 °C)
4. High grade metamorphism
   (>600 °C)

Thermal metamorphism results in mineral assemblages stable at the temperatures of metamorphic grades 1 to 4 (pressure under 2 kbar), burial metamorphism is restricted to the metamorphic grade of 1 (very-low grade) and regional metamorphism to metamorphic grades of 2 to 4 at pressures above 2 kbar (Fig. 12).

The division into the four metamorphic grades is done according to mineral reactions occurring in the natural and polynary rock systems which have been calibrated according to pressure and temperature in the laboratory. The result is a **petrogenetic grid of univariant reaction** curves (which are connected via **invariant points**) dividing the pressure-temperature-composition-space into specific subsystems (equivalent to specific mineral assemblages) with thermodynamic relevance. Such a petrogenetic grid for specific and metamorphic minerals reactions is displayed in Fig. 13. The displayed reactions are either solid-solid or solid-solid-vapor equilibria. The latter reveal a continuous dehydration and/or decarbonation with increasing temperature toward the (hydrous) solidi. Univariant mineral reactions in metamorphism degenerate into a higher variancy (that is with one more degree of freedom) if the vapor phase consists of $H_2O + CO_2$. Thus, mineral reactions in which $H_2O + CO_2$ participate are not very well suited for a calibration of the natural temperature-pressure-composition-space, because the mol fraction of $CO_2/(H_2O + CO_2)$ can hardly be reckoned during metamorphism. At a given pressure any specific temperature requires a specific mol fraction of $CO_2/(H_2O + CO_2)$. For example, bivariant reactions occurring within the $CaO—MgO—Al_2O_3—SiO_2—H_2O—CO_2$ system - modelling the prograde

1. Sehr schwach-gradige Metamorphose,
   ungefährer Temperaturbereich: 190 bis 360 °C
2. Schwach-gradige Metamorphose,
   ungefährer Temperaturbereich: 360 bis 510 °C
3. Mittel-gradige Metamorphose,
   ungefährer Temperaturbereich: 510 bis 600 °C
4. Hoch-gradige Metamorphose,
   Temperaturbereich: >600 °C.

Thermometamorphose äußert sich in Mineralparagenesen, die die Metamorphosegrade 1 bis 4 umfassen (<2 kbar). Die Regionale Versenkungsmetamorphose ist dagegen auf Mineralparagenesen begrenzt, die dem Metamorphosegrad 1 angehören. Regionalmetamorphose umfaßt bei Drucken von mehr als 2 kbar die metamorphen Grade 2 bis 4. Die Unterteilung in vier Metamorphosegrade erfolgt auf Grund von Mineralreaktionen, die im Vielkomponenten-System der Gesteine in einzelnen Konzentrationsräumen in der Natur ablaufen und im Laborexperiment thermometrisch und barometrisch kalibriert werden. Das Resultat ist ein **petrogenetisches Netz** von **univarianten Mineralreaktionen** (über **invariante Punkte** verknüpft), die den Druck-Temperatur-Konzentrations-Raum in diskrete Mineralparagenesen unterteilen, denen thermodynamische Relevanz zuzuordnen ist. Ein petrogenetisches Netz mit ausgewählten univarianten Modellreaktionen zeigt Fig. 13. Es sind (bis der Solidus bestimmter Konzentrationsverhältnisse im Vielkomponenten-System Gestein erreicht wird) Festkörper-Reaktionen oder Festkörper-Gas-Reaktionen die mit Zunahme der Temperatur eine fortschreitende Dehydration und Dekarbonisation erkennen lassen. Bei Vorliegen eines Gasgemisches ($H_2O + CO_2$) als Bestandteil des thermodynamischen Gleichgewichtes degenerieren die univarianten Reaktionen in Zustände höherer Varianzen (Di- und Trivarianz). Die Fugazitäten der Gasspezies in der polynären Gasphase bedingen (nach dem Gibbs'schen Phasengesetz thermodynamisch) zusätzliche Freiheitsgrade, die den Systemzustand unbestimmter werden lassen (bivariant, trivariant). Reaktionen, bei denen Gasgemische teilnehmen, sind deshalb für die Kalibrierung des Druck-Temperatur-Zusam-

Fig. 13a···c. Portions of the petrogenetic grid illustrating mineral reactions with increasing temperature from very-low to high grade metamorphism (from [Win76] slightly modified). Pressure p, temperature T.
a) very-low grade; curves labelled GW, BA, and UM occur at compositions of graywackes, basalts, and peridotites;
b) low grade; PG, pelites and graywackes;
c) medium and high grade; MARL, marl; DOL, siliceous dolomites and limestones.

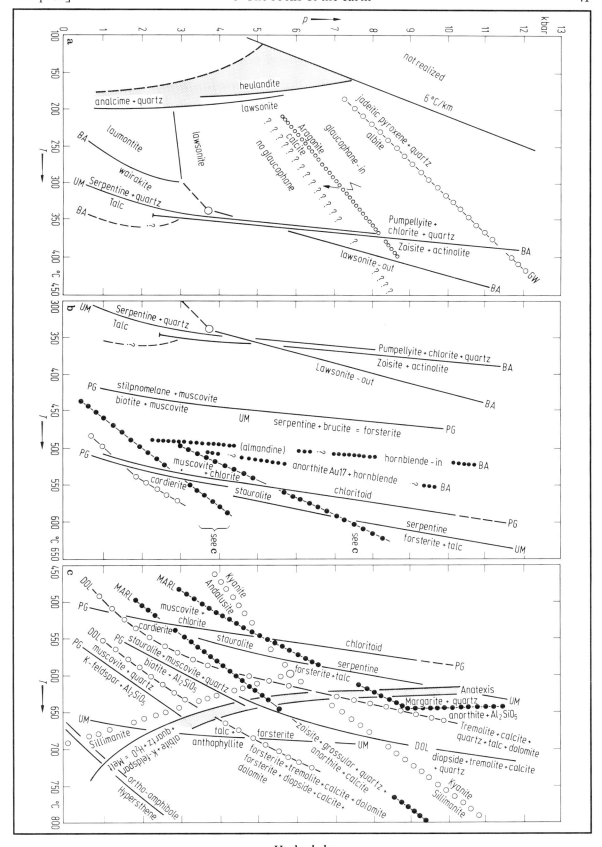

metamorphism of siliceous limestones – are presented in an isobaric (5 kbar) temperature vs. $CO_2/(H_2O+CO_2)$ diagram in Fig. 14.

mensetzungs-Raumes schlecht geeignet. Ihre Anwendung auf natürliche Mineralparagenesen setzt die Kenntnis der $CO_2/(CO_2+H_2O)$-Fraktion in der Gasphase zur Zeit der Metamorphose voraus; denn zu jeder bestimmten Temperatur gehört ein ganz bestimmtes $CO_2/(H_2O+CO_2)$. Solche bivarianten Reaktionen für das $CaO-MgO-SiO_2-H_2O-CO_2$ System – das die progressive Metamorphose von kieseligen Kalksteinen veranschaulicht – sind in einem isobaren (5 kbar) Temperatur-vs. $CO_2/(CO_2+H_2O)$-Schnitt in Fig. 14 dargestellt.

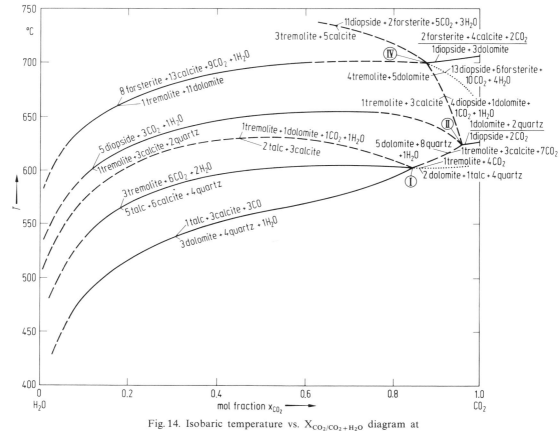

Fig. 14. Isobaric temperature vs. X_{CO_2/CO_2+H_2O} diagram at 5 kbar illustrating (isobaric) univariant mineral reactions which will occur in siliceous limestones during metamorphism (from [Win76]). Isobaric invariant points are labelled I, II, and IV having talc+tremolite+quartz+calcite+dolomite+vapor (H_2O+CO_2), diopside+tremolite+quartz+calcite+dolomite+vapor (H_2O+CO_2), or forsterite+diopside+tremolite+calcite+dolomite+vapor, respectively, in equilibrium. Vapor is H_2O+CO_2.

The change from diagenesis to metamorphism is documented by sedimentary assemblages which have been converted into those of the very-low grade metamorphic type. Not all mineral assemblages, however, are able to exhibit the sedimentary/metamorphic-boundary. Very often, the chemical bulk composition is not suitable for a documentation of certain metamorphic grades. As an example, chlorite + illite + quartz- or calcite + quartz-assemblages of sedimentary origin persist at low or medium metamorphic grade before they are converted into metamorphic mineral assemblages.

Conventionally metamorphism begins – a suitable bulk chemistry assumed – if *albite* + H_2O form from sedimentary assemblages of *analcite* + *quartz* or if sedimentary *heulandite* decomposes to *laumontite* + *quartz* + H_2O. This indicates temperatures between 180° and 200 °C. Increasing temperature during metamorphism breaks down *laumontite* to *wairakite* + H_2O and *prehnite*- and *pumpelleyite*-bearing assemblages occur gradually. At pressures above 3 kbar assemblages containing *lawsonite*, *glaucophane*, and *jadeite* become stable during prograde metamorphism and *calcite* is transformed into *aragonite*. The change from very-low grade to low grade metamorphism is shown by the disappearance of *lawsonite*, *prehnite*, and *pumpelleyite* and by the appearance of *zoisite*/*clinozoisite* and *actinolite*. *Chloritoid* and *epidote* in assemblages of Fe^{2+}- and Al-rich bulk compositions are critical mineral phases. The transition of low grade to medium grade metamorphism is documented by the disappearance of Fe-rich *chlorites* in the presence of *muscovite*. Assemblages containing *chloritoid* are replaced by those of *staurolite*. *Staurolite* + *quartz* react to form *cordierite* + *andalusite* + H_2O or *almandine* + *sillimanite* + H_2O under the conditions of medium grade metamorphism. The *garnets* are almandine-rich solid solutions.

The reactions of *muscovite* + *quartz* changing to a Al_2SiO_5-*polymorph* + *K-feldspar* + H_2O approximately defines the boundary of medium to high grade metamorphism. *Andalusite* is the Al_2SiO_5-polymorph occurring below 3 kbar, and *sillimanite* that occurring at higher pressure. The range of high grade metamorphism is restricted to pressures of 3.5 kbar ($P_{total} = P_{H_2O}$) and temperatures between 740° and 660 °C. Above 3.5 kbar the hydrous solidi of *K-feldspar* + *muscovite* + *quartz* and of *albite* + *K-feldspar* + *quartz* intersect the *muscovite* + *quartz* + *sillimanite* + *K-feldspar* + H_2O – reaction

Der Schritt von der Diagenese zur Metamorphose zeigt sich an sedimentären Paragenesen, die sich in sehr schwach-gradige metamorphe Paragenesen umwandeln. Nicht alle Mineralgemenge sind aber von ihrem Pauschalchemismus in der Lage die Grenze Diagenese/Metamorphose kritisch zu belegen. Sedimentäre Chlorit + Illite + Quarz- und Calcit + Quarz-Paragenesen persistieren z. B. unter höher-gradigen Metamorphosebedingungen, ehe sie sich in (kritische) metamorphe Mineralparagenesen umwandeln.

Konventionell ist der Schritt von der Diagenese zur sehr schwach-gradigen Metamorphose vollzogen, wenn sich *Albit* + H_2O aus der diagenetischen Paragenese *Analzim* + *Quarz* oder *Laumontit* + H_2O aus diagenetischem *Heulandit* gebildet haben. Das Erscheinen der metamorphen Paragenesen impliziert Temperaturen zwischen 180 bis 200 °C. Bei Ansteigen der Temperatur wandelt sich *Laumontit* in *Wairakit* + H_2O um und es erscheinen nacheinander *Prehnit* sowie *Pumpellyit*-führende Paragenesen. Bei Drucken über 3 kbar werden mit aufsteigender Metamorphose Paragenesen mit *Lawsonit*, *Glaukophan* und *Jadeit* stabil und *Calcit* durch *Aragonit* ersetzt. Der Wechsel von sehr-schwach-gradiger zu schwach-gradiger Metamorphose wird durch das Verschwinden von *Lawsonit*, *Pumpellyit*, *Prehnit* und durch die Bildung von *Zoisit*/ *Klinozoisit* und *Aktinolith* belegt. *Chloritoid* und *Epidot* in Fe^{2+} und Al-reichen Gesteinen sind aber noch für die schwach-gradige Metamorphose kritisch. Der Übergang schwach-gradiger zur mittel-gradigen Metamorphose wird durch das Verschwinden von Fe-reichen *Chloriten* in Gegenwart von *Muskovit* angezeigt. Paragenesen mit *Chloritoid* werden durch solche mit *Staurolith* ersetzt. Paragenesen mit *Staurolith* + *Quarz* wandeln sich innerhalb der mittel-gradigen Metamorphose bereits in *Cordierit* + *Andalusit* + H_2O oder *Granat* + *Sillimanit* + H_2O um. Der *Granat* ist ein Almandin-betonter Mischkristall.

Die Grenze mittel-gradige/hoch-gradige Metamorphose ist durch die Reaktion *Muskovit* + *Quarz* $\rightleftharpoons$ *Kalifeldspat* + Al_2SiO_5-*Modifikation* + H_2O gekennzeichnet. Die Al_2SiO_5-Modifikation ist dabei *Andalusit* im Druckbereich unter 3 kbar, aber *Sillimanit* bei höheren Drucken. Der Bereich der hoch-gradigen Metamorphose ist bei $P_{total} = P_{H_2O}$ auf Drucke unter 3,5 kbar und Maximaltemperaturen von etwa 740 °C begrenzt. Oberhalb dieser Drucke und zwischen 740° bis 660 °C werden aus Alkalifeldspat-führenden Paragenesen bereits granitische Schmelzen (anatektisch) gebildet; denn

curve (see Fig. 3) and a granitic magma can be formed by partial melting of the high grade metamorphic gneiss host (anatexis). Deficiency of H_2O within the rock system – that is $P_{total} \gg P_{H_2O}$ – does not cause partial melting. Anhydrous mineral assemblages with *feldspars + quartz + kyanite* or *sillimanite* of the granulite-rock family are formed from metamorphites with a granitic bulk chemistry. *Eclogite* with *omphacite + pyrope-rich garnet* is the anhydrous product of basalt transformation.

Mineral zonations in nature from the low grade to the high grade type imply prograde metamorphism, i.e. with increasing temperature under the continuous release of H_2O and CO_2. A subsequent decrease in temperature under exclusion of these volatiles preserves mineral assemblages formed during the culmination of metamorphic conditions. Access of volatiles, however, initiates reactions in a reversed order during the metamorphic cooling process (retrograde metamorphism).

die hydrischen Solidi von *K-Feldspat + Muskovit + Quarz* und von *Albit + K-Feldspat + Quarz* schneiden die *Muskovit + Quarz + Sillimanit + K-Feldspat + H_2O* Reaktionskurve (Fig. 3). Wasseruntersättigung im Gesteinssystem, d.h. $P_{total} \gg P_{H_2O}$, führt nicht zur Anatexis. Anhydrische Mineralparagenesen mit *Feldspäten + Quarz + Disthen* oder *Sillimanit* bilden die metamorphen Gesteine der Granulit-Familie. Ebenso erzeugt anhydrische Metamorphose von basaltischen Gesteinen *Eklogit* mit *Omphacit + Pyrop-betontem Granat*.

In der Natur implizieren Mineralzonen von schwach-gradig- zu hoch-gradig-metamorph eine progressive Metamorphose bei Temperaturanstieg unter H_2O- und CO_2-Abgabe. Erfolgt die rückläufige Abkühlung unter Ausschluß von H_2O und CO_2, so bleiben die typischen Mineralparagenesen der progressiven Metamorphose (weitgehend) erhalten. Zutritt der leichtflüchtigen Komponenten bei der Abkühlung läßt die Mineralreaktionen der progressiven Metamorphose in umgekehrter Reihenfolge ablaufen, sodaß die höher-gradigen Stadien verwischt werden (retrograde Metamorphose).

0.4 The sedimentary rocks — Die sedimentären Gesteine
0.4.1 Nomenclature — Nomenklatur

Sedimentary rocks or sediments are the product of sedimentation. Igneous and metamorphic (as well as previously sedimentary) rocks are disintegrated physically and chemically by weathering. The residual material is transported as debris and/or in solution by wind, water, and ice. It accumulates at the locus of deposition and/or precipitation. Occurrence and origin distinguish two main types of sedimentary rocks:

Sedimentäre Gesteine oder Sedimente (Sedimentgesteine, Absatzgesteine, Ablagerungsgesteine, Schichtgesteine) sind die Produkte der Sedimentation. Magmatische und metamorphe (aber auch alte sedimentäre) Gesteine werden physikalisch und chemisch durch die Verwitterung zerkleinert. Die Residuen werden als Schutt und/oder als Lösungen von Wind, Wasser und Eis transportiert. Sie bilden Akkumulationsprodukte am Ort ihrer Ablagerung und/oder Ausfällung. Nach Auftreten und Herkunft lassen sich generell zwei Gruppen der Sedimentgesteine unterscheiden. Dies sind die

i) the clastic or fragmental sedimentary rocks or sediments and

ii) the chemical and biochemical sedimentary rocks or sediments.

i) Klastischen oder Trümmer-Sedimente und die

ii) Chemischen und biochemischen Sedimente.

Clastic sediments make up about 92% of the total, chemical and biochemical sediments the rest. More than 50% of the earth's surface is covered by a (thin) blanket of sediments which makes up about 6% of the (total) earth's crust. Sedimentary mineral deposits of economic importance comprise much less than 1% of the total sedimentary bulk composition.

Klastische Sedimente machen etwa 92 % aller sedimentären Gesteine aus, chemische und biochemische Sedimente umfassen den Rest. Mehr als 50 % der Erdkruste sind mit einer (dünnen) Decke von sedimentären Gesteinen bedeckt, die etwa 6 % der Erdkruste ausmachen. Minerallager oder -Lagerstätten sedimentärer Entstehung, die vielfach wirtschaftliche Bedeutung besitzen, machen weniger als 1 Prozent aller sedimentären Gesteine aus.

0.4.2. Mineral composition of sedimentary rocks — Die mineralische Zusammensetzung der sedimentären Gesteine

The average mineral composition of the earth's crust is characterized by silicate minerals (feldspars + quartz >80%). Sediments, as the result of weathering of igneous and metamorphic rocks, thus consist mainly of the residual silicate debris and/or of newly formed silicate-, hydroxide-, and carbonate-minerals. The most abundant sedimentary rock on earth has the composition of a shale (solidified) or a clay (not solidified). Its average mineral composition was extrapolated by geochemical calculation ([Cor49]) and is presented in Table 15. This **average shale** or **clay** contains large amounts of quartz, phyllosilicates (mica, clay minerals), and feldspars.

One of the most resistant (physically and chemically) minerals on the earth's surface is quartz. Thus, detrital quartz becomes enriched in the sedimentary debris during the process of weathering and transportation. Partial or complete disintegration of the feldspars, the amphiboles, the pyroxenes, the mica and olivines enab-

Die mineralische Zusammensetzung der Erdkruste ist durch Silikatminerale (Feldspat + Quarz > 80 %) gekennzeichnet. Sedimente als Produkte der Verwitterung bestehen deshalb aus den silikatischen Verwitterungsresten und Neubildungen neben Hydroxyden und Karbonaten. Das häufigste Sediment-Gestein der Erdoberfläche hat die Zusammensetzung eines Tonschiefers (verfestigt) oder Tones (unverfestigt). Seine mittlere Mineralzusammensetzung wurde durch geochemische Überschlagsrechnung von [Cor49] ermittelt und ist in Tab. 15 zusammengestellt. Dieser **mittlere Tonschiefer** oder **Ton** enthält wesentliche Gehalte an Quarz, Phyllosilikaten und Feldspäten.

Eines der widerstandsfähigsten Minerale (physikalisch und chemisch) an der Erdoberfläche ist Quarz. Daher wird detrischer Quarz im sedimentären Schutt (Detritus) durch den Prozeß der Verwitterung und des Transportes angereichert. Feldspäte, Glimmer, Amphibole, Pyroxene und Olivine werden (fast immer) voll-

Table 15. Mineral composition of the *average sedimentary rock* (in per cent by weight) calculated on the basis of the average sedimentary chemistry as given in Table 22, column 1 (nach [Cor48] slightly modified).

	wt-%
Quartz	30
Feldspar	9
plagioclase	(1)
alkali feldspar	(8)
Mica	23
muscovite, illite	(15.5)
biotite, glauconite	(7.5)
Clay Minerals	17.5
kaolinite, halloysite,	
montmorillonite	(17.5)
Chlorite	2
Calcite	6
Dolomite	2.5
Oxides	6.5
Apatite	<0.5
Carbonaceous Matter	0.5
Others	2.5

les the formation of residual remnants (**weathering residua**) or of new minerals (**weathering products**). The latter are mainly very fine-grained phyllosilicates (grains less than 0.02 mm in diameter) of the so-called clay mineral group. The most important of these are listed in Table 8. Clay minerals such as *kaolinite*, *dickite*, and *nakrite* are $Al_4Si_4O_{10}(OH)_8$-polymorphs which can be distinguished by different stacking order of their tetrahedral + octahedral-layers. *Halloysite* has the same chemical composition but its layers are bent and rolled up to fine tubes. Clay minerals of *montmorillonite* or *vermiculite* composition (montmorillonite, nontronite, saponite, hectorite, beidellite, vermiculite) are built up of alternating tetrahedral + octahedral + tetrahedral-units containing intercalated layers of water. These structures can swell or contract with the release or incorporation of water. *Illites* (dioctahedral hydromica) are mainly residua of weathered (primary) muscovite. *Calcite*, *dolomite*, and – due to the action of organisms or hot springs – *aragonite* are sedimentary carbonate minerals. Sedimentary calcite mostly incorporates Mg for Ca in the structure and, because of greater disorder, is more soluble (in seawater) than ideal calcite.

Heavy minerals (minerals with a density of >2.9) of igneous and metamorphic assemblages that are physically and chemically resistant to the activities of weathering will be enriched when transported by wind and water. Resistant minerals such as *zircon*, *tourmaline*, *sphene*, the *garnets*, members of the *epidote* group, *corundum*, *spinel*, *diamond*, *magnetite*, *ilmenite*, *straurolite*, the Al_2O_5-polymorphs as well as certain *amphiboles* and *pyroxenes* are often accumulated in heavy mineral concentrates or deposits. Noble metals as *gold*, *platinum* and *osmiridium* can also be found in these economically important mineral deposits.

kommen oder teilweise zerstört, bleiben als **Verwitterungsreste** übrig oder bilden OH-haltige (silikatische) **Verwitterungsneubildungen.** Die Produkte sind sehr feinkörnige und fein-kristalline Phyllosilikate (meist weniger als 0,002 mm Durchmesser) der Tonmineral-Gruppe. Die wichtigsten Tonminerale sind in Tab. 8 aufgeführt. Die Tonminerale *Kaolinit*, *Dickit* und *Nakrit* sind $Al_4Si_4O_{10}(OH)_8$, die sich durch unterschiedliche Aufeinanderstapelung des Tetraeder + Oktaeder-Schichtpaketes unterscheiden. *Halloysit* ist zu Röhrchen aufgerolltes $Al_4Si_4O_{10}(OH)_8$. *Montmorin*-Minerale (Montmorillonit, Nontronit, Saponit, Hectorit, Beidellit, Vermikulit) sind von alternierenden Tetraeder + Oktaeder + Tetraeder-Einheiten aufgebaut und können zwischengelagerte H_2O-Schichten enthalten. Durch Wasserabgabe oder -Aufnahme kann sich die Montmorillonit-Struktur zusammenziehen oder aufblähen. *Illite* sind sehr häufige Tonminerale und als unvollständige (dioktaedrische) Glimmer (Hydroglimmer) meist die Verwitterungsreste von Muskovit. *Kalzit*, *Dolomit* und – durch die Mitwirkung von Organismen oder heißer Quellen – *Aragonit* sind sedimentäre Karbonat-Minerale. Sedimentärer Kalzit kann Mg (für Ca) in sein Gitter aufnehmen und ist dadurch – weil ungeordneter – leichter löslich (in Meerwasser) als stöchiometrisch idealer Kalzit.

Minerale magmatischer und metamorpher Paragenesen, die chemisch und mechanisch gegen Verwitterung resistent sind und die hohe Dichten (>2,9) aufweisen, werden beim Transport und Ablagerung angereichert. Sie bilden nutzbare Lager, Seifen oder Schwermineralkonzentrate. Beteiligt sind vor allem *Zirkon*, *Turmalin*, *Titanit*, Minerale der *Granat*- und *Epidotgruppe*, *Korund*, *Spinell*, *Diamant*, *Magnetit*, *Ilmenit* aber auch *Staurolith*, Al_2SiO_5-Modifikationen sowie *Pyroxene* und *Amphibole*. Edelmetalle wie *Gold*, *Platin* und *Osmiridium* können ebenfalls in Seifen angereichert werden.

0.4.3 The texture and structure of sedimentary rocks — Das Gefüge der sedimentären Gesteine

A classification of sedimentary rocks can be achieved on the basis of grain size and mineral (modal) composition. **Psephites** (rudites; gravel, rubble, boulder, pebble) are (clastic) sediments with more than 50% (by weight) of the constituents >2 mm in diameter. **Psammites** (arenites; sand) have more than 50% of their grains within the sand size fraction (2 to 0.002 mm). **Pelites** (argillites; clay, mud) are restricted to grain sizes <0.002 mm. A transitional grain size fraction within the psammite fraction is that of **silt** (from 0.06 to 0.002 mm).

The determination of the grain size and its distribution (grain size distribution) is obtained by sifting of the loose material (gravel, sand) and by sedimentation analysis for (loose) sediments below a grain size of 0.06 mm (silt, clay). For consolidated, solid sedimentary

Die Einteilung der Sedimente geschieht nach Korngröße und Mineralbestand. Als **Psephite** (Steine, Blockwerk) bezeichnet man (klastische) Gesteine, wenn mehr als 50 % ihrer Körner einen Durchmesser von >2 mm aufweisen. **Psammite** (Sand) sind klastische Gesteine, die über 50 % der Körner auf die Korngröße von 2 bis 0,02 mm verteilen. **Pelite** (Ton) sind klastische Gesteine mit Korngrößen <0,02 mm. Silt umfaßt die Korngrößen von 0.06 bis 0.002 mm innerhalb der Psammit-Fraktion.

Die Bestimmung der Korngrößenverteilung erfolgt beim Lockersediment (Kies, Sand) durch Sieben und ab <0,063 mm durch Schlämmen. Am verfestigten Sediment müssen die Korngrößenklassen im Dünnschliff ausgemessen werden. Im allgemeinen enthalten

rocks the grain size distribution has to be measured by means of petrographic microscopy. Sedimentary rocks usually contain a sequence of grain size fractions. A division into (suitable) fractions is made logarithmically (Brigg's logarithm) in Germany and displayed using the grain size scale of DIN 4022 (Table 16). **Gravel, sand, silt,** and **clay** are names expressing a grain size fraction or fractions and not a specific mineral composition. Thus, **sand** must not necessarily be an accumulation of loose quartz grains. A sand can also contain grains of feldspars and calcite only, or fragments of rocks (igneous, metamorphic as well as sedimentary) and (organic) shells with a size distribution which covers the fraction of sand according to the scale of Table 16. The measured grain size distribution of a sedimentary rock is graphically expressed by a grain size vs. grain weight diagram in which the ordinate either represents the specific amounts of the individual fractions participating in the distribution (Fig. 15a, b, grain size distribution curve), or the grain size distribution is expressed as summation of the fractions up to 100% (Fig. 15c, grain size summation curve). The three diagrams roughly reflect the complex relationships between transportation and deposition of particles in a specific medium. Sands with grains restricted to one or a few size fractions are (usually) transported by the action of wind or surging water (Fig. 15a). Ice does not sort the debris according to the size and the density. The resulting grain size distribution pattern is (more or less) irregular with maxima separated by minima (Fig. 15b).

die klastischen Gesteine verschiedenen Korngrößen-Klassen oder -Fraktionen. Die Einteilung nach Korngrößenklassen geschieht in Deutschland logarithmisch nach DIN 4022 (Briggsche Logarithmen), siehe Tab. 16. Bezeichnungen wie **Sand, Schluff** oder **Ton** sind reine Korngrößenbezeichnungen, die nichts über die mineralische Zusammensetzung der klastischen Sedimente aussagen. So muß **Sand** nicht gleichbedeutend mit Quarz sein; es kann sich ebenso um Feldspäte oder Kalzit, aber auch um Bruchstücke von Organismen in der Korngrößenklasse des Sandes handeln. Die experimentell am Sediment(gestein) ermittelte Korngrößenverteilung kann graphisch durch die Korngrößenverteilungs- oder -summenkurve mit den Korngrößenklassen (als Abszisse) vs. den Kornprozenten (als Ordinate) dargestellt werden (Fig. 15a, b, c). Beide Darstellungsarten eines sedimentären Gesteines spiegeln die komplexen Transport- und Absatzbedingungen der Sediment-Partikel wider. Im allgemeinen kann man annehmen, daß z.B. Sande, die nur wenige oder nur eine einzige Korngrößenfraktion aufweisen, von Wellen am Strand oder durch Wind abgelagert worden sind (Fig. 15a). Eis dagegen transportiert und setzt Material der unterschiedlichsten Korngrößenklassen ab. Eine solche Korngrößenverteilungskurve kann (mehrere) Maxima und Minima aufweisen (Fig. 15b).

Table 16. *German* grain size scale of DIN 4022 (based on Brigg's logarithm). Grain sizes are given as diameter. (Clay and boulder portions slightly modified; from [Cor68]).

Clay (Ton)	Silt (Schluff)			Sand			Gravel (Kies)			Boulder (Blockwerk, Steine)
0.002	0.0063	0.02	0.063	0.2	0.63	2.0	6.3	20	63	200 mm
	fine	medium	coarse	fine	medium	coarse	fine	medium	coarse	

The properties of a grain size distribution can be expressed quantitatively by the so-called **median diameter** (Md) and the coefficient of **sorting** (So). The median diameter is defined as the specific size of the summation curve at which 50% of the particles are greater or smaller than the Md. The coefficient of sorting, So, can also be derived from the summation curve and is defined by the equation

Die Eigenschaften der Korngrößenverteilung lassen sich zahlenmäßig und damit quantitativ durch den sogenannten **Mittleren Korndurchmesser** Md (= Medianwert) und den **Sortierungsgrad** So (= Sortierung) ausdrücken. Der Md bezeichnet diejenige Korngröße der Kornsummenkurve bei der 50% der Sedimentkörner größer oder kleiner sind. Der So ist ebenfalls aus der Summenkurve zu ermitteln und ergibt sich aus

$$\text{So} = \sqrt{Q_3/Q_1}.$$

Q_3 and Q_1 define grain sizes at which 75% and 25% of the sample, respectively, have smaller grain sizes than Q_3 and Q_1. Advanced sorting requires a close neighbourhood of Q_3 and Q_1 (with Md in between), i.e. only a few fractions make up the sediment (Fig. 15a). A clastic sediment can be called **very-well sorted** when the coefficient So is <1.4, **well sorted** with 1.4 to 1.87, and **poorly sorted** with >1.87.

Die Viertelgewichtsdurchmesser Q_1 und Q_3 (auch 1. und 3. Quartilwert genannt) sind die Korngrößen, die unter 25 bzw. 75% der Summenkurve liegen. Ein Sediment ist umso besser sortiert, je dichter Q_1 bei Q_3 (und Md) liegt, d.h. das Sedimentgestein weist nur wenige Korngrößenfraktionen auf (Fig. 15a). Werte für die So <1,4 bezeichnen **sehr gut sortierte,** bis 1,87 **gut sortierte** und >1,87 **schlecht sortierte** Sedimente.

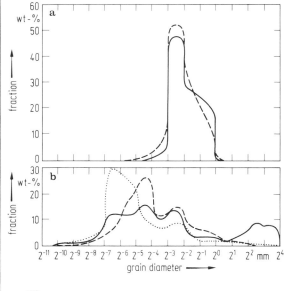
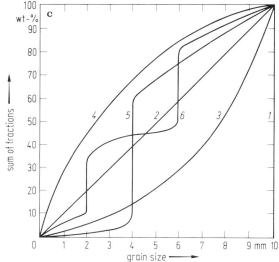

Fig. 15a···c. Graphic presentations of the grain size distribution of clastic sediments (from [Cor68]).
a) and b) grain size vs. grain weight (in %), American WENTWORTH-grain size scale. a) very-well sorted river (solid curve) and dune sand (dashed curve), respectively; b) three poorly sorted glacial tills, c) grain size summation curves, grain size scale arbitrarily divided in to mm. *1*, only one grain size of 10 mm; *2*, all grain sizes from <1 to 10 mm; *3*, few fine-grained but many coarse-grained particles; *4* many fine-grained but few coarse-grained particles; *5*, with a maximum at 4 mm; *6*, with maxima at 2 and 6 mm, respectively, and a minimum at 4 mm.

◄

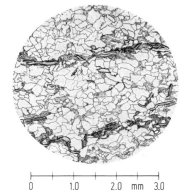

Fig. 16. Thin section drawing of a layered (clastic) sedimentary rock. Triassic sandstone from Karlshafen (Weser River) from [Cor68]. Quartz and (few) feldspar (light), mica flakes (heavy-contoured) in a matrix of fine-grained mica and (few) quartz (gray to dark).

Transport and sedimentation within a specific medium effect the **sphericity** (shape) and **roundness** (angularity) of the particles. Detrital minerals, however, are neither spheres nor ideal and isometric polyhedra. They are prisms, cubes, fibers, sheets, etc. and their shapes are not necessarily the products of abrasion. Abrasion in a specific medium smoothes the edges and corners of the detrital particles more or less, according to their physical resistance and hardness (with a given intensity and time of abrasion). The effect can be seen in the **degree of roundness** or angularity expressed as *angular*, *subangular*, *subrounded*, or *rounded* (Fig. 17).

Die Einwirkung der Transport- und Absatzmedien macht sich auch in der **Kornform** und dem **Rundungsgrad** der Sedimentkörner bemerkbar. Ist die Kornform durch die primäre Ausbildung der Mineralkomponenten (z.B. isometrisch, prismatisch, stenglig, blättchenförmig etc.) vorgezeichnet, so ist der Rundungsgrad mehr oder weniger ein Anzeichen für das mechanische Einwirken der Medien bei Absatz und Transport. Abrieb in einem (bestimmten) Medium schleift und rundet Kanten und Ecken der detritischen Minerale in Abhängigkeit ihrer physikalischen Resistenz und Härte (bei einer bestimmten Stärke und in einer bestimmten Zeit). Der Abrasionseffekt drückt sich im **Grad der Rundung** mit Klassen wie *sehr eckig*, *eckig*, *gerundet* und *gut gerundet* aus (Fig. 17).

In a sand, for example, grains of quartz and calcite can reveal different degrees of roundness. The soft calcite (hardness of 3) may exhibit excellent roundness whereas the hard and resistant quartz (hardness of 7) still remains angular.

Sedimentation causes loosely-packed arrangements of grains. If the sediment is very-well or well sorted, the porosity (percentage of pore space) is high and most of the pores are accessible and permeable to media (liquids, solutions, gases, etc.). Poorly sorted sediments have smaller porosities than well-sorted ones. Their interstices are filled by the debris of the smaller grain sizes, thereby reducing the pore space capacity.

Textural properties such as grain size distribution, median diameter, sorting, sphericity, roundness, porosity and permeability can be used to classify a sedimentary rock and an example is given in Table 17. The most obvious structural property of all sedimentary rocks, however is the (stratal) **layering** or **stratification** of their particles (Fig. 16). Layering or stratification implies differences in the amount and kind of material making up a layer, stratum or bed. The amount of material carried by a river can alternate because of sea-

So können Kalzit- und Quarzkörner eines Sandes unterschiedliche Rundungsgrade aufweisen. Kalzit mit Härte 3 wird beim Transport und Ablagerung wesentlich besser gerundet (d.h. abgeschliffen) als der härtere und mechanisch sehr resistente Quarz mit Härte 7.

Sedimentation bewirkt eine lockere Packung der Körner. Bei sehr gut sortierten oder gut sortierten (klastischen) Sedimenten ist die Porosität (Prozentanteil des Porenraumes am Gesamtgestein) hoch und die Poren sind (meist) für bestimmte Medien (Flüssigkeiten, Lösungen, Gase, etc.) zugänglich. Schlecht sortierte (klastische) Sedimente weisen im allgemeinen kleinere Porositäten auf als gut sortierte. Die Zwischenräume zwischen den groben Körnern sind mit feinerem Material gefüllt, sodaß ihr Porenraum stark reduziert ist.

Korngrößenverteilung mit Sortierungsgrad und mittlerem Korndurchmesser, Kornform und Kornrundung sind strukturelle Eigenschaften des sedimentären Gefüges. Diese dienen vor allem zur (strukturellen) Kennzeichnung der klastischen Gesteine, als Beispiel mögen die Daten von Tab. 17 dienen. Die auffälligste Gefügeeigenschaft aller Sedimentgesteine ist jedoch die schichtige Anordnung der sedimentbildenden Bestandteile im Gesteinsverband (Fig. 16). Diese texturelle Gefügeeigenschaft ist so typisch, daß die

Table 17. Textural and structural properties of a conglomeratic (psephitic) graywacke from the Harz Mountains, Germany [Huc66].

Grain size fractions (diameter in mm)	wt-%	
11.240···6.320	5.7	Median diameter: 0.710 mm
6.320···3.560	5.4	Sorting: 2.24
3.560···2.000	7.7	Total mineral density:
2.000···1.124	27.8	2.71 g cm^{-3}
1.124···0.632	15.9	Rock density: 2.63 g cm^{-3}
0.632···0.356	11.4	Porosity: 1.5%
0.356···0.200	9.5	Permeability:
0.200···0.112	7.2	$\perp$ layering $9 \cdot 10^{-4}$ d
0.112···0.063	3.4	$\parallel$ layering $45 \cdot 10^{-4}$ d
0.063···0.036	1.0	Roundness for grains of
0.036···0.020	1.0	2···0.2 mm: subrounded to subangular
<0.020	4.0	Length/width-ratio for grains of 2···0.2 mm: ≈1.5

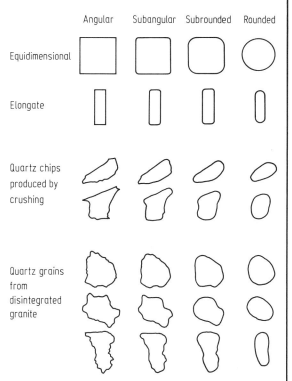

Fig. 17. Roundness and sphericity scale for grains from psammitic (and psephitic) rocks. From [Wil55].

sonal oscillations; the kind of material can vary if the source area – producing and delivering the weathered debris – has changed or the salt solution precipitates different minerals. Slurries (fine-grained particles, usually clay minerals, dispersed in water) deposit their sheet-like minerals parallel or sub-parallel to the floor. After solidification, these minerals are responsible for the (well developed) fissility of the slates and shales. Streaming and flowing media orientate particles with anisotropic shapes. Ellipsoidal bodies are arranged like (roofing) tiles in the flow direction or assymmetric ripple-marks – with a flat wind-ward but a steep lee-side – are produced in sands. Symmetric ripple-marks are the result of waves or surging water. Alternating changes in the flow direction in cooperation with sedimentation of material causes structures known **as cross-lamination** or **cross-bedding.** Slurries often produce a water-solid system which has higher densities than pure (sea) water. These are **turbidity currents** with densities of about 1.2 and the ability to flow and carry denser (and larger) components on slopes with very small inclinations. Turbidity currents deposit sediments with **graded bedding** (rhythmic alternations of units with the coarse material at the bottom and grading into finer material toward the top within a macro-layer) very similar in appearance to laminated clays which are the product of seasonal oscillations during glacial deposition.

Sedimentgesteine auch **Schichtgesteine** genannt werden. **Schichtung** ist das Abbild einer mechanischen Anlagerungstextur, die Unterschiede bezüglich des Materials in den Schichten, bzw. Lagen anzeigt. Solche Materialunterschiede können jahreszeitliche bzw. klimatische Schwankungen der Suspensionszufuhr bedeuten; sie können aber auch einen Wechsel der Materiallieferungen aus dem Verwitterungsgebiet anzeigen. Feinste Trübe, die im wesentlichen eine Suspension von feinsten Phyllosilikaten in Wasser darstellt, setzt die blättchenförmigen Minerale parallel zum Boden ab. Nach der Kompaktion solcher Sedimente bewirken die Phyllosilikate die gute Teil- und Spaltbarkeit der Schiefertone und Tonschiefer. Strömende Medien regeln anisotrope Materialien nach der Form ein. Ellipsoide Körper werden mit ihren Längsachsen (dachziegelartig) zur Strömungsrichtung angeordnet. In Sanden werden asymmetrische Strömungsrippeln mit flacher Luv- aber steiler Lee-Seite erzeugt. Wellenschlag bedingt dagegen symmetrische Oszillationsrippeln. Azimutal-räumlicher Wechsel der Strömungsrichtung mit gleichzeitiger Sedimentschüttung erzeugt Sedimenttexturen, die im Anschnitt diagonal bzw. kreuzartig angeordnet sind: **Diagonal-** und **Kreuzschichtung.** Feinste Trübe kann aber auch ein System Wasser-Feststoff bilden, das physikalisch höhere Dichten als reines (Meer-)Wasser aufweist. Solche **Suspensionsströme** mit Dichten um 1,2 sind als **Trübströme** (turbidity currents) auf schon schwach geneigter Unterlage fließfähig und transportieren gröbere und spezifisch schwere Komponenten. Bei Absatz kommt es zur Ausbildung einer korngrößenmäßig **gradierten Schichtung** (graded bedding), die der jahreszeitlich bedingten gradierten Schichtung der eiszeitlichen Bändertone mit ihrer typischen Warventextur im Aussehen gleicht.

0.4.4 Clastic or fragmental sedimentary rocks — Klastische oder Trümmer-Sedimente

(Classification and modal composition)

(Klassifikation und modale Mineralzusammensetzung)

a) **Psephites** (rudites; gravel, rubble, boulder, pebbles) are clastic sediments with grain sizes above 2 mm. Not solidified or consolidated psephites with rounded components are called **gravels. Rubbles** are loose psephites containing angular components. **Conglomerates** (with rounded) and **breccias** (with angular) components are the solidified psephites. In general, the components of the psephites are rock pebbles and boulders as (weathered) residua with the finer debris filling the interstices. Individual mineral fragments will occur if the source area bears rocks with grain sizes greater than 2 mm. These are mainly quartz and feldspars. Psephite nomenclature depends on the participating fragmental material. A *granite conglomerate* or a *chert breccia* are rocks which consist pre-

a) **Psephite** (Rudite; Blockwerk, Steine, Gerölle) sind klastische Sedimente, deren **Md** > 2 mm beträgt. Nichtverfestigte Psephite, deren Bestandteile gerundet sind, nennt man **Schotter.** Sind die Bestandteile eckig, spricht man von **Schutt. Konglomerate** mit gerundeten und **Breccien** mit eckigen Bestandteilen sind die verfestigten Produkte der Psephite. Im allgemeinen bestehen die Hauptbestandteile der Psephite aus Mineralaggregaten, d.h. aus Geröllen oder Bruchstücken von präexistenten Gesteinen neben Sand- und Tonkomponenten, die die Zwischenräume ausfüllen. Stehen im Gebiet der Verwitterung grobkörnige Gesteine mit > 2 mm Korndurchmesser zur Verfügung, so können in den Psephiten auch Gerölle oder Bruchstücke von Einzelmineralen (z.B. Feldspäte und Quarz) vor-

dominantly of (rounded) granite pebbles or (angular) chert rubbles, respectively. If the fragments are individual minerals, e.g. quartz or feldspar, the psephite is called a quartz or feldspar conglomerate (or breccia), respectively. An exact and quantitative determination of the psephite-forming components is extremely difficult because the fragments very often exceed the dimension of an ordinary thin section (2 by 2 cm). Psephites become normally solidified by a silicate- and/or carbonate matrix or cement during the process of diagenesis.

b) **Psammites** (arenites; sand) are clastic sediments with components in the grain size fractions of 2 to 0.002 mm, thus incorporating the **silt** fraction of 0.06 to 0.002 mm. Loose psammites are **sands**, solidified products are **sandstones**. A mineral or modal classification can be achieved on the basis of quartz, the feldspars, the phyllosilicates and the carbonates (mainly calcite and less abundant dolomite). Thus, the mineral composition of a sandstone (or sand) can be graphically presented in a Q (quartz) + F (feldspar) + P (phyllosilicate) + k (carbonate) tetrahedron (Fig. 18a). Many sandstones (and sands), however, contain less than 10% carbonate minerals (unless the sediment has carbonate cement or consists of clastic calcite or dolomite grains) and a projection on the Q—F—P basis appears to be reasonable. The main sandstone types according to this projection, such as normal sandstone, arkoses, and graywackes are given in Fig. 18b. Average and detailed modal compositions of members of the sandstone group are listed in Table 18.

The presence of (polygenetic) rock fragments in a sandstone or sand is a function of the grain size of the exposed rock in the source area and its resistance during the weathering and transportation process. Coarse-grained granites (with grain sizes of above 2 mm) in the source area will contribute fragments of feldspar and quartz. Sand-sized rock fragments can be derived from fine-grained (volcanic) igneous rocks (grain sizes less than 2 mm). Fig. 20 gives an impression of how the various components making up a sandstone (graywacke) can be allocated to its grain size pattern.

kommen. Die qualitative Bezeichnung der Psephite erfolgt nach dem Gesteins- bzw. Mineralinhalt. So zeigt der Name *Granitkonglomerat* oder *Kieselschieferbreccie*, daß das Konglomerat (vorwiegend) aus Granitgeröllen und die Breccie (vorwiegend) aus den Bruchstücken von Kieselschiefer besteht. Bestehen die Gerölle aus Einzelmineralen z.B. Feldspäten oder Quarz, so ist die Bezeichnung Feldspatkonglomerat oder Quarzkonglomerat angebracht. Eine quantitative Bestimmung der gesteinsbildenden Komponenten ist mit Schwierigkeiten verbunden, oft gehen die Gerölle über die Dimension eines Dünnschliffes (2 × 2 cm) hinaus. Psephite werden gewöhnlich durch eine Phyllosilikat + Quarz + Feldspat-Matrix (vorwiegend die Fein-Sand- und Ton-Fraktion in den Zwickeln zwischen den Geröllen und Bruchstücken) und/oder durch karbonatisches Bindemittel während des Prozesses der Diagenese verfestigt.

b) **Psammite** (Arenite: Sand) sind klastische Gesteine, deren Md zwischen 2 und 0,002 liegt und die das Gebiet des **Schluffes** oder **Siltes** von 0,063 und 0,002 mm mit einschließen. Unter **Sanden** versteht man die Lockerprodukte der Psammite; **Sandsteine** sind durch Diagenese und Zementation verfestigte Psammite. Eine Einteilung der Psammite geschieht quantitativ nach den Anteilen an Quarz, Feldspäten, Phyllosilikaten und Karbonatmineralen (vorwiegend Kalzit, selten Dolomit). Die Mineralzusammensetzung kann in einem Konzentrations-Tetraeder mit den Komponenten Q (Quarz) + F (Feldspat) + P (Phyllosilikate) + k (Karbonate) veranschaulicht werden. Der Anteil an Karbonatmineralen liegt – wenn nicht Sandsteine mit karbonatischem Bindemittel oder klastischen Karbonaten vorliegen – im allgemeinen unter 10%. Gewöhnliche Sande und Sandsteine können deshalb direkt auf die Q—F—P-Grundfläche des Konzentrations-Tetraeders (Fig. 18a) projiziert werden und eine Unterteilung der Psammite in Normal-Sandsteine (mit Untergruppen), Arkosen (mit Untergruppen) und Grauwacken (mit Untergruppen) ist möglich (Fig. 18b). Mittlere und spezielle Modalbestände von Psammiten sind in Tab. 18 aufgeführt.

Das Vorhandensein von (polygenetischen) Gesteinsbruchstücken in Sanden oder Sandsteinen im Korngrößenintervall der Psammite hängt von der primären Korngröße der (verwitternden) Magmatite und Metamorphite im Liefergebiet und von deren Resistenz gegen Transport und Absatzbedingungen ab. So liefern grobkörnige Granite (mit Korngrößen oberhalb des Limits der Sandsteine) vorwiegend Einzelminerale von Feldspäten und Quarz. Feinkörnige Vulkanite im Verwitterungsgebiet (deren Mineralkorngröße innerhalb oder unterhalb des Limits der Sandsteine liegt) liefern fast immer Gesteinsbruchstücke in der Korngrößenklasse der Sande und Sandsteine.

Table 18. Modal composition of selected sedimentary rocks.

	Psephitic (conglomeratic) graywacke, Harz Mountains [Huc66]	Average of 295 sandstone samples [Huc63a]	Average of normal sandstones [Huc63a]	Average of quartz sandstones [Huc63a]	Average of feldspar sandstones [Huc63a]	Average of phyllosandstones [Huc63a]	Average of arkoses [Huc63a]	Average of quartz arkoses [Huc63a]	Average of feldspar arkoses [Huc63a]	Average of graywackes [Huc63a]	Average of quartz graywackes [Huc63a]	Average of normal graywackes [Huc63a]	Average of quartz-poor graywackes [Huc63a]	Average of shales [Wed69]
	vol-%	vol-%	vol-%	vol-%	vol-%	vol-%	vol-%	vol-%	vol-%	vol-%	vol-%	vol-%	vol-%	vol-%
Quartz (+SiO$_2$ minerals)	37.4	59	82	96	76	79	52	62	39	46	61	37	21	20
Plagioclase	34.2 }	19	5	1	16	3	38	28	50	22	12	28	41	10–15
Alkalifeldspar	4.4 }													
Muscovite, illite	7.5 }	15	8	2	2	13	4	5	3	25	21	29	29	45–55
Chlorite	11 }													
Kaolinite, montmorillonite	trace	1	+	+	+	+	1	1	1	1	1	+	+	14
Calcite	1 }	3	3	1	1	4	2	1	4	3	4	3	+	3
Dolomite														
Accessories (Poligenetic rock fragments)	4.5	3 (0–63)	2 (0–48)	+ (0–9)	5 (0–14)	1 (0–14)	3 (0–63)	3 (0–31)	3 (0–63)	3 (0–15)	1 (0–53)	3 (0–50)	8 (0–55)	1

Accessories are: opaque minerals, apatite tourmaline, zircon, etc., and in the average of quartz-poor graywackes:

c) **Pelites** (argillites; clay, mud) are clastic sediments of particles with a median diameter of less than 0.002 mm. They usually deposit from slurries. Quartz, feldspars and micas (hydromica, illite) in the grain-size of clay (and silt) occur (mainly) as weathering residua. The most abundant components in clays, however, are the sheet-like and fine-grained phyllosilicates of the clay mineral group as formed by weathering (and the remnants of the decay of organic material). Because of the very fine-grained nature of the pelite-forming particles, X-ray diffraction methods and electron microscopy are the predominant tools in clay and clay mineral analyses.

c) **Pelite** (Argillite; Ton, Schlamm) sind klastische Gesteine, deren Md <0,002 mm beträgt. Sie werden als Tone bezeichnet und sind der Absatz aus feinster Trübe oder Suspensionen. Neben den Verwitterungsresten von Quarz, Feldspäten und Glimmern (Hydroglimmer, Illite) enthalten sie als Verwitterungsneubildungen vor allem die blättchenförmigen, sehr feinen Tonminerale aber auch die Reste organischer Substanz oder deren Zersatz. Wegen der Feinheit der Pelitbildenden Partikel kann die Bestimmung der Mineralkomponenten (meist) nur röntgenographisch und elektronenmikroskopisch durchgeführt werden.

continued p. 55

Table 18 (continued)

Ceramic clay, Miocene, Rohrhof, Bavaria [Cor68]		Clay, Lias, Göttingen [Cor68]	Carbonate-bearing shale, Devonian, near Goslar [Cor68]	Carbonate-bearing shale, Devonian, near Goslar [Cor68]	Solidified tuff, Carboniferous [Cor68]	Average of limestones [Wed69]	Triassic limestone, near Göttingen [Cor68]	Limestone, Devonian, near Goslar [Cor68]	Marl, Triassic, near Göttingen [Cor68]	Marl, Cretaceous, near Göttingen [Cor68]	Marl, Triassic, near Göttingen [Cor68]	Permian dolomite, Bad Lauterberg, Southern Harz Mountains [Cor68]	Carbonate-bearing chert, Carboniferous, Rheinisches Schiefergebirge [Cor68]	Carboniferous chert, Harz Mountains [Cor68]	Carboniferous chert, Harz Mountains [Cor68]	Chert, Cretaceous, Lower Saxony [Cor68]
wt-%	vol-%	wt-%	wt-%	wt-%	vol-%	wt-%	wt-%	wt-%	wt-%	vol-%	wt-%	wt-%	wt-%	wt-%	wt-%	
17.3	17	28.0	22.1	36.0		1.2	4.5	6.2	52	26	1.0	28.5	84.5	66.9	86	
} 0.3		5.3	0.8	48.5			0.4	+	5.0	+	2	} 0.2	1.7	1.1	5.0	+
		1.8														
	49.1	25.4	34.6		14	6.0	4.0	62.7	5	3	1.0	4.0	8.2	22.0	8	
18.3	7.6	27.2	21.3	6.5			1.5		+	1	0.2		3.5	2.7	1	
60.6	23.5								5	6	0.1				5	
} 0.7		12.3	21.2	3.0	84	92.0	89.6	26.6	38		5.1	64.5			+	
				5.0	2					62	89.8					
3.8	1.7			1.0				0.4	+		1.6	1.4	2.7	3.4	+	

amphiboles and pyroxenes.

Table 19. Estimated abundance of sandstone types consisting the sandstone family. A, from Krynine (1948); B, from Pettijohn (1960, 1963); C, from Middleton (1960); A, B, and C are listed in [Wed69]; D is from [Huc63a].

Sandstones	A	B	C	D
(quartz) sandstones	22.5	34	34	36
arkoses	32.5	15	16	15
graywackes	45	46	50	49

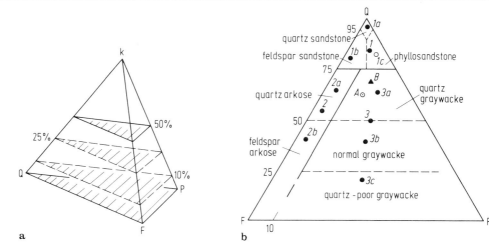

Fig. 18a, b. a) The Q (quartz) – F (feldspars) – P (phyllosilicates) – k (carbonate) tetrahedron for the classification of psammitic sedimentary rocks. Sandstones with 0 to 10 % k, sandstones *free* of or *poor* in carbonate minerals; with 10 to 25 % k, sandstones *containing* carbonate minerals; with 25 to 50 % k, sandstone *rich* in carbonate; with > 50 % k, *impure carbonate rock* and *carbonate rock*. The most abundant carbonate mineral is calcite, dolomite is mostly subordinate [Huc63a].

b) The essential sandstone types projected on the Q–F–P basis. *A* is the average of 295 sandstones [Huc63a]; *B* the average of sandstones after KRYNINE, from [Huc63a]. Full circles are projection points for averages of: *1*, normal sandstone; *1a*, quartz sandstone; *1b*, feldspar sandstone; *1c*, phyllosandstone; *2*, arkose; *2a*, quartz arkose; *2b*, feldspar arkose; *3*, graywacke; *3a*, quartz graywacke; *3b*, normal graywacke; *3c*, quartz-poor graywacke. Averages are from [Huc63a].

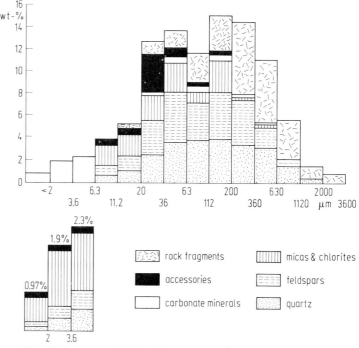

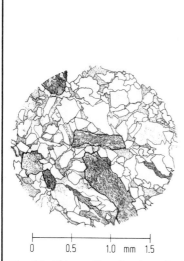

Fig. 19. Thin section drawing of a sandstone (from [Cor68]), Carboniferous, Eder River, Rheinisches Schiefergebirge. Rock fragments (heavy-contoured), quartz (light) and mica (dark). Interstices are filled with a micaceous matrix.

Fig. 20. Average grain size distribution and mineral composition of 8 Devonian and Carboniferous graywackes from the Harz Mountains [Huc63b]. The percentages scale of the interval < 6.3 μm is enlarged four times.

Diagenesis solidifies clay to **slate** and **shales** exhibiting excellent fissility. **Mudstones** (claystones) are more compact, their particles are cemented by calcite. Pelites are named according to the most abundant (clay) mineral. In *kaoline* kaolinite is the major clay mineral. *Bentonite* is a montmorillonite-bearing clay and usually the product of the weathering of (vitric volcanic) tuffs. Detailed and average mineral (modal) compositions are given in Table 18.

d) **Pyroclastic rocks** or **tuffs** are, strictly speaking, not sediments. Their abundant material is usually produced by volcanic eruptions which splash lava (magma) mingled with country rocks into the atmosphere or, if submarine, into the (sea) water. The mingled material is transported by wind and water and (usually) settles as a layered sediment. Its components can cover the grain size fractions of gravel, sand and/or clay. Because of the variability of the lava/country rock ratio, the nomenclature of the pyroclastic sediments is complex. The grain size classification of pyroclastic material is – as yet – not identical with that of the real sediments. Size fractions of <0.25 mm are called **fine ash, dust** or **tuff** (if solidified), between 0.25 and 4 mm **coarse ash** or **tuff** (if solidified), between 4 and 32 mm **lapilli** or **lapilli tuff** (if solidified), and >32 mm **bombs, blocks** or **agglomerates** and **volcanic breccias** (if solidified).

Ashes and tuffs may be further classified by their contents of glassy material, crystals, and rock debris. Those composed mainly of glassy particles are called **vitric ashes** or **vitric tuffs**. Hot glass lumps are able to weld together again forming the so called **welded tuffs** (ignimbrites). Glass is not very stable under the conditions of the earth's surface. In a hydrous environment, it devitrifies quickly. *Palagonite* is a (devitrified) basaltic, *bentonite* a more acid vitric tuff. Pyroclastic rocks made up chiefly of crystals are called **crystal ashes** or **crystal tuffs**. Those in which country rock fragments predominate are called **lithic ashes** or **lithic tuffs**. When volcanic material (ashes, lapilli) falls into basins where sedimentation is already going on, intimate mingling with the (real) sedimentary particles occurs. Materials formed in this way are **ashy sediments** or **tuffaceous sedimentary rocks.**

Durch Diagenese verfestigte Pelite sind **Schiefertone** und **Tonschiefer.** Enthalten sie kalzitisches Bindemittel, sind sie kompakt und fehlt ihnen die typische schiefrige Spaltbarkeit der Tonschiefer und Schiefertone, so spricht man von **Tonsteinen.** Die petrographische Bezeichnung eines Pelites geschieht in erster Linie nach dem quantitativ vorherrschendem (Ton-) Mineral. *Kaoline* sind Kaolinit-Tone, *Bentonite* dagegen Montmorillonit-Tone. Letztere sind meist das Ergebnis der Verwitterung von (vulkanischen) Tuffen. Detaillierte und mittlere modale Zusammensetzungen sind in Tab. 18 aufgeführt.

d) **Pyroklastische Gesteine** oder **Tuffe** (Pyroklastica, pyroklastische Gesteine) sind keine Sedimente im Sinne der Definition. Sie werden durch vulkanische Eruptionen produziert, die – subaerisch oder submarin – Lava vermischt mit Nebengestein erumpieren. Das vermischte Material wird durch Wind und Wasser transportiert und lagert sich als (geschichtetes) Sediment ab. Seine Komponenten können die Korngrößen-Fraktionen von Kies, Sand und Ton überdecken. Durch sehr unterschiedliche Mischungsverhältnisse von Lava und Nebengestein wird eine Einteilung der Tuffe erschwert. Außerdem ist ihre Korngrößen-Einteilung (bis jetzt) nicht identisch mit der der (wirklichen) Sedimente. So werden Korngrößenklassen <0,25 mm als **feine Asche, Staub** oder **Tuff** (verfestigt), zwischen 0,25 und 4 mm als **grobe Asche** oder **Tuff** (verfestigt), zwischen 4 und 32 mm als **Lapilli** oder **Lapilli-Tuff** (verfestigt) und >32 mm als **Bomben, Blöcke, vulkanische Breccien** oder **Agglomerate** (verfestigt) bezeichnet.

Aschen und Tuffe können weiter nach ihrem Gehalt an Glas, Kristallen und Gesteinsfragmenten eingeteilt werden. Solche, die vorwiegend Glas-Komponente enthalten, nennt man **vitrische Aschen** oder **Tuffe** (Glasaschen, Glastuffe). Heiße Glasfladen und -batzen können wieder zusammenkleben und sich verschweißen: **Schmelz-** oder **Schweißtuffe** (Ignimbrite). Unter den (klimatischen) Bedingungen der Erdoberfläche ist Glas nicht unbegrenzt beständig und devitrifiziert (in hydrischer Umgebung) ziemlich schnell. So ist *Palagonit* ein (entglaster) Basalt-, *Bentonit* ein (entglaster) Andesit- oder Rhyolith-Tuff. Pyroklastische Gesteine, die vorwiegend kristallines Material in Form von Kristallen enthalten, nennt man **Kristall-Aschen** oder **-Tuffe.** Pyroklastische Gesteine mit großen Anteilen an Nebengestein werden als **lithische Aschen** oder **Tuffe** bezeichnet. Fällt vulkanisches Material (Aschen, Lapilli) in (Wasser-) Becken, in denen die (normale) Sedimentation bereits im Gange ist, so vermischen sich Aschen und Sedimentmaterial, es entstehen **aschenhaltige** oder **tuffogene** (tuffige) **Sedimente.**

0.4.5 Chemical and biochemical sedimentary rocks — Chemische und biochemische Sedimente

(Classification and modal composition) (Klassifikation und modale Zusammensetzung)

a) **Limestone, dolomite,** and **marl.** The most abundant chemical sediments are carbonate rocks including those formed by biochemical processes. **Limestones** make up about 6 and **dolomites** about 2 per cent of the total of chemical sediments. Limestones predominantly contain calcite, dolomites predominantly dolomite as the major carbonate phase. Limestones, however, can also consist of detrital grains of calcite (or dolomite) covering the sand-or gravel-size fractions. Such clastic rocks are called **lime sandstone, lime conglomerates, lime breccias, lime siltstones, calcirudites, calcarenites,** etc. Layering, cross-lamination and graded bedding can be displayed in these rocks but commonly diagenetic recrystallization has overprinted the primary (clastic) texture and structure.

Sedimentary carbonate rocks very commonly contain clastic components with the grain size of sand or silt. These are predominantly silicates. In the case of quartz (or other polymorphs of SiO_2), the carbonate rock is called a **siliceous limestone** or **dolomite.** The clastic components, however, can also be residual feldspar and mica or newly formed clay minerals which were deposited together with the (precipitated) carbonate minerals. According to the mixing ratios between a pure carbonate rock and a clay (or silt), transitional clay-carbonate sediments, the so called **marls,** are produced. The main carbonate phase is calcite. Dolomite-bearing marls are not very abundant. A possible classification scheme is shown in Table 20.

a) **Kalksteine, Dolomite** und **Mergel.** Die häufigsten chemischen Sedimente sind die Karbonatgesteine einschließlich derjenigen, die durch biochemische Prozesse gebildet worden sind. So sind **Kalksteine** mit $\approx 6\%$ und **Dolomite** mit $\approx 2\%$ an der (mittleren) Zusammensetzung des Sedimentmittels beteiligt. Kalksteine enthalten fast ausschließlich Kalzit, Dolomite fast ausschließlich Dolomit als Karbonat-Phase. Kalksteine können jedoch auch aus detritischen Partikeln von Kalzit (und im Falle von Dolomit aus Dolomit) im Korngrößenbereich des Sandes oder Kieses bestehen. Solche klastischen Gesteine nennt man **Kalk-Sandsteine, Kalk-Konglomerate, Kalk-Breccien, Kalk-Silt-** oder **-Schluffsteine, Calcirudite, Calcarenite,** etc. Solche Gesteine können alle sedimentären Gefügeeigenschaften der klastischen Gesteine (Schichtung, Diagonalschichtung, gradierte Schichtung) aufweisen. Diagenetische Rekristallisations-Vorgänge haben diese (klastischen) Eigenschaften jedoch meistens überprägt.

Sedimentäre (chemische und biochemische) Karbonat-Gesteine enthalten gewöhnlich klastische Komponenten in der Korngröße des Sandes oder Schluffes. Meistens sind dies Silikate. Ist dies Quarz, (oder eine Modifikation von SiO_2) so spricht man von **kieseligen Kalksteinen** oder **Dolomiten** (auch Kieselkalk oder Kieseldolomit ist gebräuchlich). In den meisten Fällen besteht die klastische Komponente aus Verwitterungsresten von Feldspat und Glimmer oder aus (neugebildeten) Verwitterungsneubildungen der Tonmineral-Gruppe, die zusammen mit den ausgefällten Karbonat-Mineralen abgesetzt wurden. Entsprechend den Mischungsverhältnissen zwischen dem reinen Karbonat-Gestein und einem Ton (oder Silt), gibt es Übergangsgesteine, die als **Mergel** bezeichnet werden. Ein mögliches Einteilungsschema von Karbonat-Mergel-Ton ist in Tab. 20 veranschaulicht.

Table 20. Classification of sediments transitional between carbonate rocks and pelites. Carbonate is (mainly) calcite and (subordinate) dolomite; the pelite fraction consists (mainly) of silicate minerals that are clay minerals, mica, feldspar, and quartz. Sediment names are given for the solidified rocks.

carbonate content				
[wt-%] 90	65	35	10	
carbonate rock (Karbonat-Gestein)	argillaceous carbonate rock (toniges (mergeliges) Karbonat-Gestein)	marl (Mergel)	carbonate-bearing shale (Karbonatführender Tonschiefer oder Schieferton)	shale (Tonschiefer oder Schieferton)

Present-day precipitation and sedimentation of $CaCO_3$ can be found on the earth's surface in shallow and warm seawater oversaturated with $CaCO_3$. Wave action keeps the newly formed $CaCO_3$-nuclei afloat and concentric spheres – the **oolites (ooids)** – are produced. The oolites will sink after they have reached accretive sizes and accumulate to an **oolite sediment** which is converted by diagenesis into an **oolitic limestone**. The Bahama platforms provide unique examples of this kind of carbonate precipitation and sedimentation. Very similar conditions must have prevailed in Central Europe during Lower Triassic times.

$CaCO_3$ formation and accumulation is also triggered by biochemical processes. In shallow (sea) water $CaCO_3$-modifications (calcite, aragonite, Mg-calcite) are incorporated in shells and skeletons of organisms, their residua form (clastic) biochemical carbonate sediments. Fragments of **foraminifera shells** cover as **globigerina ooze** vast areas of the present-day ocean floors. Most $CaCO_3$ shells, however, become dissolved while descending to ocean depths because the abyssal seawater is undersaturated with $CaCO_3$. The non-carbonate leftovers are deposited as **pelagic clay** on the abyssal ocean floors far from the continents and contain authigenic zeolites (e.g. phillipsite) as minor constituents. This kind of ocean floor sediment has previously been called **red clay**. The abyssal surface of pelagic clay sediments is often (sparsely) dotted by accretive **manganese nodules** which have been formed by coating (detrital) sediment particles with MnO_2.

Primary dolomite precipitation from seawater requires a high salinity which can only be produced in (shallow) lagoons. The major sedimentary dolomite rocks are formed, however, by exchange reactions caused by Mg-bearing solutions which react with the $CaCO_3$ of the limestone. The longer the solutions can circulate in the pore system of the limestone, the more calcite is replaced by dolomite. Thus, a sequence of decreasing dolomite contents can be found in limestones of the Paleozoic to Tertiary and Quaternary ages.

Karbonat-Sedimentation findet (rezent) auf der Erde vorwiegend im flachen, warmen Wasser statt, das mit $CaCO_3$ übersättigt ist. Wellenbewegung hält $CaCO_3$-Keime (oft) längere Zeit in Schwebe. Es bilden sich sphärische Körper, die sogenannten **Ooide**, die nach Absatz und Verfestigung die **oolithischen Kalksteine** oder **Oolithe** bilden. Rezent werden oolithische Kalksteine an den Bänken der Bahamas gebildet. Ähnliche Bedingungen müssen aber auch zur Zeit des Mittleren Buntsandsteins geherrscht haben, als die sogenannten **Rogensteine** (oolithische Kalksteine) entstanden.

Außer durch reine anorganische Ausfällung kann $CaCO_3$ auch auf biochemischem Wege über bestimmte Organismen gebildet werden. Im Flachwasserbereich werden die $CaCO_3$-Modifikationen von Pflanzen und Tieren als Gerüste und Schalen ausgeschieden und bilden nach dem Absterben klastische, biochemische Karbonatsedimente. Die Reste von kalkschaligen **Foraminiferen** bedecken als **Globigerinen**-Schlamm weite Gebiete der heutigen Ozeanböden. Die meisten Kalkschalen und (-skelette) werden aber aufgelöst, wenn sie in größere Tiefe hinabsinken, da ozeanische Tiefenwässer an $CaCO_3$ untersättigt sind. Die nichtkarbonatischen Überbleibsel der Schalen und Skelette sedimentieren als **pelagischer Ton** im Tiefwasserbereich fern von den Kontinenten und wurden früher als **roter Ton** bezeichnet. Zeolithe (z. B. Phillipsit) sind (akzessorische) authigene Bestandteile. Die Oberfläche des Sediments ist sehr häufig mit konkretionären **Mangan-Knollen** (bei unterschiedlicher Besetzungsdichte) belegt. Die Knollen sind das Ergebnis von MnO_2-Ausscheidungen um (detritische) Sedimentpartikel.

Dolomitausscheidung direkt aus Seewasser erfordert hohe Elektrolyt-Gehalte. Diese können nur in seichtem Wasser unter teilweisem oder vollständigem Abschluß vom offenen Meer z.B. in Lagunen erzeugt werden. Die überwiegende Menge der Dolomitgesteine wird durch Austauschreaktionen von Mg-haltigen Lösungen mit dem $CaCO_3$ von Kalksteinen gebildet. Je länger solche Lösungen im Porensystem der Kalksteine zirkulieren können, desto mehr Kalzit wird durch Dolomit ersetzt. Daher zeigen Kalksteine vom Paläozoikum bis zum Tertiär und Quartär eine Folge abnehmender Dolomitgehalte.

Table 21. Estimated areas covered by deep sea sediments. A, from Vaughan (1924); B, from Sverdrup et al. (1942); C, from Ronov (1968), all listed in [Wed69].

	A	B	C
pelagic clay	50	37.8	39
foraminifera ooze (+ pteropod ooze)	38	48.7	44
diatomaceous and radiolaria ooze	10	13.5	17

A large proportion of the carbonate grains comprising a limestone or a dolomite rock are not the phases precipitated primarily. Most of the chemical and biochemical precipitates are converted during the process of diagenesis because of their high solubility. The original proportions of $CaCO_3$ (calcite, aragonite) from inorganic precipitation and (clastic) shell debris can usually not be determined with certainty. Properties of diagenetic conversion, such as resolution and recrystallization mainly expressed in "**granoblastic**" fabrics and stylolitic seams are not always visible in chemical and biochemical sedimentary rocks. Preserved organic remnants (such as coral reefs, algae, stromatolites, shells and their accumulations) and (critical) inorganic precipitation products (e.g. oolitic textures), however, are suitable indicators to unravel the (primary) sedimentary environment.

b) Rare chemical and biochemical sedimentary rocks and sediments. Iron-rich sedimentary rocks (chamositic-, sideritic-, cherty-, residual ironstones), bauxites (boehmite-, diaspor-, and hydrargillite-bearing (soil) sediments), bedded gypsum, bedded anhydrite, bedded halite, the evaporites of desert lakes and precipitates of heavy metal sulfides from stagnant water by bacterial sulfate reduction (sapropelites) are rather rare chemical and biochemical sediments ($\ll 1\%$ of the total sediments). They can, however, be of economic importance.

Phosphate rocks (phosphatic sediments, phosphorite, collophane) can be of biochemical and of inorganic chemical origin. (Fluid) animal excrements react with limestones to form (the present-day) **guano**. Calciumphosphate can precipitate non-biogenetically from solutions in (continental) non-marine deposits.

Cherts are silica-rich sedimentary rocks. Silica precipitation is caused by organisms. Abyssal seawater containing 2.5 to 5 ppm Si and (Si-undersaturated) surface seawater cannot precipitate any form of SiO_2 or colloidal silica. Silica originates mainly in the skeletons of *radiolaria*, *diatoma* and/or *sponges*. Accumulation of biogenetically derived silica residua produces **radiolarian-** and **diatomaceous oozes** or **earths**. Chert is the consolidated rock usually bearing no indication of the origin of silica. Reconstitution of the material due to high solubility of silica in pore solutions extinguishes the organic remnants (skeletons, needles, etc.) during the solidification process in most cases.

In gewöhnlichen Kalksteinen oder Dolomiten besteht der größte Anteil der Karbonat-Körner nicht mehr aus den primär ausgeschiedenen Phasen. Die meisten chemischen und biochemischen Ausfällungs-Produkte werden während der Diagenese infolge der (sehr) großen Löslichkeit ihrer Karbonat-Komponenten umkristalliert. Der ehemalige Anteil an (primärem und) anorganischem Kalzit (und Aragonit) und (klastischen) organischen Komponenten (Schalen, Skeletten, etc.) kann deshalb in der meisten Zahl der Fälle nicht mit Sicherheit bestimmt werden. Aber auch typische Eigenheiten der diagenetischen Umwandlung – wie zum Beispiel „**granoblastische**" Korngefüge oder die Tonhäutchen von Stylolithen – sind in (diagenetischen) chemischen und biochemischen Sedimentgesteinen nicht immer erkennbar. Erhaltene organische Strukturen (z.B. die Reste von Korallenriffen, Algen, Stromatolithe) und anorganische Ausfällungserscheinungen (z.B. die Ooide der Oolithe und Rogensteine) sind oft die einzigen Hinweise, die Herkunft und Entstehung des Sedimentes zu enträtseln.

b) Seltene chemische und biochemische sedimentäre Gesteine und Sedimente. Eisen-führende Sedimentgesteine (chamositische, sideritische, kieselige, residuale Eisensteine), Bauxite (Böhmit-, Diaspor-, Hydrargillit-führende (Boden-) Sedimente), die Gips-, Anhydrit- und Salzlager, die Salzausscheidungen der Wüsten (-Seen), sowie die (Schwermetall-) Sulfidausfällungen stagnierender Gewässer durch bakterielle Sulfatreduktion (Sapropelite) sind seltene chemische und biochemische Sedimente ($\ll 1\%$ der gesamten Sedimente), die aber ökonomische Wirtschaftlichkeit besitzen können.

Phosphat-Gesteine (Phosphat-Sedimente, Phosphorite, Kollophan) können sowohl auf biochemischen Wege als auch anorganisch gebildet werden. (Flüssige) Tierexkremente reagieren mit Kalksteinen und bilden den (heutigen) **Guano**. Calciumphosphat kann aber auch auf anorganischem Wege aus Lösungen in nicht-mariner Umgebung ausgefällt werden.

Kieselgesteine (Hornsteine, Flinte und Feuersteine, (schichtige) Lydite und Lydite im Schichtverband als Kieselschiefer) sind SiO_2-reiche Gesteine. Die Kieselsäureausscheidung wird durch Organismen bewirkt. Die Tiefenwässer der Ozeane mit 2,5 bis 5 ppm Si und gewöhnliche (Si-untersättigte) Oberflächenwässer sind nicht in der Lage kristallines oder kolloidales SiO_2 auszuscheiden. Planktonische Organismen wie *Radiolarien* und *Diatomeen*, aber auch die *Schwämme*, bauen SiO_2 in ihre Schalen und Gerüstsubstanzen ein. SiO_2-Anreicherung erfolgt durch Absterben der Organismen und Absinken der Schalen oder Gerüstsubstanz. Die resultierenden Sedimente sind die **Radiolarien-** und **Diatomeen**-Schlicke der Ozeane oder die **Kieselgur** der Süßwasser-Seen. Verfestigte Radiolaren-Schlicke nennt man Radiolarite; in den meisten Fällen werden aber in solchen Gesteinen durch die hohe Lös-

The decay and degradation of plant and animal tissue exclusively produces **biochemical** or **biogenetic sediments**. In the first place, protein, sugars and starches oxidize and break down, followed by cellulose and fats. Chitin, resins, and waxes can persist, however, even over geologic periods. **Peat** is predominantly made up of cellulose. **Coalification** follows burial by (partial) decomposition of plant remnants and prevents the access of air under an (slight) increase of temperature and pressure. Carbonaceous constituents are widely abundant in sedimentary rocks. **Coal** intercalations within a pile of sediments, however, do rarely exceed one per cent of the total rock unit. Hydrocarbons are formed by the decomposition of phyto- and zooplankton. The products are added to the (worldwide) carbon cycle, remain incorporated in the (organic and inorganic) sediment and are more abundant than carbonaceous matter. The decomposition products of phyto- and zooplankton are richer in fats and proteins and more mobile than carbonaceous matter. Thus, hydrocarbons as **bitumen** (hydrocarbons soluble in carbon bisulfide) can move from the locus of primary deposition and accumulate in reservoirs with reasonable porosities and permeabilities (oil sandstones, **bituminous sandstone**).

0.4.6 The chemical composition of sedimentary rocks — Die chemische Zusammensetzung der sedimentären Gesteine

The chemical composition of the **average sedimentary rock** is given in Table 22 (first column). The composition was extrapolated from geochemical calculations by [Cor48] and contains about 8% limestones, about 15% sandstones (sandstones+graywackes+arkoses) and about 77% shales and clays. Thus, this average sedimentary rock is a shale containing essential amounts of quartz, phyllosilicates (mica, clay minerals) and feldspars (Table 15).

Psephites as conglomerates and breccias do not reveal their real chemical composition even in a large handspecimen taken for chemical analysis. Coarse grain size and poor sorting prevent an average chemical analysis, even of a large hunk of (that conglomeratic) rock in most cases. One available sample, however, is presented in Table 22 (column 7). The chemistry of shales, clays, sandstones, graywackes, limestones, and cherts is given as average. Selected samples of these rocks and supplementary chemical compositions of arkose, clays, marls, dolomite, and tuff are also listed in Table 22.

Table 22. Chemical composition of average and selected sedimentary rocks. For comparison with igneous and metamorphic rocks the CIPW-norm as normative composition is given.

	Average of all sedimentary rocks [Cor49]	Average of 253 sandstones in [Wed69]	Average of 3700 sandstones from platforms in [Wed69]	Average of 61 graywackes in [Wed69]	Average of 24 Devonian and Carboniferous graywackes from the Harz Mountains [Huc66]	Arkose, Massif Central, Auvergne [Huc63b]	Psephitic (conglomeratic) graywacke from the Harz Mountains [Huc66]	Average of 277 shales mainly from synclines in [Wed69]	Average of 6800 shales from platforms in [Wed69]	Ceramic clay, Miocene, Rohrhof, Bavaria [Cor68]	Clay, Lias, Göttingen [Cor68]	Carbonate-bearing shale, Devonian, near Goslar [Cor68]	Carbonate-bearing shale, Devonian, near Goslar [Cor68]
chemical composition [wt-%]													
SiO_2	55.64	78.7	70.0	66.7	69.7	76.6	69.6	58.9	50.7	51.89	53.63	51.79	43.15
TiO_2	0.69	0.25	0.58	0.6	0.5	0.6	0.4	0.78	0.78	1.31	1.92	0.70	0.58
Al_2O_3	14.44	4.8	8.2	13.5	13.9	12.4	14.6	16.7	15.1	30.76	20.11	15.55	14.58
Fe_2O_3	} 6.87	1.1	2.5	1.6	1.5	0.7	2.4	2.8	4.4	2.13	3.16	0.80	2.11
FeO		0.3	1.5	3.5	2.9	0.2	2.0	3.7	2.1		1.46	4.70	7.49
MnO		0.03	0.06	0.1	0.2	trace	0.2	0.09	0.08			0.13	0.71
MgO	2.93	1.2	1.9	2.1	1.8	0.3	0.9	2.6	3.3	0.47	4.82	4.46	4.10
CaO	4.69	5.5	4.3	2.5	1.1	0.4	1.8	2.2	7.2		2.72	6.53	9.29
Na_2O	1.21	0.45	0.58	2.9	2.9	0.3	3.8	1.6	0.8	0.12	0.25	0.83	0.14
K_2O	2.87	1.3	2.1	2.0	1.5	3.8	1.0	3.6	3.5	1.97	3.57	3.34	3.81
H_2O^+	5.54	1.3	3.0	2.4	2.5	2.7	2.4	5.0	5.0	7.10	7.10	3.80	4.09
P_2O_5	0.17	0.08	0.10	0.2	0.1	0.2	0.1	0.16	0.10		0.24	0.17	0.16
CO_2	3.80	5.0	3.9	1.2	1.0	trace	0.4	1.3	6.1	11.4[a])	0.21	5.40	9.3
C	0.65		0.26	0.1	0.2	trace	trace	0.6	0.67			1.5	0.24
S	0.32			0.1	0.2	(1.7 BaO)	0.1	0.24			(0.46)		
										0.6			
SO_3		0.07	0.07	0.3									
normative composition [wt-%]													
qz (quartz)	33.0	70.6	60.1	37.1	43.4	58.3	39.4	30.0	31.3	42.9	26.8	25.4	19.1
cor (corundum)	9.2	2.7	5.9	5.7	7.5	6.4	5.2	9.6	10.8	28.4	12.0	10.8	10.3
or (K-feldspar)	16.6	7.7	12.5	11.9	8.9	22.5	5.9	21.3	20.7	11.6	21.2	20.1	22.6
ab (albite)	10.0	3.3	0.3	22.6	24.6	2.5	32.3	13.6	2.8	1.1	2.1	7.2	1.2
an (anorthite)				3.5		3.8	5.8	1.7			10.6		
hy (hypersthene)	6.5	1.2	3.3	9.4	7.2	0.7	3.4	9.4	7.0	1.2	12.0	17.4	17.7
mt (magnetite)	6.2	0.3	3.4	2.3	2.2		3.5	4.1	4.8			1.2	3.1
hm (hematite)	2.5	0.9	0.2			0.7			1.1	2.2	3.2		
ilm (ilmenite)	1.3	0.5	1.1	1.1	1.1	0.4	0.8	1.5	1.5		2.0	1.4	1.1
ap (apatite)	0.4	0.2	0.2	0.5	0.2	0.5	0.2	0.4	0.2		0.6	0.4	0.4
cc (calcite)	7.8	9.6	7.5	2.7	1.7		0.9	3.0	12.6		0.5	11.5	16.3
mc (magnesite)		1.5	1.2		0.5				1.1			0.9	4.2
py (pyrite)	0.6			0.2	0.4		0.2	0.5			0.9		
ru (rutile)										1.3			

[a]) Loss on ignition.
[b]) Plus H_2O^- and loss on ignition.
[c]) Not calculable.

Table 22 (continued)

Tillite (solidified glacial till) in [Wed69]	Average of 92 limestones in [Wed69]	Average of 1500 to 8300 limestones from platforms in [Wed69]	Triassic limestone, near Göttingen [Cor68]	Limestone, Devonian, near Goslar [Cor68]	Marl, Triassic, near Göttingen [Cor68]	Marl, Cretaceous, near Göttingen [Cor68]	Marl, Triassic, near Göttingen [Cor68]	Permian dolomite, Bad Lauterberg, Southern Harz Mountains [Cor68]	Average of 10 cherts in [Wed69]	Carbonate-bearing chert, Carboniferous, Rheinisches Schiefergebirge [Cor68]	Carboniferous chert, Harz Mountain [Cor68]	Carboniferous chert, Harz Mountain [Cor68]	Average of 430 pelagic clays in [Wed69]	Solidified tuff, Carboniferous, Harz Mountains [Cor68]	Chert, Cretaceous, Lower Saxony [Cor68]
58.9	6.9	8.2	4.41	6.90	37.64	47.27	22.7		89.9	30.9	90.6	80.8	54.9	71.8	27.7
0.79	0.05		0.02	0.07	0.09	0.08	0.1		0.2	0.08	0.1	0.10	0.78	0.1	0.1
15.9	1.7	2.2	1.75	1.47	15.10	5.44	2.7		3.7	1.80	4.0	9.7	16.6	11.4	2.7
3.3	0.98	1.0	0.38	0.36	4.36	1.60	0.6		2.3	0.60	1.8	2.9	} 7.7	0.5	0.6
3.7	1.3	0.68	0.28	0.80	0.95	0.17	0.2							1.3	0.2
0.10	0.08	0.07	0.02	0.36	0.04	0.03	0.2		0.1	0.33	0.6	0.16	2.0	0.48	0.2
3.3	0.97	7.7	0.97	1.84	3.13	1.37	18.1	19.50	0.5	trace	1.0	0.8	3.4	2.2	18.1
3.2	47.6	40.5	50.26	47.41	14.34	21.20	18.6	30.18	0.3	35.7	0.4	0.1	0.72	1.5	18.6
2.1	0.08		0.09	0.07	0.4	0.18	0.2		0.7	0.2	0.1	0.5	1.3	5.5	0.2
3.9	0.57		0.45	0.40	3.16	1.25	0.3		0.7	0.4	0.6	1.8	2.7	0.6	0.3
3.0	0.84		0.91	0.41	9.15[b]	3.40[b]			1.2	0.5	1.3	2.8	(9.2)	2.0	
0.21	0.16	0.07	0.15	0.03	0.04	0.06	0.05		0.9	0.14	0.06	0.05	0.72	0.07	0.05
0.6	38.3	35.5	40.3	39.30	11.75	16.94	29.3	45.83		28.0				2.3	29.3
		0.23	(0.03 Cl)	0.32	(0.13 Cl)	0.24	2.4	(4.1)[c]	0.3	0.6	0.06	0.05		trace	2.4
0.08	0.11														
0.09	0.02	3.1													
21.5	[c]	[c]	2.3	[c]	20.6	40.3	18.9	[c]	[c]	[c]	85.1	69.9	[c]	34.7	62.4
4.5			1.2		11.6	3.8	2.1				2.6	6.7		1.7	11.8
23.2			2.7		19.3	7.5	1.8				3.5	10.7		3.6	
17.3			0.5		2.5	1.5	1.7				0.8	4.2		46.7	
10.8											1.6	0.2			
11.1			0.3		6.8	2.6	11.7				2.5	2.0		5.5	0.7
4.8			0.6		3.0	0.4	0.9				1.7	0.2		0.7	
					2.4	1.3					0.6	2.7			16.8
1.5			<0.1		0.2	0.2	0.2				0.2	0.2		0.2	
0.5			0.4		0.1	0.1	0.1				0.1	0.1		0.2	
1.4			89.3		26.4	38.1	33.7							2.5	
			1.9		1.1	0.7	28.8							2.3	
0.2															

Chemical compositions of evaporites can be derived from the (predominantly monomineralic) composition of their salt minerals (Tables 23, 24 and 25). Marine evaporites are mainly alkali chlorides (and sulfates). Evaporites of the continents usually contain alkali (sodium) carbonates, sulfates, and borates. Phosphate rocks contain P_2O_5 in the form of apatite.

Die chemische Zusammensetzung von Ausfällungssedimenten kann aus ihren (vorwiegend monomineralisch) zusammengesetzten Salz-Mineralen abgeleitet werden (Tab. 23, 24 und 25). Marine Ausfällungssedimente sind vorwiegend Chloride (und Sulfate). Die der Kontinente enthalten meistens Alkali (Natrium)-Karbonate, -Sulfate und -Borate. Phosphat-Gesteine haben P_2O_5 in Form von Apatit gebunden.

Table 23. The most common minerals of the marine salt deposits.

chlorides:	
halite (rocksalt, Steinsalz)	$NaCl$
sylvite (Sylvin)	KCl
bischofite	$MgCl_2 \cdot 6H_2O$
carnallite	$KMgCl_3 \cdot 6H_2O$
sulfates:	
anhydrite	$CaSO_4$
gypsum	$CaSO_4 \cdot 2H_2O$
kieserite	$MgSO_4 \cdot H_2O$
epsomite (reichardtite)	$MgSO_4 \cdot 7H_2O$
polyhalite	$Ca_2K_2Mg(SO_4)_4 \cdot 2H_2O$
bloedite (astrakhanite)	$Na_2Mg(SO_4)_2 \cdot 4H_2O$
loeweite	$3[Na_{12}Mg_7(SO_4)_{13}] \cdot 15H_2O$
vanthoffite	$Na_6Mg(SO_4)_4$
langbeinite	$K_2Mg_2(SO_4)_3$
carbonates:	
calcite	$CaCO_3$
dolomite	$CaMg(CO_3)_2$
others:	
kainite	$K_4Mg_4Cl_4(SO_4)_4 \cdot 11H_2O$
boracite (stassfurtite)	$Mg_3B_7O_{13}Cl$

Table 24. The major components of the seawater in gram per kilogram at a salt content of 35‰ (from [Cor68]).

Cations	g/kg	Anions	g/kg
sodium (Na)	10.75	chlorine (Cl)	19.345
potassium (K)	0.39	bromine (Br)	0.065
magnesium (Mg)	1.295	sulfate (SO_4)	2.701
calcium (Ca)	0.416	hydrocarbonate (HCO_3)	0.145
strontium (Sr)	0.008	borate (BO_3)	0.027

Table 25. Common salt minerals of arid areas or desert lakes from the continents.

trona	$Na_3H(CO_3)_2 \cdot 2H_2O$
soda (Natrit)	$Na_2CO_3 \cdot 10H_2O$
thermonatrite	$Na_2CO_3 \cdot H_2O$
gaylussite	$Na_2Ca(CO_3)_2 \cdot 5H_2O$
borax	$Na_2B_4O_7 \cdot 10H_2O$
kernite	$Na_2B_4O_7 \cdot 4H_2O$
ulexite	$NaCaB_5O_9 \cdot 8H_2O$
thenardite	Na_2SO_4
glaserite	$NaKSO_4$
sodaniter (Nitronatrit, Natronsalpeter)	$NaNO_3$
niter (Kalisalpeter)	KNO_3
lautarite	$Ca(IO_3)_2$
gypsum	$CaSO_4 \cdot 2H_2O$
halite (Steinsalz)	$NaCl$
carbonates	$CaCO_3$, $CaMg(CO_3)_2$

0.4.7 The origin and genesis of sedimentary rocks — Die Herkunft und Entstehung der sedimentären Gesteine

Minerals of the igneous and metamorphic rock cycles are not stable under the climatic conditions of the earth's surface. Alternating temperatures, the action of wind, water and ice crush the material of the rocks – mainly assemblages of silicate minerals – physically. Atmospheric oxygen, carbon dioxide, water, etc. attack chemically and dissolve the (physically disintegrated) debris partially or completely. The results of these activities are weathered residua and weathered products which are mainly of silicate composition. *In situ* accumulation of the remnants produces **soils**; transportation, precipitation and deposition form **sediments**. Sedimentation in a specific medium (e.g. water) separates and sorts the particles according to their size and density (Stokes' law).

Accumulation rates for sediments usually depend on the geotectonic environment. Weathering on (stable) platforms delivers more mobile material to the (continental) run off than do orogenic zones. Clay accumulation in the Central South Atlantic is in the order of a millimeter per thousand years; shelf sediments are deposited more quickly with rates of a centimeter per thousand to hundred years.

The primary and loose sediment is buried in sinking areas (basins, troughs, synclines). The load of the overlying (sedimentary) rock pile compacts the texture of the sediment. The grains are pressed (closer) together. Pore solutions (mainly primary or converted seawater) are trapped and react with the minerals attached to the walls of the pores. The temperature will increase slightly but remains below 200 °C. The pores become (partly) filled by newly formed minerals crystallizing from the trapped solutions. (Some of) the pore capillaries are clogged. Limestones of organic as well as inorganic origin can reprecipitate their carbonate material entirely.

The (very) loosely packed sedimentary material, e.g. a sand or a clay, solidifies under the reduction of its pore space (and its permeability). The result is a solid sedimentary rock, e.g. a sandstone and a shale, respectively. The process of changing a loosely packed sediment into a solid (and massive) sedimentary rock is called **diagenesis**. The distinction between a diagenetic sedimentary rock and a very-low grade metamorphite is conventionally defined by mineral assemblages which are of critical diagenetic or critical metamorphic origin.

Minerale des magmatischen und metamorphen Gesteins-Kreislaufs sind (überwiegend) unter den klimatischen Bedingungen der Erdoberfläche nicht stabil. Temperaturwechsel, die Tätigkeit von Wind, Wasser und Eis zerkleinern das Gesteinsmaterial physikalisch. Atmosphärischer Sauerstoff, Kohlensäure, Wasser und andere Atmospherilien greifen chemisch an und lösen den (physikalisch zerkleinerten) Gesteinsschutt partial oder vollkommen auf. Das Ergebnis dieser Tätigkeiten sind Verwitterungsreste und Verwitterungsneubildungen von vorwiegend silikatischer Zusammensetzung. *In situ*-Anreicherung dieser Materialen erzeugt **Böden**, ihr Transport, Ausfällung und Ablagerung erzeugt (echte) **Sedimente**. Ablagerung oder Sedimentation in einem (spezifischen) Medium (z. B. Wasser) separiert und sortiert die Partikel entsprechend ihrer (Korn-) Größe und Dichte (Stokessches Gesetz).

Ablagerungsraten für Sedimente hängen meistens von den geotektonischen Bedingungen des sedimentliefernden Gebietes ab. Verwitterung auf (stabilen) Schilden erzeugt mehr mobilen Detritus als in orogenen Zonen. Tonablagerung im zentralen Südatlantik geschieht in Größenordnungen von Millimeter pro tausend Jahre, während Schelf-Sedimente Raten von Zentimeter pro tausend bis hundert von Jahren aufweisen.

Das primäre und lockere Sediment wird in sinkenden Zonen (Becken, Tröge, Synklinen) begraben. Das Gewicht des auflagernden (sedimentären) Gesteins-Stapel verdichtet das lockere Sedimentgefüge. Die Körner werden enger zusammengepreßt. Porenlösungen (vorwiegend primäres oder verändertes Seewasser) werden eingeschlossen und reagieren mit den Mineralen an den Porenwänden. Die Temperatur steigt gering an, bleibt aber unter 200 °C. Die Poren werden mit neugebildeten Mineralen (teilweise) gefüllt, die in den eingeschlossenen Lösungen kristallisieren. Einige der Poren-Kapillaren werden verschlossen. Kalksteine organischer und anorganischer Entstehung können ihren (gesamten) Karbonatgehalt umlösen und rekristallisieren.

Das (sehr) locker gepackte Sediment-Material – ein Sand oder Ton – wird unter Verringerung seines Porenraumes (und seiner Permeabilität) verdichtet. Das Resultat ist ein festes und dichtes Sedimentgestein – ein Sandstein oder ein Tonschiefer. Der Vorgang, der ein lockeres Sediment in ein festes und dichtes Sedimentgestein umwandelt, wird **Diagenese** genannt. Die Unterscheidung zwischen einem diagenetisch-veränderten Sedimentgestein und einem sehr schwachgradigen Metamorphit geschieht konventionell auf der Grundlage von Mineralparagenesen, die als kritisch diagenetisch und kritisch metamorph angesehen werden.

Organic and inorganic sedimentary rock formation at the earth's surface is, geochemically speaking, a fractionation process under the conditions of the atmosphere. Less soluble elements are enriched in the weathering residua, more soluble elements are incorporated in the weathering products. The (highly) soluble elements are transported as the (continental) run off into the oceans and precipitate in the sequence carbonate→sulfate→chloride as marine salt deposits. Calcium carbonate precipitation depends on the solubility of carbon dioxide in (sea) water. Isobaric temperature increase decreases, isothermal pressure increase raises the solubility of carbon dioxide in (sea) water. Increasing salt concentration at a given pressure and temperature increases the solubility of calcium carbonate. Thus, tropical (warm) ocean waters are oversaturated with calcium carbonate at the surface and precipitate $CaCO_3$. Abyssal waters of the oceans are under higher pressure ($\gg 1$ atm) and are cold. They are undersaturated with calcium carbonate and, therefore, organic debris of calcium carbonate (shell, skeletons, etc.) descending to greater depths is dissolved.

Organische und anorganische Sedimentbildung an der Erdoberfläche ist geochemisch gesehen ein Fraktionierungsprozess unter den Bedingungen der Atmosphäre. Schwer lösliche Elemente werden in den Verwitterungsresten angereichert, löslichere Elemente gehen in die Verwitterungsneubildungen. Lösungen transportieren sehr leicht lösliche Elemente in die Meere. Hier erfolgt Ausscheidung in der Folge Karbonat→Sulfat→Chlorid. Kalziumkarbonat-Ausfällung hängt in erster Linie von der Löslichkeit der Kohlensäure im Meerwasser ab. Isobarer Temperaturanstieg erniedrigt, isothermer Druckanstieg erhöht die Löslichkeit von Kohlensäure in Wasser. Ansteigende Salzgehalte bei gegebenem Druck und gegebener Temperatur erhöhen die Löslichkeit von $CaCO_3$. Daher sind tropische Oberflächenwasser an Calciumkarbonat übersättig und scheiden $CaCO_3$ aus. Tiefenwässer der Ozeane sind unter höherem Druck ($\gg 1$ atm) und kälter; sie sind untersättigt an Calciumkarbonat, und organischer Detritus (Schalen, Skelette, u. anderes) wird beim Absinken in größere Tiefe meistens aufgelöst.

Table 26. Chemical composition of selected salt deposits from the continents (from [Cor68]).

	Old Walker Lake, North Branch	Alkali crust from Westminster, Orange, California	Perth Lake, Nevada	Bottom of Lake Altai, Sibiria	Desert lake, east of the Sandia Mts., New Mexico	West of Black Rock, Nevada
Na	39.15	31.16	15.27	31.03	34.71	36.26
K	0.62	12.98				0.73
Mg			9.75	0.07	1.50	
Ca			1.31	0.16		
CO_3	41.15	38.08		0.16		5.13
SO_4	11.83	11.98	72.96	64.84	4.66	18.29
Cl	2.10	6.41	0.45	0.17	54.47	36.65
B_4O_7	3.2					0.77
SiO_2	1.96					2.18
Fe_2O_3				0.11		
H_2O					4.66	
residua				3.46		

0.5 References for 0.1···0.4 — Literatur zu 0.1···0.4

Bar52	Barth, T.F.W.: Theoretical petrology. A textbook on the origin and evolution of rocks. New York: John Wiley & Sons, Inc.; London: Chapman & Hall, Limited **1952**.
Cor48	Correns, C.W.: Naturwissenschaften **35** (1948).
Cor49	Correns, C.W.: Einführung in die Mineralogie, Berlin-Heidelberg-Göttingen: Springer **1949**.
Cor68	Correns, C.W.: Einführung in die Mineralogie (Kristallographie und Petrologie). 2nd edition. Berlin-Heidelberg-New York: Springer **1968**.
Cro02	Cross, Ch.W., Iddings, J.P., Pirson, L.V., Washington, H.S.: J. Geol. **10** (1902) 555–690.
Daw62	Dawson, J.B.: Bull. Geol. Soc. Am. **73** (1962) 545–560.
Huc63a	Huckenholz, H.G.: Geol. Fören. i Stockholm Förh. **85** (1963) 156.
Huc63b	Huckenholz, H.G.: J. Sedi. Petrol. **33** (1963) 914–918.
Huc66	Huckenholz, H.G.: Contr. Mineral. Petrol. **14** (1966) 65–71.
Huc71	Huckenholz, H.G., Yoder, H.S., Jr.: N. Jb. Miner., Abh. **114** (1971) 246–280.
Meh69	Mehnert, K.R., in: Handbook of Geochemistry, Vol. I, K.H. Wedepohl editor, Berlin-Heidelberg-New York: Springer **1969**.
Mue77	Mueller, R.F., Saxena, S.K.: Chemical Petrology, New York-Heidelberg-Berlin: Springer **1977**.
Nes76	Nesbitt, R.W., Shen-Su Sun: Earth Planetary Lett. **31** (1976) 433–453.
Noc54	Nockolds, S.R.: Bull. Geol. Soc. Am. **65** (1954) 1007–1032.
Str67	Streckeisen, A.: N. Jahrb. Mineral. Abhandl. **107** (1967) 144–240.
Trö35/38	Tröger, E.: Spezielle Petrographie der Eruptivgesteine, Berlin **1935**. Nachtrag **1938**.
Wed69	Wedepohl, K.H.: Handbook of geochemistry. Vol. I. Berlin-Heidelberg-New York: Springer **1969**.
Wil55	Williams, H., Turner, F.J., Gilbert, Ch.M.: Petrography. An introduction to the study of rocks in thin sections, San Francisco: W.E. Freeman & Co. **1955**.
Win76	Winkler, H.G.F.: Petrogenesis of metamorphic rocks, New York-Heidelberg-Berlin: Springer 4th edition **1976**.
Yod76	Yoder, H.S., Jr.: Generation of basaltic magma. National Academy of Sciences, Washington, D.C. **1976**.

1 Density — Dichte

The density is mostly indicated by ϱ and is defined as the ratio of the mass m and the unit-volume V of any matter

Die Dichte wird meist mit ϱ bezeichnet und ist als Quotient, gebildet aus der Masse m mit der Volumeneinheit V eines Stoffes definiert:

$$\varrho = \frac{m}{V} \tag{1}$$

The unit of the density using the SI-system of units is

Die Einheit der Dichte im SI-Maßsystem ist

$$\text{g cm}^{-3} = 10^3 \cdot \text{kg m}^{-3}$$

The density ϱ is related to the specific gravity σ by the relation

Über die Beziehung

$$\sigma = g \cdot \varrho \tag{2}$$

with g = gravitational acceleration. To avoid confusion regarding the notions density and specific gravity, the tendency is to replace completely the specific gravity – which correctly is a function of the geographical coordinates – by the notion density according to relation (2) [Wes70].

ist die Dichte ϱ verknüpft mit dem spezifischen Gewicht σ. Hierbei ist g die Fallbeschleunigung. Die Tendenz geht heute dahin, wegen der häufigen Verwechslungen der Begriffe Dichte und spezifisches Gewicht auf den Begriff des - genaugenommen ortsabhängigen - spezifischen Gewichtes zu verzichten und ihn stets durch den Begriff Dichte gemäß (2) zu ersetzen [Wes70].

The following tabulated densities of minerals, rocks and some natural organic materials at normal pressure and room temperature are taken from publications specified by reference keys. The literature is given in 1.1.2 and 1.2.7.

Bei den folgenden tabellarischen Darstellungen der Dichtewerte von Mineralen, Gesteinen und einigen natürlichen organischen Substanzen bei normalem Druck und Zimmertemperatur wird jeweils auf die Quelle hingewiesen. Die Angabe der Quellen erfolgt in den Abschnitten 1.1.2 und 1.2.7.

Density values and ranges are often cited from several authors to underline the variety of this parameter. Special information on methods of density determinations is given by [Koh65].

Durch die Angabe mehrerer Dichtewerte oder Dichtebereiche für das gleiche Mineral oder Gestein von verschiedenen Autoren soll deutlich auf die Schwankungsbreite der einzelnen Angaben hingewiesen werden. Angaben über die Methoden zur Bestimmung von Dichtewerten finden sich z.B. bei [Koh65].

1.1 Density of minerals — Dichte der Minerale

1.1.0 Introduction — Einleitung

The mineral densities or density ranges listed in the tables are determined by experiment. Densities based on X-ray methods can be accurately calculated according to the relation

In den Tabellen zur Dichte der Minerale werden die Zahlenwerte der im Experiment bestimmten Dichtewerte oder Dichtebereiche angegeben. Die in eckige Klammern gesetzten Dichtewerte wurden mit Hilfe röntgenographischer Untersuchungen nach der Beziehung

$$\varrho_x = \frac{M\,Z}{N\,V} \tag{3}$$

with

berechnet. Hierbei bedeuten:

M = formula weight [g mol^{-1}]
Z = number of formula units per unit cell
V = volume of the unit cell [Å^3 = 10^{-24} cm^3]
N = Avogadro's number, $6.02252 \cdot 10^{23}$ mol^{-1}

M = Formelgewicht [g mol^{-1}]
Z = Zahl der Formeleinheiten pro Elementarzelle
V = Volumen der Elementarzelle [Å^3 = 10^{-24} cm^3]
N = Avogadrosche Zahl, $6.02252 \cdot 10^{23}$ mol^{-1}

These densities ϱ_x are given in brackets. They were taken from the "Handbook of Chemistry and Physics" [Wea76]. Further detailed information on ϱ_x-densities is given in "Powder Diffraction File" [McC79].

Diese Dichteangaben wurden dem „Handbook of Chemistry and Physics" [Wea76] entnommen. Weitere detaillierte Angaben zur ϱ_x-Dichte der Mineralien finden sich im „Powder Diffraction File" [McC79].

The chemical formulas are added to the table to avoid difficulties in the identification of minerals because of non-uniform nomenclature. The chemical formulas are taken from [Str77]. Exceptions are marked by the reference keys behind the formulas.

Die Aufnahme der chemischen Formeln in das Tabellenwerk soll Schwierigkeiten bei der Identifikation von Mineralen durch uneinheitliche Namensgebung vermeiden helfen. Die chemischen Formeln sind [Str77] entnommen. Ausnahmen sind durch Angabe des Literaturschlüssels hinter der Formel gekennzeichnet.

1.1.1 Density, ϱ, and X-ray density, ϱ_X, of minerals (ϱ_X in brackets)

Mineral	Formula	ϱ, [ϱ_X] g cm^{-3}	Ref.
Acanthite, see also Argentite	Ag$_2$S	7.2···7.3 [7.248(9)]	Dan44 Wea76
Acmite (Aegirin)	NaFe^{3+}Si$_2$O$_6$	3.3···3.558 [3.576]	Spe27
Acmite-augite	(Na, Ca)(Fe^{3+}, Fe^{2+}, Mg, Al)[Si$_2$O$_6$] [Dee62]	3.40···3.55 3.26···3.42	Dee62 Spe27
Actinolite (Strahlstein)	Ca$_2$(Mg, Fe^{2+})$_5$[(OH, F)Si$_4$O$_{11}$]$_2$	2.9···3.2 2.9···3.1	Hod52/53 Klo48
Adamite	Zn$_2$[OHAsO$_4$]	4.3···4.5	Klo48
Aegirin, see Acmite			
Aenigmatite	Na$_4$Fe$^{2+}_{10}$Ti$_2$[O$_4$(Si$_2$O$_6$)$_6$]	3.732···3.758	Spe27
Aeschynite	(Ce, Th, Ca ...)[(Ti, Nb, Ta)$_2$O$_6$]	≈ 5	Klo48
Afwillite	Ca$_3$[SiO$_3$OH]$_2$ · 2H$_2$O	2.619···2.630	Spe27
Aguilarite	Ag$_4$SeS	7.586	Dan44
Aikinite (Nadelerz)	2PbS · Cu$_2$S · Bi$_2$S$_3$	6.8···7.2	Klo48
Åkermanite	Ca$_2$Mg[Si$_2$O$_7$]	2.944···2.980 2.9375(29)	Spe27 Wea76
Akrochordite	(Mn, Mg)$_5$[(OH)$_2$AsO$_4$]$_2$ · 4H$_2$O	3.194 2.9···3.2 2.9···3.1	Spe27 Hod52/53 Klo48
Alabandite	α-MnS	[4.0546(12)]	Wea76
Alaskaite	Ag$_2$S · 3Bi$_2$S$_3$	6.23 6.83(5)	Spe27 Dan44
Albite	Na[AlSi$_3$O$_8$]	2.603···2.688 [2.617(5)] 2.63	Spe27 Wea76 Dee62
Algodonite	Cu$_{6...7}$As	8.38	Dan44
Allactite	Mn$_7$[(OH)$_4$AsO$_4$]$_2$	3.8	Klo48
Allanite (Orthite)	(Ca, Ce)$_2$(Fe^{2+}, Fe^{3+})Al$_2$[OOHSiO$_4$Si$_2$O$_7$]	2.50···4.20 3.0···4.2	Spe27 Cla66
Alleghanyite	Mn$_5$[(OH)$_2$(SiO$_4$)$_2$]	4.02	Klo48
Allemontite	SbAs [Dan44]	5.8···6.2 5.80···6.34	Dan44 Spe27
Allodelphite see Synadelphite			
Allophane	Al$_2$O$_3$ · ySiO$_2$ · zH$_2$O [Klo48]	1.88···1.94	Spe27
Almandine	Fe$^{2+}_3$Al$_2$[SiO$_4$]$_3$	3.688···4.33 [4.318(1)]	Hod52/53 Wea76
Almandine-pyrope		3.80···3.95	Spe27
Alstonite	CaBa(CO$_3$)$_2$ [Dee62]	3.67···3.707	Spe27
Altaite	PbTe	8.15 [8.2459(19)]	Dan44 Wea76
Aluminite	Al$_2$[(OH)$_4$SO$_4$] · 7H$_2$O	1.7	Klo48
Alumogel	Al$_2$O$_3$ + aq. [Klo48]	2.4···2.5	Klo48

continued

1.1.1 Density of minerals (continued)

Mineral	Formula		$\varrho, [\varrho_x]$ g cm^{-3}	Ref.
Alunite	$KAl_3[(OH)_6(SO_4)_2]$		2.58···2.75	Hod52/53
			[2.822(4)]	Wea76
Alunogen	$Al_2(SO_4)_3 \cdot 18H_2O$	[Hod52/53]	1.6···1.8	Hod52/53
Amarantite	$Fe_2^{3+}[O(SO_4)_2] \cdot 7H_2O$		2.2	Klo48
Ambatoarinite	$Sr(Ce, La, Nd)[OCO_3)_3](?)$		5.25	Spe27
Amblygonite	$LiAl[(F, OH)PO_4]$		2.989···3.101	Spe27
			2.98···3.15	Hod52/53
Amesite	$Mg_{3.2}Fe_{0.8}^{2+}Al_{2.0}[(OH)_8Al_2Si_2O_{10}]$		2.77	Dee62
Aminoffite	$Ca_3Be_2[(OH)_2Si_3O_{10}]$		2.94	Klo48
Ampangabeite	$(Y, U, Ca)(Nb, Fe^{3+})_2(O, OH)_6$		3.348···4.644	Spe27
see also Samarskite				
Analcime	$Na[AlSi_2O_6] \cdot H_2O$		2.2···2.285	Spe27
			[2.258(3)]	Wea76
Anapaite	$Ca_2Fe[PO_4]_2 \cdot 4H_2O$		2.8	Klo48
Anatase,	TiO_2		3.9	Dan44
see also Xanthitane			[3.893(5)]	Wea76
			3.82···3.97	Dee62
Anauxite	$Al_2O_3 \cdot 3SiO_2 \cdot 2H_2O$	[Klo48]	2.524	Spe27
Ancylite	$Sr_3(Ce, La, Dy)_4[(OH)_4(CO_3)] \cdot 3H_2O$		3.95	Klo48
Andalusite	$Al^{[6]}Al^{[5]}[OSiO_4]$		3.118···3.29	Spe27
			[3.145(2)]	Wea76
Andesine	$(CaO, Na_2O)Al_2O_3 \cdot 4SiO_2$	[Hod52/53]	2.647···2.69	Hod52/53
Andorite	$Pb_4Ag_4Sb_{12}S_{24}$		5.35(2)	Dan44
Andradite	$Ca_3Fe_2^{3+}[SiO_4]_3$		3.50···3.85	Spe27
			[3.860(1)]	Wea76
Anemousite	(Plagioclase)		2.684	Spe27
Aneylite, Calcio-			3.82	Spe27
Angaralite	$(Mg, Ca)_2(Al, Fe)_{10}[O_5SiO_4]_6(?)$ [Klo48]		2.62	Klo48
Anglesite	$PbSO_4$		6.12···6.39	Hod52/53
			[6.324(8)]	Wea76
Anhydrite	$Ca[SO_4]$		2.899···2.985	Hod52/53
			[2.964(4)]	Wea76
Ankerite	$CaFe[CO_3]_2$		2.93···3.10	Dee62
Annite, see Lepidomelan				
Annabergite (Nickelblüte)	$Ni_3[AsO_4]_2 \cdot 8H_2O$		2.907	Spe27
			3···3.1	Klo48
Anorthite	$Ca[Al_2Si_2O_8]$		2.703···2.763	Spe27
			[2.760(1)]	Wea76
Anorthoclase	$(NaK)_2O, Al_2O_3 \cdot 6SiO_2$	[Hod52/53]	2.56···2.651	Hod52/53
Anthophyllite	$(Mg, Fe)_7[OHSi_4O_{11}]_2$		2.857···3.2	Hod52/53
			[2.953(12)]	Wea76
– ferro			3.24···3.83	Spe27
Anthraxolite			1.845	Spe27
Antigorite	$Mg_6[(OH)_8Si_4O_{10}]$		2.618	Spe27
			2.55···2.62	Hod52/53
Antimonite, see Stibnite				
Antimony	Sb		6.61···6.72	Dan44
			[6.685(4)]	Wea76
Antlerite (Stelznerite,	$Cu_3[(OH)_4SO_4]$		3.9	Klo48
Heterobrochantite)			3.757	Spe27

continued

1.1.1 Density of minerals (continued)

Mineral	Formula	$\varrho, [\varrho_x]$ g cm^{-3}	Ref.
Antofagastite (Eriochalcite)	$CuCl_2 \cdot 2H_2O$	2.4	Klo48
Apatite	$Ca_5[F(PO_4)_3]$	3.1···3.35	Dee62
−, Chlorine		3.14···3.20	Spe27
−, Fluorine		3.18···3.206	Spe27
−, Manganese		3.257	Spe27
−, Sulfate		3.196···3.207	Spe27
Aphrosiderite	Variety of Daphnite	2.959	Spe27
Aphthitalite, see Glaserite			
Apophyllite	$KCa_4[F(Si_4O_{10})_2] \cdot 8H_2O$	2.3···2.4	Hod52/53
		2.33···2.37	Dee62
Aragonite	$CaCO_3$	2.85···2.94	Hod52/53
		[2.930(4)]	Wea76
		2.94	Dee62
Arakawaite (Veszelyite)	$(Cu, Zn)_3[(OH)_3PO_4] \cdot 2H_2O$	3.09	Spe27
		3.5	Klo48
Aramayoite	$Ag(Sb, Bi)S_2$	5.447···5.602	Spe27
		5.62	Klo48
		5.602	Dan44
Arcanite	K_2SO_4	[2.661(3)]	Wea76
Ardennite (Dewalquite)	$Mn_4^{2+}(Mg, Al, Fe^{3+})_2Al_4[(OH)_6(As, V)$ $\cdot O_4(SiO_4)_2Si_3O_{10}]$	3.6	Klo48
Arduinite		2.26	Spe27
Arfvedsonite	$Na_{2.5}Ca_{0.5}(Fe^{2+}, Mg, Fe^{3+}, Al)_5[(OH, F)_2$ $\cdot Al_{0.5}Si_{7.5}O_{22}]$	3.4···3.5	Klo48
Argentite, see also Acanthite	Ag_2S	7.04	Dan44
		7.24···7.40	Hod52/53
		[7.125(35)]	Wea76
Argentopyrite	$AgFe_2S_3$	[4.269(17)]	Wea76
Argyrodite	$4Ag_2S \cdot GeS_2$	6.32	Dan44
		6.2	Klo48
Arizonite	$Fe_2Ti_3O_9$ [Klo48]	4.25	Dan44, Klo48
Armangite	$Mn_3(AsO_3)_2$	4.23	Spe27, Klo48
Armenite	$BaCa_2Al_3[Al_3Si_9O_{30}] \cdot 2H_2O$	2.77	Klo48
Arsenic	As	5.64···5.78	Hod52/53
		[5.776(4)]	Wea76
Arsenic Nickel	NiAs	7.7···7.8	Klo48
Arseniosiderite	$Ca_3Fe_4[OHAsO_4]_4 \cdot 4H_2O$	3.8···3.9	Klo48
Arsenkies, see Arsenopyrite			
Arsenobismite	$Bi_4[OHAsO_4]_3 \cdot H_2O(?)$	5.70	Spe27
Arsenoclasite	$Mn_5[(OH)_2AsO_4]_2$	4.16	Klo48
Arsenolamprite	As-Modification	5.3···5.5	Dan44
Arsenolite	As_2O_3	3.87(1)	Dan44
		[3.870(5)]	Wea76
Arsenopyrite (Arsenkies)	FeAsS	6.07(15)	Dan44
		5.89···6.20	Hod52/53
		5.88···6.14	Spe27
		[6.162(15)]	Wea76
Artinite	$Mg_2[(OH)_2CO_3] \cdot 3H_2O$	2.03	Klo48
Asbestos, Amphibole	$H_4Mg_3Si_2O_9$ [Klo48]	2.97	Spe27
Asbolan (Erdkobalt)	CO, MnO_2 [Klo48]	2.985	Spe27

continued

1.1.1 Density of minerals (continued)

Mineral	Formula	$\varrho, [\varrho_x]$ g cm^{-3}	Ref.
Ascharite, see also Camsellite and Szaibelyite	$Mg_2[B_2O_5] \cdot H_2O$	2.69	Spe27
Ashcroftine	$KNa(Ca, Mg, Mn)[Al_4Si_5O_{18}] \cdot 8H_2O$	2.61(?)	Dee62
Asowskite	$Fe_3^{3+}[(OH)_6PO_4]$	2.5	Klo48
Astrakanite, see Bloedite			
Astrolite	$(Na, K)_2Fe^{2+}(Al, Fe^{3+})_2Si_5O_{15} \cdot H_2O$	2.8	Klo48
Astrophyllite	$(K, Na)_3(Fe, Mn)_7(Ti, Zr)_2[Si_8(O, OH)_{31}]$	3.3···3.4	Klo48
Atacamite	$Cu_2(OH)_3Cl$	3.77···3.94	Hod52/53
Atelestite, see Rhagite			
Augelite	$Al_2[(OH)_3PO_4]$	2.77	Hod52/53
		2.07	Klo48
Augite	$(Ca, Mg, Fe^{2+}, Fe^{3+}, Ti, Al)_2[(Si, Al)_2O_6]$	3.2···3.6	Hod52/53
		3.23···3.52	Dee62
–, Titan-		3.29···3.39	Spe27
Aurichalcite	$(Zn, Cu)_5[(OH)_3CO_3]_2$	3.274···3.64	Spe27
Auripigment (Rauschgelb)	As_2S_3	3.4···3.5	Klo48
		3.49	Dan44
Aurosmirid	50% Ir + 25% Au + 25% Os [Klo48]	20	Dan44
Austinite	$CaZn[OHAsO_4]$	4.12	Klo48
Autunite	$Ca[UO_2PO_4]_2 \cdot 10(12···10)H_2O$	3.198	Spe27
		3.05···3.19	Hod52/53
Avogadrite	$K[BF_4]$	2.498···2.617	Spe27
Awaruite	Ni_3Fe	7.746	Spe27
Axinite	$Ca_2(Fe, Mn)AlAl[BO_3OHSi_4O_{12}]$	3.22···3.314	Spe27
		3.26···3.36	Dee62
Azurite	$Cu_3[OHCO_3]_2$	3.7···3.9	Klo48
		[3.787(5)]	Wea76
Bababudanite	Variety of Riebeckite	3.18	Spe27
Babingtonite	$Ca_2Fe^{2+}Fe^{3+}[Si_5O_{14}OH]$	3.351···3.398	Spe27
		3.4	Klo48
Baddeleyite (Zircon oxide)	ZrO_2	5.4···6.02	Dan44
		[5.8267(23)]	Wea76
		4.9···5.4	Klo48
Badenite	$\approx (Co, Ni, Fe)_2(As, Bi)_3$	7.104	Dan44
Bakerite	$Ca_8B_{10}Si_6O_{35} \cdot 6H_2O$ [Klo48]	$\approx$ 2.73	Klo48
Banalsite	$BaNa_2[Al_2Si_2O_8]_2$	3.06	Klo48
		[3.092(11)]	Wea76
Bandylite	$Cu[ClB(OH)_4]$	2.81	Klo48
Barbertonite, see Stichtite			
Bardolite		2.470	Spe27
Barite (Baryt)	$Ba[SO_4]$	4.3···4.6	Hod52/53
		[4.480(5)]	Wea76
–, Sr-		4.29	Spe27
–, Pb-		4.62	Spe27
Barkevikite	$Ca_2(Na, K)(Fe^{2+}, Mg, Fe^{3+}, Mn)_5[(OH, F)_2 \cdot Al_{1.5}Si_{6.5}O_{22}]$	3.298···3.518	Spe27
Barthite	$CaZn[(OH)AsO_4] + Cu$, variety of Austinite [Klo48]	4.19	Spe27
Barylite	$BaBe_2[Si_2O_7]$	4.027	Spe27
		4.0	Klo48
Barysilite	$Pb_8Mn[Si_2O_7]_3$	6.53···6.706	Spe27
Baryt, see Barite			

continued

1.1.1 Density of minerals (continued)

Mineral	Formula		$\varrho, [\varrho_x]$ g cm^{-3}	Ref.
Barytocalcite	BaCa[CO$_3$]$_2$		3.71	Spe27
Bassetite	Fe[UO$_2$PO$_4$]$_2 \cdot 10 \cdots 12$H$_2$O		3.10	Spe27
Bastnaesite (Weibyite)	Ce[FCO$_3$]		4.948	Spe27
			4.78	Don73
			[5.02]	Wea76
Baumhauerite	Pb$_5$As$_9$S$_{18}$		5.329	Dan44
Bauxite	Al$_2$O$_3 \cdot 2$H$_2$O	[Hod52/53]	2.4$\cdots$2.5	Spe27
			2.55	Hod52/53
Bavalite			3.20	Spe27
Bavenite	Ca$_4$Al$_2^{[4]}$Be$_2^{[4]}$[(OH)$_2$Si$_9$O$_{26}$]		2.7	Klo48
Bayldonite	PbCu$_3$[OHAsO$_4$]$_2$		5.21$\cdots$5.50	Spe27
Bazzite	Sc$_2$Be$_3$[Si$_6$O$_{18}$]		2.80	Spe27
Beckelite	(Ca, Ce, La, Nd)$_5$[(O, OH, F)(SiO$_4$)$_3$]		4.1	Klo48
Becquerelite	6[UO$_2$(OH)$_2$]$\cdot$Ca(OH$_2$)$\cdot 4$H$_2$O		4.967	Spe27
			5.2	Dan44
Beegerite	Pb$_6$Bi$_2$S$_9$	[Klo48]	7.27	Dan44
Bellingerite	Cu$_3$[IO$_3$]$_6 \cdot 2$H$_2$O		4.89	Klo48
Bellite	(Pb, Ag)$_5$[Cl(CrO$_4$, AsO$_4$, SiO$_4$)$_3$]		≈ 5	Klo48
Bementite	Mn$_8$[(OH)$_{10}$Si$_6$O$_{15}$]		3.106	Spe27
Benitoite	BaTi[Si$_3$O$_9$]		3.7	Klo48
			3.64$\cdots$3.65	Hod52/53
Benjaminite	Pb$_2$Ag$_2$Bi$_4$S$_9$		6.34	Dan44
Beraunite (Eleonorite)	Fe^{2+}Fe$_5^{3+}$[(OH)$_5$(PO$_4$)$_4$]$\cdot 6$H$_2$O		2.85$\cdots$2.99	Spe27
			≈ 2.9	Klo48
Berlinite	Al[PO$_4$]		[2.618(6)]	Wea76
Bermanite	Mn^{2+}Mn$_2^{3+}$[OHPO$_4$]$_2 \cdot 4$H$_2$O		2.84	Klo48
Berthierite	FeS$\cdot$Sb$_2$S$_3$		4.64	Dan44
Berthonite, see also Bournonite	2PbS$\cdot$Cu$_2$S$\cdot$Sb$_2$S$_3$		5.49	Spe27, Dan44
Bertrandite	Be$_4$[(OH)$_2$Si$_2$O$_7$]		2.59$\cdots$2.604	Spe27
			2.571$\cdots$2.60	Hod52/53
Beryl (Emerald, Smaragd)	Al$_2$Be$_3$[Si$_6$O$_{18}$]		2.545$\cdots$2.910	Spe27
			[2.641(3)]	Wea76
			2.66$\cdots$2.83	Dee62
			2.648$\cdots$2.709	Spe27
Beryllonite	NaBe[PO$_4$]		2.85	Klo48
Berzelianite	Cu$_2$Se		6.71$\cdots$7.23	Dan44
			[6.835(35)]	Wea76
Berzeliite	(Ca, Na)$_3$(Mg, Mn)$_2$[AsO$_4$]$_3$		3.9$\cdots$4.4	Klo48
Betafite	(U, Ca)$_2$(Ti, Nb, Ta)$_2$O$_6$(O, OH, F)		3.75$\cdots$4.17	Spe27
			3.7$\cdots$5	Dan44
Bianchite	(Zn, Fe)[SO$_4$]$\cdot 6$H$_2$O		≈ 2.3	Klo48
Bindheimite	Pb$_{1\cdots 2}$Sb$_{2\cdots 1}$(O, OH, H$_2$O)$_{6\cdots 7}$		4.6$\cdots$5	Klo48
Biotite	K(Mg, Fe, Mn)$_3$[(OH, F)$_2$AlSi$_3$O$_{10}$]		2.692$\cdots$3.16	Spe27
			2.7$\cdots$3.3	Dee62
Bischofite	MgCl$_2 \cdot 6$H$_2$O		1.59	Klo48
Bismite	α-Bi$_2$O$_3$		10.4	Dan44
			9.70$\cdots$9.83	Hod52/53
			[9.371(11)]	Wea76
Bismoclite	BiOCl		7.5	Klo48
Bismuth	Bi		9.70$\cdots$9.83	Dan44
			[9.8071(50)]	Wea76

continued

1.1.1 Density of minerals (continued)

Mineral	Formula	$\varrho, [\varrho_x]$ g cm^{-3}	Ref.
Bismuthinite	Bi$_2$S$_3$	6.8···7.2	Klo48
(Wismutglanz)		6.55···6.73	Spe27
		[6.8081(38)]	Wea76
		6.78 (3)	Dan44
Bismuth oxide	Bi$_2$O$_3$ [Klo48]	≈9	Klo48
Bismutoplagionite, see also Galenobismutite	PbS · Bi$_2$S$_3$	5.35	Spe27
Bismutogutallite		8.26	Dan44
Bismutotantalite	Bi$_2$O$_3$(Ta, Nb)$_2$O$_5$ [Klo48]	8.15	Klo48
Bittersalz, see Epsomite			
Bityite	CaLiAl$_2$[(OH)$_2$AlBeSi$_2$O$_{10}$]	3.05	Klo48
Bixbyite	(Mn, Fe)$_2$O$_3$ (Mn:Fe=1:1)	4.945	Dan44
		[5.032(8)]	Wea76
Blei, see Lead			
Bleiglanz, see Galena			
Bleigummi, see Plumbogummite			
Bloedite (Astrakanite)	Na$_2$Mg[SO$_4$]$_2$ · 4H$_2$O	2.32	Spe27
Blomstrandine	Priorite group	4.88···5.00	Spe27
Blomstrandite (Uranpyrochlore, see also Ellsworthite)	(U, ...)$_2$(Nb, Ta, Ti)$_2$O$_6$(O, OH, F)	4.07···4.17	Spe27
Bobierrite	Mg$_3$[PO$_4$]$_2$ · 8H$_2$O	2.2	Klo48
Boehmite	γ-AlOOH	3.01···3.06	Dan44
		[3.071(4)]	Wea76
Boksputite	Pb$_6$Bi$_2$[O$_6$(CO$_3$)$_3$] [Klo48]	7.3	Klo48
Boleite	(26 PbCl$_2$ · 3AgCl) · 24 Cu(OH)$_2$ · 6AgCl · 3H$_2$O	4.74···5.155	Spe27
Bolivarite	Al$_2$[(OH)$_3$PO$_4$] · 5H$_2$O	2.05	Spe27
		2.52	Klo48
Bolivianite		4.1	Spe27
Boothite	Cu[SO$_4$] · 7H$_2$O	2.02	Spe27
Boracite	β-Mg$_3$[ClB$_7$O$_{13}$]	2.89···2.91	Spe27
Borax	Na$_2$[B$_4$O$_5$(OH)$_4$] · 8H$_2$O	1.7···1.8	Klo48
		[1.7128(13)]	Wea76
Bornite	Cu$_5$FeS$_4$	5.06···5.08	Dan44
		[5.0910(28)]	Wea76
Boulangerite	5 PbS · 2Sb$_2$S$_3$	6.274···6.407	Spe27
		6.23; 5.98(2)	Dan44
		5.8···6.2	Klo48
Bournonite, see also Berthonite	2 PbS · Cu$_2$S · Sb$_2$S$_3$	5.83(3)	Dan44
		5.7···5.9	Hod52/53
Boussingaultite	(NH$_4$)$_2$ · Mg[SO$_4$]$_2$ · 6H$_2$O	1.7	Klo48
Bradleyite	Na$_3$Mg[PO$_4$CO$_3$]	2.646	Klo48
Braggite	(Pt, Pd, Ni)S	10	Dan44
Brannerite	(U, Ca, Th, Y)[(Ti, Fe)$_2$O$_6$]	4.5···5.43	Spe27
Braunite	Mn^{2+}Mn$_6^{4+}$[O$_8$SiO$_4$] (Mn/Si = 7/1)	4.7···4.9	Klo48
Bravoite	(Ni, Fe, Co)S$_2$	4.62	Dan44
	(Fe/Ni=1/4, 28 NiS$_2$)	4.66	Dan44
	Cobalt-Nickel-Pyrite	4.716	Spe27
Brazilianite	NaAl$_3$[(OH)$_2$PO$_4$]$_2$	2.94	Klo48
			continued

1.1.1 Density of minerals (continued)

Mineral	Formula	$\varrho, [\varrho_x]$ g cm^{-3}	Ref.
Breithauptite	NiSb	8.23	Dan44
		8.1	Klo48
		[8.639(5)]	Wea76
Brewsterite	(Sr, Ba, Ca)[Al$_2$Si$_6$O$_{16}$]·5H$_2$O	2.45	Klo48
Brochantite (Kamarezite)	Cu$_4$[(OH)$_6$SO$_4$]	3.8···3.9	Hod52/53
		[3.982(8)]	Wea76
		4.0	Klo48
Bromargyrite (Bromyrite)	AgBr	5.8···6.0	Hod52/53
		[6.4772(17)]	Wea76
Bromyrite, see Bromargyrite			
Bromellite	BeO	3.017	Dan44
		3.008	Cla66
		[3.0104(12)]	Wea76
Bronzite	(Mg, Fe)$_2$[Si$_2$O$_6$]	3.2···3.5	Klo48
Brookite	TiO$_2$	3.87···4.084	Hod52/53
		[4.119(4)]	Wea76
		4.08···4.18	Dee62
Brucite	Mg(OH)$_2$	2.38···2.39	Spe27
		[2.368(6)]	Wea76
Brugnatellite	Mg$_6$Fe^{3+}[(OH)$_{13}$CO$_3$]·4H$_2$O	2.14	Dan44
		2.07	Klo48
Brushite, see also Metabrushite	CaH[PO$_4$]·2H$_2$O	2.3	Klo48
Bunsenite	NiO	6.898	Dan44
		[6.809(10)]	Wea76
Burkeite	Na$_6$[CO$_3$(SO$_4$)$_2$]	2.57	Klo48
Bustamite	(Mn, Ca)$_3$[Si$_3$O$_9$]	3.32···3.43	Dee62
		[3.326(5)]	Wea76
Buttgenbachite	Cu$_{19}$[Cl$_4$(OH)$_{32}$(NO$_3$)$_2$]+2H$_2$O	3.33	Spe27
Bytownite		≈2.75	Klo48
Cacoxenite	Fe$_4^{3+}$[OHPO$_4$]$_3$·12H$_2$O	≈2.3···2.8	Klo48
Cadmium oxide	CdO	8.1···8.2	Dan44
		6.2	Klo48
Cadmoselite	CdSe	[5.6738(28)]	Wea76
Calaverite	(Au, Ag)Te$_2$	9.24(20)	Dan44
Calcio-thomsonite	(Thomsonite without Na)	2.405	Spe27
Calcite	CaCO$_3$	2.699···2.82	Spe27
		[2.7100(11)]	Wea76
Caledonite	Pb$_5$Cu$_2$[(OH)$_6$CO$_3$(SO$_4$)$_3$]	6.4	Klo48
Calomel	α-Hg$_2$Cl$_2$	6.4···6.5	Klo48
		6.482	Hod52/53
		[7.166(16)]	Wea76
Camsellite, see also Ascharite		2.60	Spe27
		2.60	Dan44
Cancrinite, see also Vishnevite	Na$_6$Ca[CO$_3$(AlSiO$_4$)$_6$]·2H$_2$O	2.42···2.51	Dee62
Canfieldite	4Ag$_2$S·(Sn, Ge)S$_2$	6.1···6.3; 6.26; 6.27	Dan44
Cannizzarite	Pb$_3$Bi$_5$S$_{11}$(?)	6.54	Spe27
Cappelenite	(Ba, Ca, Ce, Na)(Y, Ce, La)$_2$[B$_2$O$_5$SiO$_4$]	4.4	Klo48
Caracolite	Pb$_2$Na$_3$[Cl(SO$_4$)$_3$]	5.1	Klo48
Carbonate-apatite	Ca$_5$[F(PO$_4$, CO$_3$OH)$_3$]	[3.281(8)]	Wea76
			continued

1.1.1 Density of minerals (continued)

Mineral	Formula		$\varrho, [\varrho_X]$ g cm^{-3}	Ref.
Carnallite	KMgCl$_3 \cdot$ 6H$_2$O		1.604	Klo48
Carnegieite	Na[AlSiO$_4$]		[2.513(50)]	Wea76
Carnotite	K$_2$[(UO$_2$)$_2$V$_2$O$_8$] $\cdot$ 3H$_2$O		4.5	Klo48
Carpholite (Strohstein)	MnAl$_2$[(OH)$_4$(Si$_2$O$_6$)]		2.9	Klo48
Caryinite	(Na, Ca)$_2$(Mn, Mg, Ca, Pb)$_3$[AsO$_4$]$_3$		4.25	Klo48
Caryocerite	≈ Na$_4$Ca$_{16}$(Y, La)$_3$(Zr, Ce)$_6$[F$_{12}$(BO$_3$)$_3$ $\cdot$ (SiO$_4$)$_{12}$]		4.3	Klo48
Caryopilite (Ektropite)	Mn$_6$[(OH)$_8$Si$_4$O$_{10}$]		2.46	Spe27
Cassiterite	SnO$_2$		6.8···7.1 [6.992(11)]	Hod52/53 Wea76
Catapleite	Na$_2$Zr[Si$_3$O$_9$] $\cdot$ 2H$_2$O		2.658 2.8	Spe27 Klo48
Cattierite	CoS$_2$		[4.8213(13)]	Wea76
Cebollite	Ca$_3$Al$_2$Si$_3$O$_{15} \cdot$ H$_2$O(?)	[Klo48]	2.96	Spe27
Celestite	SrSO$_4$	[Hod52/53]	3.84···3.968 [3.972(5)]	Spe27 Wea76
Celsian	Ba[Al$_2$Si$_2$O$_8$]		3.10···3.39	Dee62
–, Para-			3.400(10) [3.340(13)]	Wea76 Wea76
Centrallasite, see Gyrolite				
Cerfluorite, see Yttrocerite				
Cerianite	(Ce, Th)O$_2$		[7.216(8)]	Wea76
Cerite	(Ca, Mg)$_2$(SE)$_8$[SiO$_4$]$_7 \cdot$ 3H$_2$O		4.9	Klo48
Cerussite	PbCO$_3$		6.46···6.54 [6.582(9)]	Spe27 Wea76
Cervantite	Sb$_2$O$_4$		6.64 art. [6.641]	Dan44 Wea76
Cesarolite	H$_2$PbMn$_3$O$_8$	[Klo48]	5.29	Spe27
Ceylonite (Pleonast)	(Mg, Fe)O $\cdot$ (Al, Fe)$_2$O$_3$		4.12 3.5···3.6	Spe27 Hod52/53
Chabazite	(Ca, Na$_2$)[Al$_2$Si$_4$O$_{12}$] $\cdot$ 6H$_2$O		2.09···2.168 2.05···2.10	Spe27 Dee62
Chalcanthite (Kupfervitriol)	Cu[SO$_4$] $\cdot$ 5H$_2$O		2.12···2.30 [2.2912(46)]	Hod52/53 Wea76
Chalcedon	SiO$_2$	[Hod52/53]	2.55···2.63	Spe27, Hod52/53
Chalcoalumite	CuAl$_4$[(OH)$_{12}$SO$_4$] $\cdot$ 3H$_2$O		2.29	Spe27
Chalcocite (Chalkosin)	Cu$_2$S		5.51···5.785 [5.792]	Spe27 Wea76
Chalcomenite	Cu[SeO$_3$] $\cdot$ 2H$_2$O		3.36	Klo48
Chalcophanite	ZnMn$_3$O$_7 \cdot$ 3H$_2$O		4.00(10)	Dan44
Chalcophyllite	(Cu, Al)$_3$[(OH)$_4$(AsO$_4$, SO$_4$)] $\cdot$ 6H$_2$O		2.4···2.6	Klo48
Chalcopyrite	CuFeS$_2$		4.12 4.1···4.3; 4.283 [4.0878(25)]	Spe27 Dan44 Wea76
Chalcostibite	Cu$_2$S $\cdot$ Sb$_2$S$_3$		4.95(5)	Dan44
Chalcosin, see Chalcosite				
Chalmersite, see Cubanite				
Chalybite, see Siderite				
Chamosite	(Fe^{2+}, Fe^{3+})$_3$[(OH)$_2$AlSi$_3$O$_{10}$](Fe, Mg)$_3$ $\cdot$ (O, OH)$_6$		3.3	Dee62

continued

1.1.1 Density of minerals (continued)

Mineral	Formula	$\varrho, [\varrho_x]$ g cm^{-3}	Ref.
Chapmanite	SbFe$_2$[OH(SiO$_4$)$_2$]	3.6	Klo48
Chenevixite	Cu$_2$Fe$_2$[(OH)$_2$AsO$_4$]$_2 \cdot$ H$_2$O	3.93	Klo48
Childrenite	(Fe^{2+}, Mn)Al[(OH)$_2$PO$_4$] $\cdot$ H$_2$O	3.2	Klo48
Chiolite	Na$_5$[Al$_3$F$_{14}$]	2.84$\cdots$3.005	Hod52/53
Chiviatite	$\approx$ Pb$_2$Bi$_6$S$_{11}$	6.92	Dan44
		7.15	Dan44
Chkalovite (Tschkalowit)	Na$_2$[BeSi$_2$O$_6$]	2.662	Klo48
Chlopinite (Eschwegeite)		5.87	Spe27
		5.24	Klo48
Chlorapatite	Ca$_5$[Cl(PO$_4$)$_3$]	[3.178(4)]	Wea76
Chlorargyrite	AgCl	5.5$\cdots$5.6	Klo48
		5.552	Hod52/53
		[5.5710(15)]	Wea76
Chlorite	(Mg, Al, Fe)$_{12}$[(Si, Al)$_8$O$_{20}$](OH)$_{16}$ [Dee62]	2.6$\cdots$3.3	Dee62
		2.386$\cdots$2.396	Spe27
Chloritoid	Fe$_2^{2+}$AlAl$_3$[(OH)$_4$O$_2$(SiO$_4$)$_2$]	3.45	Spe27
		[3.619(8)]	Wea76
		3.51$\cdots$3.80	Dee62
Chloromagnesite	MgCl$_2$	[2.333(6)]	Wea76
Chloromelanite	Jadeite-series	3.365	Spe27
Chlorophaeite	Leptochlorite, Fe-rich	1.81	Spe27
Chlorophoenicite	(Zn, Mn)$_5$[(OH)$_7$AsO$_4$]	3.55	Klo48
Chloroxiphite	Pb$_3$O$_2$Cl$_2 \cdot$ CuCl$_2$	6.763	Spe27
Chloromercury		6.4$\cdots$6.5	Klo48
Chondrodite	Mg$_5$[(OH, F)$_2$(SiO$_4$)$_2$]	3.175	Spe27
		3.16$\cdots$3.26	Dee62
		[3.136(21)]	Wea76
Chromite	Cr$_2$FeO$_4$	4.32$\cdots$4.57	Hod52/53
		5.09	Dee62
Chromohercynite	(Al, Cr)$_2$FeO$_4$	4.415	Spe27
Chrysoberyl	Al$_2$BeO$_4$	3.50$\cdots$3.84	Hod52/53
		[3.6997(25)]	Wea76
Chrysocolla	Cu$_4$H$_4$[(OH)$_8$Si$_4$O$_{10}$]	2.400$\cdots$2.417	Spe27
Chrysolite, see Olivine			
Chrysotile		2.457$\cdots$2.57	Spe27
Chubuttite, see also Lorettoite	PbCl$_2 \cdot$ 7 PbO [Klo48]	7.952	Spe27
Cinnabar (Zinnober)	HgS	8.090	Dan44
		8.0$\cdots$8.2	Hod52/53
		[8.187(4)]	Wea76
Clarkeite	Na$_2$U$_2$O$_7$	6.39	Dan44
Claudetite	As$_2$O$_3$	4.15	Dan44
		3.85$\cdots$4.151	Hod52/53
		[4.1863(25)]	Wea76
Clausthalite	PbSe	7.8	Dan44
		[8.269(20)]	Wea76
Cleveite	(UO$_2$) [Hod52/53]	7.49	Hod52/53
Clinochlore	(Mg, Al)$_3$[(OH)$_2$AlSi$_3$O$_{10}$]Mg$_3$(OH)$_6$	2.65$\cdots$2.78	Hod52/53
Clinoclase	Cu$_3$[(OH)$_3$AsO$_4$]	4.2$\cdots$4.4	Klo48
Clinoedrite	Ca$_2$Zn$_2$[(OH)$_2$Si$_2$O$_7$] $\cdot$ H$_2$O	3.33	Klo48
Clinoferrosilite	FeSiO$_3$	[4.005(2)]	Wea76
Clinohumite	Mg$_9$[(OH, F)$_2$(SiO$_4$)$_4$]	3.21$\cdots$3.35	Dee62
		[3.167(18)]	Wea76

continued

1.1.1 Density of minerals (continued)

Mineral	Formula	ϱ, $[\varrho_x]$ g cm^{-3}	Ref.
Clinoptilolite, see Heulandite			
Clinozoisite	Ca$_2$Al$_3$[OOHSiO$_4$Si$_2$O$_7$]	3.336(11)	Wea76
		3.21···3.38	Dee62
Clintonite	Ca(Mg, Al)$_{3···2}$[(OH)$_2$Al$_2$Si$_2$O$_{10}$]	3···3.1	Dee62
Cobalt-Nickel-Pyrite, see Bravoite			
Cobalt Olivine	Co$_2$SiO$_4$	[4.716(3)]	Wea76
Cobalt Pyrite	partly Pyrite with 14% Co, partly Linneite	4.965	Klo48
Cobalt Titanate	CoTiO$_3$	[4.986(3)]	Wea76
Cobalticalcite (Kobaltspat)	CoCO$_3$	4.1	Klo48
		[4.2159(20)]	Wea76
Cobaltite (Kobaltglanz)	CoAsS	6.0···6.3	Hod52/53
		[6.275(168)]	Wea76
Cocinerite	Cu$_4$AgS [Dan44]	6.14	Spe27
Coesite	SiO$_2$	[2.9110(5)]	Wea76
Coffinite	U[SiO$_4$]	[7.155(10)]	Wea76
Cohenite	Fe$_3$C	7.20···7.65	Dan44
Colemanite	Ca[B$_3$O$_4$(OH)$_3$] · H$_2$O	2.417···2.428	Hod52/53
		[2.4194(21)]	Wea76
Colerainite	Variety of Pennine or Corundophillite (Chlorite)	2.51	Spe27
Coloradoite	HgTe	8.04 (average)	Dan44
		8.0···8.1	Klo48
		[8.0855(23)]	Wea76
Columbite	(Fe, Mn)(Nb, Ta)$_2$O$_8$ [Hod52/53]	5.26···7.30	Hod52/53
		5.147···6.845	Spe27
Colusite	Cu$_3$(Fe, As, Sn)S$_4$	4.50	Dan44
Connellite	Cu$_{19}$[Cl$_4$(OH)$_{32}$SO$_4$ + 4 H$_2$O]	3.54	Spe27
		3.4	Hod52/53
Cooperite	PtS	9.5	Dan44
		[10.254(4)]	Wea76
Copiapite	(Fe^{2+}, Mg)Fe$_4^{3+}$[OH(SO$_4$)$_3$]$_2$ · 20 H$_2$O	2.087	Spe27
		2.1···2.2	Hod52/53
Copper (Kupfer)	Cu	8.62···8.80	Spe27
		8.95	Dan44
		8.8···8.9	Hod52/53
		[8.9331(37)]	Wea76
Coquimbite	Fe$_2^{3+}$[SO$_4$]$_3$ · 9 H$_2$O	2.07···2.105	Hod52/53
Cordierite	Mg$_2$[Al$_4$Si$_5$O$_{18}$]	2.53···2.78	Dee62
	orthorhombic	[2.508(1)]	Wea76
	hexagonal	[2.513(2)]	Wea76
Cornetite	Cu$_3$[(OH)$_3$PO$_4$]	4.10	Spe27
Coronadite	Pb$_{\leqq 2}$Mn$_8$O$_{16}$	5.44	Dan44
Corundophyllite	(Mg, Fe, Al)$_3$[(OH)$_2$Al$_{1.5···2}$Si$_{2.5···2}$O$_{10}$] · Mg$_3$(OH)$_6$	2.881	Spe27
Corundum (Korund)	Al$_2$O$_3$	3.9···4.1	Klo48
		[3.9869(11)]	Wea76
	α-Al$_2$O$_3$	3.98···4.02	Dee62
	β-Al$_2$O$_3$	3.30	Spe27
Corvusite	V$_2^{4+}$V$_{12}^{5+}$O$_{34}$ · n H$_2$O	2.82	Klo48
			continued

1.1.1 Density of minerals (continued)

Mineral	Formula	$\varrho, [\varrho_x]$ g cm^{-3}	Ref.
Cosalite	$2\,PbS \cdot Bi_2S_3$	6.07⋯7.13	Spe27
		6.76	Dan44
Cotunnite	$PbCl_2$	5.84	Hod52/53
		[5.906(12)]	Wea76
Couzeranite	partly decomposed Dipyr, partly Andalusite	2.625	Spe27
Covellin, see Covellite			
Covellite (Covellin)	CuS	4.6⋯4.76; 4.671 art.	Dan44
		[4.682(1)]	Wea76
Crednerite	$CuMnO_2$	4.972⋯5.03	Spe27
Creedite	$Ca_3[(Al(F, OH, H_2O)_6)_2SO_4]$	2.713⋯2.730	Spe27
		2.71	Klo48
Crestmorite	$4\,CaSiO_3 \cdot 7\,H_2O\,(?)$ [Klo48]	2.22	Spe27
Cristobalite	SiO_2	2.32⋯2.36	Spe27
	α-SiO_2	[2.3344(30)]	Wea76
	β-SiO_2	[2.194]	Wea76
Crocidolite, see Riebeckite			
Crocoite	$Pb[CrO_4]$	5.9⋯6.1	Hod52/53
Cronstedtite	$Fe_4^{2+}Fe_2^{3+}[(OH)_8Fe_2^{3+}Si_2O_{10}]$	3.45	Dee62
Crookesite	$(Cu, Tl, Ag)_2Se$	6.9	Klo48
Cryolite	$Na_3[AlF_6]$	2.95⋯3.00	Hod52/53
		[2.965(9)]	Wea76
Cryolithionite	$Na_3Li_3[AlF_6]_2$	2.78	Klo48
		2.777⋯2.778	Hod52/53
Cryptohalite	$(NH_4)_2[SiF_6]$	2.004	Spe27
Cubanite (Chalmersite)	$CuFe_2S_3$	4.03⋯4.18	Dan44
		[4.026(10)]	Wea76
		4.04	Spe27
Cumengeite	$5\,PbCl_2 \cdot 5\,Cu(OH)_2 \cdot 0.5\,H_2O\,(?)$	4.74⋯4.88	Spe27
Cummingtonite	$(Mg, Fe)_7[OHSi_4O_{11}]_2$	3.10⋯3.60	Dee62
		[2.950(5)]	Wea76
Cuprite	Cu_2O	5.85⋯6.15	Hod52/53
		[6.1047(43)]	Wea76
Cuprodescloizite, see Mottramite			
Cuprozincite		4.104	Spe27
Curite	$3\,PbO \cdot 8\,UO_3 \cdot 4\,H_2O$	7.192	Spe27
		7.26	Dan44
Curtisite (Idrialite)	Composition, mainly Picen, $C_{22}H_{14}$, and Chrysen, $C_{18}H_{12}$	1.21	Spe27
Cuspidine (Custerite)	$Ca_4[(F, OH)_2Si_2O_7]$	2.965⋯2.989	Spe27
		2.91	Klo48
Custerite, see Cuspidine			
Cyanite, see Disthen			
Cylindrite	$FePb_3Sn_4Sb_2S_{14}(?)$ [Klo48]	5.46(3)	Dan44
Dahllite	$3\,Ca_3(PO_4)_2 \cdot 2\,CaCO_3 \cdot H_2O$ [Klo48]	3.00⋯3.094	Spe27
Dakeite, see Schroeckingerite			
Danalite	$Fe_8[S_2(BeSiO_4)_6]$	3.28⋯3.44	Dee62
Danburite	$Ca[B_2Si_2O_8]$	2.93⋯3.02	Hod52/53
		[2.992(13)]	Wea76

continued

1.1.1 Density of minerals (continued)

Mineral	Formula	$\varrho, [\varrho_x]$ g cm^{-3}	Ref.
Dannemorite	Mn-Cummingtonite	3.516	Spe27
Daphnite	$(Fe^{2+}, Al)_3[(OH)_2Al_{1.2\cdots1.5}Si_{2.8\cdots2.5}O_{10}]$ $\cdot Fe_3(OH)_6$	3.2	Klo48
Darapskite	$Na_3[SO_4NO_3] \cdot H_2O$	2.20	Klo48
Darchiardite	$(K, Na, Ca_{0.5})_5[Al_5Si_{19}O_{48}] \cdot 12 H_2O$	2.17	Klo48
Datolite	$CaB^{[4]}[OHSiO_4]$	2.89$\cdots$3.00	Hod52/53
		[3.003(20)]	Wea76
		2.96$\cdots$3.00	Dee62
Daubréeite	$BiCl_3 \cdot 2Bi_2O_3$ [Klo48]	6.4	Klo48
Daubréelite	$FeCr_2S_4$	3.81(1)	Dan44
		[3.866(6)]	Wea76
Davyne	$(Na, Ca, K)_9[(Cl, SO_4, CO_3)_3(Si, Al)_{12}O_{24}]$ $\cdot 0.5 H_2O$	2.34$\cdots$2.492	Spe27
Delafossite	$CuFeO_2$	5.41; 5.52 art.	Dan44
Delessite	$(Mg, Fe^{2+}, Fe^{3+})_3[(OH)_2Al_{0\cdots0.9}Si_{4\cdots3.1}O_{10}]$ $\cdot (Mg, Fe^{2+})_3(O, OH)_6$	2.6$\cdots$2.9	Klo48
Delorenzite (Tanteuxenite)	$(Y, Ce, U, Pb, Ca)(Ta, Nb, Ti)_2(O, OH)_6$	4.7	Dan44
Delvauxite	$Fe_2^{3+}[(OH)_3PO_4] \cdot 3.5 H_2O$	1.8$\cdots$2.0	Klo48
Demantoid	$Ca_3Fe_2Si_3O_{12}$ [Klo48]	3.801	Spe27
Derbylite	$Fe_3^{2+}Ti_3SbO_{11}OH$	4.512$\cdots$4.530	Hod52/53
Descloizite	$Pb(Zn, Cu)[OHVO_4]$	5.5$\cdots$6.2	Klo48
Desmine, see Stilbite			
Destinezite	$\approx Fe_4[(OH)_4(PO_4, SO_4)_3] \cdot 13 H_2O$	2.105	Spe27
Dewalquite, see Ardennite			
Dewindtite, see also Stasite	$Pb[(UO_2)_4(OH)_4(PO_4)_2] \cdot 8 H_2O(?)$	4.8	Spe27
Diabantite	Fe-analogue to Pennine	2.77$\cdots$2.79	Spe27
		2.6$\cdots$2.9	Klo48
Diaboleite	$Pb_2[Cu(OH)_4Cl_2]$	6.412	Spe27
		5.48	Klo48
Dialogite, see Rhodochrosite			
Diamond	C	3.150$\cdots$3.525	Hod52/53
		[3,5155(3)]	Wea76
Diaphorite (Ultrabasite)	$4 PbS \cdot 3 Ag_2S \cdot 3 Sb_2S_3$	6.04	Dan44
		6.026	Spe27
Diaspore	α-AlOOH	3.2$\cdots$3.5	Dee62
		[3.378(5)]	Wea76
Diaspore (crystal)		3.44(2)	Dan44
Dickinsonite	$Na_2(Mn^{2+}, Fe^{2+})_5[PO_4]_4$	3.41	Klo48
Dickite	$Al_4[(OH)_8Si_4O_{10}]$	[2.600(2)]	Wea76
Didymolite	$(Ca, Mg, Fe)Al_2[OSi_3O_9](?)$ [Klo48]	2.71	Klo48
Dietzeite	$Ca_2[CrO_4(IO_3)_2]$	3.7	Klo48
Digenite	Cu_9S_5	5.602$\cdots$5.605	Cla66
		5.546; 5.706	Dan44
Dihydrite (Pseudomalachite)	$Cu_5[(OH)_2PO_4]_2$	4$\cdots$4.4	Klo48
		3.58	Spe27
Dimorphite	As_4S_3	2.58; 2.60 art.	Dan44
Diopside	$CaMg[Si_2O_6]$	3.22$\cdots$3.38	Dee62
		[3.277(5)]	Wea76
–, Jadeite		3.270$\cdots$3.330	Spe27
–, Chromium		3.337	Spe27
–, Fluorine		3.236	Spe27
			continued

1.1.1 Density of minerals (continued)			
Mineral	Formula	$\varrho, [\varrho_X]$ g cm^{-3}	Ref.
Dioptase	$Cu_6[Si_6O_{18}] \cdot 6H_2O$	3.05···3.35	Hod52/53
		[3.247(10)]	Wea76
		3.53···3.65	Dee62
Disthen (Kyanite)	$Al^{[6]}Al^{[6]}[OSiO_4]$	3.282···3.593	Spe27
		3.559···3.675	Hod52/53
		[3.675(6)]	Wea76
Dixenite	$Mn_5As_2^{3+}[O_6SiO_4] \cdot H_2O$	4.20	Spe27
Djalmaite	$(U, ...)(Ta, Nb, Ti)_2O_6(O, OH)$	5.75···5.88	Dan44
Dolerophanite	$Cu_2[OSO_4]$	3.3	Klo48
		[4.171(8)]	Wea76
Dolomite	$CaMg[CO_3]_2$	2.80···2.99	Hod52/53
		[2.8661(13)]	Wea76
Domeykite	Cu_3As [Klo48]	7.2···7.9	Dan44
Douglasite	$K_2[Fe^{2+}Cl_4(H_2O)_2]$	2.16	Hod52/53
Dravite (Tourmaline)	$NaMg_3Al_6[(OH)_{1+3}(BO_3)_3Si_6O_{18}]$	3.03···3.15	Dee62
		3.027	Cla66
		[3.004(6)]	Wea76
Dufrenite	$Fe_3^{2+}Fe_6^{3+}[(OH)_3PO_4]_4$	3.3···3.5	Klo48
Dufrenoysite	$Pb_8As_8S_{20}$	5.53(3)	Dan44
Duftite	$PbCu[OHAsO_4]$	6.19	Spe27
Dumortierite	$(Al, Fe)_7[O_3BO_3(SiO_4)_3]$	3.30···3.36	Spe27
Dundasite	$PbAl_2[(OH)_2(CO_3)_2] \cdot 2H_2O$ [Klo48]	3.2	Klo48
Durangite	$NaAl[FAsO_4]$	3.9···4	Klo48
Dussertite	$BaFe_3^{3+}H[(OH)_6(AsO_4)_2]$	3.75	Spe27
Dysanalyte	$(Ca, Na, Ce)(Ti, Nb, Fe)O_3$	4.02···4.26	Hod52/53
Dyscrasite	Ag_3Sb	9.74(7)	Dan44
		9.4···10	Klo48
Eakleite, see Xonotlite			
Ecdemite	$Pb_3AsO_{<4}Cl_{<2}$	7.1	Klo48
Eckermannite	$Na_{2.5}Ca_{0.5}(Mg, Fe^{2+}, Fe^{3+}, Al)_5[(OH, F)_2$ $\cdot Al_{0.5}Si_{7.5}O_{22}]$	3.00	Dee62
Edenite	$NaCa_2Mg_5[(OH, F)_2AlSi_7O_{22}]$	3.02···3.45	Dee62
Edingtonite	$Ba[Al_2Si_3O_{10}] \cdot 3H_2O$	2.7···2.8	Dee62
Eglestonite	Hg_6Cl_4O	8.327	Hod52/53
Eichbergite	$(Cu, Fe)Sb_3Bi_3S_5(?)$	5.36	Spe27
Eis, see Ice			
Ekmanite	$(Fe^{2+}, Mg, Mn, Fe^{3+})_{<3}[(OH)_2(Si, Al)Si_3O_{10}]$ $\cdot X_n(H_2O)_2$	2.671	Spe27
Ektropite, see Caryopilite			
Eläolith, see Nepheline			
Elbaite (Tourmaline)	$Na(Li, Al)_3Al_6[(OH)_{1+3}(BO_3)_3Si_6O_{18}]$	3.03···3.10	Dee62
		[3.271(6)]	Wea76
Eleonorite, see Beraunite			
Ellsworthite (Uranpyrochlore), see also Blanstrandite	$(U, ...)_2(Nb, Ta, Ti)_2O_6(O, OH, F)$	3.608···3.758	Spe27
Elpidite	$Na_2Zr[Si_6O_{15}] \cdot 3H_2O$	2.54	Klo48
–, Titano-		2.533···2.560	Spe27
Embolite	$Ag(Cl, Br)$	5.31···5.81	Hod52/53
Emerald (Smaragd), see Beryl			
			continued

1.1.1 Density of minerals (continued)

Mineral	Formula	$\varrho, [\varrho_x]$ g cm^{-3}	Ref.
Emery	Composition of Corundum, Magnetite Haematite, Quartz, and Spinel [Hod52/53]	3.75···4.31	Hod52/53
Emmonsite	$Fe_2[Te, O_3]_3 \cdot 2H_2O$	≈4.52	Klo48
Emplectite	$Cu_2S \cdot Bi_2S_3$	6.38	Dan44
Empressite	AgTe	7.510	Dan44
Enargite	Cu_3AsS_4	4.43···4.55	Hod52/53
		[4.463(6)]	Wea76
Endeiolite	$R^{2+}Nb_2O_5(OH)_2 \cdot R^{2+}SiO_3$ [Klo48]	3.44	Klo48
Enstatite	$Mg_2[Si_2O_6]$	3.10···3.43	Hod52/53
		[3.194(5)]	Wea76
–, Clino-		[3.190(5)]	Wea76
–, Proto-		[3.101(8)]	Wea76
Epiboulangerite	≈$Pb_3Sb_2S_8$ [Dan44]	6.303	Spe27
Epichlorite		2.52	Spe27
Epidesmine	Variety of Stilbite	2.152···2.16	Spe27
Epidote	$Ca_2(Fe^{3+}, Al)Al_2[OOHSiO_4Si_2O_7]$	3.07···3.50	Hod52/53
		[3.587(15)]	Wea76
		3.38···3.49	Dee62
Epigenite	$(Cu, Fe)_5AsS_6(?)$ [Klo48]	4.5	Dan44
Epinatrolite		2.235···2.24	Spe27
Epistilbite	$Ca[Al_2Si_6O_{16}] \cdot 5H_2O$	≈2.2	Dee62
Epistolite	$(Na, Ca)_2(Nb, Ti, Mg, Fe, Mn)_2[O(OH)_2Si_2O_7]$	2.9	Klo48
Epsomite (Bittersalz)	$Mg[SO_4] \cdot 7H_2O$	[1.679(3)]	Wea76
Erdkobalt, see Esbolan			
Erikite	≈$(Na, Ca, Ce)_2(Al, Ce, La, Nd)_3$ $\cdot [OH((Si, P)O_4)_3] \cdot H_2O$	3.77	Klo48
Erinite	$Cu_5[(OH)_2AsO_4]_2$ [Klo48]	4···4.1	Klo48
Eriochalcite, see Antofagastite			
Erionite	$(Ca, Na, K)_2[Al_3Si_9O_{24}] \cdot 9H_2O$	≈2.02	Dee62
Erythrite (Kobaltblüte)	$Co_3[AsO_4]_2 \cdot 8H_2O$	3.149	Spe27
		2.912···2.948	Hod52/53
Eschwegeite, see Chlopinite			
Eskolaite	Cr_2O_3	[5.225(6)]	Wea76
Eucairite	α-$Cu_2Se \cdot Ag_2Se$	7.6···7.8	Dan44
		[7.887(24)]	Wea76
Euchroite	$Cu_2[OHAsO_4] \cdot 3H_2O$	3.3	Klo48
Euclase	$Al^{[6]}Be^{[4]}[OHSiO_4]$	3.051···3.103	Hod52/53
		[3.116(7)]	Wea76
Eucolite	Nb-Eudialyte	2.97	Spe27
Eucryptite	$LiAl[SiO_4]$	2.667	Spe27
–, Pseudo-		2.365	Spe27
Eudialyte	$(Na, Ca, Fe)_6Zr[(OH, Cl)(Si_3O_9)_2]$	2.8···3.1	Hod52/53
Eudidymite	$NaBe^{[4]}[OHSi_3O_7]$	2.55	Klo48
Eulytite	$Bi_4[SiO_4]_3$	6.1	Klo48
Euxenite	$(Y, Ce, U, Pb, Ca)(Nb, Ta, Ti)_2(O, OH)_6$	4.594···4.99	Spe27
Evansite	$Al_3[(OH)_6PO_4] \cdot 6H_2O$	1.924···1.929	Spe27
Fairfieldite	$Ca_2(Mn, Fe)[PO_4]_2 \cdot 2H_2O$	3.1	Klo48
Falkmanite	$Pb_2Sb_2S_6$ [Klo48]	6.24	Klo48
Famatinite	$Cu_3(Sb, As)S_4$ [Klo48]	4.52(5); 4.50	Dan44
		[4.687(9)]	Wea76

continued

1.1.1 Density of minerals (continued)

Mineral	Formula	$\varrho, [\varrho_X]$ g cm^{-3}	Ref.
Fassaite	Ca(Mg, Fe^{3+}, Al)[(Si, Al)$_2$O$_6$]	2.96···3.34	Dee62
Faujasite	Na$_2$Ca[Al$_2$Si$_4$O$_{12}$]$_2$·16H$_2$O	≈1.92	Dee62
Fayalite	Fe$_2$[SiO$_4$]	3.91···4.34	Hod52/53
		[4.3928(88)]	Wea76
–, Mangan-		4.32	Spe27
Ferberite	FeWO$_4$	6.801···7.109	Hod52/53
		[7.520(10)]	Wea76
Fergusonite	YNbO$_4$	5.6···5.8	Dan44
		4.7···6.2	Klo48
Fermorite		3.518	Spe27
Ferrazite	(Ba, Pb)$_3$[P$_4$O$_{13}$]·8H$_2$O(?)	3.0···3.3	Spe27
Ferriannite	KFe$_3$FeSi$_3$O$_{10}$(OH)$_2$	[3.454(3)]	Wea76
Ferrierite	(Na, K)Mg[Al$_3$Si$_{15}$O$_{36}$]·9H$_2$O	≈2.15	Dee62
Ferrimolybdite	Fe$_2^{3+}$, MoO$_4$)$_3$·7H$_2$O	2.99	Spe27
Ferrinatrite (Leucoglaugite)	Na$_3$Fe^{3+}[SO$_4$]$_3$·3H$_2$O	2.6	Spe27
		≈2.56	Klo48
Ferrisymplesite	Fe$_3^{3+}$[(OH)$_3$(AsO$_4$)$_2$]·5H$_2$O	2.885	Spe27
Ferroactinolite	Ca$_2$Fe$_5^{2+}$[(OH, F)Si$_4$O$_{11}$]$_2$	3.02···3.44	Dee62
Ferrohastingsite	NaCa$_2$Fe$_4^{2+}$(Al, Fe^{3+})[(OH, F)$_2$Al$_2$Si$_6$O$_{22}$]	3.50	Dee62
Ferropallidite (Szomolnokite, Schmöllnitzite)	Fe[SO$_4$]·H$_2$O	3.19	Klo48
Ferropicotite	Fe-Spinel	3.93	Spe27
Ferroselite	FeSe$_2$	[7.134(13)]	Wea76
Ferrotremolite	Ca$_2$Fe$_5$[Si$_8$O$_{22}$](OH)$_2$	[3.434(8)]	Wea76
Fersmanite	Na$_4$Ca$_4$Ti$_4$[(O, OH, F)$_3$SiO$_4$]$_3$(?)	3.44	Klo48
Fibroferrite	Fe^{3+}[OHSO$_4$]·5H$_2$O	1.901···2.09	Spe27
Fillowite	≈Na$_2$(Mn^{2+}, Fe^{2+}, Ca, H$_2$)$_5$[PO$_4$]$_4$	3.43	Klo48
Finnemanite	Pb$_5$Cl(AsO$_3$)$_3$	7.08···7.265	Spe27
		7.62	Klo48
Fischerite, see Wavellite			
Flagstaffite	C$_{10}$H$_{18}$(OH)$_2$·H$_2$O	1.092	Spe27
Flinkite	Mn$_2^{2+}$Mn^{3+}[(OH)$_4$AsO$_4$]	3.87	Klo48
Florencite	CeAl$_3$[(OH)$_6$(PO$_4$)$_2$]	3.58	Klo48
Fluellite (Kreuzbergite)	AlF$_3$·H$_2$O	2.17	Klo48
		2.139	Spe27
Fluoborite	Mg$_3$[(F, OH)$_3$BO$_3$]	2.89	Klo48, Spe27
Fluocerite (Tysonite)	(Ce, La)F$_3$	5.73	Spe27
		6.1	Klo48
Fluorapatite	Ca$_5$(PO$_4$)$_3$F [Cla66]	[3.2007(25)]	Wea76
Fluor-Edenite	NaCa$_2$Mg$_5$[AlSi$_7$O$_{22}$]F$_2$	[3.076(5)]	Wea76
Fluor-Humite	3Mg$_2$SiO$_4$·MgF$_2$	[3.2017(39)]	Wea76
Fluorite	CaF$_2$	2.97···3.25	Hod52/53
		[3.1792(7)]	Wea76
Fluor-Norbergite	Mg$_2$SiO$_4$·MgF$_2$	[3.194(4)]	Wea76
Fluor-Richterite	Na$_2$CaMg$_5$[Si$_8$O$_{22}$]F$_2$	[3.033(5)]	Wea76
Fluosiderite	Silicate of Ca, Mg, Al, little Fe and Mn	3.13	Spe27
Forbesite	(Ni, Co)H[AsO$_4$]·3···4H$_2$O	3.08	Klo48
Forsterite	Mg$_2$[SiO$_4$]	3.222	Dee62
		3.191···3.33	Hod52/53
		[3.2136(20)]	Wea76

continued

1.1.1 Density of minerals (continued)			
Mineral	Formula	$\varrho, [\varrho_X]$ g cm^{-3}	Ref.
Foshagite	Ca$_4$[(OH)$_2$Si$_3$O$_9$]	2.36	Spe27
Foshallasite	Ca$_3$[Si$_2$O$_7$]·3H$_2$O	2.5	Klo48
Fourmarierite	8[UO$_2$(OH)$_2$]·2Pb(OH)$_2$·4H$_2$O	6.046	Spe27
Franckeite	5PbS·3SnS$_2$·Sb$_2$S$_3$	5.90	Dan44
Franklinite	Fe$_2$ZnO$_4$	5.07···5.22	Dan44, Hod52/53
		5.34	Dee62
Freieslebenite	PbAgSbS$_3$	6.04···6.23	Dan44
Freirinite (Lavendulan)	(Ca, Na)$_2$Cu$_5$[Cl(AsO$_4$)$_4$]·4···5H$_2$O	3.3	Klo48
Friedelite	(Mn, Fe)$_8$[(OH, Cl)$_{10}$Si$_6$O$_{15}$]	3.07	Klo48
Frohbergite	FeTe$_2$	8.094(11)]	Wea76
Fuchsite	Cr-Muscovite [Klo48]	2.85	Klo48
Fueloeppite	3PbS·4Sb$_2$S$_3$	5.23	Dan44
Gadolinite	Y$_2$Fe^{2+}Be$_2$[OSiO$_4$]$_2$	4.0···4.6	Spe27
Gageite	Mn$_7$[(OH)$_6$Si$_3$O$_{10}$]	3.58	Klo48
Gahnite	Al$_2$ZnO$_4$	4.478···4.602	Hod52/53
		[4.6083(34)]	Wea76
		4.62	Dee62
Gajite	Composition of Calcite and Brucite	2.619	Spe27
Galaxite	MnAl$_2$O$_4$ [Cla66]	4.04	Dee62
		[4.078(3)]	Wea76
Galena (Bleiglanz)	PbS	[7.5973(19)]	Wea76
		7.5···7.6	Dee62
		7.30···7.55	Spe27
Galenobismutite, see also Bismutoplagionite	PbS·Bi$_2$S$_3$	7.04	Dan44
Gamagarite	Ba$_2$(Fe, Mn)$_2$[VO$_4$]$_2$·0.5H$_2$O	6.42	Klo48
Ganomalite	Pb$_6$Ca$_4$[(OH)$_2$(Si$_2$O$_7$)$_3$]	5.57···5.7	Hod52/53
Ganophyllite	(Na, K, Ca)$_{1.2}$(Mn, Fe^{3+}, Al)$_{4.7}$·[(OH)$_5$(Si, Al)$_6$O$_{15}$]·2H$_2$O	2.878	Klo48
Garnet (Granat)	X$_3$Y$_2$Z$_3$O$_{12}$ [Klo48]	3.4···4.6	Klo48
Garnierite	H$_2$(Ni, Mg)SiO$_4$ [Hod52/53]	2.27···2.87	Hod52/53
Gaylussite	Na$_2$Ca[CO$_3$]$_2$·5H$_2$O	1.93···1.95	Hod52/53
Gearksutite	CaAl(F, OH)$_5$·H$_2$O	2.710···2.768	Spe27
Gedrite	(Mg, Fe)$_{6...5}$Al$_{1...2}$[OH(Al, Si)Si$_3$O$_{11}$]$_2$	2.85···3.57	Dee62
Gehlenite, see also Velardeñite	Ca$_2$Al[(Si, Al)$_2$O$_7$]	2.9···3.07	Hod52/53
		[3.0387(30)]	Wea76
		3.038	Dee62
Geikielite	MgTiO$_3$	3.98···4.0	Hod52/53
		[3.896(8)]	Wea76
		4.05	Dee62
Genthelvite	Zn$_8$[S$_2$(BeSiO$_4$)$_6$]	3.66	Klo48
		3.44···3.70	Dee62
Geocronite	5PbS·AsSbS$_3$	6.4(1)	Dan44
Georgiadesite	Pb$_3$[Cl$_3$AsO$_4$]	7.1	Klo48
Gerhardtite	Cu$_2$[(OH)$_3$NO$_3$]	3.31···3.46	Spe27
		[3.399(4)]	Wea76
Germanite	Cu$_3$(Ge, Fe)S$_4$	4.46···4.59	Dan44
Gersdorffite	NiAsS	5.6···6.2	Klo48
		5.9 (average)	Dan44
		[5.964(3)]	Wea76
			continued

1.1.1 Density of minerals (continued)

Mineral	Formula	$\varrho, [\varrho_x]$ g cm^{-3}	Ref.
Gibbsite (Hydrargillite)	γ-Al(OH)$_3$	[2.441(1)]	Wea76
Gibbsite (crystal)		2.40(2)	Dan44
Gibbsite (massive)		2.3···2.4	Dan44
Gillespite	BaFe[Si$_4$O$_{10}$]	3.33	Spe27, Klo48
Ginorite	Ca$_2$[B$_4$O$_5$(OH)$_4$][B$_5$O$_6$(OH)$_4$]$_2 \cdot$ 2H$_2$O	2.09	Klo48
Gips, see Gypsum			
Gismondine	Ca[Al$_2$Si$_2$O$_8$] $\cdot$ 4H$_2$O	≈ 2.2	Dee62
Gladite	2PbS $\cdot$ Cu$_2$S $\cdot$ 5Bi$_2$S$_3$	6.96	Dan44
Glaserite (Aphthitalite)	K$_3$Na(SO$_4$)$_2$	2.65	Klo48
		2.662	Hod52/53
Glauberite	CaNa$_2$[SO$_4$]$_2$	2.7···2.85	Hod52/53
Glaubersalz, see Mirabilite			
Glaucocerinite	(Zn, Cu)$_{10}$Al$_4$[(OH)$_{30}$SO$_4$] $\cdot$ 2H$_2$O	2.749	Klo48
Glaucochroite	CaMn[SiO$_4$]	2.216	Spe27
		[3.441(9)]	Wea76
		3.4	Klo48
Glaucodot	(Co, Fe)AsS	6.04(12)	Dan44
		[6.161(75)]	Wea76
Glauconite	(K, Na, Ca)$_{<1}$(Al, Fe^{3+}, Fe^{2+}, Mg)$_2$ $\cdot$ [(OH)$_2$Al$_{0.35}$Si$_{3.65}$O$_{10}$]	2.4···2.95	Dee62
Glaucophane	Na$_2$Mg$_3$Al$_2$[(OH, F)Si$_4$O$_{11}$]$_2$	3.08···3.30	Dee62
		2.991···3.15	Hod52/53
		[2.906]	Wea76
Glinkite	Fe-Olivine	3.463	Spe27
Gmelinite	(Na$_2$, Ca)[Al$_2$Si$_4$O$_{12}$] $\cdot$ 6H$_2$O	2.045···2.135	Spe27
		≈ 2.1	Dee62
Goethite (Nadeleisenerz)	α-FeOOH	[4.269(8)]	Wea76
		4.0···4.4	Hod52/53
Goethite (massive)		3.3···4.3	Dan44
Goethite (crystal)		4.28(1)	Dan44
		3.8···4.3	Klo48
Gold	Au	[19.282(3)]	Wea76
– amalgam		15.47	Dan44
Goldmanite	Ca$_3$V$_2$Si$_3$O$_{12}$	[3.765(5)]	Wea76
Gonnardite	(Ca, Na)$_3$[(Al, Si)$_5$O$_{10}$]$_2 \cdot$ 6H$_2$O	2.3	Klo48, Dee62
Gorceixite	BaAl$_3$H[(OH)$_6$(PO$_4$)$_2$]	3.1	Klo48
Goongarrite, see Warthaite			
Goslarite, see Zinkvitriol			
Graftonite	(Fe^{2+}, Mn, Ca, Mg)$_3$[PO$_4$]$_2$	3.67	Klo48
Grammatite, see Tremolite			
Granat, see Garnet			
Grandidierite	(Mg, Fe)Al$_3$[OBO$_4$SiO$_4$]	3.0	Klo48
Graphite	C	2.09···2.25	Hod52/53
		[2.2670(4)]	Wea76
Gratonite	9PbS $\cdot$ 2As$_2$S$_3$	6.22(2)	Dan44
Greenalite	(Fe^{2+}, Fe^{3+})$_{<6}$[(OH)$_8$Si$_4$O$_{10}$]	≈ 3.2	Klo48, Dee62
Greenockite	β-CdS	4.9; 4.820 art.	Dan44
		[4.8261(24)]	Wea76
Greigite	Fe$_3$S$_4$	[4.079(3)]	Wea76

continued

1.1.1 Density of minerals (continued)

Mineral	Formula	$\varrho, [\varrho_x]$ g cm^{-3}	Ref.
Griffithite	Variety of Saponite	2.309	Spe27
Griphite	$(Mn, Na, Ca)_3(Al, Mn)_2[PO_3(OH, F)]_3$	3.39	Klo48
Grodnolite	Variety of Kollophan with Kaoline	2.974	Spe27
Grossularite	$Ca_3Al_2[SiO_4]_3$	3.226···4.452	Spe27
		[3.595(1)]	Wea76
Groothin		3.79···3.90	Spe27
Gruenlingite	Bi_4S_3Te [Klo48]	8.08	Dan44
		8.15	Klo48
Grunerite	$(Fe, Mg)_7[OHSi_4O_{11}]_2$	3.59(10)	Cla66
		3.10···3.60	Dee62
		[3.603(6)]	Wea76
Guanajuatite	$Bi_2(Se, S)_3$	6.25···6.98	Dan44
Guarinite	$Ca_2NaZr[(F, O)_2Si_2O_7]$	3.31	Spe27
Gudmundite	FeSbS	6.72	Dan44
		[6.987(59)]	Wea76
Gumminite		5.08	Spe27
Gummite	$(Pb, Ca, Ba)SiU_3O_{12} \cdot 5H_2O(?)$ [Hod52/53]	3.9···5.16	Hod52/53
Gymnite	$Mg_4Si_3O_{10}$ + aq. [Klo48]	2.0···2.3	Klo48
Gypsum (Gips)	$Ca[SO_4] \cdot 2H_2O$	2.314···2.328	Hod52/53
		[2.305(7)]	Wea76
Gyrolite (Centrallasite)	$Ca_2[Si_4O_{10}] \cdot 4H_2O$	2.35···2.40	Spe27
		2.5	Klo48
Hafnia, see Hafnium dioxide			
Hafnium dioxide (Hafnia)	HfO_2 [Cla66]	[10.108(4)]	Wea76
Hainite	Na–Ca–Ti–Zr-Silicate	3.18	Klo48
Halite (Chlornatrium, Steinsalz, Rock salt)	NaCl	2.135···2.170	Hod52/53
		[2.1634(3)]	Wea76
		2.16···2.17	Dee62
Halloysite (Lithomarge)	$Al_4[(OH)_8Si_4O_{10}](H_2O)_4$	2.1···2.2	Klo48
		2.44···2.714	Spe27
Halotrichite	$Fe^{2+}Al_2[SO_4]_4 \cdot 22H_2O$	1.807···1.899	Spe27
Hambergite	$Be_2[(OH, F)BO_3]$	2.347···2.36	Hod52/53
		[2.3663(6)]	Wea76
Hamlinite	$SrAl_3H[(OH)_6(PO_4)_2]$	3.2	Klo48
Hancockite	$(Ca, Pb, Sr, Mn)_2(Al, Fe, Mn)_3$ $\cdot [OOHSiO_4Si_2O_7]$	4	Klo48
Hanksite	$KNa_{22}[Cl(CO_3)_2(SO_4)_9]$	2.55···2.6	Klo48
		2.562	Hod52/53
Hardystonite	$Ca_2Zn[Si_2O_7]$	3.4	Klo48
		[3.357(29)]	Wea76
Harmotome (Harmotomite)	$Ba[Al_2Si_6O_{16}] \cdot 6H_2O$	2.345···2.50	Hod52/53
		2.41···2.47	Dee62
Harmotomite, see Harmotome			
Harstigite	$MnCa_6Be_4[OOHSi_3O_{10}]_2$	3.0	Klo48
Hastingsite	$NaCa_2Fe_4^{2+}(Al, Fe^{3+})[(OH, F)_2Al_2Si_6O_{22}]$	3.16···3.426	Spe27
Hatchettolite		4.417···4.509	Spe27
Hauchecornite	$(Ni, Co)_9(Bi, Sb)_2S_8(?)$	6.4	Klo48
Hauerite	MnS_2	3.463	Dan44
		[3.4816(10)]	Wea76

continued

1.1.1 Density of minerals (continued)

Mineral	Formula	$\varrho, [\varrho_x]$ g cm^{-3}	Ref.
Hausmannite	Mn_2MnO_4	4.722···4.856	Hod52/53
		[4.873(7)]	Wea76
Hauyne	$(Na, Ca)_{8···4}[(SO_4)_{2···1}(AlSiO_4)_6]$	2.44···2.50	Dee62
Hawleyite	α-CdS	[4.835(5)]	Wea76
Heazlewoodite	Ni_3S_2	[5.867(3)]	Wea76
Hedenbergite	$CaFe[Si_2O_6]$	3.5(1)	Cla66
		[3.632(10)]	Wea76
		3.50···3.56	Dee62
Hedleyite	$Bi_{14}Te_6$	8.93	Klo48
Hellandite	$Ca_3(Y, Yb ...)_4B_4[O(OH)_2Si_2O_7]_3$	3.70	Klo48
Helvite	$(Mn, Fe, Zn)_8[S_2(BeSiO_4)_6]$	3.20···3.44	Dee62
Hematite	Fe_2O_3	4.9···5.3	Hod52/53
		[5.2749(21)]	Wea76
	α-Fe_2O_3	5.256	Dee62
Hematophanite	$4 PbO \cdot Pb(Cl, OH)_2 \cdot 2 Fe_2O_3$	7.70	Dan44
Hemimorphite (Kalamin)	$Zn_4[(OH)_2Si_2O_7] \cdot H_2O$	3.3···3.5	Klo48
		[3.482(4)]	Wea76
Hercynite	Al_2FeO_4	3.91···3.95	Hod52/53
		[4.265(5)]	Wea76
		4.40	Dee62
Herderite	$CaBe[(F, OH)PO_4]$	2.952···3.012	Hod52/53
Herzenbergite	SnS	[5.197(3)]	Wea76
Hessite	Ag_2Te	[8.405]	Wea76
Hessite (isometric)		8.24···8.45	Dan44
Hessonite (Fe-Grossularite)	$3 CaO \cdot (Al, Fe)_2O_3 \cdot 3 SiO_2$ [Hod52/53]	3.4···3.6	Hod52/53
		[3.595]	Wea76
Hetaerolite	Mn_2ZnO_4	4.6···4.85	Spe27
		5.18	Dan44
Heterobrochantite, see Antlerite			
Heterogenite	CoOOH	3.128	Spe27
Heteromorphite	$11 PbS \cdot 6 Sb_2S_3(?)$	5.73	Dan44
Heterosite	$(Fe^{3+}, Mn^{3+})[PO_4]$	3.4	Klo48
Heulandite (Clinoptilolite)	$Ca[Al_2Si_7O_{18}] \cdot 6 H_2O$	2.16···2.249	Hod52/53
		2.1···2.2	Dee62
Hewettite	$CaV_6O_{16} \cdot 9 H_2O$	2.554	Spe27
Hexahydrite	$Mg[SO_4] \cdot 6 H_2O$	1.757	Spe27
		[1.732(15)]	Wea76
Hibbenite	$Zn_7[OH(PO_4)_2]_2 \cdot 6 H_2O$	3.213	Spe27
Hibschite, see Plazolite			
Higginsite	$CuCa[OHAsO_4]$	4.33	Spe27, Klo48
Hilgardite	$Ca_2[ClB_5O_8(OH)_2]$	2.71	Klo48
Hillebrandite	$Ca_2[SiO_4] \cdot H_2O$	2.7	Klo48
Hinsdalite	$PbAl_3[(OH)_6PO_4SO_4]$	3.65	Spe27
Hintzeite, see Kaliborite			
Hisingerite	Fe_2O_3-Silicate [Klo48]	2.50	Spe27
		2.6···3	Klo48
Hjelmite	Composition of Tapiolite and Pyrochlore	5.2···5.8	Klo48
Hochschildtite	$\approx PbSnO_3 \cdot 5···6 H_2O$	4.4···4.6	Klo48
Hodgkinsonite	$Mn^{[6]}Zn_2^{[4]}[(OH)_2SiO_4]$	3.91	Spe27, Klo48

continued

1.1.1 Density of minerals (continued)

Mineral	Formula	$\varrho, [\varrho_x]$ g cm^{-3}	Ref.
Hoegbomite	Na$_x$(Al, Fe, Ti)$_{24-x}$O$_{36-x}$	3.81	Dan44
Hoelite		1.43	Spe27
Hoernesite	Mg$_3$[AsO$_4$]$_2 \cdot$ 8 H$_2$O	2.7	Klo48
		2.57	Spe27
Holdenite	(Mn, Ca)$_4$(Zn, Mg, Fe)$_2$[(OH)$_5$O$_2$AsO$_4$]	4	Klo48
Hollandite	Ba$_{\leq 2}$Mn$_8$O$_{16}$	4.95	Dan44
Holmquistite	Li$_2$Mg$_3$Al$_2$[OHSi$_4$O$_{11}$]$_2$	3.06$\cdots$3.13	Dee62
Hopeite	Zn$_3$[PO$_4$]$_2 \cdot$ 4 H$_2$O	3.03	Spe27, Hod52/53
Hornblende	(Ca, Na, K)$_{2\cdots 3}$(Mg, Fe^{2+}, Fe^{3+}, Al)$_5$ $\cdot$ [(OH, F)$_2$(Si, Al)$_2$Si$_6$O$_{22}$]	3.0$\cdots$3.5 3.02$\cdots$3.45	Hod52/53 Dee62
Hornblende, basaltic	Ca$_2$(Na, K)$_{0.5\cdots 1.0}$(Mg, Fe^{2+})$_{3\cdots 4}$(Fe^{3+}, Al)$_{2\cdots 1}$ $\cdot$ [(O, OH, F)$_2$Al$_2$Si$_6$O$_{22}$]	3.19$\cdots$3.30	Dee62
Horsfordite	Cu$_6$Sb	8.812	Dan44
Hortonolite	(Fe, Mg)$_2$[SiO$_4$]	3.98	Spe27
Howlite	Ca$_2$[(BOOH)$_5$SiO$_4$]	2.58	Klo48
Huebnerite	MnWO$_4$	7.2$\cdots$7.5 [7.228(10)]	Hod52/53 Wea76
Hulsite	(Fe^{2+}, Mg^{2+}, Fe^{3+}, Sn^{4+})$_3$[O$_2$BO$_3$]	4.31	Spe27
Humboldtilite		2.92$\cdots$2.975	Spe27
Humboldtine, see Oxalite			
Humite	Mg$_7$[(OH, F)$_2$(SiO$_4$)$_3$]	3.20$\cdots$3.32	Dee62
Huntite	CaMg$_3$[CO$_3$]$_4$	[2.880(2)] 2.696	Wea76 Dee62
Huréaulite	(Mn, Fe^{2+})$_5$H$_2$[PO$_4$]$_4 \cdot$ 4 H$_2$O	3.2	Klo48
Huronite		2.819	Spe27
Hutchinsonite	PbTlAs$_5$S$_9$	4.6	Dan44, Hod52/53
Hyalophane	(K, Ba)[Al(Al, Si)Si$_2$O$_8$]	2.90 2.58$\cdots$2.82	Spe27 Dee62
Hyalotekite	(Pb, Ca, Ba)$_4$B[Si$_6$O$_{17}$(F, OH)]	3.81	Klo48
Hydrargillite, see Gibbsite			
Hydrocalumite	2 Ca(OH)$_2 \cdot$ Al(OH)$_3 \cdot$ 3 H$_2$O	2.15	Dan44
Hydrocerussite	Pb$_3$[OHCO$_3$]$_2$	6.02$\cdots$6.80	Spe27
Hydrogiobertite	Hydromagnesite group [Dan44]	2.152	Spe27
Hydrogrossular	Ca$_3$Al$_2$[(Si, H$_4$)O$_4$]$_3$	3.594$\cdots$3.13	Dee62
Hydrohematite (Turgite)	Fe$_2$O$_3 \cdot$ 0.5 H$_2$O [Klo48]	4.33$\cdots$4.7 4.29$\cdots$5.00	Spe27 Hod52/53
Hydrohetaerolite	(Mn, H$_3$)$_2$ZnO$_4$	4.6?	Dan44
Hydromagnesite	Mg$_5$[OH(CO$_3$)$_2$]$_2 \cdot$ 4 H$_2$O	2.152$\cdots$2.16	Spe27
Hydronephelite	HNa$_2$Al$_3$Si$_3$O$_{12} \cdot$ 3 H$_2$O [Hod52/53]	2.263$\cdots$2.48	Hod52/53
Hydrophlogopite	(K, H$_2$O)Mg$_3$[(OH, H$_2$O)$_2$AlSi$_3$O$_{10}$]	2.783	Spe27
Hydroromeite, see also Stibiconite and Volgerite	SbSb$_2$O$_6$OH	3.50	Klo48
Hydrotalcite	Mg$_6$Al$_2$[(OH)$_{16}$CO$_3$] $\cdot$ 4 H$_2$O	2.04$\cdots$2.091	Hod52/53
Hydrotenorite	CuO with $\approx$ 10% H$_2$O	4.15	Klo48
Hydrotungstite	WO$_2$(OH)$_2 \cdot$ H$_2$O	4.6	Klo48
Hydroxyl-Apatite	Ca$_5$[OH(PO$_4$)$_3$]	[3.155(4)]	Wea76
Hydrozincite	Zn$_5$[(OH)$_3$CO$_3$]$_2$	3.2$\cdots$3.8	Klo48
Hypersthene	(Mg, Fe)$_2$[Si$_2$O$_6$]	3.36$\cdots$3.415 3.5	Spe27 Klo48

continued

1.1.1 Density of minerals (continued)

Mineral	Formula	$\varrho, [\varrho_x]$ g cm^{-3}	Ref.
Ice (Eis)	H_2O	0.9174	Cla66
Iddingsite		2.54···2.80	Spe27
Idrialite, see Curtisite			
Idokras, see Vesuvianite			
Illite	$K_{1...1.5}Al_4[Si_{7...6.5}Al_{1...1.5}O_{20}](OH)_4$ [Dee62]	2.6···2.9	Dee62
Ilmenite	$FeTiO_3$	4.44···4.90	Hod52/53
		[4.788(12)]	Wea76
		4.70···4.78	Dee62
Ilmenorutile	Composition of Rutile:Mossite (5:1)	5.18	Spe27
Ilvaite	$CaFe^{2+}Fe^{3+}[OHOSi_2O_7]$	3.95···4.00	Spe27
		4.1	Klo48
Imerinite, see Richterite			
Inderite	$Mg[B_3O_3(OH)_5] \cdot 5H_2O$	1.85	Klo48
Inesite (Rhodotilith)	$Ca_2Mn_7[Si_5O_{14}OH]_2 \cdot 5H_2O$	3.029···3.03	Spe27
		3.1	Klo48
Inyoite	$Ca[B_3O_3(OH)_5] \cdot 4H_2O$	1.87	Klo48
Iodargyrite (Iodyrite)	β-AgI	5.60···5.707	Hod52/53
		[5.683(4)]	Wea76
Iodyrite, see Iodargyrite			
Iridium	Ir	22.6···22.8	Klo48
Iridosmine	(Os, Ir …)	19···21	Klo48
Iron	α-Fe	7.3···7.87	Dan44
		[7.8748(41)]	Wea76
Ishikawaite		6.2···6.4	Spe27
Ixiolite	$(Ta, Nb, Sn, Mn, Fe)_2O_4$	7.0···7.3	Dan44
		7.88	Spe27
Jacobsite	Fe_2MnO_4	[4.990(4)]	Wea76
		4.87	Dee62
Jadeite	$NaAl[Si_2O_6]$	[3.347(6)]	Wea76
		3.24···3.43	Dee62
Jamesonite	$4PbS \cdot FeS \cdot 3Sb_2S_3$	5.5···5.7	Klo48
		5.63	Dan44
Janite	$(Ca, Na, K)(Fe, Al, Mg)[Si_2O_6] \cdot 2H_2O$	2.32	Klo48
Jarlite	$NaSr_2[AlF_6] \cdot [AlF_5]H_2O$	3.93	Klo48
Jarosite	$KFe_3^{3+}[(OH)_6(SO_4)_2]$	3.15···3.26	Hod52/53
Jefferisite		2.38	Spe27
Jeremejevite	$AlBO_3$	3.3	Klo48
Jezekite (Morinite)	$Ca_2NaAl_2[(F, OH)_5(PO_4)_2] \cdot 2H_2O$	2.94	Klo48
Joaquinite	$CaNa(Ti, Fe)_3[Si_4O_{15}]$	3.89	Klo48
Johannsenite	$CaMn[Si_2O_6]$	3.6	Klo48
		[3.629(23)]	Wea76
		3.44···3.55	Dee62
Johnstrupite (Mosandrite)	$(Ca, Na, Y)_3(Ti, Zr, Ce)[(F, OH, O)_2Si_2O_7]$	3.17	Spe27
		3.4	Klo48
Jordanite	$5PbS \cdot As_2S_3$	6.4	Klo48
		6.451	Spe27
Joséite	$(A)Bi_4(Te, S)_3(?)$	7.688···7.793	Spe27
		8.10	Klo48
		8.18	Dan44

continued

Mineral	Formula		$\varrho, [\varrho_x]$ g cm^{-3}	Ref.
1.1.1 Density of minerals (continued)				
Jurupaite, see Xonotlite				
Justite	$Mg_5Al_2(OH)_{12}Cl_4$	[Dan44]	2.98···3.24	Spe27
Kaemmererite	$(Mg, Cr)_{<3}[(OH)_2AlSi_3O_{10}]Mg_3(OH)_6$		2.59···2.67	Spe27
Kaersutite	$Ca_2(Na, K)(Mg, Fe^{2+}, Fe^{3+})_4Ti$ $\cdot [(O, OH, F)_2Al_2Si_6O_{22}]$		3.2···3.28	Dee62
Kainite	$KMg[ClSO_4]\cdot 2.75 H_2O$		2.132	Spe27
			2.067···2.188	Hod52/53
Kainosite	$Ca_2Y_2[CO_3Si_4O_{12}]\cdot H_2O$		3.5	Klo48
Kalamin, see Hemimorphite				
Kalialaun, see Kalinite				
Kaliborite (Hintzeite)	$KMg_2H[B_6O_8(OH)_5]_2\cdot 4H_2O$		2.1	Klo48
Kalinite (Kalialaun)	$KAl(SO_4)_2\cdot 11 H_2O(?)$		1.7···1.8	Klo48
Kaliophilite	$K[AlSiO_4]$		2.49···2.67	Hod52/53
			[2.650(2)]	Wea76
Kalkowskyn	$\approx Fe_2Ti_4O_{11}$		4.01(3)	Dan44
Kalsilite (Kaliophilite)	$K[AlSiO_4]$		2.59	Klo48
			[2.618(2)]	Wea76
			2.59···2.625	Dee62
Kamarezite, see Brochantite				
Kaolinite	$Al_4[(OH)_8Si_4O_{10}]$		2.32···2.59	Spe27
			[2.594(7)]	Wea76
			2.61···2.68	Dee62
Karelianite	V_2O_3		[5.0216(54)]	Wea76
Kasolite	$Pb_2[UO_2SiO_4]_2\cdot 2H_2O$		5.962	Spe27
			6.5	Klo48
Katangite	Variety of Chrysocolla		2.4	Spe27
Katophorite	$Na_2CaFe_4^{2+}(Fe^{3+}, Al)[(OH, F)_2AlSi_7O_{22}]$		3.20···3.50	Dee62
Katoptrite	$Mn_{14}Sb_2(Al, Fe)_4[O_{21}(SiO_4)_2]$		4.5	Spe27
Keatite	SiO_2		[2.503(3)]	Wea76
Keeleyite, see Zinckenite				
Kempite	$Mn_2(OH)_3Cl$		2.94	Spe27
Kentrolite	$Pb_2Mn_2'''[O_2Si_2O_7]$		6.2	Klo48
Kermesite	Sb_2S_2O		4.68	Dan44
			4.5	Klo48
Kernite	$Na_2[B_4O_6(OH)_2]\cdot 3H_2O$		1.935	Spe27
			[1.9038(32)]	Wea76
Kieserite	$Mg[SO_4]\cdot H_2O$		2.57	Hod52/53
Kirschsteinite	$CaFe[SiO_4]$		[3.564(7)]	Wea76
Klaprothite	$Cu_6Bi_4S_9$	[Klo48]	6.01	Dan44
			6.4	Klo48
Kleinite	$[Hg_2N](Cl, SO_4)\cdot xH_2O$		7.975···7.987	Spe27
			8.0	Klo48
			5	Klo48, Dan44
Klockmannite	$CuSe$		[6.122(20)]	Wea76
Knebelite	$(Mn, Fe)_2[SiO_4]$		3.96···4.25	Dee62
			[4.249(12)]	Wea76
Knopite	Perovskite with 4···5% Cerium		4.2	Klo48
Kobaltblüte, see Erythrite				
Kobaltglanz, see Cobaltite				
Kobaltspat, see Cobalticalcite				

continued

1.1.1 Density of minerals (continued)

Mineral	Formula	$\varrho, [\varrho_X]$ g cm^{-3}	Ref.
Kobellite	$5\,PbS \cdot 4(Bi, Sb)_2S_3$	6.535	Spe27
		6.334	Dan44
Kochite	$Al_4[SiO_4]_3 \cdot 5\,H_2O$ (?)	2.927···2.932	Spe27
Koenenite	$2\,MgCl_2 \cdot 3\,MgO \cdot Al_2O_3 \cdot 6$ (or 8)H_2O [Klo48]	1.98	Klo48
Kolbeckite	$Sc[PO_4] \cdot 2\,H_2O$	2.39	Klo48
Kollophan	CO_3 and Fluor-Apatite	2.6···2.9	Spe27
Kolskite	$3\,MgO \cdot 2\,SiO_2 \cdot 2\,H_2O$	2.4	Klo48
Koninckite	$Fe^{3+}[PO_4] \cdot 3\,H_2O$	3.08	Klo48
Kornelite	$Fe_2^{3+}[SO_4]_3 \cdot 7.5\,H_2O$	2.306	Spe27
Kornerupine, see Prismatine			
Korund, see Corundum			
Kotoite	$Mg_3[BO_3]_2$	3.11	Klo48
Krausite	$KFe^{3+}[SO_4]_2 \cdot H_2O$	2.840	Klo48
Krennerite	$(Au, Ag)Te_2$	8.4	Klo48
		8.62	Dan44
Kreuzbergite, see Fluellite			
Kupfer, see Copper			
Kupfervitriol, see Chalcanthite			
Kyanite, see Disthen			
Labradorite	$NaAlSi_3O_8 : CaAl_2Si_2O_8$, 1:1···1:3	2.686···2.718	Spe27
Lacroixite	$Na_4Ca_2Al_3[(OH, F)_8(PO_4)_3]$	3.126	Spe27
Lamprophyllite	$Na_3Sr_2Ti_3[(O, OH, F)_2Si_2O_7]_2$	3.44	Klo48
Lanarkite	$Pb_2[OSO_4]$	6.3···6.8	Hod52/53
Landsbergite, see Moschellandsbergite			
Langbanite	$Mn^{2+}Mn_6^{4+}[O_8SiO_4]$	4.9	Klo48
Langbeinite	$K_2Mg_2[SO_4]_3$	2.8	Klo48
Lanthanite	$(La, Dy, Ce)_2[CO_3]_3 \cdot 8\,H_2O$	2.6···2.74	Hod52/53
Lapislazuli, see Lasurite			
Lapparentite	$Al[OHSO_4] \cdot 4.5\,H_2O$ (?)	1.9···2.0	Klo48
Larnite	$\beta\text{-}Ca_2[SiO_4]$	[3.338(17)]	Wea76
Larsenite	$PbZn[SiO_4]$	5.9	Klo48
Lasurite (Lapislazuli)	$(Na, Ca)_8[(SO_4, S, Cl)_2(AlSiO_4)_6]$	2.38···2.45	Hod52/53
Laubanite	$Ca_2Al_2Si_5O_{15} \cdot 6\,H_2O$ [Hod52/53]	2.23	Hod52/53
Laumontite (Leonhardite)	$Ca[AlSi_2O_6]_2 \cdot 4\,H_2O$	2.2···2.3	Dee62
		2.23···2.42	Hod52/53
Laurionite	$PbOHCl$	6.24	Hod52/53
Laurite	RuS_2	6···6.99	Dan44
		[6.248(67)]	Wea76
Lautarite	$Ca[IO_3]_2$	4.6	Klo48
Lautite	$CuAsS$	4.53···4.96	Klo48
		4.9(1)	Dan44
Lavendulan, see Freirinite			
Lawrencite	$FeCl_2$	[3.212(17)]	Wea76
Lawsonite	$CaAl_2[(OH)_2Si_2O_7] \cdot H_2O$	[3.101(4)]	Wea76
		3.05···3.10	Dee62
Lazulite	$(Mg, Fe^{2+})Al_2[OHPO_4]_2$	2.958	Spe27
		3.057···3.122	Hod52/53
Lead (Blei)	Pb	11.273	Spe27
		[11.342(3)]	Wea76

continued

1.1.1 Density of minerals (continued)

Mineral	Formula		$\varrho, [\varrho_x]$ g cm^{-3}	Ref.
Leadhillite	$Pb_4[(OH)_2SO_4(CO_3)_2]$		6.45···6.55	Klo48
			6.26···6.44	Hod52/53
Lechatelierite	SiO_2		2.20	Klo48
Leifite	$Na_2[(F, OH, H_2O)_{1...2}(Al, Si)Si_5O_{12}]$		2.565···2.578	Spe27
Leightonite	$K_2Ca_2Cu[SO_4]_4 \cdot 2H_2O$		2.95	Klo48
Lengenbachite	$6PbS \cdot (Ag, Cu)_2S \cdot 2As_2S_3$		5.80···5.85	Dan44
Leonhardite, see Laumontite				
Leonhardtite	$MgSO_4 \cdot 4H_2O$		[2.0071(25)]	Wea76
Leonite	$K_2Mg[SO_4]_2 \cdot 4H_2O$		2.201	Spe27
			2.23	Klo48
Lepidocrocite	γ-FeOOH		[3.973(11)]	Wea76
			4.09	Dee62
Lepidolite	$KLi_2Al[(F, OH)_2Si_4O_{10}]$		2.80···2.90	Dee62
			[2.698(65)]	Wea76
Lepidomelan (Annite)	$KFe_3[(OH)_2AlSi_3O_{10}]$		3.151···3.294	Spe27
			[3.318(9)]	Wea76
Lessingite	$(Ca, Ce, La, Nd)_5[(O, OH, F)(SiO_4)_3]$		4.69	Klo48
Leuchtenbergite	Variety of Clinochlore		2.648···2.735	Spe27
Leucite	$K[AlSi_2O_6]$		2.45···2.51	Spe27, Klo48, Hod52/53
			[2.469(1)]	Wea76
			2.47···2.50	Dee62
–, Fe-	$K[FeSi_2O_6]$		[2.695(1)]	Wea76
Leucoglaugite, see Ferrinatrite				
Leucophoenicite	$Mn_7[(OH)_2(SiO_4)_3]$		3.8	Klo48
Levyne	$Ca[Al_5Si_4O_{12}] \cdot 6H_2O$		≈2.1	Dee62
Lewisite	$5CaO \cdot 2TiO_5 \cdot 3Sb_3O_5$	[Hod52/53]	4.950	Hod52/53
Libethenite	$Cu_2[OHPO_4]$		≈3.8	Klo48
Liebigite, see Uranothallite				
Lillianite	$3PbS \cdot Bi_2S_3$		7.0···7.2	Dan44
			7.0	Klo48
Lime	CaO		[3.3453(10)]	Wea76
Limnite	$Fe_2O_3 \cdot 3H_2O$	[Dan44]	2.8	Spe27
Limonite	$FeOOH \cdot nH_2O$	[Dan44]	2.7···4.3	Dee62, Dan44
			3.6···4.0	Hod52/53
Linarite	$PbCu[(OH)_2SO_4]$		5.23	Spe27
			5.3···5.5	Klo48
Lindgrenite	$Cu_3[OHMoO_4]_2$		4.26	Klo48
Lindstroemite	$2PbS \cdot Cu_2S \cdot 3Bi_2S_3$		7.0	Klo48
			7.01	Dan44, Spe27
Linnaeite	Co_3S_4		4.82···4.85	Spe27
			4.5···4.8	Klo48
			[4.8772(16)]	Wea76
Liroconite	$Cu_2Al[(OH)_4AsO_4] \cdot 4H_2O$		≈2.9	Klo48
Litharge	PbO		9.13	Hod52/53
			[9.334(20)]	Wea76
Lithidionite (Litidionite)	$(K, Na)_2Cu[Si_3O_7]_2$		2.56	Spe27
Lithionite			2.972	Spe27
Lithiophilite	$Li(Mn^{2+}, Fe^{2+})[PO_4]$		3.4···3.6	Klo48
			3.42···3.56	Hod52/53
				continued

1.1.1 Density of minerals (continued)

Mineral	Formula	$\varrho, [\varrho_x]$ g cm^{-3}	Ref.
Lithomarge, see Halloysite			
Litidionite, see Lithidionite			
Liveingite	$Pb_9As_{13}S_{28}$	5.3	Dan44
Livingstonite	$HgSb_4S_8$	4.8	Klo48
		5.00	Dan44
Lizardite	$(Mg_{5.494}Fe^{2+}_{0.033}Fe^{3+}_{0.288})_{\Sigma 5.815}$	≈ 2.55	Dee62
	$\cdot [(OH)_{8.283}Al_{0.154}Si_{3.834}O_{10}]$	7.22…7.23	Spe27
Loellingite	$FeAs_2$	7.40(1)	Dan44
		[7.477(5)]	Wea76
Loeweite	$Na_{12}Mg_7[SO_4]_{13} \cdot 15 H_2O$	2.374	Spe27
Loparite	$(Na, Ce, Ca)TiO_3$	4.73…4.77	Spe27
Lorandite	$TlAsS_2$	5.53	Dan44
Lorenzenite	$Na_2(Ti, Zr)_2[O(SiO_4)_2]$ [Klo48]	3.4	Klo48
Lorettoite, see also Chubuttite	$PbCl_2 \cdot 6 PbO$	7.39…7.65	Spe27
		7.6	Klo48
Loseyite	$(Mn, Zn)_7[(OH)_5CO_3]_2$	3.27	Klo48
Lovchorrite	metamict Rinkolite	3.32	Spe27
Lovozerite	$Na_2Zr[Si_6O_{12}(OH)_6] \cdot 0.5 NaOH$	2.38	Klo48
Lubeckite	$4CuO \cdot 0.5 Co_2O_3 \cdot Mn_2O_3 \cdot 4H_2O$	4.8	Spe27
Ludlamite	$Fe^{2+}_3[PO_4]_2 \cdot 4H_2O$	3.19	Spe27
Ludwigite	$(Mg, Fe^{2+})_2Fe^{3+}[O_2BO_3]$	4.0	Klo48
Luzonite	Cu_3AsS_4	[4.478(9)]	Wea76
Mackayite	$Fe_2(TeO_3)_3 \cdot x H_2O$	4.86	Klo48
Mackensite	$\approx (Fe^{2+}, Fe^{3+})_6[(OH)_8Fe_{2\ldots 1}Si_{2\ldots 3}O_{10}]$ [Klo48]	4.89	Spe27
Maghemite	$\gamma\text{-}Fe_2O_3$	4.88	Dee62
Magnalite	partly Montmorillonite, partly Saponite	2.34	Spe27
Magnesiaalaun, see Pickeringite			
Magnesiochromite (Picrochromite)	Cr_2MgO_4	4.43	Dee62
		[4.415(5)]	Wea76
Magnesioferrite	$(Fe^{3+}Mg)Fe^{3+}O_4$	4.52	Dee62
		4.56…4.65	Dan44
Magnesiokatophorite	$Na_2CaMg_4(Fe^{3+}, Al)[(OH, F)_2AlSi_7O_{22}]$	3.20…3.50	Dee62
Magnesite	$MgCO_3$	2.95…3.2	Hod52/53
		2.98	Dee62
		3.0095(14)	Wea76
Magnetite	$(Fe^{3+}Fe^{2+})Fe^{3+}O_4$	4.967…5.180	Hod52/53
		[5.2003(9)]	Wea76
Magnetkies and Magnetkies-Troilite, see Pyrrhotite			
Magnetoplumbite	$PbO \cdot 6 Fe_2O_3$	5.517	Spe27, Dan44
Malachite	$Cu_2[(OH)_2CO_3]$	3.90…4.03	Hod52/53
		[4.030(6)]	Wea76
Maldonite	Au_2Bi	15.46 art.	Dan44
		[15.891(12)]	Wea76
Manasseite	$Mg_6Al_2[(OH)_{16}CO_3] \cdot 4H_2O$	2.05(5)	Dan44
Manganite	$\gamma\text{-}MnOOH$	4.2…4.4	Hod52/53
Manganolangbeinite	$K_2Mn_2(SO_4)_3$	3.02…3.03	Spe27
Manganophyllite	Mn-Biotite, up to 18% MnO	2.743…2.954	Spe27

continued

1.1.1 Density of minerals (continued)			
Mineral	Formula	$\varrho, [\varrho_x]$ g cm^{-3}	Ref.
Manganosite	MnO	5.364; 5.0···5.4 art. [5.3653(18)] 5.18	Dan44 Wea76 Hod52/53
Marcasite	FeS$_2$	4.609···4.887 4.61···4.90 [4.8813(36)]	Spe27 Hod52/53 Wea76
Margarite	CaAl$_2$[(OH)$_2$Al$_2$Si$_2$O$_{10}$]	3···3.1 [2.975(17)]	Dee62 Wea76
Margarosanite	Pb(Ca, Mn)$_2$[Si$_3$O$_9$]	3.991···4.39	Spe27
Marialite	Na$_8$[(Cl$_2$, SO$_4$,CO$_3$)(AlSi$_3$O$_8$)$_6$]	2.50···2.692 2.50···2.62 [2.566(4)]	Spe27, Hod52/53 Dee62 Wea76
Marshite	CuI	5.68 [5.710(3)] 5.59···5.62	Klo48 Wea76 Hod52/53
Mascagnite	(NH$_4$)$_2$[SO$_4$]	1.76···1.77 [1.7693(20)]	Hod52/53 Wea76
Massicot	β-PbO	8.61 9.56 art. [9.641(12)]	Spe27 Dan44 Wea76
Matildite	AgBiS$_2$	7.07 6.9	Spe27 Dan44
Matlockite	PbFCl	7.21 [7.111]	Hod52/53, Klo48 Wea76
Maucherite	Ni$_{<3}$As$_2$	7.73···7.901 8.00; 8.03 art.	Spe27 Dan44
Maucherite	Ni$_{11}$As$_8$	8.04 [8.0343(44)]	Dan44 Wea76
Meerschaum, see Sepiolite			
Meionite	Ca$_8$[(Cl$_2$, SO$_4$, CO$_3$)$_{2(?)}$(Al$_2$Si$_2$O$_8$)$_6$]	2.70···2.815 2.78 [2.737(7)]	Hod52/53 Dee62 Wea76
Melanite	Ca$_3$Fe$_2$(SiO$_4$)$_3$ [Klo48]	3.908 3.8···4.1	Spe27 Klo48
Melanophlogite	SiO$_2$	[1.9065(17)]	Wea76
Melanotekite	Pb$_2$Fe$''_2$[O$_2$Si$_2$O$_7$]	5.7	Klo48
Melanovanadite	Ca$_2$V$_{10}$O$_{25}$ · 7 H$_2$O	3.477	Spe27
Melanterite	Fe[SO$_4$] · 7 H$_2$O	1.89···1.90 [1.8972(33)]	Hod52/53 Wea76
Melilite	(Ca, Na)$_2$(Al, Mg)[(Si, Al)$_2$O$_7$]	2.9···3.4 2.95···3.05	Hod52/53 Dee62
Melinophan (Meliphanite)	(Ca, Na)$_2$(Be, Al)$^{[4]}$[Si$_2$O$_6$F]	3.0	Klo48
Meliphanite, see Melinophan			
Mellite	Al$_2$[C$_{12}$O$_{12}$] · 18 H$_2$O	1.55···1.65	Hod52/53
Melonite	NiTe$_2$	7.35 (average) [7.575(42)]	Dan44 Wea76
Mendelejewite	Variety of Betafite	4.44···4.766	Spe27
Mendipite	PbCl$_2$ · 2 PbO	7.240 7···7.1	Spe27 Hod52/53
Mendozite	NaAl[SO$_4$]$_2$ · 11 H$_2$O(?)	1.9	Klo48
			continued

1.1.1 Density of minerals (continued)

Mineral	Formula	$\varrho, [\varrho_x]$ g cm^{-3}	Ref.
Meneghinite	$4\,PbS \cdot Sb_2S_3(?)$	6.3···6.4	Klo48
		6.36(1)	Dan44
Mennige, see Minium			
Mercallite	$KH[SO_4]$	2.31	Klo48
Mercury (Quecksilber)	Hg	13.596	Dan44
Mercury (crystal −46°)		14.26	Dan44
Merrillite	$Na_2Ca_3(PO_4)_2O(?)$ [Dan44]	3.10	Spe27
Merwinite	$Ca_3Mg[SiO_4]_2$	3.15	Spe27, Klo48
Mesitite	Magnesite group	3.375	Spe27
Mesolite	$Na_2Ca_2[Al_2Si_3O_{10}]_3 \cdot 8\,H_2O$	2.257···2.260	Spe27
		≈2.26	Dee62
Metabrushite, see also Brushite	Brushite with less H_2O	3.666	Spe27
Metacinnabar	HgS	7.23	Spe27
		7.65	Dan44
		[7.712(4)]	Wea76
Metahewettite	$CaV_6O_{16} \cdot 3\,H_2O$	2.511	Spe27
Meyerhofferite	$Ca[B_3O_3(OH)_5] \cdot H_2O$	2.120	Spe27, Klo48
Miargyrite	$AgSbS_2$	5.2	Klo48
		5.36	Spe27
		5.26	Dan44
		[5.646(9)]	Wea76
Microcline	$K[AlSi_3O_8]$	2.54···2.57	Hod52/53
		[2.560(2)]	Wea76
		2.56···2.63	Dee62
Microlite	$(Ca, Na)_2(Ta, Nb)_2O_6(O, OH, F)$	4.2···6.4	Dan44
		5.405···5.562	Hod52/53
Miersite	α-AgI	5.64	Hod52/53
		[5.688(3)]	Wea76
Milarite	$KCa_2AlBe_2[Si_{12}O_{30}] \cdot 0.5\,H_2O$	2.6	Klo48
		2.5···2.59	Hod52/53
Millerite	β-NiS	5.5(2)	Dan44
		[5.3743(20)]	Wea76
Mimetesite	$Pb_5[Cl(AsO_4)_3]$	≈7.1	Klo48
Mimetite	$9\,PbO \cdot 3\,As_2O_5 \cdot PbCl_2$ [Klo48]	7.1	Klo48
		6.98···7.25	Hod52/53
Minguetite, see Stilpnomelane			
Minium (Mennige)	Pb_3O_4	[8.926(9)]	Wea76
		8.9···9.2 art.	Dan44
		4.6	Klo48
Minnesotaite	$(Fe^{2+}, Mg, H_2)_3[(OH)_2(Si, Al, Fe^{3+})_4O_{10}]$	3.01	Klo48
		[3.239(64)]	Wea76
Minyulite	$KAl_2[(OH, F)(PO_4)_2] \cdot 4\,H_2O$	2.46	Klo48
Mirabilite (Glaubersalz)	$Na_2[SO_4] \cdot 10\,H_2O$	1.40···1.481	Hod52/53
		[1.466(3)]	Wea76
		1.49	Klo48
Mitscherlichite	$K_2[CuCl_4(H_2O)_2]$	2.42	Klo48
Mixite	$(Bi^{3+}, Fe^{3+}, ZnH, CaH)Cu_{12}$ $\cdot [(OH)_{12}(AsO_4)_6] \cdot 6\,H_2O$	3.8	Klo48
Mizzonite		2.60	Spe27
Moissanite	SiC	3.1(1)	Dan44
		3.2	Hod52/53

continued

1.1.1 Density of minerals (continued)

Mineral	Formula		$\varrho, [\varrho_X]$ g cm^{-3}	Ref.
Molybdenite	MoS_2		4.62	Spe27
			4.62⋯4.73; 5.06 art.	Dan44
			[4.9982(33)]	Wea76
			4.7⋯4.8	Hod52/53
Molybdite	MoO_3		4.0⋯4.5	Klo48
			[4.710(6)]	Wea76
Molybdophyllite	$Pb_2Mg_2[(OH)_2Si_2O_7]$		4.7	Klo48
Monazite	$Ce[PO_4]$		5.2 (4.9⋯5.3)	Hod52/53
			5.0⋯5.3	Dee62
Monetite	$CaH[PO_4]$		2.75⋯2.863	Hod52/53
Monimolite	$(Pb, Fe, Mn)_2Sb_2O_7$	[Klo48]	5.9⋯7.2	Klo48
Montbrayite	Au_2Te_3		9.94	Klo48
Montebrasite	$(Li, Na)Al(PO_4)(OH, F)$	[Dan44]	3.008	Spe27
Monteponite	CdO		[8.2386(53)]	Wea76
Monticellite	$CaMg[SiO_4]$		3.03⋯3.25	Hod52/53
			[3.046(4)]	Wea76
Montmorrilonite	$(Al_{1.67}Mg_{0.33})[(OH)_2Si_4O_{10}]^{0.33-}$ $\cdot Na_{0.33}(H_2O)_4$		1.7⋯2.7	Klo48
			2⋯3	Dee62
Montroydite	HgO		11.23; 11.2(1) art.	Dan44
			11.14	Hod52/53
			[11.21(1)]	Wea76
Mooreite	$(Mg, Zn, Mn)_8[(OH)_{14}SO_4] \cdot 4H_2O(?)$		2.47	Klo48
Mordenite (Ptilolite)	$(Ca, K_2, Na_2)[AlSi_5O_{12}]_2 \cdot 6H_2O$		2.12⋯2.15	Dee62
			2.10⋯2.30	Spe27
Morinite, see Jezekite				
Mosandrite, see Johnstrupite and Rinkolite				
Moschellandsbergite (Landsbergite)	γ-(Ag, Hg)		13.48⋯13.71	Dan44
Mossite	$(Fe, Mn)(Nb, Ta)_2O_6$		5.20	Spe27
			6.5	Klo48
Mottramite (Cuprodescloizite)	$Pb(Cu, Zn)[OHVO_4]$		5.90⋯5.93, 6.19	Spe27
			5.5⋯6.2	Klo48
Mullite	$Al_4^{[6]}Al_4^{[4]}[O_3(O_{0.5}, OH, F)Si_3AlO_{16}]$		[3.124]	Wea76
Murmanite	$Na_2MnTi_3[OSi_2O_7]_2 \cdot 8H_2O$		2.84	Klo48
Muscovite, see also Sericite	$KAl_2[(OH, F)_2AlSi_3O_{10}]$		2.77⋯2.88	Dee62
			[2.831(4)]	Wea76
Muthmannite	$\approx(Ag, Au)Te$		5.598	Dan44
Myeline (Nacrite)	$Al_4[(OH)_8Si_4O_{10}]$		2.714	Spe27
			[2.602(8)]	Wea76
Nacrite, see Myeline				
Nadeleisenerz, see Goethite				
Nadelerz, see Aikinite				
Nadorite	$PbSbO_2Cl$		7.2	Klo48
Nagyagite	$AuTe_2 \cdot 6Pb(S, Te)(?)$		7.41(5)	Dan44
			6.8⋯7.5	Klo48
Nantokite	$CuCl$		≈4	Klo48
			3.930	Hod52/53
			[4.139(7)]	Wea76
				continued

1.1.1 Density of minerals (continued)

Mineral	Formula	$\varrho, [\varrho_x]$ g cm^{-3}	Ref.
Narsarsukite	Na$_4$Ti$_2$[O$_2$Si$_8$O$_{20}$]	2.75	Klo48
Natramblygonite (Natromontebrasite)	(Na, Li)Al[(OH, F)PO$_4$]	2.98···3.11	Dan44
Natroalunite	NaAl$_3$(SO$_4$)$_2$(OH)$_6$	[2.821(4)]	Wea76
Natrochalcite	NaCu$_2$[OH(SO$_4$)$_2$]·H$_2$O	3.48	Klo48
Natrojarosite	NaFe$_3^{3+}$[(OH)$_6$(SO$_4$)$_2$]	3.11	Spe27
		3.18	Dan44
Natrolite	Na$_2$[Al$_2$Si$_3$O$_{10}$]·2H$_2$O	2.18···2.25	Hod52/53
		2.20···2.26	Dee62
		[2.245(5)]	Wea76
Natromontebrasite, see Natramblygonite			
Natronorthoclase	(Na, K)AlSi$_3$O$_8$ [Klo48]	2.58···2.59	Klo48
Natronsalpeter, see Soda niter			
Naumannite	Ag$_2$Se	6.527	Spe27
		7.0···8.0	Dan44
Neighborite	NaMgF$_3$	[3.058(1)]	Wea76
Neotocite	(Mn, Mg)$_4$(SiO$_2$)$_3$(OH)$_{10}$	2.5	Spe27
Nepheline (Nephelite, Eläolith, Pseudonepheline)	KNa$_3$[AlSiO$_4$]$_4$	[2.623(3)]	Wea76
		2.56···2.665	Dee62
		2.68	Spe27
Nephelite, see Nepheline			
Nephrite	Variety of Actinolite or Antophyllite	2.938···3.06	Spe27
Neptunite	KNa$_2$Li(Fe, Mn)$_2$Ti$_2$[OSi$_4$O$_{11}$]$_2$	3.23	Klo48
–, Mangan-		3.203	Spe27
Newberyite	MgH[PO$_4$]·3H$_2$O	2.10	Hod52/53
Niccolite (Nickeline, Rotnickelkies)	NiAs	7.33···7.67	Hod52/53
		[7.776(5)]	Wea76
Nickel	Ni	[8.9117(38)]	Wea76
Nickelblüte, see Annabergite			
Nickeline, see Niccolite			
Nickel-Iron	α-(Fe, Ni)	7.8···8.22	Dan44
Nickel-olivine	Ni$_2$SiO$_4$	[4.917(4)]	Wea76
Nickelous Carbonate	NiCO$_3$	[4.3886(20)]	Wea76
Niggliite	Pt(Sn, Te)	4	Dan44
Niobite	(Fe, Mn)(Nb, Ta)$_2$O$_6$	8.1	Klo48
Niter (Kalisalpeter, Nitrokalite)	KNO$_3$	1.9···2.1	Klo48
		[2.105(3)]	Wea76
Nitronatrite, see Soda-Niter			
Nocerine	Mg$_3$[F$_3$BO$_3$]	2.96	Spe27
Nontronite	Fe$_2^{3+}$(OH)$_2$Al$_{0.33}$Si$_{3.67}$O$_{10}$]$^{0.33-}$·Na$_{0.33}$(H$_2$O)$_4$	2.29···2.295	Spe27
Norbergite	Mg$_3$[(OH, F)$_2$SiO$_4$]	3.15···3.18	Dee62
		3.13···3.15	Spe27
Nordenskioeldine	CaSn[BO$_3$]$_2$	4.2	Klo48
Nordite	≈Na$_4$(Ce, La ...)$_{1.5}$(Sr, Mn, Ca, Mg)$_3$Si$_8$O$_{23}$	3.43	Klo48
Norsethite	BaMg[CO$_3$]$_2$	[3.838(9)]	Wea76
Nosean (Noselite)	Na$_8$[SO$_4$(AlSiO$_4$)$_6$]	2.30···2.40	Dee62
Noselite, see Nosean			
Okenite	Ca$_{1.5}$[Si$_3$O$_6$(OH)$_3$]·1.5H$_2$O	2.206···2.332	Spe27
		2.3	Klo48

continued

1.1.1 Density of minerals (continued)

Mineral	Formula		$\varrho, [\varrho_x]$ g cm^{-3}	Ref.
Oldhamite	CaS		2.58	Dan44
			[2.602(8)]	Wea76
Oligoclase	NaAlSi$_3$O$_8$ + CaAl$_2$Si$_2$O$_8$	[Hod52/53]	2.612···2.672	Spe27
–, K$_2$Co$_3$ (Potasche)			2.615···2.625	Spe27
Olivenite	Cu$_2$[OHAsO$_4$]		4.1···4.4	Hod52/53
Olivine	(Mg, Fe)$_2$[SiO$_4$]		3.301···3.56	Spe27
(Peridot, Chrysolite)			3.26···3.40	Hod52/53
–, Lime-	α-Ca$_2$SiO$_4$		3.190···3.341	Spe27
			[2.914(9)]	Wea76
Omphacite	(Ca, Na)(Mg, Fe^{2+}, Fe^{3+}, Al)[Si$_2$O$_6$]		3.29···3.37	Dee62
Opal	SiO$_2$ + aq.		2.06···2.22	Spe27
			2.1···2.3	Hod52/53
Orangite	Variety of Thorite		5.2···5.4	Klo48
Orientite	Ca$_4$(Mn^{3+}, Mn^{2+}, Al)$_6$[(OH)$_6$(Si, H$_4$)O$_4$ · (SiO$_4$)$_2$Si$_3$O$_{10}$](?)		3.05	Spe27
Orpiment	As$_2$S$_3$		[3.490(13)]	Wea76
Orthite, see Allanite				
Orthoclase	K[AlSi$_3$O$_8$]		[2.570(4)]	Wea76
			2.55···2.63	Dee62
Orthochlorite, see Zebedassite				
Orthoferrosilite	Fe^{2+}[SiO$_3$]	[Dee62]	3.96	Dee62
			[3.998(4)]	Wea76
Oruetite	Bi$_8$S$_4$Te	[Klo48]	7.6	Spe27
Otavite	CdCO$_3$		[5.0265(22)]	Wea76
Overite	Ca$_3$Al$_8$[(OH)$_3$(PO$_4$)$_4$]$_2$ · 15H$_2$O		2.53	Klo48
Owyheeite	5PbS · Ag$_2$S · 3Sb$_2$S$_3$		6.03	Dan44
Oxalite (Humboldtine)	Fe[C$_2$O$_4$] · 2H$_2$O		2.28	Spe27
			2.25	Klo48
Oxammite	(NH$_4$)$_2$C$_2$O$_4$ · H$_2$O	[Klo48]	1.48	Klo48
Oxi-Apatite, see Voelckerite				
Pachnolite	CaNa[AlF$_6$] · H$_2$O		2.976	Spe27
Paigeite	(Fe, Mg)$_2$Fe^{3+}BO$_5$	[Dan44]	4.7	Dan44
			4.78	Spe27
Palaite	Mn$_5$H$_2$[PO$_4$]$_4$ · 3H$_2$O	[Klo48]	3.15	Klo48
			3.14···3.20	Spe27
Palladium	Pd		11.3···11.8	Klo48
Palladiumamalgam, see Potarite				
Palmierite	PbK$_2$[SO$_4$]$_2$		4.50	Spe27
Parabayldonite	(Cu, Pb)$_2$(AsO$_4$)(OH)(?)	[Dan44]	5.44···5.512	Spe27
Paradoxite	Variety of Potassium-feldspar		2.425···2.430	Spe27
Paragonite	NaAl$_2$[(OH, F)$_2$AlSi$_3$O$_{10}$]		2.8···2.9	Klo48
			2.85	Dee62
			[2.893(28)]	Wea76
Parahopeite	Zn$_3$[PO$_4$]$_2$ · 4H$_2$O		3.3	Klo48
			3.21···3.236	Spe27
Paramelaconite (Paratenorite)	CuO		6.04	Dan44
Pararammelsbergite	NiAs$_2$		7.12	Dan44
			[7.244(22)]	Wea76

continued

1.1.1 Density of minerals (continued)

Mineral	Formula	$\varrho, [\varrho_x]$ g cm^{-3}	Ref.
Paratenorite, see Paramelaconite			
Paraurichalcite	$(Cu, Zn)_2(OH)_2(CO_3)$ [Dan44]	4.137···4.201	Spe27
Paravauxite	$Fe^{2+}Al_2[OHPO_4]_2 \cdot 8H_2O$	2.291···2.30	Spe27
Paredrite	Variety of Rutile with 0.6% H_2O	3.97···4.08	Spe27
Pargasite	$NaCa_2Mg_4(Al, Fe^{3+})[(OH, F)_2Al_2Si_6O_{22}]$	3.069···3.181	Spe27
		3.05	Dee62
Parisite	$CaCe_2[F_2(CO_3)_3]$	4.320···4.42	Hod52/53
Parsettensite	$(K, H_2O)(Mn, Fe^{3+}, Mg, Al)_{<3}$ $\cdot [(OH)_2Si_4O_{10}]X_n(H_2O)_2$	2.590···2.681	Spe27
Parsonsite	$Pb_2[UO_2(PO_4)_2]$	6.23	Spe27
Pascoite	$Ca_3[V_{10}O_{28}] \cdot 16H_2O$	2.46	Klo48
Paternoite	$MgB_8O_{13} \cdot 4H_2O$ [Klo48]	2.11	Spe27
Pearceite	$8(Ag, Cu)_2S \cdot As_2S_3$	6.15(2)	Dan44
		6.1	Klo48
Pectolite	$Ca_2NaH[Si_3O_9]$	2.74···2.88	Hod52/53
		2.86···2.90	Dee62
		[2.876(7)]	Wea76
Pennine	$(Mg, Al)_3[(OH)_2Al_{0.5···0.9}Si_{3.5···3.1}O_{10}]$ $\cdot Mg_3(OH)_6$	2.6···2.85	Hod52/53
Penroseite (Blockite)	$(Ni, Co, Cu)Se_2$	6.93	Spe27
		6.9(2)	Dan44, Klo48
Pentlandite	$(Ni, Fe)_9S_8$	4.638	Spe27
	Fe:Ni = 1:1	4.6···5.0	Dan44
Penwithite	weathered Rhodonite	2.20	Spe27
Percylite	$30\,PbCl_2 \cdot 30\,Cu(OH)_2$	4.675···4.71	Hod52/53
Periclase	MgO	[3.5837(13)]	Wea76
		3.56···3.68	Dee62
Peridot, see Olivine			
Perovskite	$CaTiO_3$	[4.0439(12)]	Wea76
		3.98···4.26	Dee62
Petalite	$Li[AlSi_4O_{10}]$	2.414	Spe27
		2.386···2.465	Hod52/53
		2.412···2.422	Dee62
		[2.385(9)]	Wea76
Petzite	Ag_3AuTe_2	7.53···8.735	Spe27
		8.7···9.02	Dan44
		[9.214(53)]	Wea76
Pharmacolite	$CaH[AsO_4] \cdot 2H_2O$	2.6	Klo48
Pharmacosiderite	$KFe_4^{3+}[(OH)_4(AsO_4)_3] \cdot 6···7H_2O$	2.9···3.0	Hod52/53, Klo48
Phenakite	$Be_2[SiO_4]$	2.944···3.041	Spe27, Hod52/53
		[2.960(3)]	Wea76
Phillipsite	$KCa[Al_3Si_5O_{16}] \cdot 6H_2O$	2.2	Klo48, Dee62
Phlogopite	$KMg_3[(F, OH)_2AlSi_3O_{10}]$	2.76···2.90	Dee62
		2.737···2.869	Spe27, Hod52/53
		[2.784(7)]	Wea76
–, Fluor-		[2.878(4)]	Wea76
Phoenicochroite	$Pb_3[O(CrO_4)_2]$	5.7	Klo48
		5.75	Dan44

continued

1.1.1 Density of minerals (continued)

Mineral	Formula	$\varrho, [\varrho_x]$ g cm^{-3}	Ref.
Phosgenite	Pb$_2$[Cl$_2$CO$_3$]	6.0···6.305	Hod52/53
Phosphoferrite	(Fe^{2+}, Mn)$_3$[PO$_4$]$_2 \cdot$ 3H$_2$O	3.0···3.2	Dan44
		3.156	Spe27
Phosphophyllite	Zn$_2$Fe[PO$_4$]$_2 \cdot$ 4H$_2$O	3.08; 3.13	Dan44
		3.081···3.082	Spe27
Phosphorochalcite (Pseudomalachite)	Cu$_5$[(OH)$_2$PO$_4$]$_2$	3.6	Klo48
Phosphosiderite	Fe^{3+}[PO$_4$] $\cdot$ 2H$_2$O	2.726	Spe27
Pickeringite (Magnesiaalaun)	MgAl$_2$[SO$_4$]$_4 \cdot$ 22H$_2$O	1.84	Spe27
Picotit	(Al, Cr, Fe)$_2$(Fe, Mg)O$_4$	4.08	Hod52/53
Picrochromite, see Magnesiochromite			
Picrotephroite	Mg-Tephroite	3.72	Spe27
Piemontite	Ca$_2$(Mn^{3+}, Fe^{3+})Al$_2$[OOHSiO$_4$Si$_2$O$_7$]	3.4	Klo48
		3.45···3.52	Dee62
		[3.810(21)]	Wea76
Pigeonite	(Mg, Fe, Ca)$_2$[Si$_2$O$_6$]	3.30···3.46	Dee62
Pinite		2.780	Spe27
Pinnoite	Mg[B$_2$O(OH)$_6$]	2.292	Spe27
Pirssonite	Na$_2$Ca[CO$_3$]$_2 \cdot$ 2H$_2$O	2.35	Klo48
Pisanite	(Fe, Cu)[SO$_4$] $\cdot$ 7H$_2$O	1.950	Spe27
Pisekite	Oxide of Nb, Ta, Ti ...	4.032	Spe27
Plagionite	5 PbS $\cdot$ 4Sb$_2$S$_3$	5.4···5.6	Klo48
		5.58(2)	Dan44
Plancheite	Cu$_8$[(OH)$_2$Si$_4$O$_{11}$]$_2 \cdot$ H$_2$O	3.37	Spe27
Platinum	Pt	16.4···19.0	Spe27
		14···19	Dan44
		13.35···19.00	Hod52/53
		[21.460(8)]	Wea70
Platiniridium	Pt, Ir	22.65···22.84	Dan44
Plattnerite	PbO$_2$	9.42(2); 8.9···9.36 art.	Dan44
Platynite	Pb$_4$Bi$_7$Se$_7$S$_4$	7.98	Spe27, Dan44
Plazolite (Hibschite)	Ca$_3$Al$_2$[(Si, H$_4$)O$_4$]$_3$	3.1	Klo48
Pleonast, see Ceylonite			
Plumboferrite	PbO $\cdot$ 2Fe$_2$O$_3$	6.07	Dan44
		6.0	Klo48
Plumbogummite (Bleigummi)	PbAl$_3$H[(OH)$_6$(PO$_4$)$_2$]	4···5	Klo48
Poechite	FeSi-gel with Mn	3.693···3.721	Spe27
Pollucite	(Cs, Na)[AlSi$_2$O$_6$] $\cdot$ H$_2$O	2.9	Klo48
		2.868···2.901	Hod52/53
Polyargyrite	Ag$_{24}$Sb$_2$S$_{15}$	6.974	Dan44
Polybasite	8(Ag, Cu)$_2$S $\cdot$ Sb$_2$S$_3$	6.1(1)	Dan44
		6···6.2	Klo48
Polyhalite	K$_2$Ca$_2$Mg[SO$_4$]$_4 \cdot$ 2H$_2$O	2.77	Klo48
Polydymite	Ni$_3$S$_4$	[4.7458(15)]	Wea76
Polymignite	(Ce, La, Y, Th, Mn, Ca)(Ti, Zr, Nb, Ta)$_2$O$_6$	4.77···4.85	Spe27
Portlandite	Ca(OH)$_2$	2.230(3)	Dan44
		[2.2415(11)]	Wea76
Potarite (Palladiumamalgam)	PdHg	13.48···16.11	Dan44
		13.33···15.82	Spe27

continued

1.1.1 Density of minerals (continued)

Mineral	Formula	$\varrho, [\varrho_X]$ g cm^{-3}	Ref.
Powellite	Ca[MoO$_4$]	4.22	Spe27
		4.356···4.526	Hod52/53
		[4.256(9)]	Wea76
Prehnite	Ca$_2$Al$^{[6]}$[(OH)$_2$AlSi$_3$O$_{10}$]	2.80···2.95	Hod52/53
		2.90···2.95	Dee62
Priceite	5CaO · 6B$_2$O$_3$ · 9H$_2$O	2.43···2.433	Spe27
		2.4	Klo48
Priorite	(Y, Ce, Th, Ca, Na, U)[(Ti, Nb, Ta)$_2$O$_6$]	4.95(10)	Dan44
Prismatine (Kornerupine)	Mg$_4$Al$_6$[(O, OH)$_2$BO$_4$(SiO$_4$)$_4$]	3.345	Spe27
		3.3	Klo48
Probertite	NaCa[B$_5$O$_7$(OH)$_4$] · 3H$_2$O	2.14	Klo48
Prochlorite (Rhipidolite)	(Mg, Fe, Al)$_3$[(OH)$_2$Al$_{1.2···1.5}$Si$_{2.8···2.5}$O$_{10}$] · Mg$_3$(OH)$_6$	2.60···2.936; 2.975	Spe27
Prosopite	CaAl$_2^{[6]}$[F, OH]$_8$	2.89	Klo48
Protolithionite	z · K$_2$LiFe$_4$Al(F, OH)$_4$[AlSi$_3$O$_{10}$]$_2$ [Klo48]	3.148···3.305	Spe27
Proustite	Ag$_3$AsS$_3$	5.51···5.64	Hod52/53
Proustite (pure)		5.57	Dan44
		[5.595(1)]	Wea76
Pseudobrookite	Fe$_2^{3+}$TiO$_5$	4.60	Spe27
		4.4···4.9	Hod52/53
		4.39	Dan44
Pseudomalachite, see Dihydrite and Phosphorochalcite			
Pseudonepheline, see Nepheline			
Pseudophite	partly Clinochlore, partly Pennine	2.693···2.695	Spe27
Psilomelan	(Ba, Mn^{2+} ···)$_3$(O, OH)$_6$Mn$_8$O$_{16}$	4.71(1)	Dan44
		3.7···4.7	Hod52/53
Ptilolite, see Mordenite			
Pucherite	Bi[VO$_4$]	6.2	Klo48
Pufahlite	Composition of Sphalerite and Teallite	5.4	Spe27
Pumpellyite	Ca$_2$(Mg, Fe, Mn, Al)(Al, Fe, Ti)$_2$ · [(OH, H$_2$O)$_2$SiO$_4$Si$_2$O$_7$] (?)	3.18···3.23	Dee62
		3.2	Spe27
Purpurite	(Mn^{3+}, Fe^{3+})[PO$_4$]	3.4	Klo48
Pyrargyrite	Ag$_3$SbS$_3$	5.790	Spe27
		5.85	Dan44
		[5.8506(25)]	Wea76
		5.77···5.86	Hod52/53
Pyrite	FeS$_2$	5.018	Dan44
		4.95···5.17	Hod52/53
		[5.0116(14)]	Wea76
Pyroaurite	Mg$_6$Fe$_2$[(OH)$_{16}$CO$_3$] · 4H$_2$O	2.12(2)	Dan44
Pyrobelonite	PbMn[OHVO$_4$]	5.4	Klo48
Pyrochlorite	RNb$_2$O$_6$ · R(Ti, Th)O$_3$ [Hod52/53]	4.2···4.36	Hod52/53
Pyrochroite	Mn(OH)$_2$	3.25(2)	Dan44
		3.258	Hod52/53
Pyrolusite	β-MnO$_2$	4.73···4.86	Hod52/53
		[5.234(8)]	Wea76
Pyrolusite (crystal)		5.06(2)	Dan44
Pyrolusite (massive)		4.4···5.0	Dan44

continued

1.1.1 Density of minerals (continued)

Mineral	Formula	$\varrho, [\varrho_x]$ g cm^{-3}	Ref.
Pyromorphite	Pb$_5$[Cl(PO$_4$)$_3$]	6.50···7.12	Hod52/53
Pyrope	Mg$_3$Al$_2$[SiO$_4$]$_3$	3.510···3.75	Spe27
		[3.559(1)]	Wea76
		3.582	Dee62
Pyrophanite	MnTiO$_3$	4.54	Dee62/Klo48
		[4.605(10)]	Wea76
Pyrophyllite	Al$_2$[(OH)$_2$Si$_4$O$_{10}$]	[2.863(14)]	Wea76
		2.65···2.90	Dee62
Pyrosmalite	(Mn, Fe)$_8$[(OH, Cl)$_{10}$Si$_6$O$_{15}$]	3···3.2	Klo48
Pyrostilpnite	Ag$_3$SbS$_3$	5.94	Dan44
Pyroxmangite (Sobralite)	(Fe, Mn)$_7$[Si$_7$O$_{21}$]	3.61···3.80	Dee62
		[3.817(20)]	Wea76
		3.60	Spe27
Pyrrhotite (Magnetkies)	Fe$_7$S$_8$–FeS [Dee62]	4.6	Dee62, Klo48
		4.53···4.66	Hod52/53
		[4.830(9)]	Wea76
		4.58···4.65	Dan44
(Magnetkies, Troilite)		4.74	Dan44
Quartz	SiO$_2$	2.649···2.697	Spe27
		2.59···2.660	Hod52/53
		[2.6483(1)]	Wea76
Quartz glass		2.194···2.213	Spe27
Quecksilber, see Mercury			
Quenselite	PbO · MnOOH	6.842	Spe27, Dan44
Racewinite		1.94···1.98	Spe27
Radiophyllite (Zeophyllite)	Ca$_4$[F$_2$(OH)$_2$Si$_3$O$_8$] · 2H$_2$O	2.45···2.60	Spe27
		2.5	Klo48
Ralstonite	Al$_2$(F, OH)$_6$ · H$_2$O	2.614	Spe27
		2.4	Klo48
Ramdohrite	Pb$_6$Ag$_4$Sb$_{10}$S$_{23}$	5.43	Dan44
Rammelsbergite	NiAs$_2$	6.734···7.02	Spe27
		7.1(1)	Dan44
		[7.091(7)]	Wea76
Ramsayite	Na$_2$Ti$_2$[O$_3$Si$_2$O$_6$]	3.43···3.47	Spe27
Rancieite	(Ca, Mn^{2+})Mn$_4^{4+}$O$_9$ · 3H$_2$O (?)	3.25···3.30	Spe27
Rathite	Pb$_6$As$_{10}$S$_{20}$	5.453	Spe27, Klo48
		5.37(4)	Dan44
Rauschgelb, see Auripigment			
Realgar	As$_4$S$_4$	3.56; 3.477	Dan44, Hod52/53
		[3.591(29)]	Wea76
Réaumurite, see Rivaite			
Reddingite	(Mn, Fe^{2+})$_3$[PO$_4$]$_2$ · 3H$_2$O	2.96···3.10	Spe27
Retgersite	NiSO$_4$ · 4H$_2$O	[2.076(3)]	Wea76
Reyerite, see Truscottite			
Rézbanyite	2 PbS · Cu$_2$S · n Bi$_2$S$_3$	6.24(15); 6.89	Dan44
Rhagite (Atelestite)	Bi$_2$[OOHAsO$_4$]	6.8	Klo48
Rhipidolite, see Prochlorite			

continued

1.1.1 Density of minerals (continued)

Mineral	Formula	$\varrho, [\varrho_x]$ g cm^{-3}	Ref.
Rhodizite	(Cs, K, Rb)Al$_4$Be$_4$[B$_{11}$O$_{26}$(OH)$_2$]	3.344	Spe27
		3.4	Klo48
Rhodochrom	Cr-Orthochlorite	2.644	Spe27
Rhodochrosite (Dialogite)	MnCO$_3$	3.30···3.76	Hod52/53
		[3.6992(17)]	Wea76
–, Zinco-		3.86	Spe27
Rhodolite	Composition of Pyrope-Almandine	3.75···3.837	Dan44
Rhodonite	CaMn$_4$[Si$_5$O$_{15}$]	3.40···3.68	Hod52/53
		[3.727(2)]	Wea76
		3.57···3.76	Dee62
Rhodotilith, see Inesite			
Richterite (Imerinite)	Na$_2$Ca(Mg, Fe^{2+}, Mn, Fe^{3+}, Al)$_5$	2.97···3.45	Dee62
	·[(OH, F)Si$_4$O$_{11}$]$_2$	3.02	Spe27
Rickardite	≈ Cu$_3$Te$_2$	7.54	Dan44
		7.6	Klo48
Riebeckite	Na$_2$Fe$_3^{2+}$Fe$_2^{3+}$[(OH, F)Si$_4$O$_{11}$]$_2$	3.02···3.42	Dee62
		[3.407(11)]	Wea76
–, Mg-	Na$_2$Mg$_3$Fe$_2$Si$_8$O$_{22}$(OH)$_2$ [Cla66]	[3.102(8)]	Wea76
Rinkolite (Mosandrite)	(Ca, Na, Y)$_3$(Ti, Zr, Ce)[(F, OH, O)$_2$Si$_2$O$_7$]	3.40	Spe27
Rinneite	K$_3$Na[FeCl$_6$]	2.35	Klo48
Rivaite (Réaumurite)	Composition of Wollastonite and glass	2.55···2.56	Spe27
Riversideite	Ca$_5$H$_2$[Si$_3$O$_9$]$_2$·2H$_2$O	2.64	Spe27
Rock salt, see Halite			
Roeblingite	PbCa$_3$H$_6$[SO$_4$(SiO$_4$)$_3$]	3.43	Klo48
Roemerite	Fe^{2+}Fe$_2^{3+}$[SO$_4$]$_4$·14H$_2$O	2.1	Klo48
Roméite, see also	(Ca, NaH)Sb$_2$O$_6$(O, OH, F)	4.7···5.4	Dan44
Schneebergite		5.044···5.074	Spe27
Rosasite	(Zn, Cu)$_2$[(OH)$_2$CO$_3$]	4.09	Spe27
Roscherite	(Ca, Mn, Fe)Be[OHPO$_4$]·$\frac{2}{3}$H$_2$O	2.92	Dan44, Spe27
Roselite	α-Ca$_2$Co[AsO$_4$]$_2$·2H$_2$O	3.7	Klo48
Rosenbuschite	(Ca, Na)$_6$Zr(Ti, Mn, Nb, ...)[(F, O)$_2$Si$_2$O$_7$]$_2$	3.3	Klo48
Rossite	Ca[V$_2$O$_6$]·4H$_2$O	2.45	Klo48
Rotnickelkies, see Niccolite			
Roweite	CaMn[B$_2$O$_5$]·H$_2$O	2.92	Klo48
Rumpfite, see also Sheridanite		2.666	Spe27
Russellite	(Bi$_2$, W)O$_3$	7.35(2)	Dan44
Rutile	TiO$_2$	4.23···5.5	Dee62
		4.123···4.272	Spe27
		4.18···5.13	Hod52/53
		[4.2453(17)]	Wea76
		4.23(2)	Dan44
Rutile (Fe)		4.2···4.4	Dan44
Rutile (Nb-Ta)		4.2···5.6	Dan44
Safflorite	CoAs$_2$	7.20(25)	Dan44
		[7.461(6)]	Wea76
Sahlinite	Pb$_{14}$[Cl$_4$O$_9$(AsO$_4$)$_2$]	7.95	Klo48
Sal ammoniac (Salmiak)	α-NH$_4$Cl	1.53	Klo48
Salesite	Cu[OHIO$_3$]	4.77	Klo48
Salmiak, see Sal ammoniac			
Salmonsite	(Mn, Fe^{3+})$_5$H$_2$[PO$_4$, (OH)$_4$]$_4$·4H$_2$O	2.88	Spe27

continued

1.1.1 Density of minerals (continued)

Mineral	Formula	$\varrho, [\varrho_x]$ g cm^{-3}	Ref.
Samarskite, see also	$(Y, U, Ca)(Nb, Fe^{3+})_2(O, OH)_6$	5.69	Dan44
Ampangabeite		4.2···6.04	Spe27
Samiresite		5.24	Spe27, Klo48
Sampleite	$CaNaCu_5[Cl(PO_4)_4] \cdot 5H_2O$	3.20	Klo48
Samsonite	$2Ag_2S \cdot MnS \cdot Sb_2S_3$	5.51	Dan44
Sanmartinite	$ZnWO_4$	[7.872(8)]	Wea76
Sanidine	$K[AlSi_3O_8]$	[2.552(2)]	Wea76
		2.56···2.62	Dee62
Saponite	$(Mg_{3\ldots2.25}Fe_{0\ldots0.75})_{\Sigma3}[(OH)_2Al_{0.33}Si_{3.67}$ $\cdot O_{10}]^{0.33}(0.5Ca, Na)_{0.33}(H_2O)_4$	2.3	Klo48
Sapphirine	$Mg_2Al_4[O_6SiO_4]$	3.40···3.58	Dee62
		3.486	Dan44
		[3.464(25)]	Wea76
Sarcolite	$(Ca, Na)_8[O_2(Al(Al, Si)Si_2O_8)_6]$ (?)	2.54	Klo48
Sarcopside	$(Fe^{2+}, Mn, Ca)_3[PO_4]_2$	3.64	Spe27
Sarkinite	$Mn_2[OHAsO_4]$	4.08···4.18	Dan44
		4.2	Klo48
Sartorite	$PbAs_2S_4$	5.10(2)	Dan44
		5.0	Klo48
Sassolite	$B(OH)_3$	1.48(2)	Dan44, Hod52/53
Scacchite	$MnCl_2$	[2.988(12)]	Wea76
Scawtite	$Ca_6[Si_3O_9]_2 \cdot CaCO_3 \cdot 2H_2O$	2.77	Klo48
Schafarzikite	$FeSb_2O_4$	4.3	Spe27
Schairerite	$Na_3[(F, Cl)SO_4]$	2.6	Klo48
Schallerite	$(Mn, Fe)_8[(OH)_{10}(Si, As)_6O_{15}]$	3.368	Spe27, Klo48
Scheelite	$Ca[WO_4]$	5.88···6.14	Hod52/53
		[6.120(12)]	Wea76
Scheteligite	$(Ca, Fe, Mn, Sb, Bi, Y)_2(Ti, Ta, Nb, W)_2$ $\cdot (O, OH)_7$ [Klo48]	≈ 4.7	Klo48
Schirmerite	$PbS \cdot 2Ag_2S \cdot 2Bi_2S_3$	6.737	Dan44, Klo48
Schmöllnitzite, see Ferropallidite			
Schneebergite, see also Roméite		5.41	Spe27
Schoenite	$K_2Mg[SO_4]_2 \cdot 6H_2O$	2.1	Klo48
Schoepite	$8[UO_2(OH)_2] \cdot 8H_2O$	5.685	Spe27
		4.8	Dan44
		4.83	Klo48
Schoerl (Tourmaline)	$NaFe_3^{2+}Al_6[(OH)_{1+3}(BO_3)_3Si_6O_{18}]$	3.10···3.25	Dee62
		[3.297(6)]	Wea76
Schorlomite	$3CaO \cdot (Fe, Ti)_2O_3 \cdot 3(Si, Ti)O_2$	3.783···3.88	Hod52/53
	[Hod52/53]		
Schreibersite	$(Fe, Ni)_3P$ [Dan44]	7.0···7.3	Dan44
Schreibersite (28,68 % Ni)		7.44	Dan44
Schroeckingerite (Dakeite)	$NaCa_3[UO_2FSO_4(CO_3)_3] \cdot 10H_2O$	2.51	Klo48
Schultenite	$PbH[AsO_4]$	5.943	Spe27, Klo48
Schwartzembergite	$Pb_3[Cl_2OOHIO_3]$	6.2	Klo48
		7.39	Spe27
Seamanite	$Mn_3[(OH)_2PO_4B(OH)_4]$	3.13	Klo48
Searlesite	$Na_2B_2[(OH)_4Si_4O_{10}]$	2.45	Spe27, Klo48
Seladonite	$K_{0.8}(Fe_{1.4}^{3+}Mg_{0.7})[(OH)_2Al_{0.4}Si_{3.6}O_{10}]$	2.8···2.9	Klo48
			continued

1.1.1 Density of minerals (continued)

Mineral	Formula	$\varrho, [\varrho_x]$ g cm^{-3}	Ref.
Selenium	Se	4.80···4.84	Dan44
		[4.8088(19)]	Wea76
Selenquecksilber, see Tiemannite			
Selenolite	SeO$_2$	[4.161(13)]	Wea76
Seligmannite	2PbS · Cu$_2$S · As$_2$S$_3$	5.44···5.48	Spe27
Sellaite	MgF$_2$	2.972···3.170	Hod52/53
		[3.177(2)]	Wea76
Semseyite	9PbS · 4Sb$_2$S$_3$	5.84···6.05	Spe27
		6.08	Dan44
		6.1	Klo48
Senaite	Pb(Ti, Fe, Mn, Mg)$_{24}$O$_{38}$	5.301	Dan44, Klo48
Senarmontite	Sb$_2$O$_3$	5.50	Dan44
		5.22···5.30	Hod52/53
		[5.5837(45)]	Wea76
Sepiolite (Meerschaum)	Mg$_4$[(OH)$_2$Si$_6$O$_{15}$] · 2H$_2$O + 4H$_2$O	2.02	Spe27, Hod52/53
Serendibite	(Ca, Mg)$_5$(AlO)$_5$[BO$_3$(SiO$_4$)$_3$]	3.4	Klo48
Sericite, see also Muscovite		2.798	Spe27
		2.78···2.88	Klo48
Serpentine	Mg$_3$[Si$_2$O$_5$](OH)$_4$ [Dee62]	2.55···2.6	Dee62
		2.50···2.65	Hod52/53
Shandite	β-Ni$_3$Pb$_2$S$_2$	[8.867(33)]	Wea76
Shattuckite	Cu$_5$[(OH)$_2$(SiO$_3$)$_4$]	3.8	Klo48
Sheridanite, see also Rumpfite	(Mg, Al)$_3$[(OH)$_2$Al$_{1.2···1.5}$Si$_{2.8···2.5}$O$_{10}$]Mg$_3$ · (OH)$_6$	2.702	Spe27
Shortite	Na$_2$Ca$_2$[CO$_3$]$_3$	2.60	Klo48
		[2.610(7)]	Wea76
Sicklerite	Li$_{<1}$(Mn^{2+}, Fe^{3+})[PO$_4$]	3.45	Spe27
Siderazot	Fe$_5$N$_2$ [Klo48]	3.147	Dan44
Siderite (Spateisenstein)	FeCO$_3$	3.63···3.96	Spe27
		3.00···3.88	Hod52/53
		[3.9436(18)]	Wea76
		3.96	Dee62
		3.7···3.9	Klo48
Sideronatrite	Na$_2$Fe^{3+}[OH(SO$_4$)$_2$] · 3H$_2$O	2.3	Klo48
Sikersite		19···21	Dan44
Sikersite (for pure Os)		22.69	Dan44
Silicamagnesiofluorite	Ca$_4$Mg$_3$H$_2$F$_{10}$Si$_2$O$_7$ (?)	2.9	Klo48
Silicon	Si	[2.3296(4)]	Wea76
Sillenite	Bi$_2$O$_3$	8.80	Dan44
Sillimanite	Al$^{[6]}$Al$^{[4]}$[OSiO$_4$]	[3.248(3)]	Wea76
		3.23···3.27	Dee62, Hod52/53
Silver	Ag	8.33···9.83	Spe27
		10.1···11.1	Dan44, Hod52/53
Silver (pure)		[10.5001(20)]	Wea76
Simpsonite	Al$_4$Ta$_3$(O$_{13}$OH)	5.92···6.27	Dan44
Sincosite	Ca[V(OH)$_2$PO$_4$]$_2$ · 3H$_2$O	2.84	Spe27, Klo48
Sjoegrenite	Mg$_6$Fe$_2$[(OH)$_{16}$CO$_3$] · 4H$_2$O	2.11(3)	Dan44
Skapolite (Wernerite)	(Na, Ca, K)$_4$[Al$_3$(Al, Si)$_3$Si$_6$O$_{24}$] · (Cl, F, OH, CO$_3$SO$_4$) [Dee62]	2.50···2.78	Dee62

continued

1.1.1 Density of minerals (continued)			
Mineral	Formula	$\varrho, [\varrho_x]$ g cm^{-3}	Ref.
Sklodowskite	MgH$_2$[UO$_2$SiO$_4$]$_2$ · 5H$_2$O	3.54···3.74	Spe27
Skolecite	Ca[Al$_2$Si$_3$O$_{10}$] · 3H$_2$O	2.16···2.4	Hod52/53
		2.25···2.29	Dee62
Skorodite	Fe^{3+}[AsO$_4$] · 2H$_2$O	2.70···3.235	Spe27
		3.1···3.3	Hod52/53
Skutterudite (Smaltine)	CoAs$_3$	6.8	Klo48
		6.1···6.9	Dan44, Spe27
		[6.7298(25)]	Wea76
Slavikite	MgFe$_3^{3+}$[(OH)$_3$(SO$_4$)$_4$] · 18H$_2$O	1.905	Dan44
		1.95	Klo48
Smaltine, see Skutterndite			
Smaragd (Emerald), see Beryl			
Smithite	Ag$_3$As$_3$S$_6$	4.88	Dan44, Klo48
Smithsonite	ZnCO$_3$	4.30···4.45	Hod52/53
		[4.4343(21)]	Wea76
Sobralite, see Pyroxmangite			
Soda	Na$_2$CO$_3$ · 10H$_2$O	1.42···1.47	Klo48
		2.24···2.290	Hod52/53
		[2.260]	Wea76
Soda-Niter (Nitronatrit, Natronsalpeter)	NaNO$_3$	2.2···2.3	Klo48
Sodalite	Na$_8$[Cl$_2$(AlSiO$_4$)$_6$]	2.14···2.40	Hod52/53
		2.27···2.33	Dee62
Soddyite	(UO$_2$)$_{15}^{2+}$[(OH)$_{20}$Si$_6$O$_{17}$] · 8H$_2$O (?)	4.63	Spe27, Klo48
Sodium Melilite	NaCaAlSi$_2$O$_7$	[2.462(3)]	Wea76
Soumansite, see Wardite			
Spangolite	Cu$_6$Al[(OH)$_{12}$ClSO$_4$] · 3H$_2$O	3.1	Klo48
Spateisenstein, see Siderite			
Spencerite	Zn$_2$[OHPO$_4$] · 1.5H$_2$O	3.123···3.142	Spe27
Sperrylite	PtAs$_2$	10.58···10.73	Spe27
		10.58	Dan44
		[10.778(27)]	Wea76
Spessartine	Mn$_3$Al$_2$[SiO$_4$]$_3$	4.0···4.3	Hod52/53
		[4.190(1)]	Wea76
Sphalerite (Zinkblende)	ZnS [Dee62]	3.9···4.1	Dan44, Hod52/53
		4.1	Dee62
		[4.0885(11)]	Wea76
	α-ZnS	3.935···4.09	Spe27
Sphen	CaTi[SiO$_4$](O, OH, F) [Dee62]	3.45···3.55	Dee62
		[3.523(11)]	Wea76
Spinel	Al$_2$MgO$_4$	3.5···4.1	Hod52/53
		[3.583(3)]	Wea76
		3.55	Dee62, Dan44
Spodumene	LiAl[Si$_2$O$_6$]	3.03···3.22	Dee62
Spodumene, α-		2.997···3.301	Spe27
		[3.188(1)]	Wea76
Spodumene, β-		2.327···2.463	Spe27
		[2.379(1)]	Wea76
		2.644···2.649	Hod52/53
Spodumene, γ-		2.313	Spe27
Spurrite	Ca$_5$[CO$_3$(SiO$_4$)$_2$]	3.01	Klo48
			continued

1.1.1 Density of minerals (continued)

Mineral	Formula	$\varrho, [\varrho_x]$ g cm^{-3}	Ref.
Stainierite	CoOOH	4.13···4.47	Dan44
		4.32	Klo48
Stannite (Zinnkies)	Cu_2FeSnS_4	4.3···4.5	Dan44, Klo48
Stasite, see also Dewindtite		5.03	Spe27
Staszicite	$CaCu(AsO_4)(OH)$ [Dan44]	4.227	Spe27
Staurolite	$AlFe_2O_3(OH) \cdot 4Al_2[OSiO_4]$	3.65···3.77	Hod52/53
		[3.825(90)]	Wea76
		3.74···3.83	Dee62
Steinsalz, see Halite			
Stelznerite, see Antlerite			
Stephanite	$5Ag_2S \cdot Sb_2S_3$	6.25(3)	Dan44
		6.2···6.3	Klo48
Stercorite	$(NH_4)NaH[PO_4] \cdot 4H_2O$	1.615	Hod52/53, Klo48
Sternbergite	$AgFe_2S_3$	4.101···4.213	Dan44
		[4.303(12)]	Wea76
Sterrettite	$Al_6[(OH)_6(PO_4)_4] \cdot 5H_2O$ [Klo48]	2.36	Klo48
Stevensite	$Mg_{2.88}Mn_{0.02}Fe^{2+}_{0.02}[(OH)_2Si_4O_{10}]^{0.16-} \cdot X_{0.16}(H_2O)_4$	2.15···2.20	Spe27
Stewartite	$MnFe^{3+}_2[OHPO_4]_2 \cdot 8H_2O$	2.94	Spe27
Stibiconite, see also Volgerite and Hydroroméite	$SbSb_2O_6OH$	5.58	Dan44
Stibiocolumbite	$Sb(Nb, Ta)O_4$ [Dan44]	5.68	Dan44
Stibiopalladinite	Pd_3Sb	≈9.5	Dan44
Stibiotantalite	$Sb(Ta, Nb)O_4$	7.34	Dan44
		6.6···7.9	Hod52/53
Stibnite (Antimonite)	Sb_2S_3	4.651···4.654	Spe27
		4.63(2)	Dan44
		[4.6276(26)]	Wea76
		4.52···4.62	Hod52/53
Stichtite (Barbertonit)	$Mg_6Cr_2[(OH)_{16}CO_3] \cdot 4H_2O$	2.161	Spe27
		2.10(5)	Dan44
Stiepelmannite	Variety of Florencite	3.7	Klo48
Stilbite (Desmine)	$Ca[Al_2Si_7O_{18}] \cdot 7H_2O$	2.1···2.2	Dee62
		2.09···2.24	Hod52/53
Stilleite	ZnSe	[5.2630(14)]	Wea76
Stilpnomelane (Minguetite)	$(K, H_2O)(Fe^{2+}, Fe^{3+}, Mg, Al)_{<3} \cdot [(OH)_2Si_4O_{10}]X_n(H_2O)_2$	2.59···2.96	Dee62
		2.86	Spe27
Stishovite	SiO_2	[4.2874(26)]	Wea76
Stokesite	$Ca_2Sn_2[Si_6O_{18}] \cdot 4H_2O$	3.2	Klo48
Stolzite	β-Pb[WO$_4$]	7.9···8.2	Klo48
		[8.4110(99)]	Wea76
Strahlstein, see Actinolite			
Strengite	$Fe^{3+}[PO_4] \cdot 2H_2O$	2.84···2.87	Hod52/53, Spe27
Strohstein, see Carpholite			
Stromeyerite	$Cu_2S \cdot Ag_2S$	6.127···6.260	Spe27
		6.194···6.845	Cla66
		[6.194(5)]	Wea76
Strontianite	$SrCO_3$	3.680···3.714	Hod52/53
		[3.785(5)]	Wea76
		3.72	Dee62

continued

1.1.1 Density of minerals (continued)

Mineral	Formula	$\varrho, [\varrho_x]$ g cm^{-3}	Ref.
Strueverite	Composition of Rutile-Tapiolite	4.91···5.036	Spe27
		5.25···5.29	Klo48
Struvite	$NH_4Mg[PO_4] \cdot 6H_2O$	1.66···1.75	Klo48
Stylotypite		4.79; 5.18	Dan44
Sulfoborite	$Mg_3[OHFSO_4(B(OH)_4)_2]$	2.4	Klo48
Sulfohalite	$Na_6[FCl(SO_4)_2]$	2.5	Spe27
Sulfur (monocline)	α-S	2.074	Spe27
		[2.065]	Wea76
			Dan44
Sulfur (monoclineprismatic)	β-S	1.982···1.985	Dan44
		[1.943]	Wea76
Sulvanite	Cu_3VS_4	4.00; 3.86	Dan44
Sursassite	$Mn_2(H_3)Al_2[OOHSiO_4Si_2O_7]$	3.252	Spe27
Svabite	$Ca_5[F(AsO_4)_3]$	3.695	Spe27
Svanbergite	$SrAl_3[(OH)_6SO_4PO_4]$	3.14	Spe27
Swedenborgite	$NaSbBe_4O_7$	4.285	Spe27
Sylvanite	$AgAuTe_4$	8.161	Dan44
		8.0···8.3	Klo48
Sylvite	KCl	1.9···2	Klo48
		1.9868(2)	Wea76
Symplesite	$Fe_3^{2+}[AsO_4]_2 \cdot 8H_2O$	3.01	Klo48
Synadelphite (Allodelphite)	$Mn_4[(OH)_5AsO_4]$	3.46···3.65	Klo48
Syngenite	$K_2Ca[SO_4]_2 \cdot H_2O$	2.579	Spe27
		2.5707(41)	Wea76
Szaibelyite, see also Ascharite	(Mg, Mn)BO$_3$H	2.76	Spe27
		≈ 2.65	Klo48
Szomolnokite (Schmöllnitzit), see Ferropallidite			
Tabergite	Composition of Clinochlore and Biotite	2.803	Spe27
Tachyhydrite	$CaCl_2 \cdot 2MgCl_2 \cdot 12H_2O$	1.664···1.669	Spe27
		1.66	Klo48
Taeniolite	$KLiMg_2[F_2Si_4O_{10}]$	2.82	Klo48
Tagilite	$Cu_2[OHPO_4] \cdot H_2O$ (?)	4	Klo48
Talc	$Mg_3[(OH)_2Si_4O_{10}]$	2.832	Spe27
		2.784(5)	Wea76
		2.58···2.83	Dee62
Tantalum	Ta	11.2	Dan44, Spe27
Tantalite	(Fe, Mn)(Ta, Nb)$_2$O$_6$	6.5···8.20	Spe27, Hod52/53
		7.95(5)	Dan44
Tanteuxenite, see Delorenzite			
Tapiolite	(Fe, Mn)(Ta, Nb)$_2$O$_6$	7.90···7.91	Spe27
		7.90(5)	Dan44
		7.3···7.8	Hod52/53
Taramite	Variety of Katophorite	3.439···3.476	Spe27
–, Fluo-		3.231···3.318	Spe27
Tarapacaite	$K_2[CrO_4]$	2.74	Klo48
Tarbuttite	$Zn_2[OHPO_4]$	4.15	Klo48

continued

1.1.1 Density of minerals (continued)

Mineral	Formula	$\varrho, [\varrho_x]$ g cm^{-3}	Ref.
Teallite	PbSnS$_2$	6.36	Dan44
		6.4	Klo48
		6.501(7)	Wea76
Teepleite	Na$_2$[ClB(OH)$_4$]	2.08	Klo48
Teinite	Cu[TeO$_3$]·2H$_2$O	3.80	Klo48
Tellurite	TeO$_2$	5.90(2)	Dan44, Klo48
		[5.7514(50)]	Wea76
–, Para-		[6.018(5)]	Wea76
Tellurium	Te	6.1···6.3	Dan44
		[6.2316(25)]	Wea76
Tellurobismuthite	Bi$_2$Te$_3$	7.815(150)	Dan44
		[7.862(7)]	Wea76
Tellurwismut, see Tetradymite			
Tennantite	Cu$_3$AsS$_{3.25}$	4.576···4.746	Spe27
		4.62	Dan44
		[4.642(6)]	Wea76
Tenorite	CuO	5.8···6.4; 6.45 art.	Dan44
		[6.509(14)]	Wea76
Tephroite	Mn$_2$[SiO$_4$]	3.78···4.1	Dee62
		4.044	Spe27
		[4.1545(58)]	Wea76
Terlinguaite	2HgO·Hg$_2$Cl$_2$	8.723···8.728	Spe27, Hod52/53
Tetradymite (Tellurwismut)	Bi$_2$Te$_2$S	7.3(2)	Dan44
		7.2···7.9	Klo48
Tetraedrite	Cu$_3$SbS$_{3.25}$	4.597···5.079	Spe27
		4.97	Dan44
		4.4···5.1	Hod52/53
		[5.024(6)]	Wea76
Thalenite	Y$_2$[Si$_2$O$_7$]	4.454	Spe27
		4.4	Klo48
Thaumasite	Ca$_3$H$_2$[CO$_3$SO$_4$SiO$_4$]·13H$_2$O	1.83···1.877	Hod52/53
Thenardite	α-Na$_2$[SO$_4$]	2.67	Spe27
		2.68···2.69	Hod52/53
		[2.663(3)]	Wea76
Thermonatrite	Na$_2$CO$_3$·H$_2$O	1.5···1.6	Hod52/53
Thomsenolite	NaCa[AlF$_6$]·H$_2$O	2.982	Spe27
		2.93···3.0	Hod52/53
Thomsonite	NaCa$_2$[Al$_2$(Al, Si)Si$_2$O$_{10}$]$_2$·6H$_2$O	2.10···2.39	Dee62
Thoreaulite	Sn[(Ta, Nb)$_2$O$_7$]	7.6···7.9	Dan44
Thorianite	ThO$_2$	9.7	Dan44
		9.32···9.33	Hod52/53
		[10.012(3)]	Wea76
Thorite	Th[SiO$_4$]	4.4···4.8	Klo48
		4.5···5 (black)	Hod52/53
		5.2···5.4 (yellow)	Hod52/53
		[6.668(8)]	Wea76
Thortveitite	Sc$_2$[Si$_2$O$_7$]	3.55···3.571	Spe27
		≈3.6	Klo48
Thuringite	(Fe^{2+}, Fe^{3+}, Al)$_3$[(OH)$_2$Al$_{1.2···2}$Si$_{2.8···2}$O$_{10}$]·(Mg, Fe, Fe^{3+})$_3$(O, OH)$_6$	≈3.2	Klo48
		3.07	Spe27

continued

1.1.1 Density of minerals (continued)

Mineral	Formula	$\varrho, [\varrho_x]$ g cm^{-3}	Ref.
Tiemannite (Selenquecksilber)	HgSe	8.19	Dan44
		8.30···8.47	Dan44
		[8.239(20)]	Wea76
Tilasite	CaMg[FAsO$_4$]	3.76···3.77	Spe27
		3.7	Klo48
Tilleyite	Ca$_5$[(CO$_3$)$_2$Si$_2$O$_7$]	2.84	Klo48
Tin (Zinn)	β-Sn	7.31	Dan44
		[7.2867(24)]	Wea76
Tinzenite	CaMn$_2$Al$_2$[BO$_3$OHSi$_4$O$_{12}$]	3.286···3.416	Spe27
		3.29	Klo48
Tirolite	Ca$_2$Cu$_9$[(OH)$_{10}$(AsO$_4$)$_4$]·10H$_2$O	≈3.1	Klo48
Titanite	CaTi[OSiO$_4$]	3.40···3.56	Hod52/53
Titanium sesquioxide	Ti$_2$O$_3$	[4.574(5)]	Wea76
Toddite	Composition of Columbite + Samarskite	5.011	Spe27
		5.041	Dan44
Todorokite	(H$_2$O, ...)$_{\leq 2}$(Mn, ...)$_{\leq 8}$(O, OH)$_{16}$	3.67	Dan44
Toernebohmite	(Ce, La, Al)$_3$[OH(SiO$_4$)$_3$]	4.94	Spe27
Topaz	Al$_2$[F$_2$SiO$_4$]	3.49···3.57	Dee62
		3.4···3.65	Hod52/53
		[3.563(5)]	Wea76
Torbernite	Cu[UO$_2$PO$_4$]$_2$·10(12···8)H$_2$O	3.22···3.60	Hod52/53
–, Meta-	Cu[UO$_2$PO$_4$]$_2$·8H$_2$O	3.67···3.68	Spe27
Torendrikite		3.153···3.21	Spe27
Tourmaline, see Dravite, see Schoerl, see Elbaite			
Tremolite (Grammatite)	Ca$_2$Mg$_5$[(OH, F)Si$_4$O$_{11}$]$_2$	2.9···3.4	Hod52/53
		[2.977(8)]	Wea76
		3.02···3.44	Dee62
–, Fluor-	Ca$_2$Mg$_5$[FSi$_4$O$_{11}$]$_2$	[3.018(5)]	Wea76
Trevorite	(Fe^{3+}Ni)Fe^{3+}O$_4$	4.67···5.165	Spe27
		[5.370(6)]	Wea76
		5.164	Dan44
		5.26	Dee62
Tridymite	SiO$_2$	2.267···2.270	Spe27
		2.28···2.33	Hod52/53
		[2.265]	Wea76
		2.27	Dee62
Trigonite	Pb$_3$MnH[AsO$_3$]$_3$	8.28	Spe27, Klo48
Trimerite	CaMn$_2$Be$_3$[SiO$_4$]$_3$	3.404	Spe27
		3.5	Klo48
Triphylite	Li(Fe^{2+}, Mn^{2+})[PO$_4$]	3.34···3.58	Dan44
		3.531	Spe27
Triplite	(Mn, Fe^{2+})$_2$[FPO$_4$]	3.5···3.9	Klo48
Triploidite	(Mn, Fe^{2+})$_2$[OHPO$_4$]	3.66···3.83	Dan44
		3.7	Klo48
Tritomite	(Ce, La, Y, Th, Zr, ...)$_{3.5···3.7}$(Si, B)$_{2.8···3}$·(O, OH, F)$_{13}$	≈4.2	Klo48
Troegerite	(H$_3$O)$_2$[UO$_2$AsO$_4$]$_2$·6H$_2$O	3.3	Klo48, Hod52/53
Trogtalite	CoSe$_2$	[7.1618(37)]	Wea76

continued

1.1.1 Density of minerals (continued)

Mineral	Formula	$\varrho, [\varrho_x]$ g cm^{-3}	Ref.
Troilite, see Pyrrhotite (Magnetkies)			
Trona	$Na_3H[CO_3]_2 \cdot 2H_2O$	2.14	Spe27
		2.11···2.147	Hod52/53
Trudellite	$Al_{10}[(OH)_4Cl_4SO_4]_3 \cdot 30H_2O$	1.93	Spe27
Truscottite (Reyerite)	$Ca_2[Si_4O_{10}] \cdot H_2O$	2.47	Spe27
Tschermakite, Ferro-	$Ca_2Fe_3^{2+}(Al, Fe^{3+})_2[(OH, F)_2Al_2Si_6O_{22}]$	3.02···3.45	Dee62
Tschkalowit, see Chkalovite			
Tschermigite	$NH_4Al[SO_4]_2 \cdot 12H_2O$	1.645	Spe27, Klo48
Tsumebite	$Pb_2Cu[OHSO_4PO_4]$	6.1	Klo48
Tucholite		1.78	Klo48
Türkis, see Turquoise			
Tuhualite	$(Na, K, Mn)_2Fe^{2+}(Fe^{3+}, Al, Mg, Ti)$ $\cdot H(Si, AlH)_8O_{20}$	2.87	Klo48
Tungstenite	WS_2	7.4; 7.5 art.	Dan44
		[7.7325(55)]	Wea76
Turgite, see Hydrohematite			
Turquoise (Türkis)	$CuAl_6[(OH)_2PO_4]_4 \cdot 4H_2O$	2.84	Spe27
		2.60···2.89	Hod52/53
		[2.927(7)]	Wea76
Tysonite, see Fluocerite			
Tyuyamunite	$Ca(UO_2)_2(VO_4)_2 \cdot nH_2O$ [Dan44]	3.67···4.35	Spe27
Uhligite	Variety of Perovskite	4.15(10)	Dan44
Ulexite	$NaCa[B_5O_6(OH)_6] \cdot 5H_2O$	1.91	Spe27
		1.7	Klo48
Ullmannite	$NiSbS$	6.70	Spe27
		6.65(4)	Dan44
Ullmannite (Kallilite)		6.66	Dan44
Ullmannite (10.28%As)		6.488	Dan44
Ullmannite (Willyamite)		6.76(3)	Dan44
Ultrabasite, see Diaphorite			
Ulvite, see Ulvospinel			
Ulvospinel (Ulvite)	$(Fe^{2+}Ti^{4+})Fe^{2+}O_4$	4.78	Dee62
Umangite	Cu_3Se_2	5.620	Dan44
		[6.604(26)]	Wea76
Uralite	$Ca_4Mg_6Fe_{3\cdots4}^{3+}[(OH, O)_4(Al, Fe)_2Si_{14}O_{44}]$	3.118	Spe27
Uraninite	UO_2	7.126···9.787	Spe27
		6.5···9.7	Hod52/53
		[10.969(6)]	Wea76
Uranocircite	$Ba[UO_2PO_4]_2 \cdot 10H_2O$	3.5	Klo48
Uranospathite	$Cu[UO_2(As, P)O_4]_2 \cdot \approx 12H_2O$	2.50	Spe27
Uranosphaerite	$[UO_2(OH)_2BiOOH]$	6.36	Dan44
Uranothallite (Liebigite)	$Ca_2[UO_2(CO_3)_3] \cdot 10H_2O$	2.1	Klo48
Uranotile	$CaU_2[(OH)_3SiO_4]_2 \cdot 4H_2O$ [Klo48]	3.8···3.9	Klo48
Uranpyrochlore, see Blomstrandite and Ellsworthite			
Ureyite	$NaCr(SiO_3)_2$	[3.605(9)]	Wea76
Ussingite	$Na_2[OHAlSi_3O_8]$	2.495	Spe27, Klo48

continued

1.1.1 Density of minerals (continued)

Mineral	Formula	$\varrho, [\varrho_x]$ g cm^{-3}	Ref.
Uvarovite	$Ca_3Cr_2^{3+}[SiO_4]_3$	3.418···3.81	Spe27, Hod52/53
		[3.848(2)]	Wea76
Uvite	$CaMg_3(Al_5Mg)[(OH)_{1+3}(BO_3)_3Si_6O_{18}]$	[3.095(6)]	Wea76
Vaesite	NiS_2	[4.4350(12)]	Wea76
Valentinite	Sb_2O_3	5.76	Dan44
		5.566	Hod52/53
		[5.8292(54)]	Wea76
Valleriite	$CuFeS_2$	≈ 4.2	Klo48
Vanadinite	$Pb_5[Cl(VO_4)_3]$	6.46	Spe27
		6.7···7.2	Hod52/53
Vandenbrandeite	$[UO_2(OH)_2] \cdot Cu(OH)_2$	5.03	Dan44
		4.96	Klo48
Vanthoffite	$Na_6Mg[SO_4]_4$	2.69	Klo48
		[2.6730(25)]	Wea76
Variscite	$Al[PO_4] \cdot 2H_2O$	2.47···2.54	Hod52/53
Varulite	$(Na, Ca)_2(Mn, Fe)_3[PO_4]_3$	≈ 3.45	Klo48
Vaterite	$CaCO_3$	[2.653(4)]	Wea76
Vauxite	$Fe^{2+}Al_2[OHPO_4]_2 \cdot 6H_2O$	2.375···2.57	Spe27
Veatchite	$Sr_2[B_5O_8(OH)]_2 \cdot B(OH)_3 \cdot H_2O$	2.69	Klo48
Velardeñite (Variety of Gehlenite)	$Ca_2Al_2SiO_7$ [Spe27]	3.038···3.039	Spe27
Vermiculite	$Mg_{2.36}Fe_{0.48}^{3+}Al_{0.16}[(OH)_2Al_{1.28}$ $\cdot Si_{2.72}O_{10}]^{0.64}Mg_{0.32}(H_2O)_4$	≈ 2.3	Dee62
Vesuvianite (Idokras)	$Ca_{10}(Mg, Fe)_2Al_4[(OH)_4(SiO_4)_5(Si_2O_7)_2]$	3.35···3.45	Hod52/53
		3.3···3.43	Dee62
		3.32···3.47	Spe27
Veszelyite, see Arakawaite			
Villamaninite	$(Cu, Ni, Co, Fe)(S, Se)_2$	4.433···4.523	Spe27
Villiaumite	NaF	2.79	Klo48, Hod52/53
		[2.8021(9)]	Wea76
Violarite	$FeNi_2S_4$	[4.725(8)]	Wea76
Viridine	Cr-Kyanite with 12.86% Cr_2O_3	3.202···3.238	Spe27
Viridite	Leptochlorite series	2.89	Spe27
Vishnevite, see also Cancrinite	$(Na, Ca, K)_{6\cdots7}[(SO_4, CO_3, Cl)_{1\cdots1.5}(AlSiO_4)_6]$ $\cdot 1\cdots5H_2O$	2.32···2.42	Dee62
Vivianite	$Fe_3^{2+}[PO_4]_2 \cdot 8H_2O$	2.58···2.693	Hod52/53
Voelckerite (Oxi-Apatite)	$Ca_{10}[O(PO_4)_6]$	3.06···3.10	Spe27
Vogtite	Metasilicate of Fe, Ca, Mn, Mg	3.39	Spe27
Volgerite, see also Stibiconite		3.082	Spe27
Voltaite	$K_2Fe_5^{2+}Fe_4^{3+}[SO_4]_{12} \cdot 18H_2O$	2.6···2.8	Klo48
Voltzine, see Voltzite			
Voltzite (Voltzine)	Zn_5S_4O [Klo48]	3.66	Klo48
		3.7···3.8	Dan44
Vonsenite	$(Fe^{2+}, Mg)_2Fe^{3+}[O_2BO_3]$	4.21	Spe27, Klo48
Vrbaite	$Tl_4Hg_3Sb_2As_8S_{20}$	5.30(3)	Dan44, Spe27
Wad	MnO_2	2.8···4.4	Dan44
Wadeite	$K_2Zr[Si_3O_9]$	3.10	Klo48

continued

1.1.1 Density of minerals (continued)

Mineral	Formula		$\varrho, [\varrho_x]$ g cm^{-3}	Ref.
Wagnerite	$Mg_2[FPO_4]$		3···3.15	Klo48, Hod52/53
Walpurgite (Waltherite)	$[(BiO)_4UO_2(AsO_4)_2] \cdot 3H_2O$		5.3···5.7	Klo48
Waltherite, see Walpurgite				
Wardite (Soumansite)	$NaAl_3[(OH)_4(PO_4)_2] \cdot 2H_2O$		2.8	Klo48
			2.87	Spe27
Warthaite (Goongarrite)	$Pb_4Bi_2S_7$	[Klo48]	7.163	Spe27
			7.12(5)	Dan44
			7.29	Klo48, Spe27, Dan44
Warwickite	$(Mg, Fe)_3Ti[OBO_3]_2$		≈ 3.31	Klo48
Wavellite (Fischerite)	$Al_3[(OH)_3(PO_4)_2] \cdot 5H_2O$		2.325	Spe27
			2.316···2.356	Hod52/53
			2.46	Klo48
Weberite	$Na_2Mg[AlF_7]$		2.96	Klo48
Wehrlite	Bi_3Te_2	[Klo48]	8.41(3)	Dan44
Weibullite	$PbS \cdot Bi_2Se_3$ (?)		6.97; 7.145	Dan44
Weibyeite, see Bastnaesite				
Weinschenkite	$(Y, Er)[PO_4] \cdot 2H_2O$		3.14	Klo48
Wernerite, see Skapolite				
Weslienite	Roméite group		4.964···4.971	Spe27
Whewellite	$Ca[C_2O_4] \cdot H_2O$		2.23	Klo48, Hod52/53
Whitneyite	(Cu, As)		8.3···8.7	Klo48
Wilkeite	$Ca_5[(F, O)(PO_4, SiO_4, SO_4)_3]$		3.234	Spe27
			3.1	Dan44
Willemite	$Zn_2[SiO_4]$		3.89···4.19	Hod52/53
			[4.251(6)]	Wea76
Wismutglanz, see Bismuthinite				
Witherite	$BaCO_3$		4.28···4.35	Hod52/53
			[4.308(6)]	Wea76
Wittite	$Pb_5Bi_6(S, Se)_{14}$ (?)	[Dan44]	7.12	Dan44, Spe27
Woehlerite	$Ca_2NaZr[(F, OH, O)_2Si_2O_7]$		3.48	Spe27
			3.44	Klo48
Wolfachite	$Ni(As, Sb)S$ (?)	[Dan44]	6.372	Dan44
Wolframite	$(Mn, Fe)WO_4$		6.93···7.49	Spe27
			7.14···7.54	Hod52/53
			[7.376(10)]	Wea76
Wollastonite	$Ca_3[Si_3O_9]$		2.897···2.992	Spe27
			2.80···2.92	Hod52/53
			2.87···3.09	Dee62
			[2.909(8)]	Wea76
–, Para-	$CaSiO_3$	[Cla66]	[2.915(2)]	Wea76
–, Pseudo-	$CaSiO_3$		[2.899(10)]	Wea76
Wuestite	$Fe_{0.953}O$	[Cla66]	[5.7471(12)]	Wea76
Wulfenite	$Pb[MoO_4]$		6.7···7.0	Hod52/53
			[6.816(13)]	Wea76
Wurtzite	β-ZnS		4.087	Spe27
			3.9; 4.1 art.	Dan44
			[4.0859(22)]	Wea76
Xanthitane, see also Anatase			3.04	Spe27

continued

1.1.1 Density of minerals (continued)

Mineral	Formula	$\varrho, [\varrho_x]$ g cm^{-3}	Ref.
Xanthoconite	Ag_3AsS_3	5.54(14)	Dan44
Xanthophyllite	$Ca(Mg, Al)_{3...2}[(OH)_2Al_2Si_2O_{10}]$	3.081	Spe27
		3...3.1	Dee62
Xanthoxenite	$Ca_2Fe^{3+}[OH(PO_4)_2] \cdot 1.5 H_2O$	2.844	Spe27
Xenotime	$Y[PO_4]$	4.55	Spe27
		4.45...4.56	Hod52/53
		[4.307(8)]	Wea76
Xonotlite (Eakleite, Jurupaite)	$Ca_6[(OH)_2Si_6O_{17}]$	2.655...2.705	Spe27
		2.75	Klo48
Yeatmanite	$(Mn, Zn)_{16}Sb_2[O_{13}(SiO_4)_4]$	4	Klo48
Yenerite	$Pb_{11}Sb_8S_{23}$ [Klo48]	6.05	Klo48
Yttrocerite (Cerfluorite)	$(Ca, Ce)F_{2...2.17}$	3.61	Spe27
Yttrofluorite	$(Ca, Y)F_{2...2.17}$	3.319...3.557	Spe27
Yttrotantalite	$(Y, U, Ca)(Ta, Fe^{3+})_2(O, OH)_6$	5.7(2)	Dan44
		5.4...5.9	Klo48
Zaratite	$Ni_3[(OH)_4CO_3] \cdot 4 H_2O$	2.6	Klo48
Zebedassite (Orthochlorite)	Variety of Saponite	2.194	Spe27
Zeophyllite, see Radiophyllite			
Zeunerite	$Cu[UO_2AsO_4]_2 \cdot 10(16...10)H_2O$	3.28	Spe27, Hod52/53
Zinc	Zn	6.9...7.2	Dan44
		[7.134(6)]	Wea76
Zincaluminite	$Zn_3Al_3[(OH)_{13}SO_4] \cdot 2 H_2O$	2.26	Klo48
Zincite	ZnO	5.675	Cla66
		5.66(2); 5.684	Dan44
		[5.6750(180)]	Wea76
		5.43...5.70	Hod52/53
Zinckenite (Keeleyite)	$6 PbS \cdot 7 Sb_2S_3$	5.30(5)	Dan44
		5.21	Spe27
		5.3	Klo48
Zinkblende, see Sphalerite			
Zinkosite	$Zn[SO_4]$	[3.883(6)]	Wea76
Zinkvitriol (Goslarite)	$Zn[SO_4] \cdot 7 H_2O$	2.0	Klo48
		1.9...2.1	Hod52/53
		[1.9723(15)]	Wea76
Zinn, see Tin			
Zinnkies, see Stannite			
Zinnober, see Cinnabar			
Zinnwaldite	$KLiFe^{2+}Al[(F, OH)_2AlSi_3O_{10}]$	2.90...3.02	Dee62, Spe27
Zircon	$Zr[SiO_4]$	4.6...4.7	Dee62
		4.02...4.86	Hod52/53
		[4.669(8)]	Wea76
Zirconoxide, see Baddeleyite			
Zirkelite	$(Ca, Fe, Th, U)_2(Ti, Zr)_2O_5$ (?)	4.3...5.22	Spe27
		4.7	Klo48
Zirklerite	$\approx 9 FeCl_2 \cdot 4 AlOOH$ with Mn, Ca, Fe^{3+}	2.6	Klo48
Zoisite	$Ca_2Al_3[OOHSiO_4Si_2O_7]$	3.35	Spe27
		[3.328(5)]	Wea76
		3.15...3.365	Dee62
Zunyite	$Al_{12}[AlO_4(OH, F)_{18}ClSi_5O_{16}]$	2.9	Klo48

1.1.2 References for 1.1 — Literatur zu 1.1

Cla66	Clark jr., S.P., (Ed.): Handbook of Physical Constants, New York: The Geol. Soc. Am., Inc. **1966**.
Dan44	Dana, J.D., Dana, E.S.: System of Mineralogy, 7th ed., New York: J. Wiley and Sons, Inc. **1944**.
Dee62	Deer, W.A., Howie, R.A., Zussman, J.: Rock-Forming Minerals, London: Longmans, Green and Co. Ltd. **1962**.
Don73	Donnay, J.D.M., Ondik, M.M.: Crystal Data (Determination Tables), U.S.Dept. of Commerce, Washington, **1973**.
Hod52/53	Hodgman, C.D., (Ed.): Handbook of Chemistry and Physics, 34th ed., Chemical Rubber Publ. Comp. **1952-1953**.
Klo48	Klockmann's Lehrbuch der Mineralogie, P. Ramdohr (Ed.), Stuttgart: F. Enke Verlag **1948**.
Koh55	Kohlrausch, W.: Lehrbuch der praktischen Physik, 2 Bde, 20. Aufl. **1955**.
McC79	McClune, W.F., Mrose, M.E., Post, B., Weissmann, S., McMurdie, H.F., Morris, M.C., Zwell, L., (Eds.): Powder Diffraction File, Alphabetical Index Inorganic Materials **1979**; Publ. SMA-29, JCPDS, International Centre for Diffraction Data, Penn. U.S.A.
Rob78	Robie, R.A., Hemingway, B.S., Fisher, J.R.: Geol. Survey Bull. 1452 (1978), U.S. Government Printing Office, Washington, USA.
Spe27	Spencer, L.J.: Mineralog. Mag. **119** (1927) 337.
Str70	Strunz, H.: Mineralogische Tabellen, 6. Aufl. **1977**, Leipzig: Akademische Verlagsgesellschaft Geest und Portig KG.
Wea76	Weast, R.C., (Ed.): Handbook of Chemistry and Physics, 56th ed. **1976**, Cleveland, Ohio: CRC-Press.
Wes70	Westphal, W.H.: Physik – ein Lehrbuch, 25./26. Aufl. Berlin-Göttingen-Heidelberg: Springer 1970.

1.2 Density of rocks — Dichte der Gesteine
1.2.0 Introduction — Einleitung

The bulk density of rocks strongly depends on their mineral components and on their content of enclosed cavities. Natural cavities often make up a considerable portion of the total volume. With the percentage Φ of enclosed cavities related to the total volume V, the density ϱ of the solid components, and the density ϱ_0 of the cavity fillings the total mass M of the volume V results in

Die natürliche Dichte von Gesteinen hängt wesentlich von deren mineralischer Zusammensetzung und von den im Gestein vorhandenen Hohlräumen ab, die beträchtliche Teile ihres Gesamtvolumens ausmachen können. Bezeichnet man mit Φ den prozentualen Anteil der Hohlräume am Gesamtvolumen V, mit ϱ die Dichte der festen Gesteinsmatrix und mit ϱ_0 die Dichte der Hohlraumfüllung, dann ist die Gesamtmasse M im Volumen V:

$$M = V \cdot \varrho \left(1 - \frac{\Phi}{100}\right) + V \cdot \varrho_0 \frac{\Phi}{100} \tag{4}$$

and the bulk density in

und die natürliche Dichte

$$\varrho_m = \left[\varrho - (\varrho - \varrho_0)\frac{\Phi}{100}\right] \tag{5}$$

In igneous and most metamorphic rocks the influence of cavities (pore volume, micro fissures etc.) on the density usually is small and can be neglected ($\varrho_m \rightarrow \varrho$). In this case the variation of density indicates variation of the mineral composition of these rock types. For non-consolidated and porous rock cavities and their fillings strongly influence the bulk density (see also: Chapter 2.1, Porosity of rocks).

In Tiefengesteinen und in den meisten metamorphen Gesteinen ist der Anteil der Hohlräume (Porenvolumina, Haarrisse etc.) am Gesamtvolumen im allgemeinen gering und zu vernachlässigen ($\varrho_m \rightarrow \varrho$). Hier charakterisieren angegebene Dichteintervalle die Variationsbreite der mineralischen Zusammensetzung des Gesteins. In lockeren und porösen Gesteinen ist der Einfluß der Hohlräume und ihrer Füllungen von wesentlicher Bedeutung für die Dichtewerte (siehe hierzu auch Kapitel 2.1, Porosität der Gesteine).

For reductions and interpretation work of gravity data the use of the bulk density ϱ_m is recommended.

It has to be pointed out that most of the reviewed references give no information on the method of determining the given density values. Therefore it is often impossible to decide wether the given densities are ϱ_m or ϱ according to (5). The listed values from Reich [Rei31/33] give the density ϱ of the solid rock components. Density ranges are given in parentheses and are preceded by their mean value.

Bei der Reduktion und Interpretation von Schweremessungen ist immer die natürliche Dichte ϱ_m zu verwenden.

Es muß darauf hingewiesen werden, daß den verwendeten Quellen in den seltensten Fällen zu entnehmen ist, auf welche Art und Weise die Dichten der Gesteine ermittelt wurden. Daher ist oft nicht zu entscheiden, ob es sich bei den angegebenen Dichten um ϱ_m oder ϱ gemäß (5) handelt. Die Werte aus den Zusammenstellungen von Reich [Rei31/33] geben die Dichte ϱ der festen Gesteinsmatrix an. Dichte-Bereiche stehen in runden Klammern. Ihnen vorangestellt ist der Mittelwert dieses Bereiches.

1.2.1 Density of intrusive rocks

See Fig. 1

Material	Density ϱ g cm^{-3}	Ref.
Granite	2.667 (2.516···2.809)	Cla66
	2.780	Rzh71
	2.60 (2.53···2.67)	Kry57
	2.65 (2.56···2.74)	Rei31/33
Biotite Muscovite Alcaline Granite	2.66	Woe63
Granodiorite	2.70	Kry57
	2.716 (2.668···2.789)	Cla66
Quartz-Diorite	2.79 (2.62···2.90)	Rei31/33, Haa53
	2.806 (2.680···2.960)	Cla66
Augite-Diorite	3.01 (2.99···3.08)	Rei31/33
Biotite Muscovite Quartz Monzonite	2.65	Woe63
Diorite	2.80	Rzh71
	2.839 (2.721···2.960)	Cla66
	2.86 (2.72···2.99)	Rei31/33
Syenite	2.757 (2.630···2.899)	Cla66
	2.71	Rzh71
	2.74 (2.60···2.95)	Rei31/33
Quartz Syenite	2.63	Kry57
Nepheline Syenite	2.62 (2.53···2.70)	Rei31/33, Haa53
Gabbro	2.976 (2.850···3.120)	Cla66
	2.90	Rzh71
	3.0 (2.89···3.09)	Rei31/33

Material	Density ϱ g cm^{-3}	Ref.
Quartz Gabbro	2.99	Woe63
Hornblende Gabbro	3.05 (2.98···3.18)	Rei31/33, Haa53
Olivine Gabbro	2.95 (2.85···3.06)	Rei31/33, Haa53
Norite	2.93 (2.70···3.24)	Rei31/33, Haa53
	2.984 (2.720···3.020)	Cla66
Anorthosite	2.734 (2.640···2.920)	Cla66
	2.75	Woe63
	2.73 (2.64···2.94)	Rei31/33
Dunite	3.277 (3.204···3.314)	Cla66
	3.22 (2.93···3.34)	Haa53
Peridotite	2.8	Rzh71
	3.234 (3.152···3.276)	Cla66
	3.06 (2.78···3.37)	Rei31/33, Haa53
Pyroxenite	3.231 (3.10···3.318)	Cla66
	3.22 (2.93···3.34)	Rei31/33, Haa53
Serpentinised Peridotite	2.67	Woe63
Essexite	2.95 (2.69···3.14)	Rei31/33
Aplite	2.50	Kry57
	2.70 (2.50···2.80)	Haa53
Lamprophyre	2.9 (2.8···3.2)	Haa53

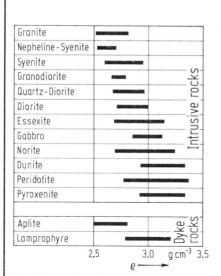

Fig. 1. Densities of intrusive rocks and dyke rocks

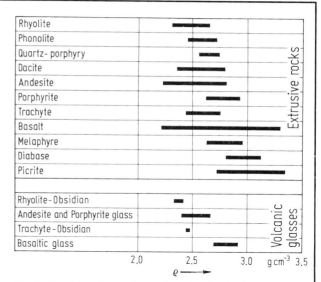

Fig. 2. Densities of extrusive rocks and volcanic glasses.

1.2.2 Density of extrusive rocks

See Fig. 2

Material	Density ϱ g cm^{-3}	Ref.	Material	Density ϱ g cm^{-3}	Ref.
Phonolite	2.56 (2.45⋯2.71)	Rei31/33, Haa53	Melaphyry	2.77 (2.63⋯2.95)	Rei31/33
Rhyolite, Liparite	2.5 (2.35⋯2.65)	Rei31/33	Diabase	2.965 (2.804⋯3.110)	Cla66
	2.05, 2.39	Woe63		2.94 (2.73⋯3.12)	Rei31/33
	2.49	Kry57		2.95, 2.82	Kry57
	2.4 (2.3⋯2.5)	Haa53	Basalt	2.75 (2.21⋯2.77)	Kry57
Porphyry	2.67 (2.60⋯2.89)	Rei31/33, Haa53		2.90 (2.74⋯3.21)	Rei31/33
				2.7⋯3.3	Jak50
Quartz-Porphyry	2.63 (2.55⋯2.73)	Rei31/33		2.860	Rzh71
Dacite	2.35⋯2.79	Rei31/33	Olivine Basalt	2.83, 3.0	Woe63
	2.46	Kry57	Cellular Olivine Basalt	2.44	Woe63
Latite (Trachytic Andesite)	2.45	Woe63	Dolerite	2.85 (2.45⋯3.09)	Haa53
Quartz Latite	2.46	Woe63	Pikrite	2.97 (2.73⋯3.35)	Rei31/33, Haa53
Andesite	2.62 (2.44⋯2.8)	Rei31/33	Vesicular Lava	2.8⋯3.0	Jak50
	2.70 (2.22⋯2.79)	Kry57	Basaltic Lava	2.9 (2.8⋯3.0)	Haa53
Hornblende Andesite	2.32	Woe63	Trachytic Lava	2.35 (2.0⋯2.70)	Haa53
			Tuffs	1.38 (1.30⋯2.40)	Haa53
Vesicular Andesite	2.57	Woe63	Trachytic Tuff	2.42	Woe63
Porphyrite	2.74 (2.62⋯2.93)	Rei31/33, Haa53	Pumice (Bimsstein)	0.64 (0.36⋯0.91)	Haa53
			Volcanic Breccia	2.19	Woe63
Trachyte	2.58 (2.44⋯2.76)	Rei31/33	Andesitic Breccia	2.73	Woe63
	2.62	Woe63	Basaltic Scoria	2.23	Woe63

1.2.3 Density of volcanic glasses
See Fig. 2, p. 115

Material	Density ϱ g cm^{-3}	Ref.
Obsidian	2.21···2.42	Rei31/33
	2.35	Woe63
	2.2···2.4	Jak50
Basaltic Glasses	2.772 (2.704···2.851)	Cla66
	2.81 (2.75···2.91)	Rei31/33
Andesite- and Porphyrite Glasses	2.474 (2.40···2.573)	Cla66
	2.50···2.66	Rei31/33
Vitrophyre	2.36···2.53	Rei31/33
Rhyolite-Glass	2.26 (2.20···2.28)	Rei31/33, Haa53
Rhyolite-Obsidian	2.370 (2.330···2.413)	Cla66
Trachyte-Obsidian	2.450 (2.435···2.467)	Cla66
Leucite-Tephrite Glass	2.55 (2.52···2.58)	Cla66
Pitchstone (Pechstein)	2.338 (2.321···2.37)	Cla66
	2.40 (2.36···2.53)	Haa53

1.2.4 Density of metamorphic rocks
See Fig. 3

Material	Density ϱ g cm^{-3}	Ref.	Material	Density ϱ g cm^{-3}	Ref.
Schist (Schiefer)	2.51···2.72	Rzh71	Hypersthene Granulite	2.93 (2.67···3.10)	Cla66
	2.7···2.9	Jak50	Graywacke	2.6···2.7	Jak50, Rei31/33
	2.39···2.87	Rei31/33	Haelleflinta	2.7···2.86	Rei31/33
Chloritic Slate (Chlorit Schiefer)	2.87 (2.75···2.98)	Rei31/33	Quartzite	3.0	Rzh71
Staurolite Garnet Mica-Schist	2.76	Cla66		2.5···2.6	Jak50
				2.64	Kry57
Chlorite-Sericite-Schist	2.82 (2.73···3.03)	Cla66	Amphibolite	3.00 (2.91···3.04)	Rei31/33
				2.99 (2.79···3.14)	Cla66
Quartzitic Slate	2.68 (2.63···2.91)	Rei31/33	Serpentine	2.95 (2.80···3.10)	Rei31/33
Quartz-Mica-Schist	2.82 (2.70···2.96)	Cla66		2.4···2.8	Jak50
Mica-Schist	2.73 (2.54···2.97)	Rei31/33, Haa53	Phyllite	2.74 (2.68···2.80)	Rei31/33
				2.7···2.8	Jak50
Two-Mica-Schist (Zweiglimmerschiefer)	2.76	Cla66		2.70 (2.60···2.90)	Haa53
			Eclogite	3.35 (3.20···3.54)	Rei31/33
				3.392 (3.338···3.452)	Cla66
Alum-Schist (Alaunschiefer)	2.46 (2.34···2.59)	Haa53	Siliceous Lime (Kalksilikatgestein)	2.67···3.11	Rei31/33
				2.7	Jak50
Marl-Schist (Mergelschiefer)	2.67 (2.5···2.75)	Haa53	Skarn	2.8···3.2	Rzh71
Gneiss	2.66 (2.61···3.12)	Kry57	Marble (Marmor)	2.88	Rzh71
	2.75 (2.59···3.0)	Rei31/33		2.61 (2.49···2.73)	Kry57
	2.4···3.0	Jak50		2.78 (2.63···2.87)	Rei31/33
Granite gneiss	2.61 (2.59···2.63)	Cla66		2.5···2.9	Jak50
Feldspar gneiss	2.67	Cla66	Dolomitic Marble	2.84	Kry57
Granulite	2.73 (2.63···2.85)	Cla66	Jadeite	3.27···3.36	Rei31/33
	2.64 (2.57···2.73)	Rei31/33, Haa53			

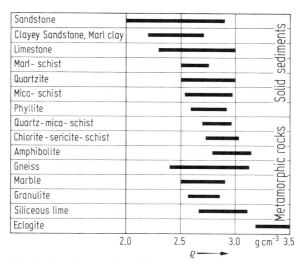

Fig. 3. Densities of solid sediments and metamorphic rocks

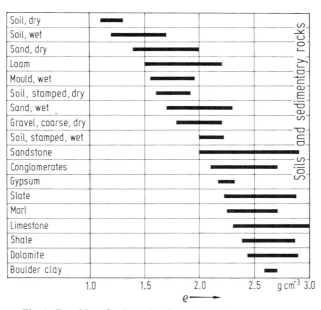

Fig. 4. Densities of soils and sedimentary rocks.

1.2.5 Density of sedimentary rocks and soils

See Fig. 4, p. 117

Material	Density ϱ g cm^{-3}	Ref.	Material	Density ϱ g cm^{-3}	Ref.
Soil, dry	1.1···1.3	Rei31/33	Shale (Schieferton)	2.59 (2.39···2.87)	Rei31/33
	1.20 (1.12···1.28)	Haa53	Bentonite	2.44···2.78	Spe27
Soil, wet	1.2···1.7	Rei31/33	Clayey Sandstone (Toniger Sandstein)	2.48	Kry57
Mould (Humus)	1.45 (1.22···1.68)	Haa53			
Mould, wet	1.73 (1.55···1.95)	Haa53			
Soil, stamped, dry	1.6···1.9	Rei31/33	Flint (Feuerstein, Hornstein)	2.48	Kry57
	1.8 (1.6···1.92)	Haa53		2.80	Rzh71
Soil, stamped, wet	2.1···2.2	Rei31/33		2.70 (2.58···2.80)	Haa53
	2.10 (2.00···2.22)	Haa53	Sandstone	2.1···2.9	Rzh71
Loess	2.64	Haa53		2.65 (2.59···2.72)	Rei31/33
Sand, dry	1.4···2.0	Rzh71		2.0···2.6	Jak50
	1.4···1.7	Rei31/33, Haa53	Variegated Sandstone	2.3 (2.3···2.5)	Haa53
Sand, wet	1.7···2.3	Rei31/33	Calcareous Sandstone	2.31	Kry57
Rubble with sand	2.0···2.5				
Moulding Sand (Formsand)	2.63 (2.54···2.63)	Haa53	Limestone (Kalkstein)	2.3···3.0	Rzh71
				2.49 (2.34···2.58)	Jak50
Loam (Lehm)	1.50···2.20	Rzh71		2.73 (2.68···2.84)	Rei31/33
Loam, wet, sandy	1.7···2.2	Rei31/33		2.67 (2.54···2.72)	Kry57
Loam, chalky	1.5	Hei68	Calcareous Tuffs (Kalktuff)	1.64 (1.57···1.71)	Haa53
Clay (Ton)	2.2···2.7	Mud60	Dolomite	2.69	Kry57
	2.02 (1.78···2.31)	Jak50		2.8	Hei68
	2.46 (2.35···2.64)	Rei31/33		2.75 (2.44···2.90)	Haa53
Marl (Mergel)	2.3···2.5	Hei68	Coral (Koralle)	2.66	Kry57
	2.5 (2.3···2.7)	Haa53	Oolitic Limestone (Kalkoolith)	2.66	Kry57
	2.25···2.6	Jak50			
Boulder Clay (Geschiebemergel)	2.66 (2.60···2.71)	Rei31/33, Haa53	Diatomite	2.0···2.35	Mud60
			Travertine	2.63	Kry57
Potter's Clay (Letten)	1.70	Haa53	Gypsum	2.20	Rzh71
				2.26 (2.17···2.31)	Haa53
Gravel, coarse, dry	2.0···2.2	Rei31/33	Anhydrite	2.96 (2.92···2.98)	Haa53
	1.90 (1.8···2.0)	Haa53	Cement	2.9···3.15	Mud60
Conglomerates (Nagelfluh)	2.40 (2.10···2.70)	Haa53	Concrete (Beton)	2.0 (1.51···2.54)	Haa53
			Bricks	1.45 (1.40···1.50)	Haa53
Syenite Breccia	2.10	Kry57	Ice (0 °C)	0.917 (0.88···0.92)	Haa53
Lime Breccia	2.28	Kry57	Snow, loose	0.125	Haa53
Slate (Tonschiefer)	2.42 (2.22···2.56)	Jak50			
	2.7 (2.51···2.83)	Rei31/33			
	2.80 (2.76···2.88)	Haa53			

1.2.6 Density of some common metallic ores, mineral ores and substances of organic deposites

Material	Density ϱ g cm^{-3}	Ref.	Material	Density ϱ g cm^{-3}	Ref.
Uranium Pitchblende (Uran-Pechblende)	8.0···9.7	Rei31/33	Mineral Coal (Hard coal)	1.26···1.33 1.2···1.8	Rei31/33 Jak50
Arseno pyrite (Arsenkies)	6.0···6.2	Rei31/33	Brown Coal (Lignite)	1.10···1.25 1.0···1.5 1.30 (1.10···1.44)	Rei31/33 Jak50 Haa53
Galena (Bleiglanz)	7.3···7.6	Rei31/33	Peat (Torf)	1.05	Rei31/33, Haa53
Magnetopyrite (Troilite, Magnetkies)	4.5···6.4	Rei31/33	Asphalt	1.1···1.2 1.2 (1.1···1.5)	Rei31/33 Haa53
Pyrite (Eisenkies)	4.9···5.2	Rei31/33	Oil (Erdöl)	0.6···0.9 0.9 (0.60···1.20)	Rei31/33 Haa53
Manganese ore	3.9···4.1	Jak50	Rock Salt	2.2 (2.1···2.3) 2.2 (2.1···2.4)	Jak50 Haa53
Hematite	4.9···5.3	Rei31/33	Kainite	2.13	Rei31/33
Siderite (Spateisen)	3.7···5.9	Rei31/33	Sylvite	1.95 (1.9···2.0)	Rei31/33, Haa53
Limonite (Raseneisenerz)	2.60	Haa53	Carnallite	1.60	Rei31/33
Barite (Schwerspat)	4.3···4.7	Rei31/33	Kaoline	2.25 (2.20···2.40)	Haa53
Magnesite	2.9···3.1	Rei31/33	Asbestos	2.5···3.2	Mud60
Bauxite	2.4···2.5 2.5 (2.0···3.0)	Rei31/33 Haa53	Gypsum	2.20 2.25 (2.17···2.31)	Rzh71 Haa53
Graphite	2.16 2.1···2.3 2.30 (2.17···2.32)	Woe63 Rei31/33 Haa53	Alabaster (Gypsum, dense)	2.66 (2.25···2.87)	Haa53
Schungite	1.122	Spe27	Chalk (Kreide)	2.20 (1.80···2.60)	Haa53
Anthracite	1.34···1.46 1.40 (1.30···1.80)	Rei31/33 Haa53	Phosphates	2.2···3.2	Haa53
			Retinite (amber-like resin)	1.03···1.051	Spe27

1.2.7 References for 1.2

Cla66	Clark jr., S. P., (Ed.): Handbook of Physical Constants, New York: The Geol. Soc. Am., Inc. **1966**.
Haa53	Haalck, H.: Lehrbuch der angewandten Geophysik, 2. erweiterte verbesserte Aufl., Berlin: Gebr. Borntraeger **1953**.
Hei68	Heiland, C.A.: Geophysical Exploration, New York and London: Hafner Publ. Co. **1968**.
Jak50	Jakovky, J.J.: Exploration Geophysics, 2nd ed. (Trija.) **1950**.
Kry57	Krynine, D.P., Judd, W.R.: Principles of Engineering Geology and Geotechnics, New York: McGraw Hill, **1957**, XIII, McGraw Hill Eng. Ser.
Mud60	Mudd, S.W., Series, Industrial Mineral and Rocks, Gillson, J.L., (Ed.), New York: Am. Inst. Min. Met. Petrol. Eng. XI, **1960**, p. 934.
Rei31/33	Reich, H.: Angewandte Geophysik für Bergleute und Geologen, Leipzig **1933**; Eigenschaften der Gesteine, in: B. Gutenberg, Handbuch der Geophysik 6, Berlin **1931**.
Rzh71	Rzhevsky, V., Nouik, G.: The Physics of Rocks, MIR Publ. **1971**.
Spe27	Spencer, L.J.: Mineralog. Mag. 119 (1927) 337.
Woe63	Woeber, A.F., Katz, S., Ahrens, T.J.: Geophys. 28 (1963) 661.

1.3 Density of minerals and rocks under shock compression — Dichte von Mineralen und Gesteinen bei Stoßwellenkompression

1.3.1 Theory of shock waves in solids — Theorie der Stoßwellen in Festkörpern

1.3.1.1 Basic thermodynamics — Grundlegende thermodynamische Beziehungen

Rapid acceleration of a solid results in transient compression by a stress wave due to the inertial response of the solid. This compression, usually transmitted by a compressional wave, becomes a shock wave travelling with supersonic velocity when a certain magnitude of the wave amplitude is exceeded [Cou48]. Neglecting dissipative effects, the shock wave or shock front represents a discontinuity where the pressure P_0, volume V_0 (or density ϱ_0), and internal energy E_0 of the unshocked material ahead of the shock increases suddenly to P_1, V_1 (or ϱ_1) and E_1, respectively (Fig. 1). The Material engulfed by a shock moves in the direction of the shock front with a certain particle velocity u. The shock wave velocity U is always greater than the sound velocity in the unshocked material ahead of the shock front, but lower than the bulk sound velocity in the shocked material behind the shock front. Therefore, in all practical cases where the shock transition is non-uniform and has a finite extent or "thickness", the shock is followed by a rarefaction wave (Fig. 2) which travels faster than the shock wave. It gradually overtakes the shock wave and thereby the peak pressure P_1 decreases to zero with increasing distance of propagation. Mathematically, a shock transition can be treated by applying the basic equations of the conservation of mass, momentum, and energy, provided that the shock wave can be considered steady and of infinite plane geometry, and that the compressed solid behaves hydrodynamically. As a consequence of the conservation laws [Duv63, Ric58], we obtain the following equations if the unshocked material is considered to be at rest:

Eine kurzzeitige, sehr schnelle Beschleunigung erzeugt in einem Festkörper auf Grund seines Trägheitsverhaltens einen vorübergehenden Druck, welcher durch eine Kompressionswelle übertragen wird. Diese Kompressionswelle wird zur Stoßwelle, die sich mit Überschallgeschwindigkeit fortpflanzt, wenn eine gewisse Größe der Wellenamplitude überschritten wird [Cou48]. Bei Vernachlässigung von dissipativen Effekten stellt die Stoßwelle oder Stoßfront eine Diskontinuität dar, durch welche der Druck P_0, das Volumen V_0 (oder die Dichte ϱ_0), und die innere Energie E_0 der nicht komprimierten Materie vor der Stoßfront sprunghaft auf P_1, V_1 (oder ϱ_1) und E_1 ansteigen (Fig. 1). Die Materie, die von einer Stoßwelle erfaßt wird, bewegt sich mit der Partikelgeschwindigkeit u parallel zur Fortpflanzungsrichtung. Die Stoßwellengeschwindigkeit U ist immer größer als die Schallgeschwindigkeit in der nicht komprimierten Materie vor der Stoßfront aber geringer als die Schallgeschwindigkeit in der komprimierten Materie hinter der Stoßfront. Deswegen folgt der Stoßfront in allen praktischen Fällen, in denen die Stoßwelle nicht stetig ist und eine endliche Ausdehnung oder „Dicke" hat, eine Verdünnungswelle (Fig. 2), welche sich schneller bewegt als die Stoßwelle. Diese überholt die Stoßwelle und erniedrigt dabei den Druck P_1 mit zunehmender Fortpflanzung auf Null. Mathematisch kann die Stoßfront durch die Anwendung der grundlegenden Gleichungen der Erhaltung der Masse, des Impulses und der Energie behandelt werden, vorausgesetzt daß die Stoßwelle als stetig betrachtet wird, eine unendliche ebene Geometrie besitzt, und der komprimierte Festkörper sich hydrodynamisch verhält. Als Folge der Erhaltungssätze der Thermodynamik [Duv63, Ric58], erhält man die folgenden Gleichungen, wenn man davon ausgeht, daß die nicht komprimierte Materie sich in Ruhe befindet:

$$\varrho_0 U = \varrho_1 (U - u) \tag{1}$$
$$P_1 - P_0 = \varrho_0 U u \tag{2}$$
$$P_1 u = \tfrac{1}{2} \varrho_0 U u^2 + \varrho_0 U (E_1 - E_0) \tag{3}$$

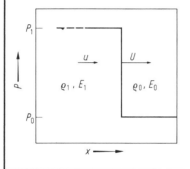

Fig. 1. Pressure-distance profile for a uniform shock front at hydrodynamic conditions [Stö72].

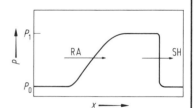

Fig. 2. Pressure-distance profile for a non-uniform shock front (SH) followed by a rarefaction wave (RA).

Combining equations (1)···(3), we get the Rankine-Hugoniot equation

$$E_1 - E_0 = \tfrac{1}{2}(P_1 + P_0)(V_0 - V_1) \qquad (4)$$

where V is the specific volume $1/\varrho$.

If the initial conditions of a solid (P_0, ϱ_0, E_0) are known, the measurement of any pair of variables of the above equations is sufficient to determine a shock state (P_1, ϱ_1, E_1). The graphical representation of any pair of variables is called the Hugoniot curve. The U-u-, the P-V-, and the P-u-plane are most commonly used for the representation of the Hugoniot curve (Figs. 3 and 4). This curve describes the thermodynamic locus for all shock states achievable in a given solid by plane, steady shock waves of variable intensity. As to be seen from the U-u-Hugoniot curves of many solids, the shock velocity U appears to be linearly related to the particle velocity u according to

$$U = a + b\,u,$$

where a and b are constants [Ric58]. This relation, which enables the definition of the Hugoniot equation of state by the three parameters ϱ_0, a, and b, is useful for a number of thermodynamic calculations [McQ70]. The P-V-relation on the Hugoniot curve would then read for any peak pressure P_1:

$$P_1 = \frac{a^2(V_0 - V_1)}{b^2(V_1 - V_0 + V_0/b)^2}.$$

A shock transition is a non-isentropic process and yields a certain amount of irreversible work which appears as heat upon pressure release. The thermodynamic path for a material shocked to a final state P_1, V_1, E_1 is given by the Raleigh line AB (Fig. 3). The release path can be considered to follow an adiabat as shown in Fig. 3.

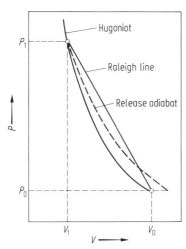

Fig. 3. Hugoniot curve plotted in the P-V-plane.

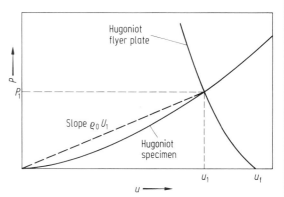

Fig. 4. Hugoniot curve plotted in the P-u-plane; $\varrho_0 U$ (= shock impedance) is the slope of the Raleigh line (compare Fig. 3) according to equation (2). Specimen and flyer plate are of different shock impedance.

The post-shock temperature of the released material increases with increasing peak pressure due to the upward concavity of the Hugoniot curve.

1.3.1.2 Calculation of equation of state data from shock wave data — Berechnung der Zustandsgleichung aus Stoßwellendaten

It is of considerable theoretical interest to use measured shock wave data for calculating equation of state data along the Hugoniot curve (temperature, entropy) and in regions off the locus of the Hugoniot curve, e.g. along adiabats or isotherms centered on the Hugoniot curve. Application of the Mie-Grüneisen equation of state to measured shock data yields the desired thermodynamic quantities as well as hydrodynamic parameters such as sound and particle velocities.

According to [McQ63] and [McQ70], the Rankine-Hugoniot equation (4) can be combined with the thermodynamic relations

$$dE = T\,dS - P\,dV \tag{5}$$

and

$$T\,dS = c_v\,dT + T\frac{\gamma}{V}c_v\,dV \tag{6}$$

where T = temperature, S = entropy, γ = Grüneisen ratio, and c_v = specific heat at constant volume.

Combining (4) and (5) yields

$$T\,dS = \tfrac{1}{2}(V_0 - V)\,dP + \tfrac{1}{2}(P - P_0)\,dV \tag{7}$$

The following differential equation is obtained if $T\,dS$ is eliminated from (6) and (7):

$$\frac{dT}{dV} = -T\left(\frac{\gamma}{V}\right) + \left[\frac{dP}{dV}(V_0 - V) + (P - P_0)\right]\frac{1}{2c_v} \tag{8}$$

from which temperatures along the Hugoniot curve (shock temperature) can be calculated by integration.

The calculation of thermodynamic states along release adiabats, the equation of state loci of rarefaction waves, may be achieved by combining the Grüneisen parameter $\gamma = V\left(\frac{\partial P}{\partial E}\right)_V$, which is assumed to be independent of pressure, and the relation $dE = P\,dV$ ($dS = 0$) with the pressure P_H, volume V_H and the energy E_H along the known Hugoniot curve.

If P_a, V_a, E_a and P_a', V_a', E_a' are thermodynamic states on the release adiabat and V_a equals V_H (see Fig. 5), an equation for the pressure on the adiabat, P_a, is obtained [McQ70]:

$$P_a = \frac{P_H - (\gamma/V_a)(P_a'\,\Delta V/2 + E_H - E_a')}{1 + (\gamma/V_a)(\Delta V/2)}.$$

Using the relation

Benützt man die Beziehung

$$E_a = E'_a - (P_a + P'_a)\, \Delta V/2,$$

P, V, E states on the adiabats can be derived. Temperatures along the adiabats may be calculated from equation (6) in which the term $T\,dS$ becomes zero, so that

so können P-, V-, E-Zustände auf den Adiabaten abgeleitet werden. Temperaturen auf den Adiabaten können aus der Gleichung (6), in welcher der Ausdruck $T\,dS$ Null wird, berechnet werden, sodaß sich ergibt

$$\frac{dT}{dV} = \frac{\gamma}{V}\, T.$$

Usually the experimental determination of Hugoniot curves requires the measurement of velocities (shock and particle velocities). This is done most commonly on the basis of impedance match techniques using a "flyer plate" which impacts a "driver plate" with known Hugoniot equation of state data, in contact with an unknown specimen. This situation is presented graphically in Fig. 6, where the Hugoniot curves for the driver plate and the specimen are designated I and II. A plane shock P_1 with a particle velocity u_1 travelling through the driver plate enters the interface to the specimen which in Fig. 6 has a lower shock impedance $\varrho_0 U$ than the driver plate. A release wave is reflected back into the driver plate increasing the particle velocity to u_2 and decreasing the shock pressure to P_2. The new shock state P_2, u_2 must have its locus on the Hugoniot curve

In der Regel müssen zur experimentellen Bestimmung der Hugoniotkurve Geschwindigkeiten (Stoßwellen- und Partikelgeschwindigkeit) gemessen werden. Dies wird meistens auf der Grundlage der sogenannten „Impedanz-Methode" durchgeführt, bei welcher eine fliegende Platte mit bekannter Hugoniotkurve auf eine planparallele Platte gleicher Zusammensetzung („Schub-Platte") impaktiert wird, die auf die ebene Oberfläche der Probe montiert ist. Diese Situation ist graphisch in Fig. 6 dargestellt, in welcher die Hugoniotkurve für die Schub-Platte und für die Probe mit I bzw. II bezeichnet sind. Eine ebene Stoßfront P_1 mit einer Partikelgeschwindigkeit u_1, die sich durch die Schub-Platte bewegt, trifft auf die Grenzfläche zur Probe, welche in Fig. 6 eine geringere sogenannte Stoßwellenimpedanz $\varrho_0 U$ als die Schub-Platte besitzt. Eine Verdünnungs-

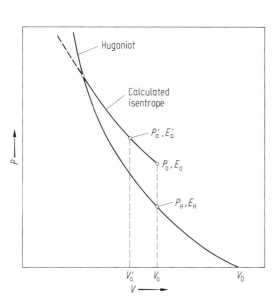

Fig. 5. Schematic representation of a Hugoniot curve and a release isentrope centered at the Hugoniot (see text).

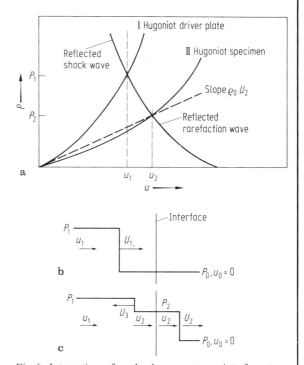

Fig. 6. Interaction of a shock wave at an interface to a medium of lower shock impedance
a) pressure-particle velocity plot
b) shock profile before interaction
c) shock profile after interaction [Stö72].

of the specimen and on the rarefaction wave curve of the driver plate because pressure and particle velocity have to be constant across the interface. If the shock velocity U_2 is measured, P_2 can be calculated from the magnitude of the slope of the Raleigh line $\varrho_0 U_2$.

If a shock wave reaches a free surface toward vacuum or air, the velocity U_{fs} of the surface is

$$U_{fs} = u_1 + \Delta u$$

For most solids Δu was found to equal u_1 ("free surface approximation"). This means that the particle velocity can be derived from the more easily measured velocity of the free surface. The free surface velocity can also be used to describe the conditions of a flyer plate impacting a specimen (Fig. 4). In Fig. 4, the specimen is accelerated to the right upon collision with a flyer plate travelling at a velocity u_f. The flyer plate decelerates when it hits the specimen. At the time of collision the flyer plate and the specimen are in the same shock state P_1, u_1. This state is given by the intersection of the specimen Hugoniot curve and the mirror image of the flyer plate Hugoniot curve. This offers the possibility of determining the particle velocity u_1 from the velocity of the impacting flyer plate ($u_f = 2u_1$) if the flyer plate is made of the same material as the specimen.

1.3.1.3 The influence of material strength and phase transitions on the shock wave structure — Der Einfluß der Materialfestigkeit und der Phasenumwandlungen auf die Struktur der Stoßwelle

The application of the thermodynamic calculation of shock transitions in solids discussed in the foregoing sections is strictly valid only if the solid behaves as an inviscid, compressible fluid ("hydrodynamic model") under shock compression. In practice, this condition holds for most solids only at very high shock pressures, especially for rocks and for silicate minerals. Therefore, the shock pressure can be taken as a hydrostatic pressure only at very high pressures. In all cases where the material strength cannot be neglected shock compression takes place under conditions of linear strain in which the shock pressure, as defined by the Hugoniot equations, represents the component of compressive stress which is normal to the shock front. For a given volume of the compressed solid, the shock pressure exceeds the

hydrostatic pressure by a factor, which is related to the yield stress of the solid (Fig. 7, for details see [Duv63, McQ70, Mur74, Gra77]).

in der Definition der Hugoniotgleichung diejenige Komponente des Drucks darstellt, welche senkrecht zur Stoßwellenfront liegt. Bei einem gegebenen Volumen, auf welches ein Festkörper komprimiert ist, übersteigt der Stoßwellendruck den hydrostatischen Druck um einen Faktor, der in Beziehung zur elastischen Grenze des Festkörpers steht (Fig. 7; Genaueres siehe [Duv63, McQ70, Mur74, Gra77]).

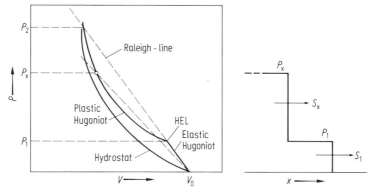

Fig. 7. Hugoniot curve of a solid with a dynamic elastic limit (HEL) (left) and two-wave profile resulting from the HEL (right); S_1 = elastic shock, S_x = plastic shock [Stö72].

Most solids exhibit dynamic yield points (Hugoniot elastic limit = HEL) and one or more phase transitions. At the critical pressure of both types of transformations the Hugoniot curve displays a cusp (Figs. 7 and 8). The HEL divides the curve into an elastic and a plastic Hugoniot curve (elastic-plastic model). As a consequence, a single front becomes unstable and two shock waves, called elastic wave or elastic precursor and plastic wave, are formed as long as the final peak pressure P_x fulfills the condition $P_1 < P_x < P_2$ (see Fig. 7). Since the shock wave velocity is proportional to the slope of the Raleigh line, the elastic wave which has a peak pressure P_1, travels faster than the plastic shock wave with pressure P_x. More details about the effects

Die meisten Festkörper haben eine dynamische elastische Grenze (Hugoniot elastic limit = HEL) und erfahren mit steigendem Druck eine oder mehrere Phasenumwandlungen. Beim kritischen Druck beider Arten von Transformationen zeigt die Hugoniotkurve einen Knickpunkt (Fig. 7 und 8). Die dynamische elastische Grenze teilt die Kurve in eine elastische und eine plastische Hugoniotkurve (elastisch-plastisches Modell). Als Folge wird eine einzelne Stoßfront instabil und teilt sich in zwei Wellenfronten, welche die „elastische Welle oder der elastische Vorläufer" und die „plastische Welle" genannt werden. Zwei Wellenfronten bilden sich für alle Spitzendrucke P_x aus, wenn die Bedingung $P_1 < P_x < P_2$ erfüllt ist (s. Fig. 7). Da die Stoßwellengeschwindig-

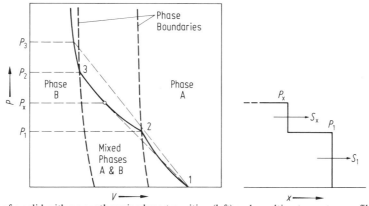

Fig. 8. Hugoniot curve of a solid with an exothermic phase transition (left) and resulting two-wave profile (right); S_1 = plastic shock, S_x = transformational.

of material strength on the Hugoniot equation of state and a discussion of the relations between the elastic constants and the shock wave data of solids are given by [McQ70, Mur74, Gra77].

Another two-wave structure caused by a phase transition is shown in Fig. 8. If $P_1 < P_x < P_3$, the peak pressure P will be achieved by two waves. A region of high compressibility exists between P_1 and P_2 which is called the mixed-phase- or two-phase-regime of the Hugoniot curve. Peak pressures $P_x > P_2$ represent shock states along the high-pressure-phase Hugoniot curve, pressures $P_x < P_1$ are connected with states on the low-pressure-phase Hugoniot curve. For solids displaying a phase change, it is sometimes necessary to reduce the high-pressure-phase Hugoniot data to a metastable Hugoniot curve centered at zero pressure which yields the zero-pressure density of the high-pressure-phase in the metastable state. Details of the calculation are given by [McQ70].

keit proportional zur Steigung der Raleigh-Linie ist, bewegt sich die elastische Welle, die einen Spitzendruck von P_1 hat, schneller als die plastische Welle mit dem Spitzendruck P_x. Einzelheiten über die Effekte der Materialfestigkeit auf die Hugoniot-Zustandsgleichung und eine Diskussion der Beziehungen zwischen den elastischen Konstanten und der Stoßwellendaten der Festkörper werden bei [McQ70, Mur74, Gra77] diskutiert.

Eine weitere, zweifache Wellenstruktur, welche durch eine Phasenumwandlung verursacht wird, ist in Fig. 8 dargestellt. Wenn $P_1 < P_x < P_2$ ist, wird der Spitzendruck P_x durch zwei Wellenfronten erreicht. Zwischen P_1 und P_2 existiert ein Gebiet hoher Kompressibilität, welches das Zweiphasengebiet der Hugoniotkurve genannt wird. Spitzendrucke $P_x > P_2$ stellen Stoßwellenzustände entlang der Hugoniotkurve der Hochdruckphase dar. Drucke $P_x < P_1$ sind mit Zuständen auf der Hugoniotkurve der Niederdruckphase verknüpft. Für Festkörper, welche eine Phasenumwandlung besitzen, ist es manchmal erforderlich, die Hugoniotkurve der Hochdruckphase in eine metastabile Hugoniotkurve zu überführen, aus welcher die Dichte der Hochdruckphase in ihrem metastabilen Zustand beim Nulldruck abgelesen werden kann. Einzelheiten der Berechnung sind bei [McQ70] zu finden.

1.3.2 Experimental techniques — Experimentelle Methoden

The scope of this article does not allow a detailed discussion of experimental techniques which are currently used to produce shock waves and to measure parameters of shock wave propagation. The reader is referred to review articles by [Duv63, McQ63, Kee71, Mur74, Gra77]. Generally, a plane shock wave is produced by the impact of a plane flyer plate against a stationary plate which either represents the specimen itself or is attached to the surface of a plane specimen. Two types of devices are most commonly used for accelerating the impacting flyer plate: high explosive plane wave generators [Duv63] and light gas guns (e.g. [Jon66, Cab70]).

Shock wave and particle velocities in the specimen may be determined by measuring the free surface motion or by means of "in-material" gauges which include electromagnetic gauges and laser interferometry. Free surface velocity measurements use optical and electrical methods or flash X-ray radiography [Mur74]. The accuracy of modern velocity measurements in shock experimentation is generally such that the pressure determination is precise to within $\pm 2\%$ or less.

Der Rahmen dieses Artikels erlaubt keine detaillierte Diskussion der experimentellen Methoden, die gegenwärtig benützt werden, um Stoßwellen zu erzeugen und um die Parameter der Stoßwellenfortpflanzung in Festkörpern zu messen. Der Leser wird auf die Überblicksartikel von [Duv63, McQ63, Kee71, Mur74, Gra77] hingewiesen. Im allgemeinen wird eine ebene Stoßwelle durch den Aufschlag einer ebenen fliegenden Platte gegen eine feststehende Platte erzeugt, welche entweder die Probe selbst darstellt oder auf der Oberfläche einer ebenen Probe befestigt ist. Zwei Typen von Vorrichtungen werden am häufigsten für die Beschleunigung der fliegenden Platte verwendet: sogenannte Hochexplosivgeneratoren zur Erzeugung ebener Stoßwellen [Duv63] und Leichtgaskanonen (z.B. [Jon66, Cab70]).

Die Stoßwellen- und Partikelgeschwindigkeiten in der Probe können durch die Messung der Geschwindigkeit der freien Oberfläche oder mit Hilfe von Meßzellen bestimmt werden, welche in die Probe eingebaut sind und die im allgemeinen auf der Basis elektromagnetischer Prozesse oder der Interferometrie mit Laserstrahlen arbeiten. Die Geschwindigkeit der freien Oberfläche kann mit Hilfe optischer und elektrischer Methoden oder mit Hilfe der Röntgenblitz-Methodik bestimmt werden [Mur74]. Mit modernen Geschwindigkeitsmessungen in der Stoßwellentechnik wird im allgemeinen erreicht, daß die Druckbestimmungen genauer als $\pm 2\%$ sind.

Direct measurement of the density of shock compressed solids has been achieved by flash X-ray techniques, but the precision of this method is unsatisfactory [Sch50, Dap57].

In recent years methods have been successfully developed to measure shock stresses directly by pressure transducers which operate on the basis of piezoelectric (quartz gauge), piezoresistive (e.g. manganin), electromagnetic, thermoelectric and polarization effects [Mur74].

Acknowledgment — Danksagung

I should like to thank the following members of our working group for most helpful technical assistance:

Gudrun Grant, Franziska Möllers, Elisabeth Bierhaus, Monika Liening, Heinz-Dieter Knöll, Udo Maerz.

1.3.3 Tables — Tabellen

For any set of shock wave data the following information is given:

a) chemical, mineralogical, and textural composition and source of the sample;
b) physical properties of the sample;
c) experimental and data reduction techniques;
d) author of the data.

Most data are taken from a compilation by van Thiel (1977) [Van77]. The meaning of the symbols and the dimensions used are summarized in Table 1.

The data are listed in the following order:

- 1.3.3.1 Standard metals — Standard-Metalle
- 1.3.3.2 Minerals — Minerale
- 1.3.3.2.1 Elements — Elemente
- 1.3.3.2.2 Oxides — Oxide
- 1.3.3.2.3 Carbonates — Carbonate
- 1.3.3.2.4 Halides — Halogenide
- 1.3.3.2.5 Sulfides — Sulfide
- 1.3.3.2.6 Silicates — Silikate
- 1.3.3.3 Rocks — Gesteine
- 1.3.3.3.1 Igneous rocks — Magmatische Gesteine
- 1.3.3.3.2 Metamorphic rocks — Metamorphe Gesteine
- 1.3.3.3.3 Sedimentary rocks and sediments — Sedimentäre Gesteine und Sedimente
- 1.3.3.4 Glasses — Gläser

Note added in proof: Recently a compilation of Hugoniot data from the Los Alamos Scientific Laboratory (LASL) has been published by Marsh [Mar80]. It contains data for the following materials not listed in this volume: Platinum, sulfur, Fe–Si alloys, gros-

sularite, kyanite, mullite, tourmaline, wollastonite, baddeleyite, gabbro, and shale. The data of [McQ66] listed in this volume are supplemented and improved in [Mar80].

Platin, Schwefel, Fe-Si-Legierungen, Grossular, Disthen, Mullit, Turmalin, Wollastonit, Baddeleyit, Gabbro und Tonschiefer. Die in diesem Band aufgeführten Daten von [McQ66] sind in [Mar80] ergänzt und verbessert worden.

Table 1 — Tabelle 1 (For details, see author of data or [Van77])

a) Symbols and dimensions — Symbole und Dimensionen

Symbol	Dimension	Explanation
ϱ	g cm^{-3}	Density under shock compression — Dichte bei Stoßwellenkompression
ϱ_0	g cm^{-3}	Density of the sample measured at zero conditions (P_0, T_0) — Dichte der Probe bei Normalbedingungen (P_0, T_0)
ϱ_{01}	g cm^{-3}	Density of the sample calculated from crystallographic information or handbook data (without porosity in the case of polycrystalline samples) — Dichte der Probe berechnet aus kristallographischen Daten oder Handbuchangaben (ohne Porosität im Falle polykristalliner Proben)
c_l	km s^{-1}	longitudinal wave velocity — Longitudinalwellen-Geschwindigkeit,
c_s	km s^{-1}	shear wave velocity — Scherwellen-Geschwindigkeit
c_0	km s^{-1}	bulk sound velocity *) — Schallgeschwindigkeit *) $\left[(c_l)^2 - \tfrac{4}{3}(c_s)^2\right]^{\tfrac{1}{2}}$
c_b	km s^{-1}	bulk sound velocity — Schallgeschwindigkeit $\left(\dfrac{\partial P}{\partial \varrho}\right)_s^{\tfrac{1}{2}}$
T_0	°C	temperature of the sample at zero shock pressure — Temperatur der Probe beim Nulldruck
U	km s^{-1}	shock wave velocity — Stoßwellen-Geschwindigkeit
u	km s^{-1}	particle velocity — Partikel-Geschwindigkeit
U_{fs}	km s^{-1}	free surface velocity or impact velocity of standard — Partikel-Geschwindigkeit an der freien Oberfläche oder Impakt-Geschwindigkeit des Standards
P	GPa	peak shock pressure — Stoßwellenspitzendruck
No.		identification number of experiment — Identifikationsnummer des Experiments
σ		standard deviation — Standardabweichung

*) Seismic parameter — seismischer Parameter: $\varphi = c_0^2$, see 3.1.3.2.

b) Experimental techniques – Experimentelle Methoden

Symbol	Technique – Methode
A	electrical pins – elektrische Kontakte
B	optical flash – optischer Blitz
C	reflecting mirror – reflektierender Spiegel / turning mirror – drehender Spiegel / total reflection of a prism – Totalreflektion eines Prismas
D	optical lever – optischer Hebel
E	knife edge – Messerschneide
F	resistance wire – Widerstandsdraht
G	capacitor – Kapazität
H	polarization – Polarisation
I	transducer gauges – Druckübertrager
J	radiographic methods – radiographische Methoden
K	magnetic induction – magnetische Induktion

c) Data reduction techniques — Methoden der Datengewinnung

Symbol	Technique — Methode
A	impedance matching (same composition of projectile and sample) — Impedanz-Methode (gleiche Zusammensetzung von Projektil und Probe)
B	impedance matching (driver plate with known Hugoniot curve) — Impedanz-Methode (fliegende Platte mit bekannter Hugoniotkurve)
C	P measured by a transducer gauge — P durch ein Druckübertragungsmedium gemessen
D	free surface approximation (U and U_{fs} measured) — Annäherung über Geschwindigkeit der freien Oberfläche (U und U_{fs} gemessen)
E	radiographic method (U and ϱ measured) — radiographische Methode (U und ϱ gemessen)
F	U and u measured — U und u gemessen
G	impact of a transducer of known Hugoniot curve (impact velocity and P measured) — Aufschlag eines Druckübertragers mit bekannter Hugoniotkurve (Impaktgeschwindigkeit und Druck gemessen)

1.3.3.1 Standard metals — Standard-Metalle

The precision of Hugoniot data of rocks and minerals as measured on the basis of the impedance match technique is very much dependent on the quality of equation of state data of some standards used as flyer plates in shock experiments. Most precisely determined Hugoniot data are available from the work of [McQ70]. The data of these authors are given in Table 2 and 3 and in Fig. 9. Table 2 contains the coefficients of the Hugoniot equation in the U-u-plane.

Die Genauigkeit der Hugoniot-Daten von Gesteinen und Mineralen, die auf der Grundlage der „Impedanz-Methode" gemessen werden, hängt sehr stark von der Qualität der Zustandsgleichungen einiger Standard-Materialien ab, die als fliegende Platte in den Stoßwellenexperimenten benützt werden. Die genauesten Hugoniotdaten für Standard-Materialien sind von [McQ70] veröffentlicht worden. Die Daten dieser Autoren sind in den Tab. 2 und 3 und in Fig. 9 zusammengestellt. Tab. 2 enthält die Koeffizienten der Hugoniotgleichung in der U-u-Ebene für die angeführten Standards.

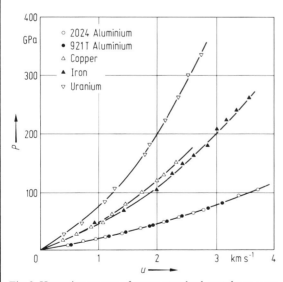

Fig. 9. Hugoniot curves of some standard metals; u = particle velocity, P = shock pressure [McQ70].

Table 2. Coefficients of the Hugoniot equation $U = a + bu + cu^2$ for some standard metals (least-squares fit) [McQ70]

Metal	a	b	c
2024 Al	5.328	1.338	0
921-T Al	5.041	1.420	0
Cu	3.940	1.489	0
Fe	3.574	1.920	−0.068
U-Mo-alloy	2.565	1.531	0

Table 3. Hugoniot data for standard metals [McQ70]

ϱ_0 g cm^{-3}	U km s^{-1}	u km s^{-1}	P GPa	ϱ g cm^{-3}	ϱ_0 g cm^{-3}	U km s^{-1}	u km s^{-1}	P GPa	ϱ g cm^{-3}
Al (2024)					Al (921T)				
2.785	6.02	0.50	8	3.04	2.810	5.72	0.50	8	3.08
2.785	6.10	0.50	9	3.03	2.810	5.74	0.53	8	3.09
2.785	6.05	0.51	9	3.04	2.810	5.98	0.63	11	3.14
2.782	6.16	0.67	12	3.12	2.811	5.88	0.65	11	3.16
2.785	6.35	0.77	14	3.17	2.828	6.72	1.19	23	3.43
2.782	6.49	0.86	16	3.21	2.828	7.04	1.40	28	3.53
2.785	6.52	0.87	16	3.22	2.828	7.18	1.49	30	3.57
2.782	6.61	0.97	18	3.26	2.828	7.21	1.57	32	3.61
2.789	6.86	1.15	22	3.35	2.828	7.42	1.63	34	3.63
2.789	6.89	1.16	22	3.35	2.828	7.73	1.86	41	3.73
2.780	6.91	1.16	22	3.34	2.828	8.06	2.13	49	3.84
2.789	6.86	1.21	23	3.38	2.828	8.09	2.13	49	3.84
2.789	7.27	1.43	29	3.47	2.828	8.35	2.28	54	3.89
2.789	7.33	1.58	32	3.55	2.828	8.42	2.40	57	3.96
2.779	7.51	1.62	34	3.54	2.828	8.51	2.44	59	3.96
2.789	7.68	1.72	37	3.60	2.828	9.07	2.87	74	4.13
2.789	7.60	1.73	37	3.61	2.828	9.38	3.09	82	4.22
2.789	7.66	1.77	38	3.63	2.828	9.40	3.10	82	4.22
2.789	7.77	1.81	39	3.64	Fe				
2.789	7.69	1.85	40	3.67	7.870	4.59	0	0	7.87
2.779	7.97	1.95	43	3.68	7.850	5.33	0.95	40	9.55
2.789	8.13	2.13	48	3.78	7.843	5.46	1.00	43	9.60
2.789	8.33	2.13	50	3.76	7.850	5.56	1.06	46	9.71
2.789	8.23	2.21	51	3.81	7.847	6.20	1.43	70	10.20
2.779	8.40	2.31	54	3.83	7.850	6.24	1.44	71	10.22
2.789	8.53	2.45	58	3.91	7.850	6.58	1.65	85	10.48
2.789	8.83	2.59	64	3.95	7.850	6.66	1.73	91	10.61
2.789	8.76	2.60	64	3.97	7.847	7.07	1.96	109	10.85
2.789	8.80	2.65	65	3.99	7.850	7.08	1.98	110	10.90
2.789	9.14	2.82	72	4.03	7.831	7.56	2.27	135	11.20
2.789	9.40	2.99	78	4.09	7.847	7.81	2.44	150	11.42
2.789	9.67	3.22	87	4.19	7.832	7.90	2.46	152	11.37
2.789	9.64	3.29	88	4.23	7.809	8.09	2.61	165	11.54
2.778	9.87	3.42	94	4.25	7.850	8.01	2.62	165	11.66
2.778	10.19	3.72	105	4.37	7.847	8.24	2.72	176	11.70
2.789	10.37	3.75	108	4.37	7.850	8.23	2.72	176	11.70
					7.850	8.23	2.72	176	11.73
					7.868	8.31	2.82	184	11.91
					7.840	8.49	2.89	192	11.88
					7.842	8.81	3.06	211	12.01
					7.850	8.94	3.12	219	12.05
					7.850	8.99	3.18	224	12.14
					7.850	9.10	3.19	228	12.08
					7.860	9.24	3.35	243	12.34
					7.860	9.28	3.36	245	12.32
					7.855	9.51	3.59	268	12.61

ϱ_0 g cm^{-3}	U km s^{-1}	u km s^{-1}	P GPa	ϱ g cm^{-3}	ϱ_0 g cm^{-3}	U km s^{-1}	u km s^{-1}	P GPa	ϱ g cm^{-3}
Cu					U—Mo-alloy (U + 3 wt-% Mo)				
8.930	4.48	0.41	16	9.82	18.450	3.17	0.41	24	21.15
8.931	4.57	0.62	17	9.82	18.450	3.33	0.49	30	21.63
8.928	4.89	0.62	27	10.23	18.450	3.67	0.72	49	22.97
8.930	4.91	0.64	28	10.26	18.450	3.73	0.75	52	23.08
8.930	4.91	0.73	32	10.49	18.450	4.24	1.11	87	24.97
8.930	5.07	0.74	33	10.45	18.450	4.54	1.29	108	25.80
8.930	4.89	0.74	32	10.53	18.450	4.53	1.32	110	26.01
8.930	5.18	0.85	39	10.67	18.480	5.19	1.76	168	27.95
8.930	5.36	0.95	46	10.86	18.450	5.31	1.80	177	27.93
8.930	5.84	1.25	65	11.37	18.450	5.36	1.84	182	28.08
8.930	6.19	1.48	82	11.74	18.450	5.83	2.11	227	28.90
8.930	6.51	1.71	99	12.10	18.450	6.06	2.36	265	30.25
8.928	6.55	1.74	102	12.15	18.450	6.27	2.38	275	29.75
8.930	6.91	1.99	123	12.54	18.450	6.51	2.54	305	30.24
8.930	6.88	2.04	125	12.68	18.450	6.76	2.75	343	31.07
8.930	7.03	2.10	132	12.73					
8.928	7.46	2.37	158	13.08					
8.930	7.47	2.37	158	13.08					

1.3.3.2 Minerals — Minerale

The shock wave data measured on single crystals or polycrystalline specimens of minerals are listed in the following sections. The minerals are grouped according to the common chemical classification of minerals. In each group the minerals are ordered alphabetically.

In den folgenden Abschnitten sind die Stoßwellendaten aufgeführt, die an Einkristallen oder an polykristallinen Proben von Mineralen gemessen wurden. Die Minerale sind nach der allgemein üblichen chemischen Klassifizierung von Mineralen angeordnet. Innerhalb einer Mineralgruppe ist die Reihenfolge alphabetisch.

1.3.3.2.1 Elements — Elemente

Antimony (Antimon) Sb

$\varrho_0 = 6.698$ g cm^{-3}; $\varrho_{01} = 6.734$ g cm^{-3}; $c_0 = 2.37$ km s^{-1}

Sample					Standard	
ϱ_0 g cm^{-3}	U km s^{-1}	u km s^{-1}	P GPa	ϱ g cm^{-3}	Composition	P GPa
6.7	3.61	1.03	24.8	9.36	Al	22.0
6.7	3.59	1.03	24.8	9.40	Al	22.0
6.7	3.63	1.02	24.9	9.31	Al	22.0
6.7	4.30	1.39	40.0	9.88	Al	33.9
6.7	4.33	1.38	40.1	9.84	Al	33.9
6.7	5.12	1.96	67.3	10.86	brass	86.9
6.7	5.06	1.88	63.7	10.65	brass	81.9
6.7	5.64	2.19	82.8	10.97	brass	104.0
6.7	5.72	2.19	83.8	10.84	brass	104.2
6.7	5.71	2.20	84.3	10.91	brass	105.1
6.7	6.31	2.70	114.2	11.71	brass	142.2
6.7	6.34	2.73	115.8	11.78	brass	144.2
6.7	6.43	2.73	117.8	11.63	brass	145.2

◄ $U = (1.989 + 1.634 u)$ km s^{-1}; $\sigma_U = 0.2\%$
Experimental technique B
Data reduction technique B
Reference: [McQ 60]

Bismuth (Wismut) Bi

$\varrho_0 = 9.794 \cdots 9.843 \text{ g cm}^{-3}$; $\varrho_{01} = 9.807 \text{ g cm}^{-3}$;
$c_1 = 2.18 \text{ km s}^{-1}$; $c_0 = 1.771 \text{ km s}^{-1}$; $c_s = 1.10 \text{ km s}^{-1}$;
$c_b = 1.86 \text{ km s}^{-1}$

Sample					
ϱ_0 g cm^{-3}	U km s^{-1}	u km s^{-1}	P GPa	ϱ g cm^{-3}	
9.807	2.300	0.009	0.2	9.84	first wave
9.807	1.790	0.009	0.2	9.84	second wave
9.807	1.860	0.040	0.76	10.01	second wave
9.807	1.950	0.080	1.55	10.22	second wave
9.807	2.050	0.126	2.55	10.44	second wave, transition point
9.807	1.755	0.300	5.5	11.69	third wave
9.807	2.183	0.500	10.7	12.72	second wave
9.807	2.718	0.750	20.0	13.55	first wave
9.807	3.253	1.000	31.9	14.15	first wave
9.807	3.660	1.200	43.1	14.59	first wave
9.807	4.730	2.000	92.8	17.00	first wave
9.807	6.070	3.000	178.7	19.38	first wave
9.807	7.410	4.000	290.8	16.35	first wave
9.807	8.080	4.500	356.7	22.14	first wave

$U = ((1.771 + 2.21 u) \pm 0.02) \text{ km s}^{-1}$;
$0.009 < u < 0.126 \text{ km s}^{-1}$;
$U = (1.113 + 2.14 u)$; $\sigma_U = 0.03 \text{ km s}^{-1}$;
$0.3 < u < 1.1 \text{ km s}^{-1}$;
$U = (2.05 + 1.341 u)$; $\sigma_U = 0.07 \text{ km s}^{-1}$;
$1.1 < u < 4.5 \text{ km s}^{-1}$
Reference: [Van 77]

Copper (Kupfer) Cu

$\varrho_0 = 8.929 \text{ g cm}^{-3}$; $\varrho_{01} = 8.937 \text{ g cm}^{-3}$;
$c_1 = 4.76 \text{ km s}^{-1}$; $c_0 = 3.93 \text{ km s}^{-1}$; $c_s = 2.33 \text{ km s}^{-1}$

Sample					Standard	
ϱ_0 g cm^{-3}	U km s^{-1}	u km s^{-1}	P GPa	ϱ g cm^{-3}	Composition	U km s^{-1}
8.920	4.31	0.21	8.1	9.38	Al	5.79
8.925	4.21	0.21	7.9	9.39	Al	5.79
8.900	4.22	0.22	8.3	9.39	Al	5.81
8.928	4.34	0.28	10.8	9.54	Al	5.94
8.925	4.35	0.28	10.9	9.54	Al	5.94
8.920	4.32	0.29	11.2	9.54	Al	5.95
8.920	4.35	0.29	11.3	9.56	Al	5.96
8.925	4.38	0.30	11.7	9.58	Al	5.98
8.930	4.30	0.30	11.5	9.60	Al	5.98
8.925	4.51	0.39	15.7	9.77	Al	6.18
8.928	4.50	0.40	16.1	9.80	Al	6.18
8.933	4.47	0.43	17.2	9.88	Al	6.25
8.933	4.50	0.44	17.7	9.90	Al	6.26
8.900	4.71	0.49	20.5	9.93	Al	6.38
8.900	4.69	0.49	20.5	9.94	Al	6.38

Copper (continued)

Sample					Standard	
ρ_0 g cm^{-3}	U km s^{-1}	u km s^{-1}	P GPa	ρ g cm^{-3}	Composition	U km s^{-1}
8.899	4.71	0.52	21.8	10.00	Al	6.44
8.924	4.73	0.52	21.9	10.03	Al	6.45
8.925	4.80	0.53	22.7	10.03	Al	6.47
8.920	4.75	0.53	22.5	10.04	Al	6.48
8.925	4.70	0.54	22.7	10.08	Al	6.47
8.900	4.71	0.55	23.1	10.08	Al	6.49
8.930	4.77	0.56	23.9	10.12	Al	6.53
8.918	4.71	0.58	24.4	10.17	Al	6.55
8.933	4.85	0.66	28.6	10.34	Al	6.72
8.900	5.01	0.69	30.8	10.32	Al	6.79
8.900	4.95	0.71	31.3	10.39	Al	6.82
8.918	5.04	0.73	32.8	10.43	Al	6.87
8.924	5.06	0.75	33.9	10.48	Al	6.91
8.930	4.94	0.76	33.5	10.55	Al	6.91
8.933	5.07	0.79	35.8	10.58	Al	6.98
8.933	5.19	0.80	37.1	10.56	Al	7.03
8.933	5.18	0.81	37.5	10.59	Al	7.04
8.933	5.22	0.83	38.7	10.62	Al	7.09
8.933	5.15	0.87	40.0	10.75	Al	7.14
8.933	5.33	0.94	44.8	10.85	Al	7.30
8.933	5.33	0.94	44.8	10.85	Al	7.31
8.899	5.44	0.99	47.9	10.88	Al	7.41
8.900	5.44	1.01	48.9	10.95	Al	7.43
8.918	5.51	1.03	50.6	10.97	Al	7.48
8.924	5.50	1.06	52.0	11.05	Al	7.54
8.930	5.53	1.06	52.3	11.05	Al	7.55
8.930	6.03	1.39	74.8	11.60	Al	8.21
8.930	6.02	1.40	75.3	11.64	Al	8.21
8.924	5.98	1.40	74.7	11.65	Al	8.21
8.918	6.92	2.00	123.4	12.54	Al	9.40
8.918	7.23	2.17	139.9	12.74	Al	9.74
8.929	7.17	2.21	141.5	12.91	Al	9.79
8.924	7.65	2.41	164.5	13.03	Al	10.23
8.924	7.60	2.46	166.8	13.20	Al	10.29
8.924	7.73	2.56	176.6	13.34	Al	10.47

$U = (A_0 + A_1 \cdot u) \text{ km s}^{-1}$; $A_0 = 3.91 \text{ km s}^{-1}$;
$A_1 = 1.51 \text{ km s}^{-1}$; $\sigma_{A_0} = 0.012 \text{ km s}^{-1}$;
$\sigma_{A_1} = 0.011 \text{ km s}^{-1}$; $\sigma_U = 0.049 \text{ km s}^{-1}$;
Experimental technique B
Data reduction technique B
Reference: [McQ 70]

Note to graphite, page 133 below
$U = (4.74 + 1.53 u) \text{ km s}^{-1}$; $0.8 < u < 2.3 \text{ km s}^{-1}$;
$\sigma_U = 0.067 \text{ km s}^{-1}$;
$U = (3.9 + 2.3 u) \text{ km s}^{-1}$; $3.25 < u < 4.0 \text{ km s}^{-1}$;
$\sigma_U = 0.22 \text{ km s}^{-1}$
Experimental technique B
Data reduction technique B
Reference: [McQ 68]

Diamond (Diamant) C

$\varrho_0 = 3.185 \text{ g cm}^{-3}$; $\varrho_{01} = 3.516 \text{ g cm}^{-3}$;
$c_0 = 11.22 \text{ km s}^{-1}$

Sample					Standard	
ϱ_0 g cm^{-3}	U km s^{-1}	u km s^{-1}	P GPa	ϱ g cm^{-3}	Composition	U km s^{-1}
3.20	9.91	1.37	43.4	3.71	Al	7.65
3.17	12.10	2.94	112.8	4.19	Al	9.93
3.20	12.49	3.19	127.5	4.30	Al	10.31

$U = (7.976 + 1.409\,u) \text{ km s}^{-1}$; $\sigma_U = 0.027 \text{ km s}^{-1}$

Experimental technique B
Data reduction technique B
Reference: [McQ 68]

The diamond samples were pressed with graphite binder (≈ 5 wt%) into $9 \cdots 12$ mm diameter discs, porosity 6%.

Gold Au

$\varrho_0 = 19.231 \cdots 19.305 \text{ g cm}^{-3}$; $\varrho_{01} = 19.286 \text{ g cm}^{-3}$;
$c_0 = 2.93 \text{ km s}^{-1}$; $c_l = 3.24 \text{ km s}^{-1}$; $c_s = 1.20 \text{ km s}^{-1}$;
$c_b = 3.05 \text{ km s}^{-1}$

Sample				
ϱ_0 g cm^{-3}	U km s^{-1}	u km s^{-1}	P GPa	ϱ g cm^{-3}
19.29	3.576	0.3	20.7	21.06
19.29	4.032	0.6	46.7	22.66
19.29	4.489	0.9	77.9	24.14
19.29	4.641	1.0	89.5	24.60
19.29	5.401	1.5	156.3	26.72
19.29	6.162	2.0	237.7	28.58
19.29	6.922	2.5	333.8	30.19
19.29	7.683	3.0	444.6	31.62
19.29	8.443	3.5	570.1	32.97

$U = (3.120 + 1.521\,u) \text{ km s}^{-1}$; $\sigma_U = 0.056 \text{ km s}^{-1}$

Summary of data from various sources compiled by [Van 77]

Graphite, pyrolytic (Graphit, pyrolitisch) C

$\varrho_0 = 2.198 \text{ g cm}^{-3}$; $\varrho_{01} = 2.267 \text{ g cm}^{-3}$;
$c_l = 3.45 \text{ km s}^{-1}$; $c_b = 3.9 \text{ km s}^{-1}$

Sample					Standard	
ϱ_0 g cm^{-3}	U km s^{-1}	u km s^{-1}	P GPa	ϱ g cm^{-3}	Composition	U km s^{-1}
2.20	3.90	0.00	0.0	2.20	Al	0.00
2.20	5.18	0.37	4.3	2.37	Al	5.88
2.20	5.17	0.37	4.3	2.37	Al	5.88
2.20	5.14	0.37	4.2	2.37	Al	5.88
2.20	6.17	0.83	11.4	2.54	Al	6.44

Graphite (continued)

Sample					Standard	
ϱ_0 g cm^{-3}	U km s^{-1}	u km s^{-1}	P GPa	ρ g cm^{-3}	Composition	U km s^{-1}
2.20	6.05	0.84	11.2	2.55	Al	6.44
2.20	6.01	0.84	11.2	2.56	Al	6.44
2.23	6.19	0.99	13.7	2.65	Al	6.62
2.20	6.41	1.01	14.2	2.61	Al	6.64
2.22	6.76	1.30	19.6	2.75	Al	7.00
2.22	6.78	1.30	19.6	2.75	Al	7.00
2.20	6.75	1.33	19.7	2.74	Al	7.02
2.20	6.72	1.33	19.7	2.74	Al	7.02
2.20	6.66	1.33	19.5	2.75	Al	7.02
2.20	7.01	1.46	22.5	2.78	Al	7.18
2.20	6.97	1.46	22.4	2.78	Al	7.18
2.22	7.64	1.81	30.8	2.91	Al	7.64
2.20	7.64	1.82	30.5	2.89	Al	7.63
2.20	7.57	1.83	30.5	2.90	Al	7.64
2.20	7.55	1.83	30.4	2.90	Al	7.64
2.20	7.50	1.83	30.3	2.91	Al	7.64
2.23	7.85	2.00	35.1	2.99	Al	7.87
2.21	8.02	2.16	38.3	3.02	Al	8.05
2.20	8.28	2.30	42.0	3.05	Al	8.23
2.20	8.24	2.31	41.9	3.06	Al	8.23
2.20	8.19	2.31	41.7	3.06	Al	8.23
2.20	8.29	2.33	42.5	3.06	Al	8.26
2.20	8.24	2.33	42.4	3.07	Al	8.26
2.20	8.21	2.34	42.3	3.08	Al	8.26
2.22	8.11	2.42	43.6	3.16	Al	8.34
2.21	8.18	2.49	45.1	3.18	Al	8.42
2.21	8.36	2.58	47.8	3.20	Al	8.54
2.20	8.33	2.81	51.5	3.32	Al	8.77
2.20	8.26	2.82	51.3	3.34	Al	8.77
2.20	8.24	2.82	51.2	3.34	Al	8.77
2.23	8.28	3.17	58.5	3.61	Al	9.14
2.22	8.06	3.22	57.7	3.70	Al	9.16
2.21	8.21	3.37	61.2	3.75	Al	9.33
2.20	8.43	3.46	64.2	3.73	Al	9.45
2.20	8.38	3.46	64.0	3.75	Al	9.45
2.20	8.36	3.47	63.9	3.76	Al	9.45
2.20	8.37	3.55	65.2	3.82	Al	9.52
2.21	8.57	3.58	67.6	3.80	Al	9.59
2.20	8.56	3.65	68.7	3.84	Al	9.66
2.23	9.13	3.68	74.9	3.74	Al	9.79
2.21	8.86	3.68	72.0	3.78	Al	9.74
2.21	9.17	3.69	74.7	3.70	Al	9.79
2.22	9.17	3.72	75.9	3.74	Al	9.84
2.20	9.07	3.76	75.1	3.76	Al	9.85
2.24	9.21	3.77	77.7	3.79	Al	9.90
2.21	9.47	3.95	82.8	3.79	Al	10.11
2.23	9.41	3.96	83.1	3.85	Al	10.13

For techniques and reference, see previous page.

Iron-Nickel Alloy (Eisen-Nickel-Legierung)
Composition (in wt%): Fe90, Ni10.
$\varrho_0 = 7.880$ g cm^{-3}; $\varrho_{01} = 7.893$ g cm^{-3}; $c_0 = 4.46$ km s^{-1}

Sample					Standard	
ϱ_0 g cm^{-3}	U km s^{-1}	u km s^{-1}	P GPa	ϱ g cm^{-3}	Composition	U km s^{-1}
7.883	5.39	1.04	44.2	9.77	Cu	5.40
7.885	5.41	1.09	46.5	9.87	Cu	5.47
7.883	6.38	1.60	80.5	10.52	Cu	6.22
7.895	6.47	1.63	83.3	10.55	Cu	6.27
7.883	6.79	1.80	96.3	10.73	Cu	6.53
7.896	7.71	2.38	144.9	11.42	Cu	7.37
7.885	8.00	2.54	160.2	11.55	Cu	7.61
7.896	8.02	2.55	161.5	11.58	Cu	7.63
7.870	7.98	2.58	162.0	11.63	Cu	7.65
7.883	8.27	2.75	179.3	11.81	Cu	7.92

$U = (3.626 + 1.711\, u)$ km s^{-1}; $\sigma_u = 0.058$ km s^{-1}
Experimental technique B.
Data reduction technique B
Reference: [McQ70]

Iron-Nickel Alloy (Eisen-Nickel-Legierung)
Composition (in wt%): Fe 82; Ni 18.
$\varrho_0 = 7.962$ g cm^{-3}; $\varrho_{01} = 7.924$ g cm^{-3}; $c_0 = 4.40$ km s^{-1}

Sample					Standard	
ϱ_0 g cm^{-3}	U km s^{-1}	u km s^{-1}	P GPa	ϱ g cm^{-3}	Composition	U km s^{-1}
7.962	5.56	1.02	45.2	9.75	Cu	5.40
7.962	5.61	1.06	47.3	9.82	Cu	5.45
7.962	6.52	1.58	82.0	10.51	Cu	6.22
7.962	6.58	1.61	84.3	10.54	Cu	6.27
7.962	6.90	1.78	97.8	10.73	Cu	6.53
7.962	7.13	1.95	110.7	10.96	Cu	6.77
7.962	7.62	2.26	137.1	11.32	Cu	7.22
7.962	7.74	2.37	146.1	11.48	Cu	7.37
7.962	8.00	2.53	161.2	11.65	Cu	7.61
7.962	8.03	2.54	162.4	11.65	Cu	7.63
7.962	7.98	2.56	162.7	11.72	Cu	7.65
7.962	8.42	2.72	182.3	11.76	Cu	7.91
7.962	8.38	2.73	182.1	11.81	Cu	7.92
7.962	8.35	2.73	181.5	11.83	Cu	7.91
7.962	8.32	2.76	182.8	11.91	Cu	7.94

$U = (3.957 + 1.608\, u)$ km s^{-1}; $\sigma_U = 0.055$ km s^{-1}
Experimental technique B
Data reduction technique B
Reference: [McQ70]

Iron-Nickel Alloy (Eisen-Nickel-Legierung)
Composition (in wt %): Fe 74; Ni 26.
$\varrho_0 = 7.968$ g cm^{-3}; $\varrho_{01} = 7.962$ g cm^{-3}; $c_0 = 4.37$ km s^{-1}

Sample					Standard	
ϱ_0 g cm^{-3}	U km s^{-1}	u km s^{-1}	P GPa	ϱ g cm^{-3}	Composition	U km s^{-1}
7.974	5.48	1.03	45.0	9.82	Cu	5.40
7.974	5.49	1.06	46.4	9.88	Cu	5.45
7.974	5.49	1.08	47.3	9.93	Cu	5.47
7.974	6.44	1.58	81.1	10.57	Cu	6.22
7.974	6.53	1.61	83.8	10.58	Cu	6.27
7.974	6.84	1.79	97.6	10.80	Cu	6.53
7.974	7.07	1.96	110.5	11.03	Cu	6.77
7.974	7.48	2.27	135.4	11.45	Cu	7.22
7.974	7.69	2.37	145.3	11.53	Cu	7.37
7.974	7.94	2.54	160.8	11.72	Cu	7.61
7.974	7.96	2.55	161.9	11.73	Cu	7.63
7.974	8.03	2.56	163.9	11.71	Cu	7.65
7.974	8.13	2.76	178.9	12.07	Cu	7.92
7.974	8.18	2.78	181.3	12.08	Cu	7.94

$U = (3.877 + 1.591\, u)$ km s^{-1}, $\sigma_U = 0.087$ km s^{-1}
Experimental technique B
Data reduction technique B
Reference: [McQ70]

Mercury (Quecksilber) Hg
$\varrho_0 = 13.532$ g cm^{-3}; $c_0 = 1.451$ km s^{-1}

Sample					Standard
ϱ_0 g cm^{-3}	U km s^{-1}	u km s^{-1}	P GPa	ϱ g cm^{-3}	Composition
13.53	2.752	0.608	22.64	17.37	Al
13.53	3.101	0.772	32.4	18.02	Al
13.53	3.504	0.978	46.37	18.77	Al

$U = (1.524 + 2.029\, u)$ km s^{-1}; $\sigma_U = 0.013$ km s^{-1}
Experimental technique B
Data reduction technique B
$T_0 = 17 \cdots 25$ °C
Reference: [Wal57]

Silver (Silber) Ag

$\varrho_0 = 10.490$ g cm^{-3}; $\varrho_{01} = 10.490$ g cm^{-3};
$c_l = 3.6$ km s^{-1}; $c_0 = 3.09$ km s^{-1}; $c_s = 1.59$ km s^{-1};
$c_b = 3.19$ km s^{-1}

ϱ_0 g cm^{-3}	U km s^{-1}	u km s^{-1}	P GPa	ϱ g cm^{-3}
Sample				
10.49	4.037	0.5	21.2	11.97
10.49	4.842	1.0	50.8	13.23
10.49	5.633	1.5	88.6	14.29
10.49	6.411	2.0	134.5	15.25
10.49	7.928	3.0	249.5	16.86
10.49	9.393	4.0	394.1	18.28

$U = (3.22 + 1.648\, u - 0.026\, u^2)$ km s^{-1}; $\sigma_U = 0.06$ km s^{-1}
Reference: [Van77]

1.3.3.2.2 Oxides – Oxide

Calcium oxide, synthetic single crystal from [Son72]
Composition: CaO with up to 2.9% Ca(OH)$_2$
$\varrho_0 = 3.305 \cdots 3.347$ g cm^{-3}; $\varrho_{01} = 3.345$ g cm^{-3}
Shock wave propagation parallel [100]

Sample					Standard	
ϱ_0 g cm^{-3}	U km s^{-1}	u km s^{-1}	P GPa	ϱ g cm^{-3}	Composition	U_{fs} km s^{-1}
Hugoniot state:						
3.306	8.204	1.801	48.9	4.236	W	2.342
3.339	9.086	2.314	70.2	4.480[1])	Al	5.099
	8.789	2.350	69.0	4.558[2])		
3.305	8.312	2.491	68.4	4.719	Al	5.226
3.325	8.831	2.846	83.6	4.906	Al	5.992
3.345	9.068	3.285	99.6	5.245	Ta	4.510
3.307	9.630	3.782	120.5	5.446	Ta	5.197
3.319	9.706	4.151	133.7	5.798	Ta	5.680
3.347	10.253	4.038	138.6	5.521	Ta	5.608
3.338	10.579	4.174	147.4	5.513	Ta	5.817
3.312	10.159	4.421	148.7	5.864	Ta	6.075
3.315	10.577	4.402	154.3	5.678	Ta	6.101
3.320	10.877	4.465	161.3	5.630	Ta	6.219
3.328	11.233	4.628	173.0	5.659	Ta	6.473
Release state:						
3.306	5.202	2.581	29.6	3.737		
3.339	5.813	2.966	38.0	4.317[2])		
3.305	no data					
3.325	6.516	3.409	49.0	4.695		
3.345	7.334	3.924	63.4	4.952		
3.307	8.723	4.800	92.3	4.537		
3.319	7.422	4.295	75.0	5.786		
3.347	9.270	5.145	105.1	4.593		
3.338	9.467	5.269	109.9	4.686		
3.312	7.960	4.319	75.8	5.859		

Calcium oxide (continued)

Sample					Standard	
ϱ_0 g cm^{-3}	U km s^{-1}	u km s^{-1}	P GPa	ϱ g cm^{-3}	Composition	U_{fs} km s^{-1}
3.315	9.869	5.523	120.1	4.697		
3.320	9.200	5.101	103.4	5.417		
3.328	10.149	5.699	127.5	4.951		

Errors for all entries are given in the original reference.
Experimental technique C
[1]) First wave arrival
[2]) Second wave arrival
Data reduction technique B
Reference: [Jea80]

Cassiterite, San Luis Potosi, Mexico (Zinnstein) SnO$_2$
$\varrho_0 = 6.451 \cdots 6.757$ g cm^{-3}; $\varrho_{01} = 6.997$ g cm^{-3}

Sample					Standard	
ϱ_0 g cm^{-3}	U km s^{-1}	u km s^{-1}	P GPa	ϱ g cm^{-3}	Composition	U km s^{-1}
6.45	6.77	2.27	99.2	9.71	Al	9.26
6.52	6.77	2.31	102.1	9.89	Al	9.33
6.75	6.94	2.55	119.3	10.66	Al	9.75
6.74	7.32	2.73	134.8	10.75	Al	10.10
6.51	7.51	2.88	141.0	10.57	Al	10.29

$U = (3.881 + 1.248\, u)$ km s^{-1}; $\sigma_U = 0.083$ km s^{-1}
Experimental technique B
Data reduction technique B
Reference: [McQ66]

Corundum, ceramic (Korund, keramisch)
Wesgo Al-995 (Table 1)
Al_2O_3 99.5 wt% minimum; MgO major impurity;
SiO_2 major impurity.
Porosity: 3.5 ··· 4.3% maximum

Sample					Standard	
ϱ_0 g cm^{-3}	U km s^{-1}	u km s^{-1}	P GPa	ϱ g cm^{-3}	Composition	No.
first wave:						
3.814	10.07	0.26	10.0	3.92	lucite	1
3.810	10.38	0.21	8.4	3.89	lucite	2
3.814	10.32	0.20	7.9	3.89	Al	3
3.809	9.82	0.18	6.7	3.88	Al	4
3.810	10.07	0.22	8.5	3.90	Al	5
3.809	10.05	0.21	8.1	3.89	Al	6
second wave:						
3.814	6.75	0.29	10.7	3.93	lucite	1
3.810	7.83	0.31	11.3	3.94	lucite	2
3.814	7.26	0.87	26.3	4.27	Al	3
3.809	7.54	0.85	25.9	4.26	Al	4
3.810	8.62	1.28	43.2	4.45	Al	5
3.809	8.59	1.26	42.4	4.44	Al	6
3.837	11.03	2.677	113.1	5.08	brass	7
3.839	10.90	2.687	112.1	5.10	brass	8
3.808	9.88	1.96	73.6	4.75	Al	9

$U = (6.43 + 1.70\,u - 0.32\,u^2)$ km s^{-1}; $\sigma_U = 0.08$ km s^{-1}

Experimental technique C was used for all entries except for the last entry of Table 1, which was obtained with technique B.

Lucalox (Table 2)
Al_2O_3 99.8 wt%.
Porosity: 0.2% maximum
$\varrho_0 = 3.808 \cdots 3.979$ g cm^{-3}; $\varrho_{01} = 3.976$ g cm^{-3};
$c_l = 10.3$ km s^{-1}; $c_s = 6.12$ km s^{-1}; $c_0 = 7.55$ km s^{-1}

Sample					Standard	
ϱ_0 g cm^{-3}	U km s^{-1}	u km s^{-1}	P GPa	ϱ g cm^{-3}	Composition	No.
first wave:						
3.98	10.98	0.368	16.1	4.12	Al	1
3.98	10.98	0.262	11.4	4.08	Al	2
3.98	10.90	0.284	12.3	4.09	Al	3
3.98	10.98	0.253	11.1	4.07	Al	4
3.98	10.88	0.228	9.9	4.06	Al	5
second wave:						
3.98	8.80	0.477	19.9	4.17	Al	1
3.98	8.79	0.495	19.5	4.19	Al	2
3.98	8.53	0.495	19.5	4.19	Al	3
3.98	9.60	0.93	36.9	4.39	Al	4
3.98	9.36	0.96	36.6	4.42	Al	5

Data reduction technique D, $U_{fs} = 2\,u$, Table 1, first 6 entries; data reduction technique B, Table 1, last 3 entries and Table 2
Sample materials Wesgo Al-995, Table 1 (Western Gold and Platinum Co., Belmond, Calif., U.S.A.)
Lucalox, Table 2 (General Electric Co., Ohio)
Reference: [Ahr68]

Corundum, crystalline (Korund, Einkristall)
Al_2O_3
$\varrho_0 = 4.854$ g cm^{-3}; $\varrho_{01} = 3.976$ g cm^{-3}

Sample					Standard	
ϱ_0 g cm^{-3}	U km s^{-1}	u km s^{-1}	P GPa	ϱ g cm^{-3}	Composition	U km s^{-1}
3.98	9.99	1.30	51.6	4.58	Al	7.72
3.98	10.51	1.66	96.3	4.73	Al	8.32
3.99	11.02	1.98	87.1	4.86	Al	8.86
3.99	11.17	2.20	97.9	4.96	Al	9.18
3.99	11.28	2.24	100.4	4.97	Al	9.26
3.99	11.24	2.27	101.6	4.99	Al	9.29
3.99	11.03	2.32	101.9	5.05	Al	9.33
3.99	11.81	2.78	130.9	5.21	Al	10.08
3.99	11.78	2.80	131.5	5.23	Al	10.11
3.99	11.69	2.95	137.6	5.33	Al	10.29
3.98	11.92	3.11	148.0	5.39	Al	10.45

$U = (7.916 + 1.897\,u - 0.195\,u^2)$ km s^{-1}; $\sigma_U = 0.12$ km s^{-1}
Experimental technique B
Data reduction technique B
Reference: [McQ66]

Hematite, natural (Hämatit)
Fe_2O_3
$\varrho_0 = 4.902$ g cm^{-3}; $\varrho_{01} = 5.283$ g cm^{-3}

Sample					Standard	
ϱ_0 g cm^{-3}	U km s^{-1}	u km s^{-1}	P GPa	ϱ g cm^{-3}	Composition	U km s^{-1}
5.01	7.62	2.34	89.6	7.24	Al	9.15
4.90	7.67	2.39	90.0	7.13	Al	9.18
4.98	7.77	2.44	94.4	7.25	Al	9.29
5.01	7.86	2.45	96.4	7.28	Al	9.33
5.01	8.49	2.92	124.3	7.63	Al	10.08
5.02	8.47	2.92	124.3	7.68	Al	10.08
5.01	8.47	2.93	124.1	7.66	Al	10.08
5.05	8.48	2.93	125.5	7.71	Al	10.11
5.05	8.45	2.93	125.3	7.75	Al	10.11
4.97	8.84	3.23	142.1	7.84	Al	10.54

$U = (4.385 + 1.393\,u)$ km s^{-1}; $\sigma_U = 0.037$ km s^{-1}
Experimental technique B
Data reduction technique B
Reference: [McQ66]

Ilmenite, Kragerø, Norway (Ilmenit)
$FeTiO_3$
$\varrho_0 = 3.774 \cdots 3.846$ g cm^{-3}; $\varrho_{01} = 4.819$ g cm^{-3}

Sample					Standard	
ϱ_0 g cm^{-3}	U km s^{-1}	u km s^{-1}	P GPa	ϱ g cm^{-3}	Composition	U km s^{-1}
3.84	7.41	2.36	67.3	5.64	Al	8.76
3.78	7.44	2.47	69.2	5.65	Al	8.86
3.77	7.43	2.47	69.1	5.64	Al	8.86
3.77	7.93	2.72	81.4	5.74	Al	9.26
3.84	8.43	3.04	98.3	6.00	Al	9.75
3.83	8.86	3.26	110.6	6.06	Al	10.10

$U = (3.317 + 1.691\,u)$ km s^{-1}; $\sigma_U = 0.69$ km s^{-1}
Experimental technique B
Data reduction technique B
Reference: [McQ66]

Ilmenite, single twinned crystal (Ilmenit, verzwillingter Einkristall) Blafjell, Norway
$Fe_{0.98}Mn_{0.07}Ti_{0.94}O_3$, sample contains several percent of exsolved hematite
$\varrho_0 = 4.73 \cdots 4.80$ g cm^{-3}; $\varrho_{01} = 4.79$ g cm^{-3}

Table 1

Sample					Standard		Sample orientation to shock direction
ϱ_0 g cm^{-3}	U km s^{-1}	u km s^{-1}	P GPa	ϱ g cm^{-3}	Composition	U_{fs} km s^{-1}	
4.749	6.288	0.332	9.9	5.013	Al	0.919	$\perp c$-axis
4.797	7.527	0.390	14.0	5.059	Al	1.118	$\perp c$-axis
4.785	7.900	0.837	31.6	5.352	Tl	1.307	$\perp c$-axis
4.785	7.734	1.127	41.7	5.601	Tl	1.727	$\perp c$-axis
4.773	7.643	1.535	56.4	5.973	Tl	2.310	$\perp c$-axis
4.766	7.296	0.834	29.0	5.381	Tl	1.274	$\parallel c$-axis
4.774	7.239	1.032	35.7	5.568	Tl	1.557	$\parallel c$-axis
4.766	7.295	1.464	50.9	5.963	W	2.024	$\parallel c$-axis
4.761	7.323	1.588	53.7	6.020	W	2.160	$\parallel c$-axis
4.754	7.462	1.780	63.3	6.246	W	2.460	$\parallel c$-axis
4.765	7.105	1.450	47.3	5.930	W	1.922	$\parallel c$-axis

Table 2a (first-shock state)

Sample						Standard		Sample orientation to shock direction
No	ϱ_0 g cm^{-3}	U km s^{-1}	u km s^{-1}	P GPa	ϱ g cm^{-3}	Composition	U_{fs} km s^{-1}	
1	4.746	7.309	1.465	50.8	5.936	W	2.026	$\parallel c$-axis
2	4.754	7.374	1.460	50.5	5.928	W	2.000	$\parallel c$-axis
3	4.733	7.250	1.721	59.0	6.206	W	2.362	$\parallel c$-axis
4	4.772	7.247	1.457	50.4	5.973	W	2.013	$\parallel c$-axis

(continued)

Ilmenite (continued)

Table 2b (reflected-shock state calculated from tungsten shock velocity)

Sample No	U km s^{-1}	u km s^{-1}	P GPa	ϱ g cm^{-3}	Thungsten witness plate shock velocity
1	4.32	0.735	69.5	7.140	4.911
2	8.10	0.831	80.7	6.431	5.038
3	6.96	0.935	93.0	6.997	5.180

Table 2c (reflected-shock state calculated from tungsten free surface velocity)

No	U km s^{-1}	u km s^{-1}	P GPa	ϱ g cm^{-3}	Tungsten free surface velocity
1	3.62	0.712	67.0	7.468	1.424
2	4.56	0.738	69.9	7.042	1.476
4	3.41	0.745	71.1	6.995	1.489

Experimental technique C for data of Tables 1 and 2a.
Data reduction technique B for data of Table 1 and 2a. Data of Tables 2b and 2c represent shock states of a second shock wave reflected from a flat tungsten witness plate placed against the upper specimen surface. Error of shock pressure P and density ϱ in Table 1: 0.6 ··· 3% and 0.15 ··· 0.3%, respectively (see reference).
Reference: [McQ66]

Magnetite, natural (Magnetit)
Fe$_3$O$_4$
$\varrho_0 = 5.155 \cdots 5.000$ g cm^{-3}; $\varrho_{01} = 5.198$ g cm^{-3}

Sample					Standard	
ϱ_0 g cm^{-3}	U km s^{-1}	u km s^{-1}	P GPa	ϱ g cm^{-3}	Composition	U km s^{-1}
5.11	6.72	1.80	62.0	6.99	Al	8.28
5.14	6.82	1.82	63.7	7.00	Al	8.32
5.13	6.77	1.83	63.4	7.02	Al	8.32
5.14	6.74	1.83	63.4	7.06	Al	8.32
5.13	7.23	2.08	77.1	7.19	Al	8.75
5.11	7.40	2.26	85.3	7.35	Al	9.01
5.13	7.19	2.29	84.5	7.52	Al	9.02
5.13	7.43	2.35	89.4	7.49	Al	9.15
5.11	7.53	2.45	94.2	7.57	Al	9.29
5.13	7.92	2.67	108.3	7.73	Al	9.67
5.14	8.26	2.93	124.1	7.97	Al	10.08
5.01	8.37	2.93	122.7	7.70	Al	10.06
5.10	8.25	2.94	123.5	7.91	Al	10.08
5.13	8.27	2.94	124.8	7.95	Al	10.11
5.13	8.45	3.03	131.0	7.99	Al	10.25

$U = (4.259 + 1.368\, u)$ km s^{-1}; $\sigma_U = 0.082$ km s^{-1}
Experimental technique B
Data reduction technique B
Reference: [McQ66]

Periclase, single crystal (Periklas, Einkristall) MgO
$\varrho_0 = 3.577$ g cm^{-3}; $\varrho_{01} = 3.586$ g cm^{-3}

Sample					
ϱ_0 g cm^{-3}	U km s^{-1}	u km s^{-1}	P GPa	ϱ g cm^{-3}	No

first wave:

3.576	9.191	0.1215	3.99	3.62	1
3.576	9.142	0.154	3.45	3.62	2
3.576	10.068	0.2476	8.90	3.66	3

second wave:

3.576	7.016	0.626	16.8	3.91	1
3.576	6.964	0.635	16.8	3.92	2
3.576	8.593	1.35	42.3	4.22	3
3.576	6.893	0.633	16.6	3.93	4
3.576	7.085	0.632	16.6	3.92	5
3.576	7.881	1.229	39.0	4.07	6
3.576	9.61	1.92	66.0	4.46	7

$U = (6.15 + 1.85\, u(\pm 0.3))$ km s^{-1} for second wave
Experimental technique C
Data reduction technique A, where $2u = U_{fs}$
Reference: [Ahr66]

The periclase single-crystals were obtained from Norton Co., Niagara Falls, N.Y., USA

Periclase, crystalline (Periklas, polykristallin) MgO
$\varrho_0 = 3.580$ g cm^{-3}; $\varrho_{01} = 3.595$ g cm^{-3}; $c_0 = 6.58$ km s^{-1}

Sample					Standard	
ϱ_0 g cm^{-3}	U km s^{-1}	u km s^{-1}	P GPa	ϱ g cm^{-3}	Composition	U km s^{-1}
3.58	7.68	0.73	20.2	3.96	Al	6.57
3.58	7.63	0.78	21.4	3.99	Al	6.63
3.58	8.15	1.04	30.4	4.11	Al	7.05
3.58	8.93	1.51	48.4	4.31	Al	7.79
3.58	9.08	1.60	52.1	4.35	Al	7.93
3.58	9.01	1.64	52.8	4.38	Al	7.98
3.58	9.23	1.86	61.6	4.49	Al	8.31
3.58	9.45	2.02	68.3	4.56	Al	8.54
3.58	9.79	2.22	78.0	4.64	Al	8.86
3.58	9.93	2.40	85.6	4.73	Al	9.12
3.58	10.13	2.43	88.4	4.72	Al	9.18
3.58	10.14	2.49	90.4	4.75	Al	9.26
3.58	10.11	2.52	91.3	4.77	Al	9.29
3.58	10.21	2.59	95.0	4.81	Al	9.40
3.58	10.33	2.64	98.0	4.82	Al	9.49
3.58	10.59	2.90	110.2	4.94	Al	9.86
3.58	10.67	2.97	113.7	4.97	Al	9.96
3.58	10.92	3.18	124.4	5.06	Al	10.25
3.58	10.96	3.20	125.8	5.06	Al	10.29

$U = (6.535 + 1.643\, u - 0.083\, u^2)$ km s^{-1}; $\sigma_U = 0.067$ km s^{-1}
Experimental technique B
Data reduction technique B
Reference: [McQ66]

$U = (2.35 + 1.56\, u)$ km s^{-1}; $\sigma_U = 0.092$ km s^{-1}
Experimental technique not reported
Data reduction technique not reported
Reference: [Alt65]

Pyrolusite, Ironton, Minnesota (Pyrolusit) MnO$_2$
$\varrho_0 = 4.065 \cdots 4.425$ g cm^{-3}; $\varrho_{01} = 5.203$ g cm^{-3}

Sample					Standard	
ϱ_0 g cm^{-3}	U km s^{-1}	u km s^{-1}	P GPa	ϱ g cm^{-3}	Composition	U km s^{-1}
4.42	5.11	0.81	18.3	5.25	Al	6.56
4.39	5.49	1.12	27.0	5.51	Al	7.01
4.06	5.20	1.41	29.8	5.57	Al	7.28
4.37	5.91	1.48	38.1	5.83	Al	7.54
4.33	6.00	1.48	38.5	5.75	Al	7.55
4.24	5.82	1.57	38.8	5.81	Al	7.62
4.19	5.83	1.64	40.2	5.84	Al	7.70
4.24	6.10	1.70	43.9	5.87	Al	7.82
4.37	6.85	1.98	59.1	6.14	Al	8.33
4.30	7.08	2.23	68.1	6.28	Al	8.68
4.36	7.26	2.25	71.3	6.32	Al	8.76
4.32	7.49	2.31	74.7	6.25	Al	8.86
4.26	7.39	2.45	77.1	6.37	Al	9.00
4.36	7.98	2.94	102.3	6.90	Al	9.75
4.34	8.37	3.16	114.9	6.97	Al	10.10
4.29	8.90	3.31	126.3	6.83	Al	10.37
4.31	8.40	3.32	120.2	7.14	Al	10.29

$U = (3.632 + 1.52\, u)$ km s^{-1}; $\sigma_U = 0.25$ km s^{-1}
Experimental technique B
Data reduction technique B
Reference: [McQ66]

Quartz, single crystal (Quarz, Einkristall) SiO$_2$
$\varrho_0 = 2.650$ g cm^{-3}; $\varrho_{01} = 2.650$ g cm^{-3}

Sample				
ϱ_0 g cm^{-3}	U km s^{-1}	u km s^{-1}	P GPa	ϱ g cm^{-3}
2.65	7.18	3.13	59.5	4.70
2.65	8.54	3.92	88.7	4.90
2.65	12.01	6.2	197.4	5.48

Quartz, single crystal (Quarz, Einkristall) SiO$_2$
$\varrho_0 = 2.655$ g cm^{-3}; $\varrho_{01} = 2.650$ g cm^{-3}

Sample					Standard		Shock direction
ϱ_0 g cm^{-3}	U km s^{-1}	u km s^{-1}	P GPa	ϱ g cm^{-3}	Composition	No	
first wave:							
2.654	5.9	0.66	10.2	2.99	Al	1	∥ x-axis
2.654	6.27	0.54	8.9	2.90	Al	2	∥ x-axis
2.654	6.427	0.537	9.1	2.90	Al	3	∥ y-axis
2.654	7.25	0.81	15.7	2.99	Al	4	∥ z-axis

(continued)

Quartz (continued)

ϱ_0 g cm^{-3}	U km s^{-1}	u km s^{-1}	P GPa	ϱ g cm^{-3}	Standard Composition	No	Shock direction
second wave:							
2.654	5.42	2.01	29.5	4.75	Al	1	∥ x-axis
2.654	5.72	1.94	29.9	4.39	Al	2	∥ x-axis
2.654	5.661	1.93	29.8	4.41	Al	3	∥ y-axis
2.654	5.92	2.40	40.1	5.03	Al	4	∥ z-axis
single wave:							
2.654	5.635	1.815	27.2	3.91	brass	1	∥ x-axis
2.654	5.696	1.803	27.2	3.88	brass	2	∥ x-axis
2.654	5.685	1.98	29.8	4.07	brass	3	∥ x-axis
2.654	5.79	2.46	37.7	4.62	brass	4	∥ x-axis
2.654	5.78	2.48	38.0	4.65	brass	5	∥ x-axis
2.654	6.003	2.61	41.6	4.70	brass	6	∥ x-axis
2.654	5.82	2.46	37.9	4.59	brass	7	∥ y-axis
2.654	6.218	2.58	42.6	4.54	brass	8	∥ y-axis

Experimental technique C   Data reduction technique B   Reference: [Ahr68a]

Quartz, crystal (Quarz, Einkristall) SiO$_2$
$\varrho_0 = 2.653$ g cm^{-3}; $\varrho_{01} = 2.650$ g cm^{-3}; $c_0 = 3.69$ km s^{-1}

ϱ_0 g cm^{-3}	U km s^{-1}	u_{max} km s^{-1}	u_{min} km s^{-1}	u_{av} km s^{-1}	P GPa	ϱ g cm^{-3}	Composition	P GPa	No.	Shock direction
first wave:										
2.65	5.89	0.33	0.275	0.285	4.3	2.78	plexiglass	3.64	1	∥ x-axis
2.65	5.92	0.34	0.275	0.29	4.3	2.78	plexiglass	3.64	2	∥ x-axis
2.65	5.93	0.37	0.29	0.32	4.6	2.79	plexiglass	3.64	3	∥ x-axis
2.65	6.00	0.40	0.34	0.36	5.4	2.81	Al	11.2	4	∥ x-axis
2.65	6.01	0.41	0.35	0.37	5.6	2.81	Al	14.7	5	∥ x-axis
2.65	6.07	0.44	0.40	0.415	6.4	2.84	Al	16.6	6	∥ x-axis
2.65	6.10		0.44		7.1	2.86	Al	22.7	7	∥ x-axis
2.65	6.12	0.52	0.48	0.50	7.8	2.87	Al	32.9	8	∥ x-axis
2.65	5.89	0.31	0.275	0.285	4.3	2.78	Al	11.2	9	∥ x-axis
2.65	5.94	0.34	0.285	0.305	4.5	2.78	Al	14.7	10	∥ x-axis
2.65	5.98	0.41	0.39	0.40	6.2	2.83	Al	22.7	11	∥ x-axis
2.65	6.03		0.35		5.6	2.81	Al	11.2	12	∥ x-axis
2.65	6.04		0.36		5.8	2.82	Al	14.7	13	∥ x-axis
2.65	6.10		0.45		7.3	2.86	Al	22.7	14	∥ x-axis
2.65	6.21		0.57		9.4	2.92	Al	32.9	15	∥ x-axis
2.65			0.61				Al	51.5	16	∥ x-axis
2.65	6.19	0.49	0.44	0.455	7.2	2.85	Al	14.7	22	∥ y-axis
2.65	6.17	0.50	0.46	0.48	7.5	2.86	Al	14.7	23	∥ y-axis
2.65	6.24	0.60	0.58	0.59	9.6	2.92	Al	22.7	24	∥ y-axis
2.65	6.26	0.66	0.64	0.65	10.6	2.95	Al	32.9	25	∥ y-axis
2.65	6.12		0.60		9.7	2.94	Al	22.7	26	∥ y-axis
2.65	6.82	0.31	0.29	0.295	5.2	2.77	plexiglass	2.98	31	∥ z-axis
2.65	6.87	0.39	0.34	0.35	6.2	2.79	plexiglass	3.64	32	∥ z-axis
2.65	7.23	0.68	0.60	0.625	11.5	2.89	Al	11.2	33	∥ z-axis
2.65	7.21	0.64	0.57	0.59	10.9	2.88	Al	14.7	34	∥ z-axis
2.65	7.44	0.71	0.685	0.695	13.5	2.92	Al	22.7	35	∥ z-axis
2.65	7.51	0.73	0.71	0.72	14.1	2.92	Al	32.9	36	∥ z-axis

Quartz (continued)

Sample					Standard			Shock direction
ϱ_0 g cm^{-3}	U km s^{-1}	u km s^{-1}	P GPa	ϱ g cm^{-3}	Composition	P GPa	No.	
second wave:								
2.65	2.88	0.43	5.6	2.94	plexiglass	3.64	3	∥ x-axis
2.65	4.74	0.67	9.4	3.05	Al	11.2	4	∥ x-axis
2.65	5.11	0.85	12.6	3.15	Al	14.7	5	∥ x-axis
2.65	5.24	0.92	13.5	3.20	Al	16.6	6	∥ x-axis
2.65	5.64	1.24	18.9	3.38	Al	22.7	7	∥ x-axis
2.65	5.69	1.69	26.9	3.75	Al	32.9	8	∥ x-axis
2.65	5.14	0.82	11.6	3.13	Al	14.7	10	∥ x-axis
2.65	5.61	1.21	18.4	3.36	Al	22.7	11	∥ x-axis
2.65	4.74	0.71	9.9	3.07	Al	11.2	12	∥ x-axis
2.65	5.18	0.81	13.2	3.17	Al	14.7	13	∥ x-axis
2.65	5.61	1.26	20.0	3.44	Al	22.7	14	∥ x-axis
2.65	5.76	1.82	27.7	3.84	Al	32.9	15	∥ x-axis
2.65	6.12	2.55	41.4	4.53	Al	51.5	16	∥ x-axis
2.65	6.29	2.70	45.0	4.64	Al	51.5	17	∥ x-axis
2.65	6.66	2.89	51.1	4.68	Al	62.0	18	∥ x-axis
2.65	6.95	3.03	55.8	4.70	Al	66.0	19	∥ x-axis
2.65	7.70	3.52	70.8	4.92	Al	85.0	20	∥ x-axis
2.65	7.63	3.50	72.1	4.91	Al	85.0	21	∥ x-axis
2.65	4.85	0.86	12.6	3.15	Al	14.7	22	∥ y-axis
2.65	4.88	0.86	12.6	3.14	Al	14.7	23	∥ y-axis
2.65	5.47	1.25	19.0	3.38	Al	22.7	24	∥ y-axis
2.65	5.61	1.71	26.3	3.76	Al	32.9	25	∥ y-axis
2.65	5.68	1.30	19.8	3.43	Al	22.7	26	∥ y-axis
2.65	6.66	2.89	51.1	4.68	Al	62.0	27	∥ y-axis
2.65	6.95	3.03	55.8	4.70	Al	66.0	28	∥ y-axis
2.65	7.72	3.50	72.5	4.85	Al	85.0	29	∥ y-axis
2.65	7.75	3.52	71.4	4.84	Al	85.0	30	∥ y-axis
2.65	3.68	1.83	13.0	3.14	Al	14.7	34	∥ z-axis
2.60	4.71	1.23	19.6	3.38	Al	22.7	35	∥ z-axis
2.65	7.76	3.42	70.3	4.74	Al	85.0	37	∥ z-axis
2.65	7.76	3.49	71.8	4.82	Al	85.0	38	∥ z-axis

$U = (A + B u)$ km s^{-1}

Fits for U_1 versus u_1 (max) along the principle axes:

		x	y	z
A		5.61	6.01	6.36
	σ_A	0.04	0.07	0.10
B		0.89	0.32	1.36
	σ_B	0.09	0.13	0.16

Fits for U_2 versus u_2, x and y direction combined:

		$u<1$	$1<u<2$	$u>2$
A		1.31	5.36	1.83
B		4.47	0.19	1.67
	σ_U	0.13	0.07	0.04

Experimental technique B and D
Data reduction technique D, for the first wave, assuming $2u = U_{fs}$, for the second wave technique B was used.
Reference: [Wac62]

An estimate of the Hugoniot curve of stishovite was made by [McQ63a]: the Hugoniot equation for stishovite with a density of 4.35 g cm^{-3} at zero pressure is $U = (10 + 1.0 u)$ km s^{-1}.

Rutile, synthetic single crystal and from Oaxaca, Mexico, polycrystalline (Rutil) TiO$_2$

$\varrho_0 = 4.20 \cdots 4.21$ g cm^{-3} for natural rutile Oaxaca;
$\varrho_0 = 4.25$ g cm^{-3} for synthetic rutile

Sample					Standard	
ϱ_0 g cm^{-3}	U km s^{-1}	u km s^{-1}	P GPa	ϱ g cm^{-3}	Composition	U km s^{-1}
4.25	7.65	0.50	16.2	4.54	Al	6.29
4.25	7.87	0.66	22.0	4.64	Al	6.57
4.25	8.18	0.68	23.4	4.63	Al	6.62
4.25	7.86	0.71	23.7	4.67	Al	6.65
4.25	8.30	1.38	48.8	5.10	Al	7.72
4.25	8.26	2.47	86.9	6.06	Al	9.18
4.25	8.31	2.52	89.1	6.10	Al	9.26
4.25	8.33	2.91	103.3	6.54	Al	9.75
4.21	8.51	2.98	106.8	6.49	Al	9.86
4.25	8.74	3.11	115.7	6.60	Al	10.08
4.25	8.77	3.13	116.5	6.60	Al	10.10
4.25	8.76	3.13	116.5	6.61	Al	10.11
4.25	9.06	3.22	124.2	6.60	Al	10.28
4.20	8.90	3.25	121.2	6.61	Al	10.25
4.25	8.94	3.25	123.5	6.67	Al	10.29

Experimental technique B
Data reduction technique B
Reference [McQ67]

Spinel, ceramic (Spinell, keramisch) MgAl$_2$O$_4$

$\varrho_0 = 3.401 \cdots 3.425$ g cm^{-3}; $\varrho_{01} = 3.582$ g cm^{-3}

Sample					Standard	
ϱ_0 g cm^{-3}	U km s^{-1}	u km s^{-1}	P GPa	ϱ g cm^{-3}	Composition	U km s^{-1}
3.41	8.46	2.35	67.8	4.35	Al	8.76
3.42	8.54	2.53	74.0	4.49	Al	9.00
3.43	9.09	2.79	87.1	4.50	Al	9.40
3.43	9.09	2.80	87.1	4.50	Al	9.40
3.43	9.16	2.85	89.6	4.50	Al	9.45
3.43	9.13	2.86	89.4	4.50	Al	9.49
3.41	9.41	3.05	97.9	5.40	Al	9.75
3.42	9.62	3.19	104.6	5.41	Al	9.96
3.41	9.91	3.25	109.8	5.40	Al	10.08
3.40	9.76	3.29	109.3	5.41	Al	10.10
3.41	9.95	3.41	115.8	5.42	Al	10.28

$U = (4.836 + 1.511\, u)$ km s^{-1}; $\sigma_U = 0.078$ km s^{-1}
Experimental technique B
Data reduction technique B
Reference: [McQ66]

Wustite, hot-pressed, synthetic polycrystals from [Gra78] (Wüstit)
Chemical composition: Fe$_{0.94}$O with 2 vol% Fe
Porosity: 4%
$\varrho_0 = 5.484 \cdots 5.576$ g cm^{-3}; $\varrho_{01} = 5.731$ g cm^{-3}

Sample						Standard	
ϱ_0[1]) g cm^{-3}	ϱ_0[2]) g cm^{-3}	U km s^{-1}	u km s^{-1}	P GPa	ϱ g cm^{-3}	Composition	U_{fs} km s^{-1}
Hugoniot state:							
5.499	5.525	6.435	1.220	43.2	6.785	W	1.705
5.494	5.557	6.954	1.629	62.2	7.175	W	2.296
5.504	5.522	7.173	1.766	69.7	7.301	W	2.500
5.484	5.563	7.221	2.034	80.6	7.635	Al	5.101
5.514	5.555	7.592	2.414	101.1	8.085	Al	5.999
5.555	5.576	8.701	3.147	152.1	8.703	Cu	5.456
5.494	5.562	9.640	3.743	198.2	8.981	Ta	5.776
5.505	5.526	10.230	4.055	228.4	9.120	Ta	6.300

[1]) Bulk density.
[2]) Archimedean density.

Wustite (continued)

Sample						Standard	
ϱ_0 [1]) g cm^{-3}	ϱ_0 [2]) g cm^{-3}	U km s^{-1}	u km s^{-1}	P GPa	ϱ g cm^{-3}	Composition	U_{fs} km s^{-1}

Release state:

5.499	5.525	5.379	1.942	12.5	6.084		
5.494	5.557	6.117	2.412	17.7	6.530		
5.504	5.522	6.738	2.808	22.6	6.248		
5.484	5.563	5.922	3.034	39.6	6.435		
5.514	5.555	6.338	3.297	46.0	7.256		
5.555	5.576	8.298	4.532	82.9	7.012		
5.494	5.562	9.316	5.174	106.2	7.485		
5.505	5.526	10.030	5.624	124.3	7.502		

Errors for all entries are given in the original reference.   Data reduction technique B
Experimental technique C   Reference: [Jea80]

[1]) Bulk density.   [2]) Archimedian density.

1.3.3.2.3 Carbonates — Carbonate

Calcite, Iceland Spar Single Crystal (Calcit, Einkristall, Island) CaCO$_3$

$\varrho_0 = 2.711$ g cm^{-3}; $\varrho_{01} = 2.711$ g cm^{-3}; c_1(x-cut) $= 7.29$ km s^{-1}; c_1(y-cut) $= 7.35$ km s^{-1}

Sample					Standard			No.	Sample orientation to shock front
ϱ_0 g cm^{-3}	U km s^{-1}	u km s^{-1}	P GPa	ϱ g cm^{-3}	Composition	U_{fs} km s^{-1}	P GPa		

first wave:

2.711	7.13	0.031	1.6	2.74	lucite, steel	1.1741	2.5	1	∥ cleavage plane
2.711	7.134	0.1144	2.2	2.76	lucite, steel	1.1520	2.45	2	∥ cleavage plane
2.711	7.068	0.0971	1.86	2.75	lucite	2.1754	5.5	3	∥ cleavage plane
2.711	6.852	0.1050	1.95	2.75	lucite, steel	1.2824	2.7	4	⊥ x-axis
2.711	6.886	0.1089	2.03	2.76	lucite, steel	1.377	2.5	5	⊥ x-axis
2.711	7.041	0.1065	2.03	2.75	brass, lucite	1.3823	3.0	6	⊥ x-axis
2.711	6.788	0.1248	2.30	2.67	lucite	2.499	6.7	7	⊥ x-axis
2.711	7.40	0.132	2.6	2.76	Al			8	⊥ x-axis
2.711	6.955	0.1202	2.27	2.76	Al	1.478	13.0	9	⊥ x-axis
2.711	6.493	1.3948	24.55	3.45	Al	3.3187	34.2	10	⊥ x-axis
2.711	6.636	0.1179	2.21	2.76	brass, lucite	1.483	3.2	11	⊥ y-axis
2.711	6.955	0.1352	2.55	2.76	lucite, steel	1.2408	2.7	12	⊥ y-axis
2.711	7.082	0.1412	2.71	2.77	lucite	2.555	6.9	13	⊥ y-axis
2.711	7.044	0.1102	2.10	2.75	Al	1.5377	13.6	14	⊥ y-axis
2.711	7.122	0.1448	2.80	2.78	Al	1.4618	12.9	15	⊥ y-axis
2.711	6.206	1.4733	24.79	3.56	Al	3.261	34.2	16	⊥ y-axis
2.711	5.451	0.1293	1.91	2.78	lucite, steel	1.4079	3.1	17	⊥ z-axis
2.711	5.407	0.1231	1.80	2.77	lucite, steel	1.4422	3.2	18	⊥ z-axis
2.711	5.752	0.1188	1.85	2.77	lucite	2.3919	6.3	19	⊥ z-axis
2.711	5.384	0.206	3.0	2.82	Al	1.482	13.0	20	⊥ z-axis
2.711	5.758	0.1026	1.60	2.76	Al	1.4952	13.2	21	⊥ z-axis
2.711	5.563	0.1393	2.10	2.78	Al	1.4087	12.5	22	⊥ z-axis
2.711	5.981	1.5135	24.60	3.64				25	⊥ z-axis
2.711	7.363	0.13	2.59	2.76				A	⊥ x-axis
2.711	7.277		2.56	2.76				B	⊥ x-axis
2.711	7.45	0.12	2.42	2.76				D	⊥ y-axis
2.711	6.09	1.47	24.3	3.57				F	⊥ z-axis
2.711	6.702	1.87	34.0	3.76				G	⊥ z-axis

(continued)

Calcite (continued)

Sample					Standard				Sample orientation to shock front
ϱ_0 gcm^{-3}	U kms^{-1}	u kms^{-1}	P GPa	ϱ gcm^{-3}	Composition	U_{fs} kms^{-1}	P GPa	No.	
second wave:									
2.711	5.04	0.134	2.3	2.76	lucite, steel	1.1741	2.5	1	∥ cleavage plane
2.711	3.952	0.2692	3.85	2.87	lucite, steel	1.1520	2.45	2	∥ cleavage plane
2.711	3.02	0.3331	5.0	2.89	lucite	2.1754	5.5	3	∥ cleavage plane
2.711	4.227	0.2173	3.23	3.09	lucite, steel	1.2824	2.7	4	⊥ x-axis
2.711	4.441	0.2280	3.46	2.83	lucite, steel	1.377	2.5	5	⊥ x-axis
2.711	4.380	0.209	3.2	2.83	brass, lucite	1.3823	3.0	6	⊥ x-axis
2.711	4.698	0.6817	9.33	3.15	lucite	2.499	6.7	7	⊥ x-axis
2.711	5.24	0.346	5.7	2.88	Al			8	⊥ x-axis
2.711	4.740	0.6813	9.42	3.14	Al	1.478	13.0	9	⊥ x-axis
2.711	4.33	0.185	2.9	2.80	brass, lucite	1.483	3.2	11	⊥ y-axis
2.711	4.290	0.2219	3.54	2.82	lucite, steel	1.2408	2.7	12	⊥ y-axis
2.711	4.582	0.6874	9.42	3.16	lucite	2.555	6.9	13	⊥ y-axis
2.711	4.788	0.6137	8.59	3.09	Al	1.5377	13.6	14	⊥ y-axis
2.711	4.765	0.5536	8.02	3.04	Al	1.4618	12.9	15	⊥ y-axis
2.711	4.146	0.2357	3.10	2.85	lucite, steel	1.4079	3.1	17	⊥ z-axis
2.711	4.222	0.2210	2.92	2.84	lucite, steel	1.4422	3.2	18	⊥ z-axis
2.711	5.13	0.586	8.3	3.05	lucite	2.3919	6.3	19	⊥ z-axis
2.711	4.652	0.7312	9.59	3.20	Al	1.482	13.0	20	⊥ z-axis
2.711	4.807	0.8368	11.13	3.27	Al	1.4952	13.2	21	⊥ z-axis
2.711	4.822	0.7892	10.56	3.23	Al	1.4087	12.5	22	⊥ z-axis
2.711	5.13	0.900	12.6	3.28	lucite	2.3989	6.3	23	⊥ z-axis
2.711	4.96	0.853	11.5	3.28	Al	1.3996	12.25	24	⊥ z-axis
2.711	6.283	1.54	26.5	3.58				A	⊥ x-axis
2.711	6.163	1.46	24.7	3.55				B	⊥ x-axis
2.711	6.675	1.86	33.8	3.76				C	⊥ x-axis
2.711	6.06	1.435	23.9	3.53				D	⊥ y-axis
2.711	6.630	1.77	32.0	3.69				E	⊥ y-axis
third wave:									
2.711	3.86	0.195	2.9	2.81	lucite, steel	1.1741	2.5	1	∥ cleavage plane
2.711	4.639	0.6246	8.68	3.10	lucite	2.1754	5.5	3	∥ cleavage plane
2.711	3.605	0.3073	4.09	2.91	lucite, steel	1.2824	2.7	4	⊥ x-axis
2.711	3.572	0.3433	4.55	2.93	lucite, steel	1.377	2.5	5	⊥ x-axis
2.711	3.76	0.328	4.4	2.92	brass, lucite	1.3823	3.0	6	⊥ x-axis
2.711	4.42	0.664	9.4	3.13	Al			8	⊥ x-axis
2.711	4.001	0.7702		3.34	Al	1.478	10.34	9	⊥ x-axis
2.711	3.33	0.284	3.8	2.90	brass, lucite	1.438	3.2	11	⊥ y-axis
2.711	3.685	0.3180	4.48	2.90	lucite, steel	1.2408	2.7	12	⊥ y-axis
2.711	4.336	0.7486	10.14	3.20	Al	1.5377	13.6	14	⊥ y-axis
2.711	4.390	0.7481	10.29	3.20	Al	1.4618	12.9	15	⊥ y-axis
2.711	3.662	0.3480	4.20	2.95	lucite, steel	1.4079	3.1	17	⊥ z-axis
2.711	3.591	0.3499	4.19	2.96	lucite, steel	1.4422	3.2	18	⊥ z-axis
2.711	4.051	0.8651	11.36	3.31	Al	1.4087	12.5	22	⊥ z-axis
2.711	4.72	1.04		3.38	Al	1.3996	13.9	24	⊥ z-axis
fourth wave:									
2.711	3.16	0.361	4.7	2.95	brass, lucite	1.3823	3.0	6	⊥ x-axis
2.711	2.86	0.331	4.1	2.98	brass, lucite	1.438	3.2	11	⊥ y-axis
2.711	3.184	0.4051	4.67	3.01	lucite, steel	1.4079	3.1	17	⊥ z-axis

Calcite (continued)

Experimental technique C
Data reduction technique B (in entries A, B, C, E, and G) and D (for all other entries)
References: [Gre62] and for entries A···G [Ahr66a]

1.3.3.2.4 Halides — Halogenide

Halite, single crystal (Halit = Steinsalz, Einkristall)
NaCl

[100] axis: $\varrho_0 = 2.155$ g cm^{-3};
$c_1 = 4.75$ km s^{-1}; $c_0 = 3.42$ km s^{-1}; $c_s = 2.40$ km s^{-1};
[111] axis: $\varrho_0 = 2.160$ g cm^{-3}; $c_1 = 4.42$ km·s^{-1};
$c_0 = 3.42$ km s^{-1}; $c_s = 2.72$ km s^{-1}; $\varrho_{01} = 2.164$ g cm^{-3}

Sample					Standard	
ϱ_0 g cm^{-3}	U km s^{-1}	u km s^{-1}	P GPa	ϱ g cm^{-3}	Composition	P GPa
Axis [100]:						
2.159	5.94	1.73	22.2	3.04	Al	29.4
2.160	5.07	1.10	12.0	2.76	Al	16.2
2.154	6.06	1.86	24.3	3.11	Al	32.0
2.163	6.24	1.97	26.6	3.16	Al	36.8
2.153	9.03	4.07	79.1	3.92	Al	105.0
2.152	7.72	3.21	53.3	3.68	Al	68.4
2.158	6.24	2.05	27.7	3.22	Al	36.8
2.157	8.47	3.77	68.9	3.89	Al	87.7
2.156	6.34	2.28	31.3	3.37	Al	41.9
2.156	6.71	2.69	38.9	3.60	Al	51.9
2.157	8.16	3.42	60.2	3.71	Al	76.4
2.155	9.05	4.02	78.4	3.88	Al	98.6
2.147	6.36	2.33	31.8	3.39	Al	42.7
2.141	7.83	3.19	53.5	3.61	Al	68.0
Axis [111]:						
2.148	5.88	1.88	23.8	3.16	Al	32.0
2.157	5.98	2.01	26.0	3.25	Al	35.1
2.163	6.00	2.00	25.9	3.24	Al	34.8
2.153	6.04	2.09	27.2	3.29	Al	36.8
2.158	8.66	3.74	69.9	3.80	Al	87.7
2.164	6.25	2.27	30.8	3.40	Al	41.9
2.161	6.77	2.68	39.2	3.58	Al	51.9
2.157	8.11	3.43	60.0	3.74	Al	76.4
2.162	9.08	4.01	78.7	3.87	Al	96.8
Unknown:						
2.165	5.86	1.76	22.3	3.09	Al	29.6
2.151	9.00	4.15	80.3	3.99	Al	102.2
2.153	7.22	3.01	46.8	3.69	Al	61.6
2.153	6.45	2.51	34.9	3.52	Al	46.9
2.159	8.62	3.73	69.4	3.81	Al	87.4
2.155	7.07	2.80	42.7	3.57	Al	55.8
2.157	7.10	2.79	42.7	3.55	Al	55.8

$U = (3.606 + 1.334\,u)$ km s^{-1} (for [100]-axis);
$\sigma_U = 0.027$ km s^{-1}; $u = 1.1 \cdots 1.97$ km s^{-1}
$U = (3.15 + 1.446\,u)$ km s^{-1} (for [100]-axis);
$\sigma_U = 0.098$ km s^{-1}; $u = 3.2 \cdots 4.07$ km s^{-1}
$U = (3.635 + 1.164\,u)$ km s^{-1} (for [111]-axis);
$\sigma_U = 0.032$ km s^{-1}; $u = 2.0 \cdots 2.68$ km s^{-1}
$U = (2.375 + 1.675\,u)$ km s^{-1} (for [111]-axis);
$\sigma_U = 0.026$ km s^{-1}; $u = 3.74 \cdots 4.01$ km s^{-1}
$U = (3.174 + 1.357\,u)$ km s^{-1}
(for [100], [111], and unknown axes)
$\sigma_U = 0.098$ km s^{-1}; $u = 2.25 \cdots 3.01$ km s^{-1}
$U = (3.234 + 1.428\,u)$ km s^{-1}
(for [100], [111], and unknown axes)
$\sigma_U = 0.101$ km s^{-1}; $u = 3.19 \cdots 4.15$ km s^{-1}

Experimental technique B
Data reduction technique B
Reference: [Van77]

Halite, single crystal (Halit = Steinsalz, Einkristall)
NaCl
$\varrho_0 = 2.160$ g cm^{-3}; $\varrho_{01} = 2.164$ g cm^{-3}

Sample					Standard	Sample					Standard
ϱ_0 g cm^{-3}	U km s^{-1}	u km s^{-1}	P GPa	ϱ g cm^{-3}	Composition	ϱ_0 g cm^{-3}	U km s^{-1}	u km s^{-1}	P GPa	ϱ g cm^{-3}	Composition
Shock front parallel to (100):						2.16	8.22	3.56	63.2	3.81	Al
2.16	4.92	1.06	11.0	2.74	plexiglass	2.16	8.54	3.70	68.2	3.81	Al
2.16	4.94	1.08	11.3	2.75	plexiglass	2.16	8.59	3.71	68.8	3.80	Al
2.16	4.95	1.05	11.1	2.73	plexiglass	Shock front parallel to (111):					
2.16	4.96	1.09	11.5	2.76	plexiglass	2.16	4.75	0.88	9.0	2.65	plexiglass
2.16	5.07	1.10	12.0	2.76	plexiglass	2.16	5.00	1.04	11.2	2.73	plexiglass
2.16	5.14	1.17	12.8	2.78	plexiglass	2.16	5.48	1.41	16.7	2.91	plexiglass
2.16	5.21	1.26	14.1	2.84	plexiglass	2.16	5.72	1.54	19.0	2.95	plexiglass
2.16	5.29	1.32	14.7	2.86	plexiglass	2.16	5.81	1.66	20.8	3.03	plexiglass
2.16	5.40	1.34	15.6	2.87	plexiglass	2.16	5.94	1.97	25.3	3.23	plexiglass
2.16	5.43	1.42	16.6	2.93	plexiglass	2.16	5.95	1.71	22.0	3.03	plexiglass
2.16	5.46	1.44	17.0	2.93	plexiglass	2.16	5.98	1.81	23.4	3.10	plexiglass
2.16	5.48	1.41	16.7	2.91	plexiglass	2.16	6.01	2.03	26.4	3.26	plexiglass
2.16	5.52	1.41	16.8	2.90	plexiglass	2.16	6.06	2.05	26.8	3.26	plexiglass
2.16	5.61	1.56	18.9	2.99	plexiglass	2.16	5.74	1.61	20.0	3.00	Al
2.16	5.62	1.46	17.7	2.92	plexiglass	2.16	5.86	1.78	22.5	3.10	Al
2.16	5.62	1.50	18.2	2.95	plexiglass	2.16	5.99	1.95	25.2	3.20	Al
2.16	5.72	1.54	19.0	2.95	plexiglass	2.16	6.00	1.92	24.9	3.18	Al
2.16	5.82	1.69	21.2	3.04	plexiglass	2.16	6.06	1.95	25.5	3.19	Al
2.16	5.86	1.69	21.4	3.03	plexiglass	2.16	6.08	1.80	23.6	3.07	Al
2.16	5.87	1.71	21.7	3.05	plexiglass	2.16	6.25	2.20	29.7	3.33	Al
2.16	5.88	1.68	21.3	3.03	plexiglass	2.16	6.42	2.38	33.0	3.43	Al
2.10	5.88	1.73	22.0	3.06	plexiglass	2.16	6.90	2.65	39.5	3.51	Al
2.16	5.91	1.69	21.6	3.03	plexiglass	Shock front parallel to (110):					
2.16	5.98	1.73	22.3	3.04	plexiglass	2.16	4.70	0.88	8.9	2.66	plexiglass
2.16	6.10	1.85	24.4	3.10	plexiglass	2.16	5.36	1.29	14.9	2.85	plexiglass
2.16	6.20	1.95	26.1	3.15	plexiglass	2.16	5.43	1.42	16.6	2.93	plexiglass
2.16	6.24	2.02	27.2	3.20	plexiglass	Shock front parallel to (112):					
2.16	5.84	1.74	21.9	3.08	Al	2.16	5.22	1.29	14.5	2.87	plexiglass
2.16	5.93	1.68	21.5	3.01	Al	2.16	5.36	1.36	15.7	2.90	plexiglass
2.16	6.15	1.92	25.5	3.14	Al						
2.16	6.33	2.05	28.0	3.20	Al						
2.16	6.37	2.14	29.4	3.25	Al						
2.16	6.43	2.36	32.8	3.41	Al						
2.16	6.44	2.27	31.6	3.34	Al						
2.16	6.45	2.39	33.3	3.43	Al						
2.16	6.55	2.46	34.8	3.46	Al						
2.16	6.74	2.50	36.4	3.43	Al						
2.16	6.79	2.62	38.4	3.52	Al						
2.16	6.81	2.66	39.1	3.55	Al						
2.16	6.88	2.66	39.5	3.52	Al						
2.16	6.95	2.75	41.3	3.58	Al						
2.16	6.99	2.76	41.7	3.58	Al						
2.16	7.01	2.80	42.4	3.60	Al						
2.16	7.15	2.91	44.9	3.64	Al						
2.16	7.27	2.96	46.5	3.64	Al						
2.16	7.43	3.08	49.4	3.69	Al						
2.16	7.71	3.20	53.3	3.69	Al						
2.16	7.95	3.33	57.2	3.72	Al						

$U = (3.52 + 1.382\, u)$ km s^{-1}, NaCl structure;
$u < 2.06$ km s^{-1}; $\sigma_U = 0.047$ km s^{-1}
$U = (3.26 + 1.35\, u)$ km s^{-1}; $2.2 < u < 3.0$ km s^{-1},
$\sigma_U = 0.045$ km s^{-1}
$U = (2.43 + 1.65\, u)$ km s^{-1}; $u > 3.1$ km s^{-1};
$\sigma_U = 0.055$ km s^{-1}

Experimental technique H for plexiglass (composition of standard); technique C for Al (composition of standard);
Data reduction technique B
Reference: [Hau71]

Halite, Sodium chloride (Summary of data)
(Halit = Steinsalz, zusammengefaßte Daten) NaCl
$\varrho_0 = 0.990 \cdots 2.16\ \text{g cm}^{-3}$; $\varrho_{01} = 2.164\ \text{g cm}^{-3}$

ϱ_0 g cm^{-3}	U km s^{-1}	u km s^{-1}	P GPa	ϱ g cm^{-3}
Fit 1 = solid, NaCl-structure:				
2.165	4.147	0.5	4.49	2.46
2.165	4.889	1.0	10.5	2.72
2.165	5.630	1.5	18.3	2.95
2.165	5.868	1.66	21.1	3.02
2.165	6.298	1.95	26.6	3.14
Fit 2 = solid, CsCl-structure:				
2.165	6.271	2.2	29.9	3.34
2.165	6.776	2.6	38.1	3.51
2.165	7.282	3.0	47.3	3.68
Fit 3 = liquid and mixed liquid-solid phases:				
2.165	7.890	3.2	54.7	3.64
2.165	8.407	3.6	65.5	3.78
2.165	8.924	4.0	77.3	3.92
2.165	10.216	5.0	110.6	4.24
2.165	12.154	6.5	171.0	4.66
Fit 4 = liquid (?):				
2.165	11.794	6.8	133.6	5.12
2.165	13.450	8.2	238.8	5.55
2.165	15.107	9.6	314.0	5.95
2.165	16.763	11.0	399.2	6.29
Fit 5 = porous:				
1.43	4.721	2.0	13.5	2.48
1.43	6.123	3.0	26.3	2.80
1.43	7.526	4.0	43.0	3.06
1.43	10.331	6.0	88.6	3.41
0.991	3.721	2.0	7.38	2.15
0.991	5.157	3.0	15.3	2.37
0.991	8.030	5.0	39.8	2.63
0.991	10.903	7.0	75.6	2.77

$U = (3.406 + 1.483\ u)\ \text{km s}^{-1}$; $\sigma_U = 0.06\ \text{km s}^{-1}$, $0.4 < u < 2.0\ \text{km s}^{-1}$

$U = (3.49 + 1.264\ u)\ \text{km s}^{-1}$; $\sigma_U = 0.11\ \text{km s}^{-1}$, $2.2 < u < 3.0\ \text{km s}^{-1}$

$U = (3.756 + 1.292\ u)\ \text{km s}^{-1}$; $\sigma_U = 0.11\ \text{km s}^{-1}$, $3.2 < u < 6.5\ \text{km s}^{-1}$

$U = (3.75 + 1.183\ u)\ \text{km s}^{-1}$; $\sigma_U = 0.07\ \text{km s}^{-1}$, $6.8 < u < 11.0\ \text{km s}^{-1}$

$U = [3.426 + 1.346\ u - 1.82(2.165 - \varrho_0) + 0.077(2.165 - \varrho_0)\ u - 0.32(2.165 - \varrho_0)^2]\ \text{km s}^{-1}$
$\sigma_U = 0.11\ \text{km s}^{-1}$, for u between the limits of the table

Sylvite, single crystal (Sylvin, Einkristall) KCl
$\varrho_0 = 1.992\ \text{g cm}^{-3}$; $\varrho_{01} = 1.987\ \text{g cm}^{-3}$

Sample					Standard	
ϱ_0 g cm^{-3}	U km s^{-1}	u km s^{-1}	P GPa	ϱ g cm^{-3}	Composition	U km s^{-1}
1.992	5.62	2.17	24.0	3.25	Al	1.72
1.992	9.76	5.35	104.0	4.41	steel	3.60
1.992	9.96	5.74	114.0	4.70	steel	3.85
1.992	10.67	6.16	131.0	4.72	steel	4.16
1.992	10.93	6.10	133.0	4.51	steel	4.13
1.992	11.43	6.71	153.0	4.82	steel	4.56
1.992	11.29	7.10	160.0	5.35	Al	6.03
1.992	12.63	8.02	202.0	5.44	steel	5.44
1.992	16.69	11.38	379.0	6.26	Al	9.95

$U = (3.6 + 1.142\ u)\ \text{km s}^{-1}$ for $u = 9.7 \cdots 16.7\ \text{km s}^{-1}$; $\sigma_U = 0.25\ \text{km s}^{-1}$

Error in U 1% ($< 10\ \text{km s}^{-1}$), $1.5 \cdots 2\%$ ($> 10\ \text{km s}^{-1}$)

Experimental technique A
Data reduction technique B
Reference: [Kor64]

Significance of the Fits:

Fit 1: solid phase, original NaCl-structure

Fit 2: solid phase, CsCl-structure: the only evidence for placing the phase change this high, is the rate of increase of the transformation pressure in KCl when NaCl is added. Reference: [Wie60]

Fit 3: liquid phase and mixed liquid-solid phases. Reference: [Kor65]

Fit 4: very high pressure data, change in slope could be due to electronic excitation

Fit 5: useful for porous samples, porous data do not show the transitions

Reference: [Van77]

Sylvite (Sylvin) KCl
$\varrho_0 = 1.992$ g cm^{-3}; $\varrho_{01} = 1.991$ g cm^{-3}; $c_b = 3.02$ km s^{-1}

Sample					Standard	
ϱ_0 g cm^{-3}	U km s^{-1}	u km s^{-1}	P GPa	ϱ g cm^{-3}	Composition	U km s^{-1}
1.99	3.67	0.28	2.0	2.15	Cu	0.17
1.99	3.63	0.58	4.18	2.37	Cu	0.35
1.99	3.61	0.98	7.05	2.73	Al	0.69
1.99	4.40	1.51	13.2	3.03	Al	1.14
1.99	5.21	1.91	19.8	3.14	Al	1.50
1.99	5.59	2.20	24.45	3.28	Al	1.74
1.99	7.50	3.40	50.8	3.64	Al	2.82
1.99	8.56	4.22	71.5	3.92	steel	2.80

◀ $U = (2.15 + 1.54\, u)$ km s^{-1} for $u = 1.0 \cdots 4.2$ km s^{-1};
$\sigma_U = 0.1$ km s^{-1}
Experimental technique A
Data reduction technique B
Reference: [Alt63]
Data with $u < 0.98$ record only the first wave of a multiple shock wave.

1.3.3.2.5 Sulfides — Sulfide

For Potassium thioferrite, see next page.

Pyrrothite, single crystals, Potosi Mine, Santa Eulalia, Mexico (crystal A) and Crucero Mine, Columbia (crystals B and C).
Composition (in wt%): Crystal A (Fe$_{0.903}$S): Fe 61.40; S 39.01; Ni 0.03; Ti 0.01; Ca 0.01; Mg 0.01; total 100.47.
Crystal B (Fe$_{0.884}$S): Fe 61.46; S 39.90; Ni 0.02; Ti 0.001; Ca 0.0002; Mg 0.003; total 101.38.
Crystal C (Fe$_{0.858}$Ni$_{0.004}$S): Fe 58.92; S 39.41; Ni 0.29; total 98.62.

Sample						Standard	
Crystal	ϱ_0 g cm^{-3}	U km s^{-1}	u km s^{-1}	P GPa	ϱ g cm^{-3}	Composition	U_{fs} km s^{-1}
Final shock states:							
B	4.604	4.73	0.235	2.56	4.721	Al	0.453
B	4.612	3.308	0.547	9.02	5.466	Al	1.083
B	4.6085	3.41	0.494	8.91	5.294	Al	1.024
B	4.624	3.787	0.663	12.24	5.565	Al	1.367
A	4.584	4.352	0.991	19.78	5.936	W	1.229
B	4.610	4.682	1.143	24.66	6.098	W	1.435
A	4.568	4.844	1.315	29.09	6.270	W	1.655
A	4.608	5.000	1.327	30.57	6.272	W	1.683
A	4.568	5.422	1.560	38.65	6.414	W	2.000
A	4.593	5.838	1.496	40.10	6.175	W	1.950
A	4.590	5.549	1.599	40.74	6.449	W	2.061
B	4.610	5.979	1.913	52.73	6.779	W	2.491
B	4.611	5.988	1.904	52.58	6.761	W	2.481
B	4.608	5.993	1.918	52.96	6.777	W	2.498
C	4.617	8.873	3.710	151.96	7.934	Ta	5.39
C	4.625	8.930	3.820	157.90	8.080	Ta	5.55
Adiabatic release state:							
B	4.612	4.521	0.2619	2.73	4.5000	Al	1.083
B	4.6085	4.620	0.440	4.68	4.839	Al	1.024
B	4.624	4.543	0.360	3.79	4.815	Al	1.367
C	4.617	8.410	4.510	83.5	7.390	Ta	5.390
C	4.625	9.330	4.510	104.9	7.540	Ta	5.550

Errors for all entries are given in the original reference.
Experimental technique C
Data reduction technique B
Reference: [Ahr79, Kin73]

Potassium thioferrite[*]), synthetic, polycrystalline
$KFeS_2$ (purity $\geq$ 99%) (Kaliumthioferrit)
Calculated average porosity: 2.6%
$\varrho_0 = 2.552 \cdots 2.635$ g cm^{-3}; $\varrho_{01} = 2.663$ g cm^{-3}

Sample					Standard	
ϱ_0 g cm^{-3}	U km s^{-1}	u km s^{-1}	P GPa	ϱ g cm^{-3}	Composition	U_{fs} km s^{-1}
2.552	2.74	0.223	1.56	2.78	Al	0.810
2.575	3.22	0.515	4.27	3.06	Al	0.785
2.583	3.41	0.630	5.55	3.17	Al	1.930
2.598	3.49	0.635	5.76	3.18	Al	0.991
2.611	3.73	0.851	8.29	3.38	Al	1.348
2.60	4.14	1.06	11.4	3.50	Al	1.74
2.563	4.35	1.31	14.6	3.67	Al	2.13
2.607	4.54	1.45	17.15	3.83	W	1.660
2.586	4.67	1.630	19.7	3.97	W	1.872
2.600	4.84	1.79	22.5	4.13	W	2.06
2.605	5.13	2.05	27.4	4.34	W	2.37
2.635	5.27	2.160	30.0	4.47	W	2.514
2.589	6.24	2.79	45.1	4.68	Al	4.80
2.590	8.70	4.45	100.3	5.30	Ta	5.69
2.597	9.08	4.72	111.1	5.40	Ta	6.05

[*]) Not known as a natural mineral, but important for geophysical applications concerning the Earth's core.

Errors for all entries are given in the original reference.
Experimental technique C
Data reduction technique B
Reference: [Som80]

1.3.3.2.6 Silicates — Silikate

Andalusite, chiastolite, South Australia (Andalusit)
Al_2SiO_5 $\varrho_0 = 3.077$ g cm^{-3}; $\varrho_{01} = 3.145$ g cm^{-3}

Sample					Standard	
ϱ_0 g cm^{-3}	U km s^{-1}	u km s^{-1}	P GPa	ϱ g cm^{-3}	Composition	U km s^{-1}
3.08	7.74	2.49	59.4	4.54	Al	8.68
3.06	7.91	2.53	61.3	4.45	Al	8.75
3.10	7.94	2.67	65.7	4.47	Al	8.93
3.07	7.94	2.76	67.3	4.47	Al	9.02
3.06	7.94	2.88	70.1	4.48	Al	9.16
3.09	8.32	2.97	76.3	4.48	Al	9.33
3.06	8.84	3.20	86.7	4.48	Al	9.67
3.08	9.20	3.50	99.3	4.50	Al	10.08
3.07	9.36	3.63	104.2	5.40	Al	10.25
3.06	9.39	3.65	104.9	5.40	Al	10.27
3.09	9.52	3.70	108.9	5.40	Al	10.37
3.09	9.87	3.80	115.8	5.40	Al	10.54

$U = (2.869 + 1.811\,u)$ km s^{-1}; $\sigma_U = 0.12$ km s^{-1}
Experimental technique B
Data reduction technique B
Reference: [McQ66]

Anorthite, single crystal, Miyake-zima, Izu Islands, Japan (Anorthit)
Composition: $An_{95.4}Ab_{4.5}Or_{0.1}$; in [wt%]: SiO_2 44.19; Al_2O_3 35.28; CaO 19.57; Na_2O 0.51; K_2O 0.02; MgO 0.04; FeO 0.66
$\varrho_0 = 2.756 \cdots 2.783$ g cm^{-3}; $\varrho_{01} = 2.74$ g cm^{-3}
Shock wave propagation parallel to [010]

Sample					Standard	
ϱ_0 g cm^{-3}	U km s^{-1}	u km s^{-1}	P GPa	ϱ g cm^{-3}	Composition	U_{fs} km s^{-1}
Hugoniot state:						
2.783	7.400	2.911	60.0	4.588	Al	5.398
2.756	8.628	3.791	90.2	4.916	Ta	4.925
2.768	9.477	4.338	113.8	5.104	Ta	5.694
Release state:						
2.783	6.226	3.226	44.3	4.459		
2.756	7.982	4.333	76.2	4.455		
2.768	8.722	4.798	92.2	4.861		

Errors for all entries are given in the original reference.
Experimental technique C
Data reduction technique B
Reference: [Jea80a]

Augite, natural (Augit) $Ca(Mg,Fe,Al)(Al,Si)_2O_6$
Grain size: ≈ 0.5 mm; $\varrho_0 = 3.366 \cdots 3.476$ g cm^{-3}; $\varrho_{01} = 3.372 \cdots 3.397$ g cm^{-3}; $c_l = 8.18 \cdots 8.26$ km s^{-1}

Sample					Standard		
ϱ_0 g cm^{-3}	U km s^{-1}	u km s^{-1}	P GPa	ϱ g cm^{-3}	Composition	U_{fs} km s^{-1}	No.
first wave:							
3.463	6.61	0.264	6.04	3.61	Al	1.55	1
3.475	7.67	0.13	3.46	3.54	Al	1.51	2
3.473	7.70	0.177	4.73	3.55	Al	1.89	3
3.475	7.27	0.13	3.28	3.54	Al	2.11	4
3.366	7.84	0.268	7.07	3.48	Al	3.58	5
3.393	7.84	0.13	3.46	3.45	brass	3.03	6
second wave:							
3.463	5.91	0.710	15.13	3.92	Al	1.55	1
3.475	6.25	0.670	15.15	3.88	Al	1.51	2
3.473	6.74	0.825	19.85	3.94	Al	1.89	3
3.475	6.68	0.945	22.17	4.04	Al	2.11	4
3.366	7.42	1.66	41.77	4.33	Al	3.58	5
3.393	7.79	1.89	49.97	4.48	brass	3.03	6
single wave:							
3.393	7.07	0.935	22.4	3.91	Al	2.11	7
3.470	7.26	1.26	31.7	4.20	Al	2.82	8
3.475	7.49	1.42	36.9	4.29	Al	3.18	9
3.475	7.88	1.87	51.2	4.55	brass	3.03	10
3.463	7.60	2.05	54.0	4.74	Al	4.41	11
3.463	7.73	2.10	56.2	4.76	brass	3.34	12
3.393	7.92	2.16	58.0	4.67	brass	3.45	13
3.475	7.94	2.15	59.3	4.77	brass		14
3.463	7.65	2.22	58.6	4.88	Al	4.78	15
3.393	7.92	2.48	66.6	4.94	brass	3.90	16
3.393	8.31	2.44	68.8	4.81	brass		17
3.463	8.01	2.84	78.8	5.37	brass	4.45	18

Experimental technique C (inclined mirror)
Data reduction technique B and D (elastic waves)
Reference: [Ahr66a]
Exact composition and locality of sample not reported.

Diopside, natural (Diopsid) $CaMgSi_2O_6$
Grain size ≈ 0.5 mm
$\varrho_0 = 3.233 \cdots 3.283$ g cm^{-3}; $\varrho_{01} = 3.282 \cdots 3.340$ g cm^{-3};
$c_l = 5.83 \cdots 8.00$ km s^{-1}

Sample					Standard	
ϱ_0 g cm^{-3}	U km s^{-1}	u km s^{-1}	P GPa	ϱ g cm^{-3}	Composition	No.
first wave:						
3.283	7.31	0.289	6.94	3.42	Al	1
3.233	7.66	0.289	7.16	3.36	Al	2
3.279	7.92				Al	3
3.272					Al	4
3.233	8.52	0.289	7.96	3.35	brass	5
3.233	10.00	0.289	9.34	3.33	brass	6
3.233	8.90	0.289	8.32	3.34	brass	7
3.166	6.57	0.201	4.18	3.27	Al	8

Sample					Standard	
ϱ_0 g cm^{-3}	U km s^{-1}	u km s^{-1}	P GPa	ϱ g cm^{-3}	Composition	No.
second wave:						
3.283	6.30	0.68	14.97	3.66	Al	1
3.233	6.86	0.96	21.97	3.74	Al	2
3.279	7.69	1.26	31.77	3.92	Al	3
3.272	7.92	1.43	37.06	4.00	Al	4
3.233	8.03	1.89	49.44	4.22	brass	5
3.233	9.22	2.09	62.89	4.17	brass	6
3.233	8.33	2.47	66.92	4.59	brass	7

Experimental technique C (inclined mirror)
Data reduction technique B and D (elastic waves)
Reference: [Ahr66a]
Exact composition and locality of sample not reported.

Enstatite, Bamle, Norway, single crystal (Enstatit, Einkristall) $Mg_{0.86}Fe_{0.14}SiO_3$
$\varrho_0 = 3.276 \cdots 3.298$ g cm^{-3}; $\varrho_{01} = 3.325$ g cm^{-3}

Sample					Standard
ϱ_0 g cm^{-3}	U km s^{-1}	u km s^{-1}	P GPa	ϱ g cm^{-3}	U_{fs} km s^{-1}
first wave:					
3.288	7.841	0.342	7.7	3.461	0.950
3.282	7.739	0.224	5.7	3.380	1.074
3.290	7.94				1.267
3.287	7.599	0.268	6.7	3.407	1.574
3.276	7.325				1.682
3.298	7.907				1.677
3.288	7.903				1.958
	7.247	0.532	12.7	3.549	
3.287	7.165	0.600	14.1	3.587	1.907
3.278	7.470				2.065
3.291	7.468				2.189
3.294	7.726				2.220
3.287	7.560				2.350
3.282					2.443
3.293	7.995				2.480
3.288	7.950				2.520
second wave:					
3.288	6.363	0.698	15.1	3.678	0.950
3.282	6.158	0.793	17.1	3.738	1.074
3.290	6.807	0.925	21.7	3.780	1.267
3.287	6.860	1.166	26.9	3.945	1.574
3.276	6.735	1.281	28.3	4.045	1.682
3.298	7.109	1.247	29.1	3.995	1.677
3.288	6.965	1.465	34.0	4.151	1.958
3.287	7.015	1.444	33.6	4.131	1.907
3.278	6.82	1.561	34.9	4.252	2.065
3.291	6.917	1.741	39.6	4.399	2.189
3.294	7.257	1.743	39.7	4.272	2.220
3.287	7.165	1.770	41.7	4.365	2.350
3.282	7.052	1.850	42.8	4.449	2.443
3.293	7.309	1.861	44.8	4.418	2.480
3.288	7.391	1.987	48.3	4.497	2.520

Tungsten (19.3 g cm^{-3}) and tungsten alloy (16.6, 16.8, and 17.0 g cm^{-3}) were used as standards.
Direction of shock wave propagation normal to (001)
Errors for all entries are given in the original reference.
Experimental technique C
Data reduction technique B
Reference: [Ahr71]

Enstatite, ceramic (Enstatit, keramisch) $MgSiO_3$
$\varrho_0 = 2.710$ g cm^{-3}; $\varrho_{01} = 3.206$ g cm^{-3}

Sample					Standard	
ϱ_0 g cm^{-3}	U km s^{-1}	u km s^{-1}	P GPa	ϱ g cm^{-3}	Composition	U km s^{-1}
2.71	5.37	2.06	30.0	4.40	Al	7.72
2.71	6.07	2.53	41.6	4.65	Al	8.32
2.71	6.74	3.07	55.9	4.97	Al	9.00
2.72	7.03	3.33	63.7	5.17	Al	9.33

$U = (2.718 + 1.304\,u)$ km s^{-1}; $\sigma_U = 0.051$ km s^{-1}
Experimental technique B
Data reduction technique B
Reference: [McQ66]

Fayalite, Rockport, Mass., USA (Fayalit) $(Fe,Mg)_2SiO_4$
$\varrho_0 = 4.184 \cdots 4.310$ g cm^{-3}; $\varrho_{01} = 4.322$ g cm^{-3}

Sample					Standard	
ϱ_0 g cm^{-3}	U km s^{-1}	u km s^{-1}	P GPa	ϱ g cm^{-3}	Composition	U km s^{-1}
4.30	6.65	2.02	57.7	6.17	Al	8.33
4.29	7.06	2.30	69.6	6.36	Al	8.76
4.23	7.23	2.37	72.5	6.30	Al	8.86
4.18	7.56	2.65	83.9	6.45	Al	9.26
4.29	7.96	2.96	101.3	6.84	Al	9.75
4.28	8.32	3.19	113.7	6.94	Al	10.10

$U = (3.862 + 1.395\,u)$ km s^{-1}; $\sigma_U = 0.038$ km s^{-1}
Experimental technique B
Data reduction technique B
Reference: [McQ66]
Composition of sample not given.

Forsterite, ceramic (Forsterit, keramisch) Mg_2SiO_4
$\varrho_0 = 3.058$ g cm^{-3}; $\varrho_{01} = 3.223$ g cm^{-3}

Sample					Standard	
ϱ_0 g cm^{-3}	U km s^{-1}	u km s^{-1}	P GPa	ϱ g cm^{-3}	Composition	U km s^{-1}
3.07	7.63	2.84	66.4	4.88	Al	9.05
3.03	8.07	3.10	75.8	4.92	Al	9.40
3.03	8.07	3.10	75.8	4.92	Al	9.40
3.07	8.20	3.14	78.8	4.97	Al	9.49
3.04	8.64	3.50	91.9	5.11	Al	9.96
3.06	9.14	3.70	103.5	5.15	Al	10.28

$U = (2.884 + 1.674\,u)$ km s^{-1}; $\sigma_U = 0.067$ km s^{-1}
Experimental technique B
Data reduction technique B
Reference: [McQ66]

Forsterite, polycrystalline (Forsterit, polykristallin) Mg_2SiO_4
$\varrho_0 = 2.625 \cdots 3.115$ g cm^{-3}; $\varrho_{01} = 3.223$ g cm^{-3}

Sample					Standard		
ϱ_0 g cm^{-3}	U km s^{-1}	u km s^{-1}	P GPa	ϱ g cm^{-3}	Composition	U_{fs}	No.
first wave:							
2.634	5.910	0.0310	0.48	2.65	Mg	1.201	1
2.627	5.820	0.0310	0.47	2.64	Mg	1.170	2
2.633	5.830	0.0310	0.47	2.65	Mg	1.420	3
second wave:							
2.634	4.190	1.0100	11.3	3.46	Mg	1.201	1
2.627	4.210	1.0300	11.6	3.47	Mg	1.170	2
2.633	4.980	1.2100	15.9	3.49	Mg	1.420	3
single wave:							
3.093	6.830	0.8550	18.1	3.53	Mg	1.110	
3.094	6.480	0.8540	17.2	3.56	Mg	1.107	
3.087	6.740	0.8970	18.7	3.56	Mg	1.161	
3.117	7.130	0.9950	22.1	3.62	Mg	1.305	
3.119	7.280	0.9750	22.2	3.60	Mg	1.285	
3.104	7.470	1.1600	26.9	3.67	Mg	1.530	
3.102	7.330	1.3500	30.7	3.80	Mg	1.770	
3.115	7.340	1.6300	37.3	4.00	Mg	2.132	

$U = (7.28 + 0.036\, u)$ km s^{-1}, for $u = 1.35 \cdots 1.63$ km s^{-1}

Errors for all entries are given in the original reference.
Experimental technique C
Data reduction technique B
Reference: [Ahr71a]

Low density material was pressed aggregate of forsterite from Atomergic Corp. Small amounts of an Al-bearing phase, probably $MgAl_2O_4$ were inferred from microprobe analysis of this material. This impurity increases ϱ_0 by ≈ 0.02.

High density material was fused from oxides by the Muscle Shoals Electrochemical Corp. It contained 0.04 wt% Fe-Si, but less than 0.5% Fe in the forsterite phase.

Forsterite, single crystal, synthetic*) (Forsterit)
Composition: Mg_2SiO_4
$\varrho_0 = 3.222$ g cm^{-3} (average of 10 measurements at 23 °C);
$\varrho_{01} = 3.214$ g cm^{-3}
Shock propagation direction within ±3° of (010)

Sample				Standard		
U km s^{-1}	u km s^{-1}	P GPa	ϱ g cm^{-3}	Composition	U_{fs} km s^{-1}	No.
Hugoniot state:						
8.96	2.40	69.2	4.40	Al	5.15	1
9.35	3.16	95.1	4.87	Ta	4.34	2
9.50	3.19	97.8	4.85	Ta	4.40	3
10.01	3.77	121.5	5.17	Ta	5.19	4
10.60	4.22	144.0	5.35	Ta	5.83	5
10.93	4.33	152.4	5.33	Ta	6.01	6
11.12	4.61	165.1	5.50	Ta	6.39	7

Sample				Standard		
U km s^{-1}	u km s^{-1}	P GPa	ϱ g cm^{-3}	Composition	U_{fs} km s^{-1}	No.
Partially released state:						
7.03	3.72	57.5	4.68			2
7.20	3.82	60.6	4.61			3
8.60	4.70	88.9	4.54			4
8.93	4.91	96.4	5.08			5
9.42	5.21	108.0	4.87			6

*) Crystal grown with the Czochralski technique by Crystal Products Division of the Union Carbide Corporation.

Errors for all entries are given in the original reference.
Experimental technique C
Data reduction technique B
Reference: [Jac79]

Garnet, Salida, Colorado, single crystal (Granat, Einkristall) $(Fe_{0.79}Mg_{0.14}Ca_{0.04}Mn_{0.03})_3Al_2Si_3O_{12}$
$\varrho_0 = 4.175\cdots 4.186\ \mathrm{g\,cm^{-3}}$

Sample					Standard		
ϱ_0 g cm^{-3}	U km s^{-1}	u km s^{-1}	P GPa	ϱ g cm^{-3}	Composition	U_{fs} km s^{-1}	No.
first wave:							
4.186	8.27	0.33	11.4	4.36	W	0.9	1
4.185	8.38	0.22	7.7	4.298	W	1.11	2
4.180	8.44	0.28	9.9	4.323	W	1.16	3
4.182	8.20	0.20	6.9	4.287	W	1.11	5
4.184	8.23	0.31	10.7	4.348	W	1.46	8
4.158	8.30	0.29	10.0	4.308	W	1.32	9
4.182	8.75	0.19	7.0	4.274	W	1.57	10
4.181	8.82	0.38	14.0	4.369	W	1.79	11
4.176	9.07	0.15	5.7	4.246	Al	0.87	15
4.180	8.58	0.24	8.6	4.300	Al	0.82	16
4.184	8.48	0.18	6.4	4.275	W	1.22	20
second wave:							
4.186	6.82	0.65	20.5	4.586	W	0.9	1
4.185	7.02	0.81	25.0	4.706	W	1.11	2
4.180	7.13	0.84	26.2	4.708	W	1.16	3
4.182	7.03	0.75	23.1	4.662	W	1.11	5
4.175	8.56	1.61	57.6	5.142	W	2.24	7
4.184	7.37	1.06	33.7	4.865	W	1.46	8
4.158	7.20	0.90	28.3	4.725	W	1.32	9
4.182	7.46	1.14	36.6	4.916	W	1.57	10
4.181	7.74	1.29	43.5	4.985	W	1.79	11
4.185	8.36	1.46	51.2	5.071	W	2.03	14
4.176	6.09	0.32	9.8	4.371	W	0.87	15
4.180	4.56	0.29	9.3	4.350	W	0.82	16
4.184	8.64	1.67	60.0	5.183	W	2.32	17
4.186	8.86	1.78	66.3	5.238	W	2.48	18
4.179	8.74	1.80	65.7	5.263	W	2.50	19
4.184	6.49	0.45	14.1	4.466	W	1.22	20
4.186	8.70	1.69	61.5	5.195	W	2.35	23
4.186	8.81	1.79	65.6	5.253	W	2.48	24

Tungsten (19.2 g cm^{-3}), tungsten alloy (fansteel) (16.9 g cm^{-3}) and aluminium 2024 were used as standards.
Direction of shock wave propagation parallel to [100].
Errors for all entries are given in the original reference.
Experimental technique C
Data reduction technique B
Reference: [Gra73]

Jadeite, Burma (Jadeit) $NaAlSi_2O_6$
Exact composition not known
$\varrho_0 = 3.333$ g cm^{-3}; $\varrho_{01} = 3.344$ g cm^{-3};
$c_0 = 6.42$ km s^{-1}; $c_1 = 8.66$ km s^{-1}

Sample					Standard	
ϱ_0 g cm^{-3}	U km s^{-1}	u km s^{-1}	P GPa	ϱ g cm^{-3}	Composition	U km s^{-1}
3.33	7.84	1.03	26.9	3.83	Al	6.95
3.33	7.78	1.05	27.1	3.85	Al	6.97
3.33	7.86	1.19	31.3	3.93	Al	7.18
3.33	8.22	1.46	40.1	4.05	Al	7.58
3.33	8.25	1.48	40.6	4.06	Al	7.60
3.33	8.20	1.51	41.3	4.09	Al	7.64
3.33	8.80	1.91	56.0	4.26	Al	8.23
3.33	8.78	1.99	58.3	4.31	Al	8.33
3.33	9.07	2.31	69.8	4.47	Al	8.77
3.33	9.05	2.32	70.1	4.49	Al	8.78
3.34	9.33	2.81	87.4	4.77	Al	9.42
3.33	9.39	3.02	94.4	4.91	Al	9.68
3.33	9.42	3.05	95.9	4.93	Al	9.73
3.35	9.72	3.34	108.6	5.10	Al	10.13
3.34	9.83	3.50	114.7	5.18	Al	10.33

$U = (6.54 + 1.124\,u)$ km s^{-1}, for $u < 2.4$ km s^{-1};
$\sigma_U = 0.088$ km s^{-1}
$U = (6.56 + 0.939\,u)$ km s^{-1}, for $u > 3.0$ km s^{-1};
$\sigma_U = 0.021$ km s^{-1}

Experimental technique B
Data reduction technique B
Reference: [McQ66]

Microcline, Ontario, Canada (Mikroklin)
$KAlSi_3O_8$, exact composition not given
$\varrho_0 = 2.550 \cdots 2.561$ g cm^{-3}; $\varrho_{01} = 2.562 \cdots 2.568$ g cm^{-3};
$c_1(001) = 6.95$ km s^{-1}

Sample					Standard		
ϱ_0 g cm^{-3}	U km s^{-1}	u km s^{-1}	P GPa	ϱ g cm^{-3}	Composition	No.	

first wave:

2.561	7.190	0.438	8.07	2.73	Al	1	
2.550	7.090	0.438	7.92	2.72	Al	2	
2.561	7.320	0.438	8.21	2.72	Al	3	
2.561	7.560	0.438	8.48	2.72	Al	4	
2.561	7.140	3.130	57.2	4.56	brass	5	
2.561	7.190	3.130	57.6	4.53	brass	6	

second wave:

2.561	3.970	0.795	11.5	3.03	Al	1	
2.550	5.220	2.10	29.5	4.17	Al	2	
2.561	6.440	2.69	45.0	4.36	Al	3	
2.561	6.560	2.66	45.5	4.27	Al	4	

Experimental technique C (inclined mirror)
Data reduction technique B and D (elastic waves)
Reference: [Ahr69]

For Oligoclase, see next page.

Olivine (Olivin)
Composition (in wt%): Forsterite (Mg_2SiO_4) 91 ± 1; Fayalite (Fe_2SiO_4) 9 ± 1
$\varrho_0 = 3.315 \cdots 3.289$ g cm^{-3}; $\varrho_{01} = 3.291 \cdots 3.325$ g cm^{-3}

Sample					Standard		
ϱ_0 g cm^{-3}	U km s^{-1}	u km s^{-1}	P GPa	ϱ g cm^{-3}	Composition	U_{fs} km s^{-1}	No.

first wave:

3.289	8.45	0.323	9.01	3.42	Al	1.55	1
3.315	8.72	0.323	9.34	3.44	Al	3.60	2

second wave:

3.289	6.58	0.665	16.3	3.62	Al	1.55	1
3.315	8.21	1.59	43.7	4.10	Al	3.60	2

single wave:

3.289	8.450	0.2720	7.5	3.40	Al	1.550	3
3.289	8.390	2.1900	60.4	4.45	Al	4.800	4
3.289	8.480	2.8000	78.1	4.91	brass	4.400	5

Experimental technique C (inclined mirror) and D (U_{fs})
Data reduction technique B and D (for u_1)
Reference: [Ahr66a]

Oligoclase, Muskawa Lake, Canada (Oligoklas)

Composition (in mol %): Albite (NaAlSi$_3$O$_8$) 75.0; Anorthite (CaAl$_2$Si$_2$O$_8$) 19.0; Orthoclase (KAlSi$_3$O$_8$) 5.0.

Chemical analysis (in wt %): SiO$_2$ 63.02; Al$_2$O$_3$ 27.93; Na$_2$O 8.55; CaO 3.89; K$_2$O 0.05; Fe$_2$O$_3$ 0.22; H$_2$O 0.13; TiO$_2$ 0.01; MnO 0.001; MgO 0.07.

$\varrho_0 = 2.632 \cdots 2.639$ g cm^{-3}

Sample					Standard		
ϱ_0 g cm^{-3}	U km s^{-1}	u km s^{-1}	P GPa	ϱ g cm^{-3}	Composition	U_{fs} km s^{-1}	No.
first wave:							
2.64	7.30	0.196	3.8	2.70	Al	2.58	1
2.63	7.14	0.195	3.7	2.70	Al	2.50	2
2.64	7.05	0.231	4.3	2.73	Al	3.39	3
2.63	6.82	0.25	4.4	2.73	Al		4
2.63	7.24	0.20	3.8	2.71	Al		5
2.63	7.61	0.20	4.0	2.70	Al	3.68	6
2.64	7.07	0.30	5.6	2.76	Al	4.74	7
2.64	7.33	0.28	5.5	2.74	Al		8
2.64	7.24	0.29	5.5	2.75	Al		9
2.63	7.24	0.29	5.5	2.74	Al		10
2.625	7.22	0.29	5.5	2.73	Al	4.79	11
2.64	7.13	0.30	5.6	2.76	Al	5.27	12
2.63	7.76	3.30	67.4	4.57	brass	6.78	13
second wave:							
2.64	4.80	1.36	18.3		Al	2.58	1
2.63	4.91	1.28	17.7		Al	2.50	2
2.64	5.42	1.85	27.2		Al	3.39	3
2.63	5.42	1.88	27.5		Al		4
2.63	5.76	1.84	28.5		Al		5
2.63	5.38	1.88	27.5		Al	3.68	6
2.64	6.30	2.44	41.0		Al	4.74	7
2.64	6.37	2.45	41.7		Al		8
2.64	6.34	2.48	42.0		Al		9
2.63	6.43	2.47	42.2		Al		10
2.625	6.30	2.49	41.7		Al	4.79	11
2.64	6.74	2.60	46.4		Al	5.27	12

Experimental technique C and D
Data reduction technique D (u_1) and B (u_2)
Reference: [Ahr69]

Orthoclase, Strongay, Madagascar,
single crystal (Orthoklas, Einkristall)
Composition (in wt%): Albite 10.8; Orthoclase 89.0;
Anorthite 0.07
$\varrho_0 = 2.556 \cdots 2.559$ g cm^{-3}

Sample					Standard
ϱ_0 g cm^{-3}	U km s^{-1}	u km s^{-1}	P GPa	ϱ g cm^{-3}	U_{fs} km s^{-1}
first wave:					
2.557	7.25	0.242	4.5	2.645	0.868
2.556	7.64	0.228	4.4	2.635	1.25
2.557	7.63	0.212	4.1	2.630	1.29
2.558	6.82	0.523	9.1	2.77	1.68
2.558	7.18	0.312	5.7	2.674	1.69
2.555	7.28	0.188	3.5	2.62	2.01
2.559	7.34	0.427	8.0	2.717	2.32
2.56	7.37	0.365	6.9	2.693	2.33
2.556	7.41	0.434	8.2	2.715	2.35
2.557					2.39
2.56					2.40
2.556					2.59
2.559	7.11				2.74
second wave:					
2.557	4.99	0.735	10.7	2.95	0.868
2.556	5.08	1.064	15.1	3.18	1.25
2.557	5.20	1.09	15.7	3.19	1.29
2.558	5.05	1.44	20.6	3.47	1.68
2.558	5.20	1.44	20.5	3.48	1.69
2.555	5.65	1.72	25.4	3.64	2.01
2.559	5.78	1.97	30.4	3.82	2.32
2.56	5.56	1.99	29.6	3.82	2.33
2.556	5.55	2.00	30.0	3.92	2.35
2.557	6.02	2.03	31.3	3.86	2.39
2.56	5.49	2.06	29.0	4.10	2.40
2.556	5.53	2.22	31.4	4.28	2.59
2.559	5.64	2.35	34.0	4.39	2.74

Direction of shock wave propagation normal to (001).
Tungsten and tungsten alloy were used as standards.
Errors for all entries are given in the original reference.
Experimental technique C
Data reduction technique B
Reference: [Ahr73]

Serpentine, Italy (Serpentin) Mg$_6$Si$_4$O$_{10}$(OH)$_8$
$\varrho_0 = 2.801$ g cm^{-3}; $\varrho_{01} = 2.653$ g cm^{-3}

Sample					Standard	
ϱ_0 g cm^{-3}	U km s^{-1}	u km s^{-1}	P GPa	ϱ g cm^{-3}	Composition	U km s^{-1}
2.78	7.60	2.72	57.5	4.33	Al	8.80
2.79	8.44	3.25	76.6	4.54	Al	9.52
2.80	8.43	3.35	79.2	4.65	Al	9.64
2.76	8.63	3.40	80.9	4.55	Al	9.70
2.83	9.01	3.53	90.1	4.66	Al	9.95
2.84	9.12	3.63	94.0	4.72	Al	10.08

$U = (3.017 + 1.666\,u)$ km s^{-1}; $\sigma_U = 0.11$ km s^{-1}
Experimental technique B
Data reduction technique B
Reference: [McQ66]
Exact composition of sample not given

Sillimanite, Dillon, Montana (Sillimanit) Al$_2$SiO$_5$
$\varrho_0 = 3.07 \cdots 3.13$ g cm^{-3}

Sample					Standard	
ϱ_0 g cm^{-3}	U km s^{-1}	u km s^{-1}	P GPa	ϱ g cm^{-3}	Composition	U km s^{-1}
3.070	7.540	2.5200	58.4	4.61	Al	8.680
3.120	7.830	2.5200	61.6	4.60	Al	8.750
3.130	8.010	2.6500	66.4	4.67	Al	8.930
3.130	8.050	2.7000	68.0	4.71	Al	9.000
3.100	7.940	2.7500	67.6	4.74	Al	9.020
3.090	8.150	2.8400	71.5	4.74	Al	9.160
3.130	8.480	2.9300	77.8	4.78	Al	9.330
3.130	8.890	3.1600	88.1	4.86	Al	9.670
3.150	9.290	3.4500	101.2	5.02	Al	10.080
3.130	9.420	3.5900	106.0	5.06	Al	10.250
3.130	9.550	3.6800	109.9	5.09	Al	10.370

Experimental technique B
Data reduction technique B
Reference: [McQ66]
Exact composition of sample not given

1.3.3.3 Rocks — Gesteine

The shock wave data measured on rock samples are listed in the following section. The rocks are grouped according to the main petrographic classes of rocks. In each class the rocks are ordered alphabetically.

Im folgenden Abschnitt sind die Stoßwellendaten aufgeführt, die an Gesteinsproben gemessen wurden. Die Gesteine sind nach den wichtigsten petrographischen Gesteinsklassen geordnet. Innerhalb jeder Klasse sind die Gesteine alphabetisch angeordnet.

1.3.3.3.1 Igneous rocks — Magmatische Gesteine

Anorthosite, Aqua Dulce quad., California (Anorthosit)
Composition (in vol %): Andesine 98 (= Albite 69···49 wt. %; Anorthite 31···51 wt. %); Apatite, Zircon, Chlorite, Hornblende 2
Grain size: 0.01···8 mm
$\varrho_0 = 2.762 \cdots 2.667$ gcm^{-3}; $\varrho_{01} = 2.710 \cdots 2.681$ gcm^{-3}

Sample					Standard		No.
ϱ_0 gcm^{-3}	U kms^{-1}	u kms^{-1}	P GPa	ϱ gcm^{-3}	Composition	U_{fs} kms^{-1}	
first wave:							
2.662	5.66	0.286	4.3	2.80	plexiglass	1.404	1
2.662	5.67	0.293	4.4	2.81	plexiglass	1.404	2
2.662	5.86	0.317	4.9	2.81	Al	1.56	3
2.662	5.40	0.403	5.8	2.88	Al	1.97	4
2.662	5.77	0.378	5.8	2.85	Al	1.97	5
2.76	6.69	0.280	5.18	2.90	Al	1.55	6
2.76	6.56	0.158	2.86	2.83	Al	1.58	7
2.75	6.80	0.300	5.72	2.88	Al	3.57	8
2.73	7.76	0.290	6.15	2.84	Al	3.57	9
2.763	7.22	0.290	5.78	2.88	Al	3.58	10
2.75	6.59	0.290	5.25	2.88	brass	3.34	11
2.75	6.81	0.290	5.43	2.87	brass	3.34	12
2.75	7.03	0.290	5.61	2.87	Al	4.41	13
2.75	7.22	0.290	5.76	2.86	Al	4.78	14
second wave:							
2.662	3.71	0.406	5.5	2.91	plexiglass		1
2.662	5.28	0.812	11.1	3.13	Al		3
2.662	5.18	1.029	14.4	3.31	Al		4
2.662	5.38	0.852	12.6	3.15	Al		5
2.76	5.25	0.820	13.0	3.24	Al	1.12	6
2.76	5.62	0.835	13.2	3.23	Al	1.14	7
2.75	6.10	1.94	33.0	3.83	Al		8
2.73	6.50	1.88	34.4	3.68	Al		9
2.763	6.14	1.94	33.5	4.01	Al	3.84	10
2.75	6.22	2.37	40.5	4.43	brass	4.65	11
2.75	6.31	2.36	41.2	4.38	brass	4.38	12
2.75	6.56	2.38	43.2	4.30	Al	4.87	13
2.75	6.71	2.57	47.6	4.44	Al	5.24	14
high pressure regime:							
2.662	5.63	1.517	22.7	3.65	Al	2.955	
2.662	5.59	1.320	19.7	3.48	Al	2.955	
2.662	5.34	0.858			Al	1.560	
2.72	6.45	2.27	40.0	4.20	brass	3.23	
2.75	6.45	2.40	42.7	4.38	Al	4.41	
2.76	6.73	2.33	43.3	4.23	brass	3.36	
2.75	6.73	2.57	47.8	4.45	Al	4.78	
2.75	7.33	3.10	62.2	4.77	brass	4.45	
2.75	7.35	3.10	62.3	4.75	brass	4.45	

Experimental technique C
Data reduction technique B, D with $U_{fs1} = 2 u_1$ (elastic wave only)
References: [Ahr64, Ahr68a, Ahr66a]

Anorthosite, Apollo 16 rock No 60025.36, Moon, Descartes region (Anorthosit) [Dix75]
Composition (in vol%): Plagioclase 98.9 (=Albite 4 mol%, Anorthite 96 mol%); Orthopyroxene 1.1; Clinopyroxene 0.1; Opaque minerals 0.1.
$\varrho_0 = 2.199 \cdots 2.24 \, \text{g cm}^{-3}$; $\varrho_{01} = 2.75 \, \text{g cm}^{-3}$
Average porosity 19%

Sample					Standard	
ϱ_0 g cm^{-3}	U km s^{-1}	u km s^{-1}	P GPa	ϱ g cm^{-3}	Composition	U_{fs} km s^{-1}
Hugoniot state:						
2.243	5.070	2.036	23.2	3.749	W	2.312
2.239	6.75	2.964	44.8	3.990	Al	4.971
2.244	7.73	3.617	62.7	4.216	Ta	4.467
2.199	9.15	4.525	91.1	4.351	Ta	5.668
2.234	9.17	4.591	94.0	4.473	Ta	5.762
2.23	10.142	5.196	117.5	4.573	Ta	6.585
Release state:						
2.239	5.523	2.783	33.9	3.943		
2.244	6.434	3.355	47.5	4.137		
2.199	8.116	4.417	79.0	4.333		
2.234	6.952	3.682	56.4	4.074		
2.23	9.349	5.194	107.0	4.573		

Errors for all entries are given in the original reference.
Experimental technique C
Data reduction technique B
Reference: [Jea78, Jea80a]

Anorthosite, Sylmar, Pennsylvania, (originally Albitite) (Anorthosit, ursprünglich Albitit)
Composition (in vol%): Oligoclase 98 (=Albite 87 wt%, Anorthite 13 wt%); Actinolite 2
$\varrho_0 = 2.611 \, \text{g cm}^{-3}$; $\varrho_{01} = 2.650 \, \text{g cm}^{-3}$

Sample					Standard	
ϱ_0 g cm^{-3}	U km s^{-1}	u km s^{-1}	P GPa	ϱ g cm^{-3}	Composition	U km s^{-1}
2.61	5.66	0.92	13.6	3.12	Al	6.49
2.61	5.56	1.34	19.5	3.44	Al	6.95
2.61	5.82	1.89	28.7	3.87	Al	7.58
2.61	5.90	1.97	30.3	3.92	Al	7.67
2.61	5.94	2.08	32.2	4.02	Al	7.79
2.61	6.31	2.49	41.0	4.31	Al	8.28
2.61	6.96	2.88	52.4	4.45	Al	8.80
2.61	7.75	3.35	67.7	4.60	Al	9.42
2.61	8.09	3.54	74.7	4.64	Al	9.68
2.61	8.14	3.70	78.6	4.78	Al	9.86
2.61	8.65	3.95	89.2	4.83	Al	10.21
2.61	8.69	3.98	90.4	4.82	Al	10.23

Anorthosite, Tahawus, New York (Anorthosit)
Composition (in vol%): Plagioclase 90 (=Anorthite 50 wt%; Albite 50 wt%); Augite 10
$\varrho_0 = 2.703 \cdots 2.762 \, \text{g cm}^{-3}$; $\varrho_{01} = 2.747 \, \text{g cm}^{-3}$;
$c_1 = 6.94 \, \text{km s}^{-1}$

Sample					Standard	
ϱ_0 g cm^{-3}	U km s^{-1}	u km s^{-1}	P GPa	ϱ g cm^{-3}	Composition	U km s^{-1}
2.72	5.93	0.92	14.8	3.22	Al	6.52
2.71	5.94	1.31	21.0	3.48	Al	6.97
2.73	5.88	1.32	21.3	3.52	Al	6.99
2.73	5.92	2.05	33.1	4.17	Al	7.79
2.71	5.93	2.05	33.0	4.14	Al	7.80
2.73	6.63	2.45	44.2	4.33	Al	8.33
2.73	6.46	2.47	43.5	4.42	Al	8.33
2.72	7.02	2.81	53.8	4.54	Al	8.78
2.73	7.06	2.82	54.2	4.54	Al	8.79
2.75	7.84	3.27	70.4	4.71	Al	9.42
2.76	8.25	3.44	78.2	4.73	Al	9.68
2.73	8.25	3.50	78.8	4.74	Al	9.73
2.75	8.39	3.68	84.9	4.90	Al	9.96
2.79	8.68	3.76	91.3	4.93	Al	10.12
2.73	8.49	3.84	88.8	4.97	Al	10.13
2.70	8.80	3.94	93.7	4.89	Al	10.28
2.72	8.81	3.97	95.2	4.96	Al	10.33

$U = (2.775 + 1.536 \, u) \, \text{km s}^{-1}$, $u = 2.05 \cdots 4.0 \, \text{km s}^{-1}$;
$\sigma_U = 0.1 \, \text{km s}^{-1}$

Experimental technique B
Data reduction technique B
Reference: [McQ66]

The modal analysis of these samples was taken from [Bir60].

◄ $U = (5.32 + 0.28 \, u) \, \text{km s}^{-1}$, $u = 0.9 \cdots 2.1 \, \text{km s}^{-1}$;
$\sigma_U = 0.1 \, \text{km s}^{-1}$
$U = (2.404 + 1.5818 \, u) \, \text{km s}^{-1}$, $u = 2.5 \cdots 4.0 \, \text{km s}^{-1}$;
$\sigma_U = 0.07 \, \text{km s}^{-1}$

Experimental technique B
Data reduction technique B
Reference: [McQ66]

The modal analysis of these samples was taken from [Bir60].

Basalt, Apollo 12 rock No. 12063, Moon, Oceanus Procellarum [Pap76] [Wil71]

Composition (in vol%): Plagioclase 22.2 (Albite 12 mol%, Anorthite 88 mol%); Pyroxene 63.7 (Wollastonite 27 mol%, Enstatite 32 mol%, Ferrosilite 41 mol%); Olivine 2.8 (Forsterite 9 mol%, Fayalite 91 mol%); Ilmenite 8.1; Quartz 1.6; Mesostasis 1.6

Chemical composition [in wt%]: SiO_2 43.48; TiO_2 5.00; Al_2O_3 9.27; FeO 21.26; MnO 0.28; MgO 9.56; CaO 10.49; Na_2O 0.31; K_2O 0.06; P_2O_5 0.14; Cr_2O_3 0.44; S 0.09

$\varrho_0 = 3.169 \cdots 3.26$ g cm^{-3}; $\varrho_{01} = 3.352 \cdots 3.380$ g cm^{-3}

Average porosity: 4.5%

Sample						Standard	
ϱ_0 g cm^{-3}	ϱ_{01} g cm^{-3}	U km s^{-1}	u km s^{-1}	P GPa	ϱ g cm^{-3}	Composition	U_{fs} km s^{-1}
Hugoniot state:							
3.184	3.352	5.208	0.385	6.4	3.438	Al	0.777
3.260	3.368	5.518	0.730	13.1	3.758	Al	1.476
3.196	3.376	6.156	1.572	30.9	4.291	W	1.932
3.224	3.358	6.700	1.963	42.4	4.561	W	2.441
3.169	3.360	8.824	3.520	98.4	5.272	Ta	4.734
3.210	3.355	9.575	3.935	120.9	5.449	Ta	5.354
3.207	3.380	10.678	4.708	161.2	5.736	Ta	6.462
Release state:							
3.184	3.352	2.921	0.376	1.3	3.438		
3.260	3.368	3.489	0.738	3.1	3.758		
3.194	3.376	5.615	2.092	14.0	4.015		
3.224	3.358	6.147	2.431	17.9	4.382		
3.169	3.360	9.004	4.251	45.8	5.005		
3.210	3.355	8.753	4.819	93.0	4.729		
3.207	3.380	11.031	5.542	73.1	5.487		

Errors for all entries are given in the original reference.
Experimental technique C
Data reduction technique B
Reference: [Ahr80]

Basalt, Apollo 17 rock No. 70215, Moon, Mare Serenitatis

Composition (in wt%): Plagioclase 14.5 (=Albite 16 mol%; Anorthite 84 mol%); Clinopyroxene a 23.0 (=Enstatite 40.5 mol%; Ferrosilite 19 mol%; Wollastonite 40.5 mol%); Clinopyroxene b 28.7 (=Enstatite 36 mol%; Ferrosilite 24 mol%; Wollastonite 39.5 mol%); Clinopyroxene c 5.8 (=Enstatite 30.5 mol%; Ferrosilite 42 mol%; Wollastonite 27.5 mol%); Olivine 6.4 (=Forsterite 68 mol%; Fayalite 32 mol%); Ilmenite 18.3; Armalcolite 0.3; Rutile 0.01; Ulvöspinel 0.2; SiO_2 2.7; Troilite 0.14.

Chemical composition (in wt%): SiO_2 38.76, TiO_2 12.61, Cr_2O_3 0.48, Al_2O_3 7.91, FeO 17.98, MnO 0.27, MgO 9.40, CaO 12.27, Na_2O 0.31, K_2O 0.01

$\varrho_0 = 3.324 \cdots 3.342$ g cm^{-3}; $\varrho_{01} = 3.378 \cdots 3.384$ g cm^{-3}

Sample					Standard	
ϱ_0 g cm^{-3}	U km s^{-1}	u km s^{-1}	P GPa	ϱ g cm^{-3}	Composition	U_{fs} km s^{-1}
Hugoniot state:						
3.338	5.801	0.021	0.41	3.351	Al	
3.338	5.627	0.3813	7.17	3.585	Al	0.8170
3.338	5.890	0.6323	12.43	3.739	Al	1.343
3.334	5.983	0.7868	15.69	3.839	Al	1.655
3.307	6.378	1.505	31.07	4.329	W	1.873
3.318	6.782	1.913	43.0	4.621	W	2.397
3.291	8.492	3.452	96.5	5.545	Ta	4.647
3.324	9.158	3.837	116.8	5.721	Cu	5.76
3.334	9.401	3.874	121.4	5.671	Cu	5.85
3.342	9.459	3.902	123.4	5.689	Cu	5.90
Release state:						
3.307	5.354	1.926	12.3	4.164		
3.318	6.221	2.478	18.4	4.360		
3.291	7.523	4.043	67.0	5.202		

Errors for all entries are given in the original reference.
Experimental technique C
Data reduction technique B
Reference: [Ahr80] [Ahr77]
Composition of minerals are averages from [Dym75]

Basalt, Centerville, Virginia

Composition (in vol%): Plagioclase 45; Augite 45; Biotite 1.8; Quartz 1.8; Microcline 3.0

$\varrho_0 = 2.985$ g cm^{-3}; $c_l = 6.73$ km s^{-1}

Sample					Standard	
ϱ_0 g cm^{-3}	U km s^{-1}	u km s^{-1}	P GPa	ϱ g cm^{-3}	Composition	U km s^{-1}
2.99	5.96	0.89	16.0	3.51	Al	6.55
3.00	6.00	0.91	16.4	3.54	Al	6.57
2.99	5.98	1.25	22.3	3.78	Al	6.97
2.99	5.99	1.27	22.6	3.79	Al	6.99
2.99	6.16	1.75	32.2	4.18	Al	7.58
2.98	6.18	1.77	32.6	4.17	Al	7.60
2.97	6.27	1.94	36.1	4.30	Al	7.80
2.97	6.21	1.94	35.8	4.32	Al	7.79
2.98	6.81	2.29	46.6	4.49	Al	8.28
2.99	6.78	2.34	47.5	4.57	Al	8.33
2.98	7.35	2.67	58.5	4.68	Al	8.78
2.98	7.96	3.15	74.7	4.93	Al	9.42
2.99	8.31	3.32	82.5	4.97	Al	9.68
2.98	8.82	3.65	96.1	5.09	Al	10.13
3.01	8.91	3.80	101.9	5.25	Al	10.33

$U = (3.403 + 1.466\,u)$ km s^{-1}, $u = 1.9 \cdots 3.8$ km s^{-1}; $\sigma_U = 0.056$ km s^{-1}

Experimental technique B
Data reduction technique B
Reference: [McQ66]

The modal analysis of these samples was taken from [Bir60].

Basalt, Vacaville, Mt. Vaca. quad., California

Sample				
ϱ_0 g cm^{-3}	U km s^{-1}	u km s^{-1}	P GPa	ϱ g cm^{-3}
2.86	5.88	2.10	35.3	4.45
2.86	6.77	2.76	53.4	4.83
2.86	7.59	3.24	70.3	4.99
2.86	8.31	3.94	93.6	5.44
2.86	9.19	4.25	111.7	5.32
2.86	9.01	4.33	111.6	5.50
2.86	9.91	4.64	131.5	5.38
2.86	10.60	5.01	151.9	5.41
2.86	10.63	5.20	158.1	5.60
2.86	12.04	5.94	204.6	5.64

$U = (2.31 + 1.615\,u)$ km s^{-1}; $\sigma_U = 0.2$ km s^{-1}

Experimental technique A
Data reduction technique A; copper and fansteel-77 alloy as standards
Reference: [Jon68]

Basalt, Frederick, Maryland

Composition (in vol%): Augite 24; Hyperstene 25; Plagioclase: 48 (= Anorthite 68 wt%; Albite 32 wt%); Olivine 1; Biotite 1

$\varrho_0 = 3.012$ g cm^{-3}; $c_l = 6.82$ km s^{-1}

Sample					Standard	
ϱ_0 g cm^{-3}	U km s^{-1}	u km s^{-1}	P GPa	ϱ g cm^{-3}	Composition	U km s^{-1}
3.01	6.00	0.89	16.1	3.53	Al	6.55
3.01	6.11	0.96	17.7	3.57	Al	6.64
3.01	6.05	1.22	22.3	3.77	Al	6.95
3.01	6.05	1.24	22.5	3.79	Al	6.97
3.01	5.98	1.28	23.1	3.83	Al	7.02
3.01	6.15	1.63	30.1	4.10	Al	7.44
3.02	6.22	1.74	32.6	4.18	Al	7.58
3.01	6.29	1.81	34.4	4.24	Al	7.67
3.01	6.39	1.91	36.8	4.29	Al	7.80
3.01	6.89	2.27	47.2	4.50	Al	8.28
3.01	6.90	2.31	48.1	4.54	Al	8.33
3.01	7.38	2.65	59.0	4.70	Al	8.78
3.01	7.39	2.66	59.3	4.71	Al	8.80
3.02	7.99	3.13	75.4	4.96	Al	9.42
3.02	8.08	3.14	76.5	4.93	Al	9.45
3.01	8.36	3.30	83.2	4.98	Al	9.68
3.02	8.75	3.65	96.3	5.17	Al	10.13
3.01	9.04	3.78	102.9	5.18	Al	10.33

$U = (3.778 + 1.371\,u)$ km s^{-1}, $u = 1.9 \cdots 3.8$ km s^{-1}; $\sigma_U = 0.050$ km s^{-1}

Experimental technique B
Data reduction technique B
Reference: [McQ66]

The modal analysis of these samples was taken from [Bir60].

◄

Composition (in vol%): Plagioclase 53 (= Anorthite 45···53 wt%; Albite 55···47 wt%); Augite 31; Magnetite 9; Celadonite 5; Apatite 2.
Grain size: 0.3···0.003 mm; Porosity: 2.0%.
Chemical analysis (in %): SiO_2 50.3; Al_2O_3 13.8; Fe_2O_3 3.5; FeO 9.1; MgO 4.4; CaO 7.7; Na_2O 3.3; K_2O 1.6; H_2O^- 1.0; H_2O^+ 1.5; TiO_2 2.4; P_2O_5 1.2; MnO 0.24; CO_2 0.05

$\varrho_0 = 2.890$ g cm^{-3}; $\varrho_{01} = 3.040 \cdots 3.175$ g cm^{-3}; $c_0 = 2.306$ km s^{-1}

Basalt, Vacaville, Mt. Vaca quad., California
Composition (in vol%): Andesine 53 (= Anorthite 47···54 wt%; Albite 53···46 wt%); Augite 31; Magnetite-Ilmenite 9; Celadonite 5; Apatite 2.
Grain size: 0.01···0.2 mm.
$\varrho_0 = 2.817 \text{ g cm}^{-3}$; $\varrho_{01} = 3.174···3.04 \text{ g cm}^{-3}$

Sample					Standard
ϱ_0 g cm^{-3}	U km s^{-1}	u km s^{-1}	P GPa	ϱ g cm^{-3}	U_{fs} km s^{-1}
first wave:					
2.82	5.43	0.328	5.0	3.00	1.968
2.82	5.33	0.291	4.4	2.98	
2.82	5.55	0.323	5.0	2.99	
second wave:					
2.82	5.31	0.964	14.5	3.44	1.968
2.82	5.19	0.906	13.4	3.41	
2.82	5.06	0.798	11.6	3.34	1.640
2.82	5.12	0.764	11.2	3.30	
2.82	5.44	1.232	18.9	3.64	2.610
2.82	5.40	1.124	17.2	3.56	

Experimental technique C
Data reduction technique B and D with $2u = U_{fs}$ for the first wave
Reference: [Ahr64]

Basalt, (Dolerite)
Composition (in vol%): Pigeonite-Augite 55 (= Enstatite 31 wt%; Ferrosilite 37 wt%; Wollastonite 32 wt%); Labradorite 30 (= Anorthite 56 wt%; Albite 44 wt%); Titanomagnetite 10; Olivine 5 (= Forsterite 52 wt%; Fayalite 48 wt%)
$\varrho_0 = 3.049 \text{ g cm}^{-3}$

Sample					Standard	
ϱ_0 g cm^{-3}	U km s^{-1}	u km s^{-1}	P GPa	ϱ g cm^{-3}	Composition	U_{fs} km s^{-1}
3.05	5.80	0.67	11.9	3.45	Al	0.69
3.05	6.21	1.53	29.0	4.05	Al	1.50
3.05	7.95	2.84	68.8	4.74	Al	2.82
3.05	9.01	3.74	102.8	5.21	steel	2.82
3.05	12.09	5.99	220.9	6.04	steel	4.56

$U = (4.10 + 1.325\, u) \text{ km s}^{-1}$, $u = 1.5···6.0 \text{ km s}^{-1}$
Error in U 1.5% (<70 GPa) and 2% (>70 GPa).
Experimental technique A
Data reduction technique B
Reference: [Tru65]

Basalt
Composition (in vol%): Labradorite 50 (= Anorthite 61 wt%; Albite 39 wt%); Olivine 25 (= Forsterite 70 wt%; Fayalite 30 wt%); Pigeonite-Augite 15 (= Enstatite 49 wt%; Wollastonite 27 wt%; Ferrosilite 24 wt%); Titanomagnetite 10
$\varrho_0 = 3.135 \text{ g cm}^{-3}$

Sample					Standard	
ϱ_0 g cm^{-3}	U km s^{-1}	u km s^{-1}	P GPa	ϱ g cm^{-3}	Composition	U_{fs} km s^{-1}
3.13	6.83	0.61	13.1	3.44	Al	0.69
3.13	6.90	1.45	31.3	3.96	Al	1.50
3.13	8.21	2.78	71.4	4.74	Al	2.82
3.13	9.26	3.66	106.1	5.17	steel	2.80
3.13	12.32	5.92	228.0	6.03	steel	4.56

$U = (4.48 + 1.326\, u) \text{ km s}^{-1}$; $u = 1.6···6.0 \text{ km s}^{-1}$
Error in U 1.5% (<70 GPa) and 2% (>70 GPa).
Experimental technique A
Data reduction technique B
Reference: [Tru65]

Basalt
Composition (in wt%): Plagioclase (Andesine?) 70; Olivine 24.5; Magnetite and Opaques 5.5.
Porosity (measured): 4.9%
Particle size: up to 0.5 mm
$\varrho_0 = 2.674 \text{ g cm}^{-3}$; $c_l = 5.36 \text{ km s}^{-1}$; $c_s = 3.1 \text{ km s}^{-1}$.

Chemical analysis (in wt%): SiO_2 51.88; FeO 9.37; Al_2O_3 17.56; CaO 6.74; MgO 5.54; Na_2O 4.26; K_2O 2.34; TiO_2 1.32; BaO 0.42 and H_2O 0.18

Sample					Standard
ϱ_0 g cm^{-3}	U km s^{-1}	u km s^{-1}	P GPa	ϱ g cm^{-3}	Composition
2.667	7.79	3.29	68.4	4.61	Al
2.673	4.87	0.79	10.3	3.19	Al
2.668	5.24	1.67	23.3	3.92	Al
2.654	8.59	3.50	79.7	4.48	Al
2.682	5.25	1.53	21.5	3.78	Al
2.693	5.70	2.10	32.2	4.25	Al
2.688	6.93	2.86	53.3	4.57	Al

Experimental technique B
Data reduction technique B
Reference [Van64]

A more extensive discussion of this material may be found in Report UOPK B 64-2 Jan 1964 by L. Rogers, Lawrence Radiation Laboratory, Livermore, California

Dunite, Mooihoek Mine, Transvaal (Dunit)
Composition (in vol%): Olivine: 90 (= Fayalite 63 wt%; Forsterite 37 wt%); Bowlingite 9; Ore 1
$\varrho_0 = 3.676 \cdots 3.846$ g cm^{-3}; $\varrho_{01} = 3.731$ g cm^{-3};
$c_1 = 7.31$ km s^{-1}

Sample					Standard	
ϱ_0 g cm^{-3}	U km s^{-1}	u km s^{-1}	P GPa	ϱ g cm^{-3}	Composition	U km s^{-1}
3.83	6.55	0.49	12.3	4.14	Al	6.16
3.83	6.71	0.74	19.0	4.30	Al	6.54
3.78	6.96	1.03	27.2	4.44	Al	6.97
3.77	6.71	1.06	26.8	4.48	Al	6.97
3.61	6.71	1.10	26.7	4.32	Al	6.99
3.78	7.05	1.22	32.6	4.57	Al	7.23
3.78	7.05	1.26	33.6	4.60	Al	7.28
3.78	7.09	1.37	36.8	4.68	Al	7.44
3.83	7.29	1.46	40.8	4.79	Al	7.59
3.82	7.41	1.52	42.9	4.81	Al	7.68
3.80	7.45	1.53	43.4	4.78	Al	7.70
3.85	7.40	1.62	46.1	4.93	Al	7.82
3.76	7.32	1.63	44.8	4.84	Al	7.80
3.82	7.40	2.09	58.9	5.32	Al	8.41
3.83	7.50	2.23	64.0	5.45	Al	8.61
3.80	7.43	2.31	65.3	5.51	Al	8.68
3.82	7.55	2.39	68.7	5.58	Al	8.80
3.80	7.52	2.39	68.3	5.57	Al	8.79
3.81	7.62	2.49	72.1	5.65	Al	8.93
3.85	7.93	2.61	79.8	5.74	Al	9.16
3.78	8.13	2.84	87.2	5.81	Al	9.44
3.84	8.22	2.89	91.1	5.92	Al	9.54
3.82	8.47	2.96	96.0	5.88	Al	9.67
3.77	8.54	2.97	95.7	5.78	Al	9.67
3.73	8.45	3.05	96.2	5.84	Al	9.73
3.68	8.48	3.06	95.5	5.76	Al	9.73
3.82	8.69	3.11	103.1	5.94	Al	9.88
3.75	8.79	3.32	109.4	6.02	Al	10.12
3.82	9.08	3.37	116.8	6.07	Al	10.27
3.80	9.00	3.37	115.4	6.08	Al	10.25
3.77	9.09	3.47	119.0	6.10	Al	10.37

$U = (6.033 + 0.828\,u)$ km s^{-1} for $u = 0.5 \cdots 1.6$ km s^{-1};
$\sigma_U = 0.12$ km s^{-1}
$U = (4.008 + 1.477\,u)$ km s^{-1} for $u = 2.3 \cdots 3.5$ km s^{-1};
$\sigma_U = 0.079$ km s^{-1}
Experimental technique B
Data reduction technique B
Reference: [McQ66]

The modal analysis of these samples was taken from [Bir60].

Dunite, Twin Sisters Peaks, Washington (Dunit)
Composition (in vol%): Olivine 92.5 (= Forsterite 84 wt%; Fayalite 16 wt%); Pyroxene 7 (= Enstatite 83 wt%; Ferrosilite 17 wt%); Serpentine 0.5.
$\varrho_0 = 3.322$ g cm^{-3}; $\varrho_{01} = 3.356$ g cm^{-3};
$c_1 = 8.55$ km s^{-1}

Sample					Standard	
ϱ_0 g cm^{-3}	U km s^{-1}	u km s^{-1}	P GPa	ϱ g cm^{-3}	Composition	U km s^{-1}
3.32	7.31	0.76	18.5	3.71	Al	6.54
3.32	7.32	0.78	19.0	3.72	Al	6.56
3.32	7.27	0.78	18.9	3.72	Al	6.56
3.32	7.35	0.84	20.5	3.75	Al	6.64
3.32	7.32	0.84	20.4	3.75	Al	6.64
3.32	7.29	0.84	20.4	3.75	Al	6.64
3.32	7.62	1.05	26.5	3.85	Al	6.95
3.32	7.48	1.06	26.4	3.87	Al	6.97
3.32	7.61	1.09	27.6	3.87	Al	7.02
3.32	7.57	1.11	28.0	3.89	Al	7.04
3.32	7.59	1.22	30.7	3.96	Al	7.18
3.32	7.69	1.25	31.9	3.97	Al	7.23
3.32	7.94	1.54	40.5	4.12	Al	7.64
3.32	8.00	1.56	41.5	4.12	Al	7.68
3.32	7.93	1.57	41.2	4.14	Al	7.67
3.32	7.87	1.64	42.9	4.19	Al	7.76
3.32	8.10	1.65	44.2	4.17	Al	7.79
3.32	8.11	1.75	47.2	4.23	Al	7.93
3.32	8.15	1.98	53.6	4.39	Al	8.23
3.32	8.13	2.01	54.2	4.41	Al	8.26
3.32	8.17	2.12	57.6	4.48	Al	8.41
3.32	8.22	2.22	60.7	4.55	Al	8.53
3.32	8.24	2.23	61.0	4.55	Al	8.54
3.32	8.22	2.23	60.9	4.55	Al	8.54
3.32	8.33	2.38	65.7	4.65	Al	8.73
3.32	8.28	2.41	66.2	4.68	Al	8.77
3.32	8.31	2.42	66.8	4.68	Al	8.78
3.32	8.28	2.43	66.9	4.70	Al	8.79
3.32	8.24	2.44	66.8	4.72	Al	8.80
3.32	8.39	2.64	73.4	4.84	Al	9.05
3.32	8.31	2.65	73.0	4.87	Al	9.05
3.32	8.75	2.86	83.1	4.93	Al	9.38
3.32	8.66	2.90	83.2	4.98	Al	9.40
3.32	8.73	2.91	84.4	4.98	Al	9.44
3.32	8.69	2.93	84.4	5.01	Al	9.45
3.32	8.77	2.95	85.8	5.00	Al	9.49
3.32	8.75	2.95	85.7	5.01	Al	9.49
3.32	9.15	3.10	94.1	5.02	Al	9.73
3.32	9.12	3.10	94.0	5.03	Al	9.73
3.32	9.25	3.28	100.7	5.14	Al	9.96
3.32	9.55	3.37	107.0	5.13	Al	10.13
3.32	9.45	3.39	106.3	5.18	Al	10.12
3.32	9.49	3.39	106.8	5.17	Al	10.13
3.32	9.69	3.49	112.3	5.18	Al	10.28

Dunite (continued)

$U = (6.65 + 0.825\, u)$ km s^{-1}, for $u = 0.7 \cdots 1.65$ km s^{-1};
$\sigma_U = 0.06$ km s^{-1}
$U = (7.63 + 0.268\, u)$ km s^{-1}, for $u = 1.65 \cdots 2.5$ km s^{-1};
$\sigma_U = 0.03$ km s^{-1}
$U = (4.140 + 1.582\, u)$ km s^{-1}, for $u = 2.6 \cdots 3.5$ km s^{-1};
$\sigma_U = 0.07$ km s^{-1}

Experimental technique B
Data reduction technique B
Reference: [McQ66]

The modal analysis of these samples was taken from [Bir60].

Dunite (Dunit)

Composition (in vol%): Olivine 65 (= Forsterite 91 wt%; Fayalite 9 wt%); Serpentine 30; Chromite 5
$\varrho_0 = 2.959$ g cm^{-3}; $\varrho_{01} = 3.205$ g cm^{-3}

Sample					Standard	
ϱ_0 g cm^{-3}	U km s^{-1}	u km s^{-1}	P GPa	ϱ g cm^{-3}	Composition	u km s^{-1}
2.96	7.08	1.46	30.7	3.75	Al	1.50
2.96	8.51	2.80	70.6	3.96	Al	2.82
2.96	12.70	5.95	224.0	5.56	steel	4.56

$U = (4.45 + 1.370\, u)$ km s^{-1}, for $u = 2.8 \cdots 6.0$ km s^{-1}
Error in U 1.5% (< 70 GPa) and 2% (> 70 GPa).
Experimental technique A
Data reduction technique B
Reference: [Tru65]
Locality of sample not given.

Dunite (Dunit)

Composition (in vol%): Olivine 80 (= Forsterite 63 wt%; Fayalite 37 wt%); Titanomagnetite 20
$\varrho_0 = 3.690$ g cm^{-3}; $\varrho_{01} = 3.891$ g cm^{-3}

Sample					Standard	
ϱ_0 g cm^{-3}	U km s^{-1}	u km s^{-1}	P GPa	ϱ g cm^{-3}	Composition	u km s^{-1}
3.69	7.61	1.27	35.6	4.43	Al	1.5
3.69	8.40	2.56	79.3	5.31	Al	2.82
3.69	7.78	2.20	63.1	5.15	Cu	1.71
3.69	9.62	3.49	123.9	5.79	steel	2.82

$U = (4.96 + 1.324\, u)$ km s^{-1}, for $u = 2.2 \cdots 3.5$ km s^{-1}
Error in U 1.5% (< 70 GPa) and 2% (> 70 GPa)
Experimental technique A
Data reduction technique B
Reference: [Tru65]
Locality of sample not given.

Dunite (Dunit)

Composition (in vol%): Olivine 85 (= Forsterite 63 wt%; Fayalite 37 wt%); Titanomagnetite 7; Serpentine 5; Talc 3
$\varrho_0 = 3.205$ g cm^{-3}; $\varrho_{01} = 3.610$ g cm^{-3}

Sample					Standard	
ϱ_0 g cm^{-3}	U km s^{-1}	u km s^{-1}	P GPa	ϱ g cm^{-3}	Composition	u km s^{-1}
3.21	7.39	1.38	32.7	3.95	Al	1.5
3.21	8.55	2.70	74.1	4.69	Al	2.82
3.21	9.85	3.58	113.3	5.04	steel	2.80
3.21	12.56	5.85	236.0	6.01	steel	4.56

$U = (5.23 + 1.270\, u)$ km s^{-1}, for $u = 2.7 \cdots 6.0$ km s^{-1}
Error in U 1.5% (< 70 GPa) and 2% (> 70 GPa)
Experimental technique A
Data reduction technique B
Reference: [Tru65]
Locality of sample not given.

Dunite (Dunit)

Composition (in vol%): Olivine 90 (= Forsterite 87 wt%, Fayalite 13 wt%); Biotite 7; Titanomagnetite 3
$\varrho_0 = 3.311$ g cm^{-3}; $\varrho_{01} = 3.356$ g cm^{-3}

Sample					Standard	
ϱ_0 g cm^{-3}	U km s^{-1}	u km s^{-1}	P GPa	ϱ g cm^{-3}	Composition	u km s^{-1}
3.31	6.98	0.59	13.6	3.62	Al	0.69
3.31	7.40	1.00	24.5	3.83	Al	1.14
3.31	7.77	1.33	34.2	3.99	Al	1.50
3.31	8.60	2.65	75.5	4.78	Al	2.82
3.31	8.19	1.70	46.1	4.18	plexiglass	3.05
3.31	8.17	2.04	55.1	4.41	Mg	2.79
3.31	9.78	3.59	116.2	5.23	steel	2.82
3.31	12.48	5.81	240.0	6.20	steel	4.56
3.31	16.83	9.07	505.0	7.18	steel	7.11

$U = (5.08 + 1.287\, u)$ km s^{-1}, for $u = 2.2 \cdots 9.5$ km s^{-1}
Error in U 1.5% (< 70 GPa) and 2% (> 70 GPa).
Experimental technique A
Data reduction technique B
Reference: [Tru65]
Locality of sample not given

Dunite (Dunit)

Composition (in vol%): Olivine 50 (= Forsterite 42 wt%; Fayalite 58 wt%); Serpentine 40; Chromite 10

$\varrho_0 = 2.899$ g cm^{-3}; $\varrho_{01} = 3.472$ g cm^{-3}

Sample					Standard	
ϱ_0 g cm^{-3}	U km s^{-1}	u km s^{-1}	P GPa	ϱ g cm^{-3}	Composition	u km s^{-1}
2.9	6.85	1.50	29.8	3.62	Al	1.5
2.9	8.28	2.85	68.4	3.82	Al	2.82
2.9	9.39	3.75	102.1	3.98	steel	2.82

$U = (4.59 + 1.290\, u)$ km s^{-1}, for $u = 2.8 \cdots 3.8$ km s^{-1}

Error in U 1.5% (<70 GPa) and 2% (>70 GPa)

Experimental technique A
Data reduction technique B
Reference: [Tru65]
Locality of sample not given.

Gabbro (actually "norite" according to composition)

Composition (in vol%): Bronzite 60 (= Enstatite 88 wt%; Ferrosilite 12 wt%); Labradorite 35 (= Albite 29 wt%; Anorthite 71 wt%); Olivine 5 (= Forsterite 42 wt%; Fayalite 58 wt%)

$\varrho_0 = 3.155$ g cm^{-3}; $\varrho_{01} = 3.115$ g cm^{-3}

Sample					Standard	
ϱ_0 g cm^{-3}	U km s^{-1}	u km s^{-1}	P GPa	ϱ g cm^{-3}	Composition	u km s^{-1}
3.15	6.92	1.44	31.4	3.98	Al	1.5
3.15	8.39	2.74	72.4	4.68	Al	2.82
3.15	9.55	3.63	109.1	5.08	steel	2.80
3.15	12.56	5.88	232.5	5.92	steel	4.56

$U = (4.66 + 1.355\, u)$ km s^{-1}, for $u = 2.7 \cdots 6.0$ km s^{-1}

Experimental technique A
Data reduction technique B
Reference: [Tru65]
Locality of sample not known.

Gabbro (leuco-gabbro), anorthositic, recrystallized, Apollo 15, rock No. 15418 Moon, Montes Apenninus (Leuko-Gabbro)

Composition (in wt%): Plagioclase 75 (= Albite 4 mol%; Anorthite 96 mol%); Orthopyroxene and Clinopyroxene 19, Olivine 5.4 (= Forsterite 53 mol%; Fayalite 47 mol%); Iron-nickel and Troilite 0.1. Porosity: 1%

Chemical composition (in wt%): SiO$_2$ 45.26, TiO$_2$ 0.27, Cr$_2$O$_3$ 0.11, Al$_2$O$_3$ 26.90, FeO 5.41, MnO 0.08, MgO 5.42, CaO 16.21, Na$_2$O 0.31, K$_2$O 0.03

$\varrho_0 = 2.806 \cdots 2.846$ g cm^{-3}

Sample				Standard		
ϱ_0 g cm^{-3}	U km s^{-1}	P GPa	ϱ g cm^{-3}	Composition	U_{fs} km s^{-1}	No.
first wave:						
2.812	6.18	5.3		Al	0.803	1
2.823	6.04	4.8		Al	0.850	2
2.846	5.94	4.2		Al	1.139	3
2.821	5.99	7.1		W	1.020	4
2.806	6.14	6.9		W	1.108	5
2.823	6.24	5.7		W	1.17	6
2.821	5.88	7.0		W	1.618	8
2.813	6.30	6.5		W	2.166	10
second wave:						
2.812		6.3	3.03	Al	0.803	1
2.823		6.5	3.08	Al	0.850	2
2.846		8.8	3.22	Al	1.139	3
2.821		12.9	3.37	W	1.020	4
2.806		14.5	3.38	W	1.108	5
2.823		14.8	3.48	W	1.17	6
2.834		15.5	3.69	W	1.318	7
2.821		20.4	3.82	W	1.618	8
2.822		26.1	4.07	W	1.992	9
2.813		28.2	4.25	W	2.166	10

Errors for all entries are given in the original reference.

Experimental technique C   Data reduction technique B   Reference: [Ahr73a]

Granite, Hardhat, Nevada (Granit)
Composition (in wt%) [Sho66, Hou59]: Orthoclase 26; Plagioclase 37; Quartz 29; Biotite 6; Hornblende 1; Accessory minerals 1
Particle size: 0.2 to 5.0 mm
$\varrho_0 = 2.675\,\mathrm{g\,cm^{-3}}$; $\varrho_{01} = 2.660\,\mathrm{g\,cm^{-3}}$

ϱ_0 g cm^{-3}	U km s^{-1}	u km s^{-1}	P GPa	ϱ g cm^{-3}	No.
first wave:					
2.675	5.96	0.28	4.5	2.81	1
2.675	5.95	0.27	4.3	2.81	2
2.675	5.90	0.29	4.6	2.81	3
2.675	5.96	0.28	4.5	2.90	4
2.675	5.99	0.28	4.5	2.81	5
2.675	5.98	0.28	4.5	2.81	6
2.675	6.04	0.28	4.6	2.81	7
2.675	5.97	0.27	4.2	2.80	8
2.674	6.02	0.34	5.4	2.83	9
2.675	5.94	0.20	3.1	2.77	10
2.67	6.05	0.29	4.7	2.80	11
2.700	6.08	0.29	4.8	2.84	12
2.701	5.98	0.28	4.5	2.83	13
second wave:					
2.675	5.19	0.85	12.3	3.18	1
2.675	5.21	0.85	12.4	3.18	2
2.675	5.68	1.53	23.3	3.65	3
2.675	5.55	1.54	23.1	3.69	4
2.675	5.37	0.87	13.0	3.17	5
2.675	5.25	0.88	12.8	3.19	6
2.675	5.50	1.54	23.0	3.70	7
2.675	5.50	1.54	22.9	3.70	8
2.674	5.51	1.22	18.3	3.42	9
2.675	5.40	1.23	18.0	3.45	10
2.67	5.45	1.25	18.6	3.45	11
2.700	5.44	1.26	18.9	3.49	12
2.701	5.71	1.82	28.4	3.95	13
2.675	7.04	3.00	56.5	4.66	14

Experimental technique C and D (here a reflecting foil is placed a known distance from the sample surface in glycerol or ethanol).
Data reduction technique B and D (first wave)
Reference: [Pet68]

Granite, Hardhat, Nevada Test Site (Granit)
Composition (in wt%): Orthoclase 47.0; Plagioclase 26.5; Quartz 19.5; Mica 5.0; Hornblende 1.5; Opaques <1.0
Porosity (measured): 0.8%
Particle size: up to 5.0 mm
Chemical analysis (grey granite; area 15 National Test Site Nevada) (in wt%): SiO_2 71.94; FeO 4.25; Al_2O_3 13.68; CaO 2.96; MgO 0.63; Na_2O 2.94; K_2O 1.44; TiO_2 0.26; H_2O 1.38
$\varrho_0 = 2.674\,\mathrm{g\,cm^{-3}}$;
$c_l = 5.64\,\mathrm{km\,s^{-1}}$; $c_s = 3.53\,\mathrm{km\,s^{-1}}$

ϱ_0 g cm^{-3}	U km s^{-1}	u km s^{-1}	P GPa	ϱ g cm^{-3}	Standard Composition
2.669	5.38	0.99	14.2	3.27	Al
2.669	5.66	1.63	24.6	3.75	Al
2.660	5.19	0.49	6.8	2.94	Al
2.680	7.58	3.35	68.0	4.80	Al
2.690	5.52	0.96	14.3	3.26	Al
2.686	6.32	2.57	43.6	4.53	Al
2.674	5.59	1.72	25.7	3.86	Al
2.674	8.27	3.87	85.6	5.03	Al
2.679	5.66	0.49	7.4	2.93	Al
2.680	5.59	0.82	12.3	3.14	Al
2.690	5.65	1.63	24.7	3.78	Al
2.675	5.68	2.11	31.9	4.25	Al
2.672	6.92	2.88	53.3	4.58	Al
2.612	5.37	1.31	18.4	3.46	Al
2.614	5.83	2.22	33.7	4.22	Al
2.618	5.37	1.00	14.0	3.22	Al
2.614	5.38	0.49	6.8	2.88	Al

$U = (2.52 + 1.5\,u)\,\mathrm{km\,s^{-1}}$, for $u \approx 2.0\,\mathrm{km\,s^{-1}}$;
$\sigma_U = 0.06\,\mathrm{km\,s^{-1}}$
Experimental technique B
Data reduction technique B
Reference: [Van64]

The table lists the results of four types of granite obtained from area 15 at the national test site in Nevada.
The first four points correspond to samples from 1005 feet exploratory core.
The two following points correspond to samples from Hardhat Tunnel. The last four points correspond to samples that can be visually characterized as pink granite [Van77]

Granite, Shoal (Granit)

Composition (in wt%) [Van77]: Plagioclase 41.0; Quartz 26.8; Orthoclase and Microcline 17.7; Biotite 9.1; Microperthite 5.4

$\varrho_0 = 2.667 \, \text{g cm}^{-3}$

Sample					
ϱ_0 g cm^{-3}	U km s^{-1}	u km s^{-1}	P GPa	ϱ g cm^{-3}	No.
first wave:					
2.82	5.80	0.25	3.8	2.94	1
2.65	5.98	0.24	3.8	2.76	2
second wave:					
2.82	4.750	0.260	4.15	2.96	1
2.65	3.566	0.870	9.47	3.40	2
2.65	3.932	0.898	10.38	3.35	3
2.65	3.993	0.985	11.39	3.44	4
2.65	4.054	0.916	10.78	3.35	5
2.65	4.481	1.171	14.56	3.53	6
2.65	4.572	1.128	14.28	3.47	7
2.65	4.663	1.195	15.24	3.52	8
2.65	4.663	1.291	16.49	3.62	9
2.65	4.968	1.500	20.09	3.76	10
2.65	5.029	1.431	19.39	3.67	11
2.65	5.182	1.481	20.56	3.69	12
2.65	5.334	1.471	20.95	3.64	13
2.65	5.425	1.965	28.33	4.13	14
2.65	5.791	2.047	31.32	4.09	15
2.65	5.882	1.672	25.70	3.71	16
2.65	6.035	1.296	20.51	3.38	17
2.65	6.126	2.380	38.41	4.34	18

Experimental technique A for pressures below 4.2 GPa, for pressures above 5.0 GPa a shadowgraph of the free surface of a wedge shaped sample was used.
Data reduction technique A for pressures below 4.2 GPa D for pressures above 5.0 GPa
Reference: [Den 69/70]

Granite, Westerly, Rhode Island (Granit)

Composition (in vol%): Microcline 35.4; Oligoclase 31.4; Quartz 27.5; Biotite 3.2

$\varrho_0 = 2.620 \, \text{g cm}^{-3}$; $\varrho_{01} = 2.624 \, \text{g cm}^{-3}$;
$c_1 = 6.1 \, \text{km s}^{-1}$

Sample					Standard	
ϱ_0 g cm^{-3}	U km s^{-1}	u km s^{-1}	P GPa	ϱ g cm^{-3}	Composition	U km s^{-1}
2.63	5.29	1.00	13.8	3.24	Al	6.55
2.63	5.31	1.01	14.1	3.25	Al	6.57
2.63	5.45	1.35	19.3	3.50	Al	6.95
2.63	5.46	1.37	19.6	3.51	Al	6.97
2.63	5.42	1.39	19.8	3.53	Al	6.99
2.63	5.45	1.43	20.5	3.56	Al	7.04
2.63	5.60	1.91	28.1	3.99	Al	7.58
2.63	5.61	1.93	28.5	4.01	Al	7.60
2.63	5.67	1.99	29.7	4.05	Al	7.67
2.62	5.78	2.10	31.8	4.11	Al	7.80
2.62	5.69	2.11	31.5	4.17	Al	7.79
2.63	6.19	2.50	40.6	4.41	Al	8.28
2.63	6.18	2.55	41.4	4.47	Al	8.33
2.63	6.83	2.88	51.7	4.55	Al	8.78
2.63	7.58	3.37	67.1	4.72	Al	9.42
2.63	7.99	3.54	74.4	4.72	Al	9.68
2.63	8.49	3.89	86.8	4.85	Al	10.13
2.63	8.41	3.90	86.1	4.90	Al	10.12
2.63	8.62	4.05	91.9	4.96	Al	10.33

$U = (4.93 + 0.372 \, u) \, \text{km s}^{-1}$ for $1.0 < u < 2.1$
$\sigma_U = 0.03 \, \text{km s}^{-1}$
$U = (2.103 + 1.629 \, u) \, \text{km s}^{-1}$ for $2.5 < u < 4.1$
$\sigma_U = 0.074 \, \text{km s}^{-1}$

Experimental technique B
Data reduction technique B
Reference: [McQ66]

The modal analysis of this sample was taken from [Bir60].

Granite, Westerly, Rhode Island (Granit) [Bir60]
Composition (in vol%): Microcline 35.4; Oligoclase 31.4; Quartz 27.5; Biotite 3.2
Porosity: ≈1%
$\varrho_0 = 2.63 \cdots 2.65$ g cm^{-3}; $\varrho_{01} = 2.624$ g cm^{-3}

Sample					Standard
U km s^{-1}	u km s^{-1}	P GPa	ϱ [1]) g cm^{-3}	Composition	U_{fs} km s^{-1}
Dry Westerly granite:					
5.05	0.048	0.64	2.665	[2])	0.227
5.03	0.067	0.89	2.676	[2])	0.316
5.14	0.088	1.20	2.686	[2])	0.408
5.37	0.104	1.48	2.692	[2])	0.495
5.24	0.13	1.8	2.707	[2])	0.582
5.52	0.180	2.61	2.729	[2])	0.77
5.50	0.242	3.53	2.761	[2])	0.485
5.6	0.287	4.26	2.783	[2])	0.573
5.57	0.315	4.59	2.798	[2])	0.63
5.23	0.380	5.26	2.847	[2])	0.76
5.38	0.405	5.68	2.855	[2])	0.81
5.44	0.400	5.72	2.850	[2])	0.80
5.60	0.395	5.86	2.840	[2])	0.79
5.50	0.485	7.02	2.895	[2])	0.97
5.56	0.490	7.12	2.895	[2])	0.98
5.18	0.605	8.18	2.989	[2])	1.21
5.68	0.625	9.34	2.966	[2])	1.25
Water-saturated Westerly granite					
5.03	0.112	1.48	2.700	[2])	0.510
5.48	0.173	2.50	2.726	[2])	0.765
5.60	0.176	2.59	2.726	[2])	0.76
5.50	0.203	2.95	2.741	[2])	0.90
5.39	0.326	4.64	2.810	[2])	1.27

[1]) Calculated from $\mu = \frac{\varrho}{\varrho_0} - 1$ assuming $\varrho_0 = 2.64$ g cm^{-3}.
[2]) Standards: Impactors consisted of polymethylmethacrylate and of Westerly granite.

Experimental technique K
Data reduction technique F
Reference: [Lar81]

Granite (Granit)
Composition (in vol%): Plagioclase 46.7; Quartz 21.1; Orthoclase 20.3; Biotite 4.9; Chlorite 6.5; Opaques 0.5
Porosity: 0.9%
$\varrho_0 = 2.681$ g cm^{-3}

Sample					Standard
ϱ_0 g cm^{-3}	U km s^{-1}	u km s^{-1}	P GPa	ϱ g cm^{-3}	Composition
2.68	11.93	6.01	192.1	5.40	steel
2.68	10.67	5.21	148.0	5.25	steel
2.68	10.22	4.90	134.0	5.14	steel
2.68	10.11	4.86	132.7	5.16	steel
2.68	9.00	4.05	98.6	4.87	Cu
2.68	8.66	4.01	93.0	4.99	Cu
2.68	7.27	3.17	62.7	4.75	Cu

$U = (2.22 + 1.625\, u)$ km s^{-1}; $\sigma_U = 0.11$ km s^{-1}
Experimental technique A
Data reduction technique A
Reference: [Isb65]
Locality of sample not reported.

The modal analysis of rock samples from the same location given above was made by [Bar67], also private communication 1968 to van Thiel [Van77]

Peridotite (Peridotit)
Composition (in vol%): Olivine 75 (= Forsterite 68 wt%; Fayalite 32 wt%); Diallage 10 (= Enstatite 48 wt%; Wollastonite 52 wt%); Labradorite 10 (= Anorthite 61 wt%; Albite 39 wt%) Titanomagnetite 5
$\varrho_0 = 3.226$ g cm^{-3}; $\varrho_{01} = 3.497$ g cm^{-3}

Sample					Standard	
ϱ_0 g cm^{-3}	U km s^{-1}	u km s^{-1}	P GPa	ϱ g cm^{-3}	Composition	u km s^{-1}
3.22	7.08	1.40	31.9	4.01	Al	1.50
3.22	8.32	2.73	73.2	4.79	Al	2.82
3.22	9.57	3.61	111.0	5.17	steel	2.80
3.22	12.60	5.84	236.9	6.00	steel	4.56

$U = (4.5 + 1.4\, u)$ km s^{-1} for $u = 2.7 \cdots 6.0$ km s^{-1}
Error in U 1.5% (<70 GPa) and 2% (>70 GPa)
Experimental technique A
Data reduction technique B
Reference: [Tru65]
Locality of sample not given

Pyroxenite, Bushveld Complex, Transvaal (Pyroxenit)
Composition (in vol%): Plagioclase 4 (= Anorthite 81 wt%; Albite 19 wt%); Pyroxene 92 (= Enstatite 88 wt%; Ferrosilite 12 wt%); Hornblende 2
$\varrho_0 = 3.30$ g cm^{-3}; $\varrho_{01} = 3.204$ g cm^{-3};
$c_l = 7.7$ km s^{-1}

Sample					Standard	
ϱ_0 g cm^{-3}	U km s^{-1}	u km s^{-1}	P GPa	ϱ g cm^{-3}	Composition	U km s^{-1}
3.30	6.74	0.52	11.6	3.58	Al	6.16
3.30	6.68	0.80	17.6	3.75	Al	6.54
3.30	6.98	1.11	25.4	3.92	Al	6.97
3.29	7.31	1.64	39.4	4.24	Al	7.68
3.30	7.31	1.64	39.7	4.25	Al	7.69
3.30	7.35	1.70	41.3	4.29	Al	7.76
3.30	7.57	2.20	55.1	4.66	Al	8.41
3.29	7.92	2.50	64.9	4.80	Al	8.80
3.30	8.65	2.93	83.6	4.99	Al	9.44
3.30	8.90	3.14	92.3	5.10	Al	9.73

$U = (4.322 + 1.462 u)$ km s^{-1} for $u = 2.2 \cdots 3.2$ km s^{-1};
$\sigma_U = 0.056$ km s^{-1}
Experimental technique B
Data reduction technique B
Reference: [McQ66]

The modal analysis of this sample was taken from [Bir60].

$U = (6.08 + 1.03 u)$ km s^{-1} for $u = 0.8 \cdots 1.3$ km s^{-1};
$\sigma_U = 0.05$ km s^{-1}
$U = (7.06 + 0.32 u)$ km s^{-1} for $u = 1.4 \cdots 2.2$ km s^{-1};
$\sigma_U = 0.04$ km s^{-1}
$U = (5.113 + 1.204 u)$ km s^{-1} for $u = 2.3 \cdots 3.6$ km s^{-1};
$\sigma_U = 0.064$ km s^{-1}
Experimental technique B
Data reduction technique B
Reference: [McQ66]

An analysis (in wt%) was given by Hess: SiO$_2$ 54.68, Al$_2$O$_3$ 1.80, Fe$_2$O$_3$ 0.50, FeO 9.19, MgO 30.19, CaO 2.22, Na$_2$O 0.04, K$_2$O 0.04, H$_2$O 0.51, TiO$_2$ 0.11, P$_2$O$_5$ 0.02, Cr$_2$O$_3$ 0.47, MnO 0.21, total 99.97

Pyroxenite, Stillwater Complex, Montana (Pyroxenit)
Composition (in vol%): Bronzite: 94 (= Enstatite 88 wt%; Ferrosilite 12 wt%); Hornblende 4; Olivine 2 (= Forsterite 82 wt%; Fayalite 18 wt%)
$\varrho_0 = 3.279$ g cm^{-3}; $\varrho_{01} = 3.296$ g cm^{-3};
$c_l = 7.66$ km s^{-1}

Sample					Standard	
ϱ_0 g cm^{-3}	U km s^{-1}	u km s^{-1}	P GPa	ϱ g cm^{-3}	Composition	U km s^{-1}
3.28	6.84	0.52	11.7	3.55	Al	6.16
3.27	6.87	0.79	17.8	3.69	Al	6.54
3.27	6.99	0.86	19.7	3.73	Al	6.64
3.29	7.02	0.87	20.2	3.76	Al	6.66
3.28	7.24	1.09	25.9	3.86	Al	6.97
3.28	7.17	1.09	25.7	3.87	Al	6.97
3.27	7.14	1.12	26.1	3.88	Al	6.99
3.28	7.44	1.28	31.2	3.96	Al	7.23
3.28	7.45	1.32	32.1	3.99	Al	7.28
3.30	7.53	1.43	35.4	4.07	Al	7.44
3.28	7.54	1.55	38.4	4.13	Al	7.60
3.22	7.54	1.56	38.0	4.06	Al	7.59
3.27	7.56	1.59	39.4	4.14	Al	7.65
3.28	7.63	1.60	40.2	4.15	Al	7.67
3.27	7.59	1.61	40.1	4.15	Al	7.68
3.28	7.54	1.62	40.2	4.18	Al	7.69
3.28	7.61	1.63	40.6	4.17	Al	7.70
3.27	7.51	1.69	41.6	4.22	Al	7.76
3.28	7.62	1.70	42.6	4.22	Al	7.79
3.28	7.65	1.72	43.2	4.23	Al	7.82
3.28	7.64	1.89	47.2	4.36	Al	8.02
3.28	7.73	2.10	53.1	4.51	Al	8.29
3.28	7.79	2.15	54.9	4.53	Al	8.37
3.23	7.72	2.15	53.7	4.47	Al	8.34
3.28	7.76	2.19	55.7	4.57	Al	8.41
3.28	7.89	2.32	60.1	4.64	Al	8.59
3.27	7.93	2.33	60.6	4.63	Al	8.61
3.29	7.96	2.39	62.6	4.70	Al	8.68
3.28	8.04	2.44	64.3	4.71	Al	8.75
3.28	8.08	2.47	65.5	4.72	Al	8.79
3.28	8.20	2.57	69.2	4.76	Al	8.93
3.28	8.32	2.63	71.8	4.80	Al	9.02
3.28	8.60	2.89	81.7	4.95	Al	9.38
3.28	8.63	2.95	83.6	4.99	Al	9.45
3.28	8.93	3.10	90.8	5.02	Al	9.67
3.28	8.88	3.11	90.4	5.04	Al	9.67
3.28	8.94	3.15	92.2	5.06	Al	9.73
3.28	8.98	3.27	96.3	5.16	Al	9.88
3.28	9.40	3.42	105.5	5.16	Al	10.13
3.28	9.17	3.46	104.0	5.26	Al	10.13
3.27	9.33	3.54	108.2	5.28	Al	10.25
3.28	9.36	3.55	109.0	5.28	Al	10.27

Pyroxenite (Pyroxenit)

Composition (in vol%): Hyperstene 95 (=Enstatite 82 wt%; Ferrosilite 18 wt%); Olivine 3 (=Forsterite 87 wt%; Fayalite 13 wt%); Labradorite 2 (=Anorthite 61 wt%; Albite 39 wt%)

$\varrho_0 = 3.289$ g cm^{-3}; $\varrho_{01} = 3.331$ g cm^{-3}

Sample					Standard	
ϱ_0 g cm^{-3}	U km s^{-1}	u km s^{-1}	P GPa	ϱ g cm^{-3}	Composition	u km s^{-1}
3.29	6.86	0.60	13.6	3.60	Al	0.69
3.29	7.46	1.36	33.4	4.02	Al	1.50
3.29	8.70	2.66	76.1	4.74	Al	2.82
3.29	7.98	1.74	45.7	4.21	plexiglass	3.05
3.29	7.87	2.26	58.4	4.61	Cu	1.71
3.29	9.85	3.56	115.4	5.15	steel	2.80
3.29	12.60	5.80	240.0	6.58	steel	4.56

$U = (5.3 + 1.265 \, u)$ km s^{-1} for $u = 2.3 \cdots 6.0$ km s^{-1}

Error in U 1.5% (<70 GPa) and 2% (>70 GPa).
Experimental technique A
Data reduction technique B
Reference: [Tru65]
Locality of sample not given.

Pyroxenite (Pyroxenit)

Composition (in vol%): Diallage 70 (=Enstatite 62 wt%; Wollastonite 38 wt%); Serpentine 25; Titanomagnetite 5

$\varrho_0 = 3.01$ g cm^{-3}

Sample					Standard	
ϱ_0 g cm^{-3}	U km s^{-1}	u km s^{-1}	P GPa	ϱ g cm^{-3}	Composition	u km s^{-1}
3.01	7.33	1.43	31.6	3.74	Al	1.50
3.01	8.47	2.79	70.9	4.49	Al	2.82
3.01	9.39	3.72	105.2	4.98	steel	2.82

$U = (5.33 + 1.12 \, u)$ km s^{-1} for $u = 2.7 \cdots 3.8$ km s^{-1}

Error in U 1.5% (<70 GPa) and 2% (>70 GPa).
Experimental technique A
Data reduction technique B
Reference: [Tru65]
Locality of sample not given.

1.3.3.3.2 Metamorphic rocks — Metamorphe Gesteine

Eclogite, Healdsburg, California (Eklogit)

Composition (in vol%): Garnet 24 (=Almandite 37 wt%; Pyrope 6 wt%; Grossularite 43 wt%; Andradite 6 wt%; Spessartite 2 wt%); Omphacite 72 (=Diopside 57 wt%; Jadeite 30 wt%; Acmite 13 wt%); Rest 4

$\varrho_0 = 3.472 \cdots 3.39$ g cm^{-3}; $\varrho_{01} = 3.472$ g cm^{-3}; $c_1 = 7.84$ km s^{-1}

Sample					Standard	
ϱ_0 g cm^{-3}	U km s^{-1}	u km s^{-1}	P GPa	ϱ g cm^{-3}	Composition	U km s^{-1}
3.35	6.52	0.53	11.5	3.65	Al	6.16
3.45	7.03	0.76	18.3	3.87	Al	6.53
3.38	6.92	0.78	18.2	3.81	Al	6.54
3.47	7.30	1.05	26.5	4.99	Al	6.96
3.38	7.20	1.08	26.2	3.93	Al	6.97
3.45	7.62	1.23	32.4	4.11	Al	7.23
3.46	7.72	1.38	36.8	4.21	Al	7.44
3.42	7.71	1.44	38.1	4.21	Al	7.52
3.42	7.77	1.48	39.2	4.22	Al	7.56
3.43	7.84	1.56	41.8	4.28	Al	7.68
3.44	7.74	1.57	41.8	4.32	Al	7.69
3.43	7.83	1.58	42.5	4.30	Al	7.71
3.44	7.88	1.61	43.8	4.32	Al	7.76
3.38	8.29	2.09	58.7	4.52	Al	8.41
3.43	8.43	2.17	62.7	4.63	Al	8.53
3.45	8.56	2.36	69.7	4.77	Al	8.79
3.40	8.60	2.37	69.4	4.70	Al	8.80
3.40	8.41	2.38	68.1	4.74	Al	8.78
3.36	8.56	2.66	76.4	4.88	Al	9.12
3.41	8.60	2.78	81.7	5.04	Al	9.30
3.39	8.79	2.87	85.8	5.04	Al	9.44
3.40	8.79	2.94	87.7	5.10	Al	9.51
3.41	9.06	3.07	95.1	5.16	Al	9.73
3.47	9.41	3.25	106.2	5.30	Al	10.03
3.43	9.33	3.37	107.8	5.36	Al	10.13

$U = (5.86 + 1.36 \, u)$ km s^{-1} for $u = 0.5 \cdots 1.4$ km s^{-1}; $\sigma_U = 0.1$ km s^{-1}

$U = (6.56 + 0.812 \, u)$ km s^{-1} for $u = 1.4 \cdots 2.7$ km s^{-1}; $\sigma_U = 0.09$ km s^{-1}

$U = (4.808 + 1.375 \, u)$ km s^{-1} for $u = 2.78 \cdots 3.4$ km s^{-1}; $\sigma_U = 0.096$ km s^{-1}

Experimental technique B
Data reduction technique B
Reference: [McQ66]

The modal analysis of this sample was taken from [Bir60], see also [Bor56].

Eclogite, Norway (Eklogit)

Composition (in vol%): Omphacite 52; Garnet 42 (= Almandite 64 wt%; Pyrope 36 wt%); Biotite 6
$\varrho_0 = 3.584 \cdots 3.484$ g cm^{-3}; $c_l = 7.4$ km s^{-1}

Sample					Standard	
ϱ_0 g cm^{-3}	U km s^{-1}	u km s^{-1}	P GPa	ϱ g cm^{-3}	Composition	U km s^{-1}
3.59	6.80	0.50	12.1	3.88	Al	6.16
3.53	7.02	0.75	18.7	3.95	Al	6.54
3.56	7.10	0.77	19.5	3.99	Al	6.57
3.57	7.40	1.02	27.0	4.14	Al	6.95
3.60	7.39	1.03	27.3	4.18	Al	6.97
3.51	7.31	1.07	27.4	4.11	Al	6.99
3.53	7.63	1.22	32.8	4.20	Al	7.23
3.55	7.86	1.35	37.6	4.29	Al	7.44
3.56	8.04	1.51	43.2	4.38	Al	7.68
3.58	7.92	1.51	42.8	4.43	Al	7.67
3.48	7.84	1.61	43.9	4.38	Al	7.76
3.52	7.99	1.61	45.3	4.41	Al	7.80
3.51	7.83	1.63	44.8	4.43	Al	7.79
3.59	8.37	2.03	60.9	4.74	Al	8.41
3.58	8.81	2.27	71.8	4.83	Al	8.78
3.54	8.72	2.31	71.3	4.82	Al	8.79
3.56	8.88	2.55	80.5	5.00	Al	9.12
3.57	9.11	2.72	88.6	5.09	Al	9.38
3.53	9.24	2.76	90.2	5.04	Al	9.44
3.52	9.36	2.99	98.4	5.17	Al	9.73
3.56	9.22	2.99	98.3	5.27	Al	9.73
3.55	9.72	3.26	112.5	5.35	Al	10.13

$U = (6.15 + 1.21\,u)$ km s^{-1} for $u = 0.5 \cdots 1.4$ km s^{-1}; $\sigma_U = 0.07$ km s^{-1}
$U = (6.4 + 0.997\,u)$ km s^{-1} for $u = 1.5 \cdots 3.3$ km s^{-1}; $\sigma_U = 0.11$ km s^{-1}

Experimental technique B
Data reduction technique B
Reference: [McQ66]

The modal analysis of this sample was taken from [Bir60].

Exact composition of omphacite not known, Escola gives the following composition for a similar sample of the same general locality (private communication of F. Birch to Van Thiel, 1967): diopside 83.5, jadeite 7.4, enstatite 4.7, Tschermaks molecule 4.1, rest 0.7 (vol%?)

Gneiss (Gneis)

Composition (in vol%): Quartz 35.0; Plagioclase 21.0···22.0 (= Albite 90···50 mol%; Anorthite 10···50 mol%); Biotite 15.0···20.0; Muscovite 15.0···10.0; Chlorite 7; Garnet 5; Pyrite 1.0···0.5; Magnetite 1.0···0.5.
Grain size: 0.1 ··· 2.0 mm
$\varrho_0 = 2.793$ g cm^{-3}

Sample				
ϱ_0 g cm^{-3}	U km s^{-1}	u km s^{-1}	P GPa	ϱ g cm^{-3}
2.79	6.57	2.56	46.9	4.57
2.79	7.63	3.17	66.7	4.84
2.79	8.47	3.74	88.5	5.00
2.79	8.74	3.98	97.0	5.12
2.79	8.99	3.99	100.0	5.01
2.79	9.08	4.27	108.2	5.26
2.79	9.22	4.34	111.7	5.27
2.79	9.47	4.42	116.9	5.23
2.79	11.35	5.25	167.3	5.20
2.79	11.99	6.08	203.4	5.66

$U = (2.49 + 1.593\,u)$ km s^{-1}; $\sigma_U = 0.22$ km s^{-1}

Experimental technique A. The projectile velocities were determined from two timed flash X-ray shadowgraphs; standards: Cu and fansteel.
Data reduction technique A
Reference: [Isb66]
Locality of sample not given.

Marble, Vermont (Marmor)

Composition: Calcite 95%; Quarz 5%
Grain size: 0.05···0.2 mm
$\varrho_0 = 2.688$ g cm^{-3}; $\varrho_{01} = 2.710$ g cm^{-3}

Sample					
ϱ_0 g cm^{-3}	U km s^{-1}	u km s^{-1}	P GPa	ϱ g cm^{-3}	No.
first wave:					
2.688	5.262	0.086	1.2	2.73	1
2.688	4.300	0.165	2.1	2.78	2
2.688	4.718	0.123	1.5	2.76	3
2.688	3.980	0.190	2.3	2.81	4
2.688	3.730	0.352	3.9	2.97	5
second wave:					
2.688	3.791	0.378	4.3	2.96	2
2.688	3.284	0.443	4.7	3.04	5

Experimental technique C
Data reduction technique B
Reference: [Gre63]

Marble, Yule (Marmor)

Calcite, polycrystalline 100%

Particle size: 0.2···0.4 mm, a few grains as large as 1.5 mm

$\varrho_0 = 2.687 \text{ g cm}^{-3}$; $\varrho_{01} = 2.770 \text{ g cm}^{-3}$

ϱ_0 g cm^{-3}	U km s^{-1}	u km s^{-1}	P GPa	ϱ g cm^{-3}	No.
first wave:					
2.687	4.63	0.396	4.9	2.95	1
2.687	4.12	0.697	8.3	3.20	2
2.687	5.73	0.103	1.6	2.78	3
2.687	4.94	0.796	10.4	3.23	4
second wave:					
2.687	3.573	0.782	9.1	3.30	2
2.687	4.478	1.024	13.4	3.41	4
2.687	5.52	1.17	17.6	3.42	5
2.687	5.83	1.12	17.7	3.35	6
2.687	5.91	1.59	25.3	3.69	7

Experimental technique C
Data reduction technique B
$T_0 = 27 \pm 3 \,°C$
Reference: [Gre63]

Marble (Marmor)

Calcite, polycrystalline

$\varrho_0 = 2.703 \text{ g cm}^{-3}$; $\varrho_{01} = 2.717 \text{ g cm}^{-3}$

ϱ_0 g cm^{-3}	U km s^{-1}	u km s^{-1}	P GPa	ϱ g cm^{-3}
2.703	4.26	0.43	5.0	2.73
2.703	4.51	0.56	6.8	3.08
2.703	4.70	0.64	8.2	3.14
2.703	4.92	0.77	10.3	3.19
2.703	5.18	0.90	12.5	3.27
2.703	5.26	0.92	13.1	3.28
2.703	5.47	1.125	16.6	3.40
2.703	5.51	1.17	17.4	3.43
2.703	5.66	1.26	19.3	3.46
2.703	5.76	1.33	20.8	3.54
2.703	6.04	1.56	25.2	3.65
2.703	6.27	1.72	29.1	3.73
2.703	6.47	1.85	32.5	3.78
2.703	7.35	2.56	50.8	4.13

Error in U 0.01 km s^{-1}.
Experimental technique B
Data reduction technique B
Reference: [Dre59]
Composition and locality of sample not given.

Quartzite (Novaculite), Arkansas, Behr-Manning Co., Troy, N.Y. (Quarzit)

Quartz ≈ 100 vol%
Particle size: ≈ 0.01 mm

$\varrho_0 = 2.626 \cdots 2.629 \text{ g cm}^{-3}$; $\varrho_{01} = 2.650 \text{ g cm}^{-3}$

Sample					Standard	
ϱ_0 g cm^{-3}	U km s^{-1}	u km s^{-1}	P GPa	ϱ g cm^{-3}	U_{fs} km s^{-1}	No.
first wave:						
2.628	6.35	0.645	10.76	2.93	2.03	1
2.628	6.26	0.673	11.07	2.94	2.03	2
2.628	6.18	0.652	10.59	2.94	2.03	3
2.628	6.04	0.598	9.49	2.92	2.03	4
2.628	6.28	0.438	7.23	2.82	2.03	5
2.628	6.24	0.522	8.56	2.87	2.03	6
2.628	6.14	0.406	6.55	2.81	1.76	7
2.628	6.14	0.496	8.00	2.86	1.76	8
2.628	5.97	0.832	13.05	3.05	1.79	9
2.628	5.94	0.665	10.38	2.96	1.78	10
2.628	6.15	0.441	7.13	2.83	1.80	11
2.628	6.14	0.444	7.16	2.83	1.81	12
2.628	6.12	0.50	8.04	2.86		13
2.628	6.313	(0.85)	14.20	3.06	3.20	22
2.628	6.144	(0.38)	6.1	2.80	3.43	23
2.628	6.170	0.64	10.3	2.93	3.49	24
2.628	6.378		10.3	2.93	3.49	25
2.628	6.302	0.671	11.1	2.94		26

(continued)

Quartzite (continued)

Sample					Standard	
ϱ_0 g cm^{-3}	U km s^{-1}	u km s^{-1}	P GPa	ϱ g cm^{-3}	U_{fs} km s^{-1}	No.
2.628	6.36	0.68	11.4	2.94		28
2.628	6.22	0.67	11.0	2.95		29
2.628	6.155	0.41	6.7	2.82		30
2.628	6.18	0.83	13.7	3.03		31
2.628	6.17	0.453	7.4	2.83	1.41	32
2.628	6.06	0.432	6.87	2.83	1.56	33
2.628	6.04	0.422	6.70	2.83	1.55	34
2.628	6.12	0.415	6.67	2.82	1.57	35
2.628	5.94	0.665	10.38	2.96	1.78	36
2.628	6.18	0.469	7.62	2.84	1.78	37
2.628	6.14	0.406	6.55	2.81	1.76	38
2.628	6.14	0.496	8.00	2.86		39
2.628	6.15	0.441	7.13	2.83	1.80	40
2.628	6.14	0.444	7.16	2.83	1.81	41
2.628	5.97	0.832	13.05	3.05	1.79	42
2.628	6.12	0.50	8.04	2.86	(1.92)	43
2.628	6.266	(0.25)			3.49	44
2.628	6.302	0.671	11.1	2.94	3.49	45
2.628	6.36	(0.68)	11.4	2.94	3.58	46
2.628	6.22	(0.67)	11.0	2.95	3.28	47
2.628	6.15	(0.41)	6.7	2.82	4.19	48
2.628	6.18	0.83	13.5	3.03	3.33	49
second wave:						
2.628	5.16	1.13	17.0	3.28	2.03	1
2.628	4.55	1.16	16.4	3.37	2.03	2
2.628	4.86	1.14	16.7	3.32	2.03	3
2.628	4.84	1.16	16.4	3.36	2.03	4
2.628	5.12	1.15	16.5	3.33	2.03	5
2.628	5.04	1.14	16.5	3.32	2.03	6
2.628	5.45	0.97	14.4	3.17	1.76	7
2.628	4.85	0.98	14.0	3.22	1.76	8
2.628	4.37	0.985	14.7	3.19	1.79	9
2.628	3.72	1.025	13.6	3.35	1.78	10
2.628	5.09	1.00	14.5	3.22	1.80	11
2.628	5.04	1.01	14.5	3.23	1.81	12
2.628	4.96					13
2.628	5.77	1.76	27.9	3.76	3.20	22
2.628	5.694	1.95	29.5	3.98	3.43	23
2.628	5.542	2.00	29.7	4.06	3.49	24
2.628	5.596	1.98	29.8	4.02	3.49	25
2.628	5.982	1.74	27.8	3.68		26
2.628	5.236	1.98	28.7	4.13		27
2.628	5.62	2.02	31.1	4.04		28
2.628	5.86	2.37	36.9	4.38		29
2.628	5.81	2.41	37.1	4.47		30
2.628	5.91	2.40	38.2	4.39		31
2.628	4.68	0.77	11.1	3.07	1.41	32
2.628	4.62	0.875	12.05	3.17	1.56	33
2.628	4.80	0.865	12.2	3.14	1.55	34
2.628	4.88	0.87	12.35	3.14	1.57	35

(continued)

Quartzite (continued)

Sample							
ϱ_0 g cm^{-3}	U km s^{-1}	u km s^{-1}	P GPa	ϱ g cm^{-3}	U_{fs} km s^{-1}	No.	
2.628	3.72	1.025	13.6	3.36	1.78	36	
2.628	5.03	0.98	14.2	3.20	1.78	37	
2.628	5.45	0.97	14.4	3.17	1.76	38	
2.628	4.85	0.98	14.0	3.22		39	
2.628	5.09	1.00	14.45	3.22	1.80	40	
2.628	5.04	1.01	14.55	3.23	1.81	41	
2.628	4.37	0.985	14.7	3.19	1.79	42	
2.628	4.96	1.08	15.4	3.29	(1.92)	43	
2.628	5.530	(1.95)	30.6	3.93	3.49	44	
2.628	5.236	1.98	28.7	4.13	3.49	45	
2.628	5.62	2.02	30.8	4.04	3.58	46	
2.628	5.86	2.37	36.9	4.38	3.28	47	
2.628	5.81	2.41	37.1	4.47	4.19	48	
2.628	5.91	2.40	37.7	4.39	3.33	49	

single wave:

2.628	6.234	2.41	39.5	4.29	4.31	50	
2.628	6.15	2.50	40.4	4.42	4.42	51	
2.628	6.27	2.47	40.7	4.34	4.43	52	
2.628	6.25	2.59	42.6	4.48	4.61	53	
2.628	6.28	2.70	44.5	4.61	4.78	54	

Quartzite, Eureka, Confusion Mtn., Ely, Nevada, (EQ) (Quarzit)

Quartz ≈ 99 vol%; SiO$_2$-cement trace

$\varrho_0 = 2.626 \cdots 2.629$ g cm^{-3}; $\varrho_{01} = 2.650$ g cm^{-3}

Sample					Standard	
ϱ_0 g cm^{-3}	U km s^{-1}	u km s^{-1}	P GPa	ϱ g cm^{-3}	U_{fs} km s^{-1}	No.
first wave:						
2.629	6.06	0.484	7.71	2.85	1.93	14
2.629	5.83	0.471	7.22	2.86	1.93	15
2.629	5.92	0.540	8.40	2.89	1.93	16
2.629	5.98	0.431	6.78	2.83	1.93	17
2.629	5.94	0.460	7.18	2.85	1.93	18
second wave:						
2.629	5.41	1.075	15.9	3.24	1.93	14
2.629	4.97	1.10	15.3	3.32	1.93	15
2.629	4.72	1.10	15.2	3.34	1.93	16
2.629	5.19	1.085	15.6	3.28	1.93	17
2.629	4.96	1.095	15.3	3.32	1.93	18

Quartzite, Sioux, Dell Rapids, S. Dakota (Quarzit)

Quartz ≈ 99 vol%; Hematite-dust remainder; SiO$_2$-cement remainder
Particle size: ≈ 0.1 mm

$\varrho_0 = 2.626 \cdots 2.629$ g cm^{-3}; $\varrho_{01} = 2.650$ g cm^{-3}

Sample					Standard	
ϱ_0 g cm^{-3}	U km s^{-1}	u km s^{-1}	P GPa	ϱ g cm^{-3}	U_{fs} km s^{-1}	No.
first wave:						
2.626	6.31	0.409	6.78	2.81	1.88	19
2.626	5.96	0.290	4.54	2.76	1.88	20
2.626	5.95	0.305	4.77	2.77	1.88	21
second wave:						
2.626	5.53	1.03	15.7	3.20	1.88	19
2.626	5.39	1.05	15.3	3.24	1.88	20
2.626	5.20	1.06	15.0	3.27	1.88	21

For quartzites Arkansas Eureka, and Sioux (entries 1$\cdots$54):
Experimental technique C
Data reduction technique B; standard materials 2024 aluminum 356 brass; data reduction technique D with $u_1 = \frac{1}{2} U_{fs1}$ (first wave)
Uncertain values in parantheses.
References: [Ahr66b] for entries No 1$\cdots$21; [Ahr68a] for entries No 22$\cdots$31 and No 50$\cdots$54; [Ahr66a] for entries No 32$\cdots$49

1.3.3.3.3 Sediments and sedimentary rocks — Sedimente und sedimentäre Gesteine

Dolomite, Blair, Berkeley Company, Martinsburg, West Virginia (Dolomit)
Composition: 98 % $CaMg(CO_3)_2$
Porosity: ≈ 1 %
$\varrho_0 = 2.84$ g cm^{-3}; $\varrho_{01} = 2.866$ g cm^{-3}

ϱ_0 g cm^{-3}	U km s^{-1}	u km s^{-1}	P GPa	ϱ g cm^{-3}	Sample Composition	Standard U_{fs} km s^{-1}
2.84	5.30	0.017	0.24	2.848	[1]	0.08
2.84	5.92	0.030	0.44	2.852	[1]	0.153
2.84	6.12	0.040	0.63	2.856	[1]	0.200
2.84	6.40	0.100	1.70	2.885	[1]	0.200
2.84	6.69	0.196	3.40	2.929	[1]	0.392
2.84	6.62	0.290	5.33	2.974	[1]	0.584

[1] Standards: Impactors consisted of polymethylmethacrylate and of Blair dolomite.

Experimental technique K
Data reduction technique F
Reference: [Lar80]

Dolomite, Kaibab, Alpha Member (Dolomit)
Composition (in vol %): Dolomite 75; Quartz 20; Calcite, Feldspar, Clay minerals, Hematite, Goethite, and heavy minerals 5
Porosity: 20.3 ··· 23.8 %
$\varrho_0 = 2.119 \cdots 2.222$ g cm^{-3}; $\varrho_{01} = 2.882 \cdots 2.817$ g cm^{-3};
$c_0 = 2.845$ km s^{-1}

ϱ_0 g cm^{-3}	U km s^{-1}	u km s^{-1}	P GPa	ϱ g cm^{-3}
2.22	5.69	2.35	29.7	3.78
2.22	6.96	3.20	49.4	4.10
2.22	8.65	4.23	81.2	4.34
2.22	9.22	4.69	96.1	4.52
2.22	10.10	5.05	113.1	4.47

$U = (1.89 + 1.597 u)$ km s^{-1}; $\sigma_U = 0.13$ km s^{-1}
Error in U and u: 1 ··· 2 % and 0.2 %, respectively (last entry $u \pm 5$ %).
Experimental technique A
Data reduction technique A; standards: Cu and fansteel-77 alloy.
Reference: [Jon68]

Dolomite, Nevada (Dolomit)
Composition (in wt %): Dolomite 96.4; Calcite 1.0; Quartz 1.8; Clay 1.7; Goethite 0.2
$\varrho_0 = 2.826$ g cm^{-3}; $\varrho_{01} = 2.857$ g cm^{-3};
$c_0 = 5.91$ km s^{-1}

ϱ_0 g cm^{-3}	U km s^{-1}	u km s^{-1}	P GPa	ϱ g cm^{-3}	Sample Composition	Standard P GPa
2.831	7.03	1.17	23.3	3.39	Al	23.1
2.820	7.45	1.79	37.8	3.71	Al	38.1
2.829	8.07	2.38	54.3	4.01	Al	55.3
2.824	8.75	3.14	77.6	4.41	Al	80.4
2.822	6.92	1.12	21.9	3.37	Al	21.8
2.825	6.68	0.813	15.3	3.22	Al	15.1
2.824	6.44	0.495	9.0	3.06	Al	8.5

$U = (5.97 + 0.876 u)$ km s^{-1}; $\sigma_U = 0.05$ km s^{-1}
Experimental technique B
Data reduction technique B
Reference: [Hor65]

These samples were obtained from depth of 1308 ··· 1341 feet in a hole designated U10B at the approximate Nevada Central coordinates N. 880,000 − E. 670,000

An average analysis (in wt %) over the depth range 1303 ··· 1346 feet [F. Stephens, Lawrence Radiation Laboratory, Livermore, California] yielded: H_2O 0.09; SiO_2 1.22; FeO 0.04; Fe_2O_3 0.13; CaO 30.63; MgO 21.07; Al_2O_3 0.48; CO_2 46.48; Insol. 1.57.

Dolomite Banded Mountain, Nevada Test Site (Dolomit)
Composition (in wt %): $CaMg(CO_3)_2$ 94 ··· 97; $CaCO_3$ < 5; SiO_2 < 1
$\varrho_0 = 2.817$ g cm^{-3}

ϱ_0 g cm^{-3}	U km s^{-1}	u km s^{-1}	P GPa	ϱ g cm^{-3}
2.82	11.54	5.32	173.1	5.22
2.82	10.02	4.14	117.1	4.80
2.82	9.77	3.73	102.8	4.56
2.82	8.44	2.65	63.1	4.11

$U = (5.39 + 1.15 u)$ km s^{-1}; $\sigma_U = 0.11$ km s^{-1}
Experimental technique A; the velocity of fansteel or high purity copper projectiles where determined by two X-ray flash photographs.
Data reduction technique A
Reference: [Isb66]

Dolomite, Blair formation, Martinsburg, W. Virginia (Dolomit-Gestein)
Composition: 98% $CaMg(CO_3)_2$
Porosity (calculated): 0.9%
$\varrho_0 = 2.84$ g cm^{-3}; $\varrho_{01} = 2.866$ g cm^{-3}

Sample

ϱ_0 g cm^{-3}	U km s^{-1}	u km s^{-1}	P GPa	ϱ g cm^{-3}
2.84	6.51	1.00	18.5	3.356
2.84	6.50	1.15	21.2	3.448
2.84	6.82	1.11	21.5	3.390
2.84	6.78	1.49	28.7	3.636
2.84	7.40	1.84	38.6	3.774
2.84	7.65	1.94	42.0	3.802

The error in the pressure determination is ±5% except for 42.0 (±8%).
Experimental technique F (manganin stress gage)
Data reduction technique C (P and U measured)
Composition of metal flyer plate not given.
Reference: [Gra76]

Limestone, Salisbury Plain (Kalk)
Composition (in wt %): $CaCO_3$ 98.2; H_2O 0.03; Mg 0.20; SiO_2 0.62; Mn 0.02; Fe 0.07; P 0.22 as PO_4; and traces of Sr, Cl, Cu, Ti, Na, K
Porosity (calculated): 35%
$\varrho_0 = 1.742$ g cm^{-3}; $\varrho_{01} = 2.711$ g cm^{-3}

Sample

ϱ_0 g cm^{-3}	U km s^{-1}	u km s^{-1}	P GPa	ϱ g cm^{-3}
1.75	1.59	0.58	1.6	2.76
1.75	1.53	0.56	1.5	2.76
1.75	1.90	0.81	2.7	3.05
1.75	1.97	0.79	2.7	2.92
1.75	2.97	1.04	5.4	2.69
1.75	2.76	1.14	5.5	2.98
1.75	4.00	1.52	10.6	2.82
1.75	4.23	1.51	11.2	2.72
1.75	4.12	1.67	12.0	2.94
1.75	4.38	1.67	12.8	2.83
1.75	4.35	1.74	13.2	2.92
1.75	4.70	2.28	18.7	3.40
1.75	4.86	2.34	19.9	3.38
1.75	5.84	3.14	32.1	3.79
1.75	6.22	3.26	35.5	3.68
1.75	6.67	3.54	41.3	3.73
1.75	6.67	3.80	44.4	4.07

$U = (0.587 + 1.24\,u + 0.718\,u^2)$ km s^{-1};
$\sigma_U = 0.2$ km s^{-1}, for $u = 0.58 \cdots 1.52$ km s^{-1}
$U = (2.151 + 1.214\,u)$ km s^{-1};
$\sigma_U = 0.17$ km s^{-1}, for $u = 1.6 \cdots 3.8$ km s^{-1}

Limestone, Solnhofen, Germany (Kalk)
Calcite 100%
Chemical analysis (in %): CO_3 56.7; Ca 37.82; Mg <0.5; Fe 0.02; Si 0.62; Al 0.15
Average grain size: 0.01 mm; maximum grain size: 0.02 mm.
$\varrho_0 = 2.585$ g cm^{-3};
$c_1 = 5.30$ km s^{-1}; $c_s = 2.89$ km s^{-1}

Sample

ϱ_0 g cm^{-3}	U km s^{-1}	u km s^{-1}	P GPa	ϱ g cm^{-3}	Standard Composition
2.590	4.33	0.69	7.8	2.08	Al
2.597	5.33	1.17	16.2	3.33	Al
2.597	5.27	1.18	16.1	3.35	Al
2.598	5.69	1.48	21.8	3.51	Al
2.594	5.67	1.50	22.0	3.53	Al
2.560	6.41	2.11	34.7	3.82	Al
2.566	5.72	1.63	24.0	3.59	Al
2.573	9.04	3.87	90.0	4.50	Al

Experimental technique B
Data reduction technique B
Reference: [Van77]

Limestone, Solnhofen, Germany (Kalk)
Composition (in %): Calcite 96; Clay 2···3; Quartz 1···2.
Porosity: 0.2···0.3%. Grain size: 0.005···0.015 mm
$\varrho_0 = 2.584$ g cm^{-3}

Sample

ϱ_0 g cm^{-3}	U km s^{-1}	u km s^{-1}	P GPa	ϱ g cm^{-3}
first wave:				
2.583	5.330	0.073	1.0	2.62
2.583	5.808	0.108	1.3	2.64
2.583	3.585	0.163	1.8	2.69
2.583	3.419	0.223	2.4	2.74
2.583	3.342	0.387	3.8	3.06
2.583	4.572	0.820	9.7	3.14
second wave:				
2.583	3.094	0.441	4.2	2.95
2.583	3.666	1.238	13.4	3.69

Experimental technique C
Data reduction technique B
The elastic wave is unstable and shows a strong strain rate dependence.
Reference: [Gre63].

◄

Experimental technique A
Data reduction technique B
Aluminium, Iron, and steel were used as standards.
Reference: [Har65]

Limestone (Kalk)

Composition (in wt%): Calcite 96.0⋯98.0; Magnesium carbonate 2.6⋯0.4; and remainders of water, manganese carbonate, iron carbonate, organics

$\varrho_0 = 2.703$ g cm^{-3}

Sample

ϱ_0 g cm^{-3}	U km s^{-1}	u km s^{-1}	P GPa	ϱ g cm^{-3}
2.7	12.59	5.82	197.7	5.02
2.7	10.22	4.41	121.4	4.74
2.7	10.16	4.23	115.9	4.63
2.7	9.38	3.93	99.5	4.63
2.7	8.07	3.29	71.7	4.57
2.7	6.72	1.99	36.1	3.84
2.7	5.23	1.04	14.7	3.37

$U = (3.51 + 1.53\,u)$ km s^{-1}; $\sigma_U = 0.26$ km s^{-1}

Experimental technique A. The projectile (fansteel or high purity copper) velocities were determined by two timed X-Ray flashes.
Data reduction technique A
Reference: [Isb66]

Regolith, Apollo 17 sample No. 70051, impactoclastic detritus, Moon, Mare Serenitatis
(Regolith, impaktoklastischer Detritus)

Composition (in vol%): Glassy agglutinates 32; Basalt fragments 19; Plagioclase (Anorthite) 8.5; Pyroxene 21; Opaques 5.6; Glass spherules 5.7; Metal spherules 1.5
Grain size: 5⋯1000 µm
Chemical composition (in wt%): SiO_2 40, TiO_2 9, Al_2O_3 11, FeO 17, MgO 10, CaO 11
$\varrho_0 = 1.795\cdots1.810$ g cm^{-3}
Porosity: 42%

Sample

ϱ_0 g cm^{-3}	P GPa	ϱ g cm^{-3}	Standard Composition	U_{fs} km s^{-1}
1.800	1.99	3.16	Al	0.810
1.798	3.95	3.16	Tl	1.042
1.795	6.61	3.41	Tl	1.444
1.799	7.79	3.45	Tl	1.573
1.810	12.60	3.68	W	2.103

Errors for all entries are given in the original reference.
Experimental technique C
Data reduction technique B
Reference: [Ahr74]

Sand

Quartz (SiO_2) 100%
Particle size less than 0.075 mm (80%) to 0.15 mm maximum; Porosity: 41%
$\varrho_0 = 1.580$ g cm^{-3}

Sample

ϱ_0 g cm^{-3}	U km s^{-1}	u km s^{-1}	P GPa	ϱ g cm^{-3}	Composition
1.58	3.13	1.17	5.8	2.52	Al
1.58	3.23	1.16	5.9	2.46	Al
1.59	3.42	1.61	8.8	3.01	Al
1.58	3.47	1.70	9.3	3.10	Al
1.56	4.26	2.25	15.0	3.31	Al
1.62	4.27	2.23	15.3	3.42	Al

$U = (1.9 + 1.02\,u)$ km s^{-1}; $\sigma_U = 0.14$ km s^{-1}
Experimental technique A
Data reduction technique B
Reference: [Bas63]
Technique used ferroelectric transducers to measure the arrival of shock waves at sample and driver plate surfaces.

Sand

Quartz (SiO_2) 100%
Particle size: less than 0.075 mm (80%) to 0.15 mm maximum, Porosity: 22%
$\varrho_0 = 2.079$

Sample

ϱ_0 g cm^{-3}	U km s^{-1}	u km s^{-1}	P GPa	ϱ g cm^{-3}	Standard Composition
2.02	3.45	1.07	7.5	2.93	Al
2.14	3.70	1.46	11.6	3.54	Al
2.03	4.78	2.03	19.7	3.53	Al

$U = (1.8 + 1.42\,u)$ km s^{-1}; $\sigma_U = 0.24$ km s^{-1}
Experimental technique A
Data reduction technique B
Reference: [Bas63]
Technique used ferroelectric transducers to measure the arrival of shock waves at sample and driver plate surfaces.

Sand
Quartz 97 wt%
Porosity: 40%
Grain size <1 mm
$\varrho_0 = 1.600$ g cm^{-3}; $\varrho_{01} = 2.650$ g cm^{-3}

Sample				
ϱ_0 g cm^{-3}	U km s^{-1}	u km s^{-1}	P GPa	ϱ g cm^{-3}
1.6	1.88	0.58	1.7	2.31
1.6	1.97	0.57	1.8	2.25
1.6	2.22	0.82	2.9	2.54
1.6	2.15	0.86	3.0	2.67
1.6	2.94	1.07	5.0	2.52
1.6	2.98	1.08	5.1	2.51
1.6	4.04	1.54	10.0	2.58
1.6	4.12	1.54	10.2	2.56
1.6	4.07	1.72	11.2	2.77
1.6	4.12	1.71	11.3	2.74
1.6	4.26	1.76	12.0	2.73
1.6	4.70	2.31	17.5	3.15
1.6	4.78	2.40	18.4	3.21
1.6	5.60	3.21	28.8	3.75
1.6	5.91	3.34	31.6	3.68
1.6	6.40	3.65	35.3	4.04
1.6	6.46	3.88	40.1	4.01

$U = (0.441 + 2.33\,u)$ km s^{-1},
$\sigma_U = 0.16$ km s^{-1}, for $u = 0.57 \cdots 1.54$ km s^{-1}
$U = (2.32 + 1.04\,u)$ km s^{-1};
$\sigma_U = 0.08$ km s^{-1}, for $u = 1.72 \cdots 3.88$ km s^{-1}
Experimental technique A
Data reduction technique B
Aluminium, iron, and brass were used as standards.
References: [Har65, Ski65]

Sand
Quartz 100 wt%
Porosity: 37 vol%
Particle size: 74 $\cdots$ 149 μ
$\varrho_0 = 1.650$ g cm^{-3}; $\varrho_{01} = 2.650$ g cm^{-3}

Sample					Standard
ϱ_0 g cm^{-3}	U km s^{-1}	u km s^{-1}	P GPa	ϱ g cm^{-3}	Composition
1.65	4.07	2.17	14.6	3.53	Al
1.65	4.05	2.25	15.0	3.71	Al
1.65	5.31	3.13	27.4	4.02	Al
1.65	6.39	3.81	40.4	4.10	Al

$U = (0.868 + 1.44\,u)$ km s^{-1}; $\sigma_U = 0.09$ km s^{-1}
Experimental technique D
Data reduction technique B
$T_0 = -10\,°C$
Reference: [And67]

Sandstone, Coconino, Arizona (Sandstein)
Composition (in vol%): Quartz 97; K-Feldspar 3; Clay and heavy minerals traces
Porosity (calculated) 25%
Grain size: 0.06 $\cdots$ 0.7 mm
$\varrho_0 = 1.980$ g cm^{-3}; $\varrho_{01} = 2.650$ g cm^{-3};
$c_0 = 1.43$ km s^{-1}

Sample				
ϱ_0 g cm^{-3}	U km s^{-1}	u km s^{-1}	P GPa	ϱ g cm^{-3}
1.98	3.67	1.33	9.7	3.11
1.98	4.10	1.63	13.2	3.29
1.98	4.49	2.18	19.4	3.85
1.98	4.84	2.57	25.6	4.23
1.98	5.66	3.12	34.9	4.41
1.98	5.79	3.25	37.3	4.52
1.98	7.57	4.30	64.4	4.58
1.98	7.79	4.43	68.4	4.59
1.98	8.82	5.07	88.6	4.66
1.98	10.09	5.94	118.6	4.82
1.98	11.20	6.43	142.6	4.65

$U = (0.36 + 1.67\,u)$ km s^{-1}
Error in U and u 0.2 $\cdots$ 2% and 0.2%, respectively
Experimental technique A
Data reduction technique A; copper and fansteel 77 alloy as standards.
Reference: [Jon68]

Sample description: Sandstone is weakly to moderately well cemented with silica, in the form of quartz overgrowths on the grains. Subparallel laminae 5.0 $\cdots$ 17.5 mm thick separated by thin laminae 0.5 mm thick containing more than average amounts of silt and clay sized grains.

◄ The sample was Ottawa Banding Sand, obtained from the Ottawa Silica Co., Ottawa, Illinois, U.S.A.

Sandstone, Coconino (Sandstein)
Composition (in wt%): Quartz 97; Microcline 3
Porosity: 24%
Grain size: 0.12···0.15 mm
$\varrho_0 = 1.978$ g cm^{-3}; $\varrho_{01} = 2.646$ g cm^{-3}

Sample					Standard		
ϱ_0 g cm^{-3}	U km s^{-1}	u km s^{-1}	P GPa	ϱ g cm^{-3}	Composition	U_{fs} km s^{-1}	No.
first wave:							
1.961	2.622	0.156	0.8	2.08	Al	0.836	7
1.961	2.853	0.074	0.41	2.01	Al	0.836	8
1.961	2.705	0.154	0.8	2.21	plexiglass	1.076	12
1.961	3.027	0.086	0.51	2.02	plexiglass	1.076	13
2.031	3.060	0.085	0.53	2.09			22
2.031	3.007	0.090	0.55	2.09			23
second wave							
1.961	4.126	1.74	14.4	3.46	Al	2.568	1
1.961	4.039	1.53	12.4	3.21	Al	2.206	2
1.961	3.285	1.26	8.3	3.24	Al	1.77	3
1.961	3.141	1.11	7.0	3.10	Al	1.54	4
1.961	3.321	1.09	7.3	2.97	plexiglass	2.46	5
1.961	3.126	1.02	6.2	2.91	plexiglass	2.25	6
1.961	2.305	0.317	1.8	2.49	Al	0.836	7
1.961	2.385	0.371	1.8	2.49	Al	0.836	8
1.961	4.386						9
1.961	4.481						10
1.961	4.600	1.98	17.3	3.45	Al		11
1.961	2.354	0.475	2.38	2.56	plexiglass	1.076	12
1.961	2.357	0.500	2.49	2.77	plexiglass	1.076	13
1.961	4.615						14
1.961	4.599						15
1.961	4.633	2.041	18.6	3.51	Al	2.872	16
2.000	2.793	0.875	4.89	2.91			17
1.970	3.11	1.09	6.68	3.03			18
1.975	3.062	1.128	6.82	3.13			19
1.960	4.26	1.75	14.6	3.33			20
2.000	4.29	1.995	17.15	3.74			21
2.000	2.483	0.796	3.95	2.94			24
2.000	2.002	0.266	1.06	2.31			25
2.000	1.601	0.066	0.21	2.09			26
high pressure data:							
1.978	4.86	2.15	20.7	3.54			27
1.978	5.19	2.75	28.3	4.21			28
1.978	5.71	2.82	31.7	3.91			29
1.978	5.97	2.99	35.2	3.96			30

$U = (1.40 + 2.65\, u)$ km s^{-1}; $\sigma_U = 0.13$ km s^{-1}, for $u < 0.4$ km s^{-1}
$U = (1.43 + 1.56\, u)$ km s^{-1}; $\sigma_U = 0.27$ km s^{-1}, for $u = 0.8 \cdots 2.0$ km s^{-1}

Experimental technique. C, except for Nos: 18, 20, 21, 22, and 23 used method D; in Nos: 24, 25, and 26 an X-ray method was used.
Data reduction technique B and D with $2u = U_{fs}$.
References: [Gre63a, Ahr64, Ahr66a]

Sandstone, Nugget sandstone, Parley's Canyon, Salt Lake City, Utah (Sandstein)

Composition: 99% Quartz

Porosity: 4…9%

$\varrho_0 = 2.42…2.58$ g cm^{-3}; $\varrho_{01} = 2.67$ g cm^{-3}

Sample					Standard	
ϱ_0 g cm^{-3}	U km s^{-1}	u km s^{-1}	P GPa	$\varrho^1)$ g cm^{-3}	Composition	U_{fs} km s^{-1}
2.55	3.86	0.0397	0.391	2.577	$^3)$	0.141
2.42	3.85	0.069	0.645	2.464	$^3)$	0.138
2.56	4.11	0.0611	0.65	2.599	$^3)$	0.227
2.55	4.10	0.0824	0.86	2.602	$^3)$	0.321
2.55	4.43	0.103	1.17	2.610	$^3)$	0.402
2.58	4.28	0.120	1.31	2.654	$^3)$	0.488
2.55	4.42	0.150	1.69	2.640	$^3)$	0.586
	4.19	0.255	2.58	0.0648 $^2)$	$^3)$	0.51
	4.39	0.39	4.26	0.097 $^2)$	$^3)$	0.78
2.50	4.30	0.397	4.27	2.753	$^3)$	0.795
2.50	4.30	0.470	5.05	2.800	$^3)$	0.94
2.51	4.48	0.615	7.02	2.902	$^3)$	1.23
	4.67	0.79	9.23	0.204 $^2)$	$^3)$	1.58

$^1)$ Calculated from $\mu = \dfrac{\varrho}{\varrho_0} - 1$.

$^2)$ Values are compression $= \dfrac{\varrho}{\varrho_0} - 1$.

$^3)$ Standards: Impactors consisted of polymethylmethacrylate.

Experimental technique K
Data reduction technique F
Reference: [Lar81]

Sand-Water mixture (50% saturated) (Sand, wasserhaltig)

Composition (in wt%): Quartz 90; H$_2$O 10

Porosity: 16%

Particle size: 74…149 μ

$\varrho_0 = 1.873$ g cm^{-3}

Sample					Standard
ϱ_0 g cm^{-3}	U km s^{-1}	u km s^{-1}	P GPa	ϱ g cm^{-3}	Composition
1.84	3.40	1.11	6.95	2.73	Al
1.84	4.66	1.98	17.0	3.20	Al
1.84	5.32	2.79	27.3	3.87	Al
1.84	5.73	3.08	32.5	3.98	Al
1.84	6.37	3.44	40.3	4.00	Al

$U = (2.11 + 1.2\, u)$ km s^{-1}; $\sigma_U = 0.16$ km s^{-1}

Experimental technique D
Data reduction technique B
$T_0 = -10$ °C
Reference: [And67]

Sand-Water mixture (20% saturated) (Sand, wasserhaltig)

Composition (in wt%): Quartz 96; H$_2$O 4

Porosity: 30%

Particle size: 74…149 μ

$\varrho_0 = 1.721$ g cm^{-3}

Sample					Standard
ϱ_0 g cm^{-3}	U km s^{-1}	u km s^{-1}	P GPa	ϱ g cm^{-3}	Composition
1.72	2.98	1.14	5.8	2.78	Al
1.72	4.34	2.05	15.3	3.26	Al
1.72	4.24	2.12	15.6	3.44	Al
1.72	5.39	3.04	28.2	3.94	Al
1.72	5.99	3.49	36.0	4.12	Al
1.72	6.43	3.74	41.3	4.11	Al

$U = (1.56 + 1.28\, u)$ km s^{-1}; $\sigma_U = 0.10$ km s^{-1}

Experimental technique D
Data reduction technique B
$T_0 = -10$ °C
Reference: [And67]

The sample was Ottawa Banding Sand, obtained from the Ottawa Silica Co., Ottawa, Illinois, U.S.A.

Sand-Water mixture (100% saturated) (Sand, wasserhaltig)

Composition (in wt%): Quartz 81; H$_2$O 19

Particle size: 74…149 μ

$\varrho_0 = 1.961$ g cm^{-3}

Sample					Standard
ϱ_0 g cm^{-3}	U km s^{-1}	u km s^{-1}	P GPa	ϱ g cm^{-3}	Composition
1.96	3.77	1.03	7.6	2.70	Al
1.96	4.00	1.01	8.0	2.62	Al
1.96	3.86	1.05	8.0	2.69	Al
1.96	5.10	2.01	20.1	3.23	Al
1.96	5.77	2.67	30.3	3.65	Al
1.96	5.63	2.71	30.0	3.78	Al
1.96	5.96	2.85	33.3	3.75	Al
1.96	6.74	3.24	42.9	3.78	Al
1.96	6.77	3.31	44.0	3.84	Al
1.96	7.04	3.52	48.7	3.92	Al

$U = (2.56 + 1.24\, u)$ km s^{-1}; $\sigma_U = 0.16$ km s^{-1}

Experimental technique D
Data reduction technique B
$T_0 = -10$ °C
Reference: [And67]

The sample was Ottawa Banding Sand, obtained from the Ottawa Silica Co., Ottawa, Illinois, U.S.A.

The sample was Ottawa Banding Sand, obtained from the Ottawa Silica Co., Ottawa, Illinois, U.S.A.

1.3.3.4 Glasses — Gläser

Shock wave data for glasses are given in the following section.

Stoßwellendaten für Gläser sind im folgenden Abschnitt angegeben.

For Pyrex, see next page.

Silica glass, synthetic (Quarzglas)
Composition: SiO_2
$\varrho_0 = 2.204$ g cm^{-3}; $\varrho_{01} = 2.649$ g cm^{-3}; $c_1 = 5.968$ km s^{-1}; $c_0 = 4.09$ km s^{-1}; $c_s = 3.764$ km s^{-1}

Sample					Standard			No.
ϱ_0 g cm^{-3}	U km s^{-1}	u km s^{-1}	P GPa	ϱ g cm^{-3}	Composition	u km s^{-1}	P GPa	
first wave:								
2.204	5.890	0.00	0.00	2.20				0
2.204	5.751	0.037	0.469	2.22				1
2.204	5.624	0.076	0.942	2.23				2
2.204	5.503	0.120	1.450	2.25				3
2.204	5.387	0.174	2.066	2.28				4
2.204	5.275	0.235	2.732	2.31				5
2.204	5.168	0.306	3.485	2.34				6
2.204	5.15	0.40	4.54	2.39	plexiglass	0.680	2.98	7
2.204	5.17	0.49	5.58	2.44	plexiglass	0.790	3.64	8
2.204	5.22	0.74	8.56	2.57	Al	0.644	11.2	9
2.204	5.25	0.81	9.43	2.61	Al	0.818	14.7	10
2.204	5.17	0.82	9.34	2.62	Al	0.818	14.7	11
2.204	5.20	0.83	9.57	2.63	Al	0.906	16.6	12
2.204	5.23	0.84	9.68	2.63	Al	1.176	22.7	13
2.204	5.20	0.85	9.80	2.64	Al	1.176	22.7	14
2.204	5.20	0.89	10.20	2.66	Al	1.580	32.9	15
2.204	5.23	0.86	9.91	2.64	Al	1.580	32.9	16
second wave:								
2.204	4.52	1.04	11.7	2.79	Al	0.906	16.6	12
2.204	4.67	1.40	15.3	3.07	Al	1.176	22.7	13
2.204	4.70	1.41	15.7	3.08	Al	1.176	22.7	14
2.204	4.97	1.90	21.1	3.53	Al	1.580	32.9	15
2.204	4.96	1.95	21.7	3.59	Al	1.580	32.9	16
2.204	5.62	2.76	34.2	4.33	Al	2.218	51.5	17
2.204	5.53	2.76	33.7	4.40	Al	2.218	51.5	18
2.204	5.62	2.78	34.6	4.30	Al	2.218	51.5	19
2.204	6.43	3.25	46.0	4.45	Al	2.66	66.0	20
2.204	6.44	3.33	48.2	4.55	Al	2.66	66.0	21
2.204	7.28	3.81	61.1	4.62	Al	3.18	85.0	22
2.204	7.30	3.87	62.3	4.69	Al	3.18	85.0	23

$U_1 = (5.76 - 2.14\, u)$ km s^{-1} for $u = 0.0 \cdots 0.4$ km s^{-1}
$U_1 = (5.07 + 0.183\, u)$ km s^{-1} for $u = 0.4 \cdots 0.9$ km s^{-1}
$U_2 = (4.03 + 0.477\, u)$ km s^{-1} for $u = 1.0 \cdots 1.9$ km s^{-1}
$U_2 = (1.3 + 1.56\, u)$ km s^{-1} for $u = 2.5 \cdots 3.8$ km s^{-1}

Experimental technique E for entries No. 0···17 and B for the rest.
Data reduction technique D used for the first wave, assuming $2u = U_{fs}$; for the second wave technique B was used.
Reference: [Wac62]

Pyrex, silicate-borate-glass
(Pyrex, Silikat-Borat-Glas)
Composition (in wt%): SiO_2 80.4···81.4; B_2O_3 13.0···10.5; Al_2O_3 3.5···1.5; CaO 0.13···0.7; MgO 0.06···0.6; Na_2O 3.2···5.1; K_2O 0.2···1.8; As_2O_3 0.5···0.75
$\varrho_0 = 2.232$ gcm^{-3};
$c_l = 5.55$ km s^{-1}; $c_0 = 3.88$ km s^{-1}; $c_s = 3.45$ km s^{-1}

Sample					Standard	
ϱ_0 g cm^{-3}	U km s^{-1}	u km s^{-1}	P GPa	ϱ g cm^{-3}	Composition	U km s^{-1}
2.230	4.93	0.56	6.2	2.52	Al	5.96
2.230	4.91	0.78	8.5	2.65	Al	6.20
2.230	4.82	1.11	11.9	2.90	Al	6.53
2.230	4.79	1.44	15.4	3.19	Al	6.87
2.230	5.11	2.10	23.9	3.79	Al	7.56
2.230	5.18	2.19	25.3	3.86	Al	7.65
2.230	5.35	2.40	28.6	4.04	Al	7.88
2.230	5.77	2.65	34.1	4.12	Al	8.19
2.230	6.30	3.06	43.0	4.34	Al	8.67
2.230	7.03	3.35	52.5	4.26	Al	9.08
2.230	7.04	3.49	54.8	4.42	Al	9.22
2.230	7.41	3.69	61.0	4.44	Al	9.48
2.230	8.02	3.99	71.4	4.44	Al	9.88

$U = (1.353 + 1.654\, u)$ km s^{-1} for $u > 2.3$ km s^{-1};
$\sigma_U = 0.1$ km s^{-1}

Experimental technique B
Data reduction technique B
Reference: [McQ70]

Silica glass, synthetic (Quarzglas)
Composition: SiO_2
$\varrho_0 = 2.20$ gcm^{-3}

Sample				Standard	
U km s^{-1}	u km s^{-1}	P GPa	ϱ g cm^{-3}	Composition	U_{fs} km s^{-1}
6.62	3.61	52.5	4.83	Ta	4.34
6.75	3.65	54.2	4.79	Ta	4.40
7.71	4.24	71.9	4.89	Ta	5.19
8.36	4.73	87.0	5.06	Ta	5.83
8.68	4.70	89.8	4.80	Ta	5.83
8.86	4.84	94.3	4.84	Ta	6.01

Errors for all entries are given in the original reference.
Experimental technique C
Data reduction technique B
Reference: [Jac79]

1.3.4 References for 1.3.1···1.3.3 — Literatur zu 1.3.1···1.3.3

Ahr64	Ahrens, T.J., Gregson jr., V.G.: J. Geophys. Res. **69** (1964) 4839.
Ahr66	Ahrens, T.J.: J. Appl. Phys. **37** (1966) 2532.
Ahr66a	Ahrens, T.J., Rosenberg, J.T., Ruderman, M.H.: Stanford Res. Inst. Rep. No. DASA 1868 (1966).
Ahr66b	Ahrens, T.J., Duvall, G.E.: J. Geophys. Res. **71** (1966) 4349.
Ahr68	Ahrens, T.J., Gust, W.H., Royce, E.B.: J. Appl. Phys. **39** (1968) 4610.
Ahr68a	Ahrens, T.J., Rosenberg, J.T., in: French, B.M., Short, N.M. (Eds.), Shock Metamorphism of Natural Materials, Baltimore: Mono Book Corp., p. 59 (1968).
Ahr69	Ahrens, T.J., Petersen, C.F., Rosenberg, J.T.: J. Geophys. Res. **74** (1969) 2727.
Ahr71	Ahrens, T.J., Gaffney, E.S.: J. Geophys. Res. **76** (1971) 5504.
Ahr71a	Ahrens, T.J., Lower, J.H., Lagus, P.L.: J. Geophys. Res. **76** (1971) 518.
Ahr73	Ahrens, T.J., Liu, H.-P.: J. Geophys. Res. **78** (1973) 1274.
Ahr73a	Ahrens, T.J., O'Keefe, J.D., Gibbons, R.V.: Proc. Lunar Sci. Conf. 4th, **1973**, New York: Pergamon Press, p. 2575.
Ahr74	Ahrens, T.J., Cole, D.M.: Proc. Lunar Sci. Conf. 5th, **1974**, New York: Pergamon Press, p. 2333.
Ahr77	Ahrens, T.J., Jackson, I., Jeanloz, R.: Proc. Lunar Sci. Conf. 8th, **1977**, New York: Pergamon Press, p. 3437.
Ahr79	Ahrens, T.J.: J. Geophys. Res. **84** (1979) 985.
Ahr80	Ahrens, T.J., Watt, J.P.: Proc. Lunar Planet. Sci. Conf. 11th, **1980**, Pergamon Press, New York, 2059.

Alt63	Al'tshuler, L.V., Pavlovskii, M.M., Kuleshova, I.V., Simakov, G.V.: Soviet Phys.-Solid State (English Transl.) **5** (1963) 203.
Alt65	Al'tshuler, L.V., Trunin, R.F., Simakov, G.V.: Izv. Akad. Nauk SSSR, Fiz. Zemli **10** (1965) 1.
And67	Anderson, G.D.: Interim Data Report FGU-6392 (1967) Stanford Research Inst., Menlo Park, California, U.S.A.
Bar67	Barnes, H.: Tech. Lett. NTS-185, June 21 (1967).
Bas63	Bass, R.C., Hawk, H.L., Chabai, A.J.: Report Sc-4907 RR (1963), Sandia Corporation, Albuquerque, N.M., U.S.A.
Bir60	Birch, F.: J. Geophys. Res. **65** (1960) 1083.
Bor56	Borg, I.: Bull. Geol. Soc. Am. **67** (1956) 1563.
Cab70	Cable, A.J., in: Kinslow, R. (Ed.), High-Velocity Impact Phenomena, New York: Academic Press, p. 1–21 (1970).
Cou48	Courant, R., Friedrichs, K.O.: Supersonic flow and shock waves, New York: Interscience Inc., (1948).
Dap57	Dapoigny, J., Kieffer, J., Vodar, B.: Acad. Sci. Paris, **245** (1957) 1502.
Den69/70	Dennen, R.S.: J. Appl. Phys. **40** (1969) 3326; J. Appl. Phys. **41** (1970) 5309.
Dix75	Dixon, J.R. and Papike, J.J.: Proc. Lunar Sci. Conf. 6th, **1975**, New York: Pergamon Press, p. 263.
Dre59	Dremin, A.N., Adadurov, G.A.: Dokl. Akad. Nauk SSSR **128** (1959) 261; Soviet Phys.-Doklady (English Transl.) **4** (1959) 970.
Duv63	Duvall, G.E., Fowles, G.R., in: Bradley R.S. (Ed.), High Pressure Physics and Chemistry, Vol. **2**, New York: Academic Press, p. 209 (1963).
Dym75	Dymek, R.F., Albee, A.L., Chodos, A.A.: Proc. Lunar Sci. Conf. 6th, **1975**, New York: Pergamon Press, p. 49.
Gra73	Graham, E.H., Ahrens, T.J.: J. Geophys. Res. **78** (1973) 375.
Gra76	Grady, D.E., Murri, W.J., Mahrer, K.D.: J. Geophys. Res. **81** (1976) 889.
Gra77	Grady, D.E., in: Manghnani, M.H. and Akimoto, S.-I. (Eds.) High Pressure Research, Applications in Geophysics, New York: Academic Press, p. 389 (1977).
Gra78	Graham, E.R., Bonzcar, L.J.: EOS Trans. Amer. Geophys. Union **59** (1978) 373.
Gre62	Gregson, V.G., Peterson, C.F., Jamieson, J.C.: Rep. SRI-PGU-3630-STR-3 (1962), Poulter Laboratories, Menlo Park, Calif., U.S.A.
Gre63	Gregson, V.G., Peterson, C.F., Jamieson, J.C.: Rep. AFCRL 63–662 (1963).
Gre63a	Gregson, V.G., Ahrens, T.J., Peterson, C.F.: Rep. AFCRL 63–662 (1963).
Har65	Hart and Skidmore, I.C.: Private Communication (1965), to Van Thiel, M. (1977).
Hau71	Hauver, G.E., Melani, A.: Ballistic Res. Lab. Rept. BRL MR 2061 (1971).
Hor65	Hord, B.L., van Thiel, M.: Lawrence Radiation Lab. equation of state file.
Hou59	Houser, F.N., Poole, F.G.: U.S. Geol. Survey, Rept. TEM-836A (1959).
Isb65	Isbell, W.M.: Contract DA-49-146-XZ-429 (1965), General Motors Technical Center, Warren, Mich., U.S.A.
Isb66	Isbell, W.M., Shipman, F.H., Jones, A.H.: Progress Rep. 5, Contract DA-49-146-X2-429 (1966), General Motors Technical Center, Warren, Mich., U.S.A.
Jac79	Jackson, I., Ahrens, T.J.: J. Geophys. Res. **84** (1979) 3039.
Jea78	Jeanloz, R., Ahrens, T.J.: Proc. Lunar Planet. Sci. Conf. 9th, **1978**, New York: Pergamon Press, p. 2789.
Jea80	Jeanloz, R., Ahrens, T.J.: Geophys. J. Royal Atron. Soc. **62** (1980) 505.
Jea80a	Jeanloz, R., Ahrens, T.J.: Geophys. J. Royal Astron. Soc. **62** (1980) 529.
Jon66	Jones, A.H., Isbell, W.M., Maiden, C.J.: J. Appl. Phys. **37** (1966) 3493.
Jon68	Jones, A.H., Isbell, W.M., Shipman, F.H., Perkins, R.D., Green, S.J., Maiden, C.J.: Interim Rep., Contract NAS2-3427 (1968), General Motors Technical Center, Warren, Mich., U.S.A.
Kee71	Keeler, R.N., Royce, E.G., in: Caldirola, P., and Knoepfel, H. (Eds.), Physics of High Energy Density, New York: Academic Press, p. 51–150 (1971).
Kin73	King, D.A., Ahrens, T.J.: Nature (London) Phys. Sci. **243** (1973) 82.
Kin76	King, D.A., Ahrens, T.J.: J. Geophys. Res. **81** (1976) 931.
Kor64	Kormer, S.B., Sinitsyn, M.V., Funtikov, A.I., Urlin, V.D., Blinov, A.V.: Zh. Eksp. Teor. Fiz. **47** (1964) 1202; Sov. Phys. JETP (English Transl.) **20** (1965) 811.
Kor65	Kormer, S.B., Sinitsyn, M.V., Krilov, G.A., Urlin, V.D.: Zh. Eksp. Teor. Fiz. **48** (1965) 1033.
Lar80	Larson, D.B.: J. Geophys. Res. **85** (1980) 293.
Lar81	Larson, D.B., Anderson, G.D.: J. Geophys. Res., in press (1981).

Mar80	Marsh, S. P. (ed.): LASL Shock Hugoniot Data, University of California Press, Berkeley (1980).
McQ60	McQueen, R.G., Marsh, S.P.: J. Appl. Phys. **31** (1960) 1253.
McQ63	McQueen, R.G., in: Gschneider, K.A., Hepworth, M.T., and Parlee, A.D. (Eds.), Metallurgy at High Pressures and High Temperatures, New York: Gordon and Breach Sci. Publ., p. 44–133 (1963).
McQ63a	McQueen, R.G., Fritz, J.N., Marsh, S.P.: J. Geophys. Res. **68** (1963) 2319.
McQ66	McQueen, R.G., Marsh, S.P., in: Clark, S.P. (Ed.), Handbook of Physical Constants, Geol. Soc. Am. Mem. **97** (1966).
McQ67	McQueen, R.G., Jamieson, J.C., Marsh, S.P.: Science **155** (1967) 1401.
McQ68	McQueen, R.G., Marsh, S.P.: Symposium on the "Behavior of Dense Media under High Pressure", I.U.T.A.M. Paris, in: International Computation Center (ed.), High Dynamic Pressures, New York: Gordon and Breach, p. 207–216 (1968).
McQ70	McQueen, R.G., Marsh, S.P., Taylor, J.W., Fritz, J.M., Carter, W.J., in: Kinslow, R. (Ed.), High velocity impact phenomena, New York: Academic Press, p. 294 (1970).
Mur74	Murri, W.J., Curran, D.R., Petersen, C.F., Crewdson, R.C., in: Wentorf, R.H. (Ed.), Advances in High-Pressure Research, Vol. **4**, London and New York: Academic Press, p. 1–163 (1974).
Pap76	Papike, J.J., Hodges, F.N., Bence, A.E., Cameron, M., Rhodes, J.M.: Rev. Geophys. Space Phys. **14** (1976) 475.
Pet68	Petersen, C.F., Murri, W.J., Anderson, G.D., Allen, C.F.: S.R.I. Interim Tech. Report, Project PGU-6618 (1968).
Ric58	Rice, H.H., McQueeen, R.G., Walsh, J.M., in: Seitz, F. and Turnbull, D. (Eds.), Solid State Physics, Vol. **6**, New York and London: Academic Press, p. 1–63 (1958).
Sch50	Schall, R.: Z. Angew. Phys. **2** (1950) 252.
Sho66	Short, N.M.: J. Geophys. Res. **71** (1966) 1195.
Ski65	Skidmore I.C.: Appl. Mater. Res. **4** (1965) 131.
Som80	Somerville, M., Ahrens, T.: J. Geophys. Res. **85** (1980) 7016.
Son72	Son, P.R., Bartels, R.A.: J. Phys. Chem. Solids **33** (1972) 819.
Stö72	Stöffler, D.: Fortschr. Mineral. **49** (1972) 50.
Tru65	Trunin, R.F., Gon'shakova, V.I., Simakov, G.V., Galdin, N.E.: Izv. Akad. Nauk SSSR, Fiz. Zemli **9** (1965) 1.
Van64	Van Thiel, M.: S Division Rep. STN 71 (1964) Lawrence Radiation Laboratory, Livermore, Calif., U.S.A.
Van77	Van Thiel, M. (Ed.), Compendium of shock wave data, Vol. 1–3, Report No. UCRL-50108, National Techn. Inform. Service, N.B.S., U.S. Dept. of Comm., Springfield, Va., USA (1977).
Wac62	Wackerle, J.: J. Appl. Phys. **33** (1962) 922.
Wal57	Walsh, J.M., Rice, M.H.: J. Chem. Phys. **26** (1957) 815.
Wie60	Wiederhorn, S., Drickamer, H.G.: J. Appl. Phys. **31** (1960) 1665.
Wil71	Willis, J.P., Ahrens, L.H., Danchin, R.V., Erlank, A.J., Gurney, J.J., Hofmeyer, P.K., McCarthy, T.S., Orren, M.J.: Proc. Lunar Sci. Conf. 2nd, **1971**, MIT Press, Cambridge, Mass., 1123.

2 Porosity and permeability — Porosität und Permeabilität
2.1 Porosity of rocks — Porosität von Gesteinen
2.1.0 Introduction — Einleitung

Porosity is one of the most important integral material properties of many rocks, and there is hardly a physical property of rock that is not influenced – directly or indirectly – by porosity. It is especially characteristic for sedimentary rocks, but recently there is increasing evidence that traces of porosity affect properties of metamorphic and igneous rocks, too.

Die Porosität ist eine der wichtigsten integralen Materialeigenschaften vieler Gesteine, und es gibt kaum eine physikalische Gesteinseigenschaft die nicht – direkt oder indirekt – von der Porosität beeinflußt wird. Besonders charakteristisch ist die Porosität für Sedimentgesteine, aber in neuester Zeit zeigen sich auch immer mehr Hinweise auf wichtige Effekte von Porositätsspuren auf die Eigenschaften metamorpher und magmatischer Gesteine.

2.1.1 Definitions — Definitionen

The porosity ϕ is defined as the ratio of the void space V_{por} inside the rock - not filled by solid rock material, but generally by liquids or gases - to the total ('bulk') volume V_{tot} of the rock

Die Porosität ϕ ist definiert als Verhältnis des Porenvolumens V_{por} innerhalb des Gesteins, das nicht von festem Gesteinsmaterial, sondern im allgemeinen von Flüssigkeiten oder Gasen erfüllt ist, zu dem Totalvolumen V_{tot} des Gesteins

$$\phi = \frac{V_{por}}{V_{tot}}$$

(The symbol ϕ is widely used for porosity, but other symbols in use are P, ε, n, f).

(Das Symbol ϕ ist für die Porosität weit verbreitet, andere gebräuchliche Symbole sind P, ε, n, f).

The complement of the pore space, called 'matrix', 'framework' or 'solids volume' is given by

Das Komplement des Porenraumes, Matrix, Gesteinsgerüst oder Feststoffvolumen genannt, wird gegeben durch

$$V_{mat} = V_{tot} - V_{por}$$

or the ratio

oder relativ ausgedrückt

$$\frac{V_{mat}}{V_{tot}} = 1 - \phi$$

Sometimes the quantity 'pore ratio' or 'relative pore volume' is used instead of porosity:

Gelegentlich in Gebrauch an Stelle der Porosität ist das sogenannte „Porenverhältnis" oder „relative Porenvolumen", das definiert ist durch

$$\frac{V_{por}}{V_{mat}} = \frac{\phi}{1-\phi}$$

It should be pointed out here that the term 'matrix' does not have an unambiguous meaning in geosciences: There is the global meaning identifying 'matrix' as the whole solid material (in aggreement with the definition of V_{mat} above). This is generally used in petrophysics, geophysics, well logging etc. On the other hand, mineralogists and petrologists, usually devide the solid rock material - especially of clastic sediments - into 'framework' (or 'grain structure') and 'matrix'. According to this latter definition the term matrix solely encompasses fine materials (e.g. silt or clay) adjoined to the framework proper (see [Sel76]). Attention is called to this ambiguous terminology.

Es soll hier darauf hingewiesen werden, daß der Gebrauch des Wortes „Matrix" in den Geowissenschaften nicht ganz einheitlich ist. Die globale Bedeutung im Sinne der obigen Definition ist vor allem in der Geophysik, Gesteinsphysik und auf dem Gebiet der Bohrlochmessungen gültig. Dagegen unterteilen Mineralogen und Petrologen üblicherweise das feste Gesteinsmaterial - insbesondere von klastischen Sedimenten - in „Korngerüst" und „Matrix"; hierbei umfaßt der Begriff „Matrix" nur das Feinmaterial (z.B. Schluff und Ton), das das eigentliche Gesteinsgerüst umgibt (siehe [Sel76]). Auf diese Uneinheitlichkeit der Terminologie sei der Leser aufmerksam gemacht.

However, in the following the suffix mat always refers to the complete solid material of the rock.

Jedoch soll sich hier im folgenden der Index mat stets auf das gesamte Festmaterial des Gesteins beziehen.

The pore space can consist completely of interconnected pores, or partially or totally of isolated pores completely enclosed by solid material. 'Total porosity' is defined to encompass both interconnected and isolated pores. 'Effective porosity' is to encompass the interconnected part only. However the meaning of this latter term is often somewhat vague in practice, since interconnection can be gradual and furthermore the interconnected pore space might contain dead end pores too. Thus in oil and gas technology, the term 'effective porosity' sometimes refers solely to the well porous, well interconnected and transconnective portion of the total volume, from which fluids can be recovered economically. A more careful naming in this sense is 'productive porosity'.

2.1.2 Origin and different types of porosity — Entstehungsursache und unterschiedliche Art der Porosität

Porosity can originate during clastic sedimentation or organogenesis ('primary porosity'), or at a later stage of the geological development of rocks ('secondary porosity').

Porosity can exist as

a) 'intergranular porosity' (primary), i.e. void space between grains, particles or fragments of clastic materials, loosely packed, compacted or even cemented;

b) 'intercrystalline' or 'intragranular porosity' (mostly secondary) inside of grains etc.

c) 'fissure' or 'fracture porosity' (secondary) caused by mainly mechanical, partly chemical action on primarily massive rock (e.g. limestone)

d) 'vugular porosity' (primary or secondary) caused by organisms during genesis or by chemical action at a later stage. (A super-macroscopical extreme example for the latter case are limestone caves).

2.1.2.1 Intergranular porosity — Intergranulare Porosität

Intergranular porosity formed during clastic sedimentation depends on grain size, sorting of grains and grain sphericity and roundness [Eng60]. Although, for a given packing, porosity should not depend on grain size theoretically, a remarkable effect can be observed especially at small grain sizes (Figs. 1, 2). This seems to be due to the fact that the primary packing is not indepedent of the grain size, but – due to the increasing ratio of friction forces to gravity forces with decreasing grain size – is less tight for smaller grains [Eng60].

Sorting generally increases porosity. The highest porosities can be expected from monodisperse packings. By mixing, the porosity is decreased (Fig. 3). Von Engelhardt [Eng60] describes the effect of mixing of two monodisperse sands on the porosity by

$$\phi = m\phi_1 + (1-m)\phi_2 - \alpha m(1-m)$$

where m and $1-m$ are the proportions, and ϕ_1 and ϕ_2 the individual porosites of the grain size components one and two. α is a mixing coefficient which increases with the size ratio. These findings can be extrapolated properly to polydisperse sands (see [Eng60]). Qualitatively, the effect of the shape of grain size distributions on porosity is given by Fig. 4.

The influence of grain shape on porosity can be seen in Table 1. The lowest porosities are obtained with balls and spherical, well rounded sand grains. Angular particles result in higher porosities. The highest porosities result from highly anisometric particles as e.g. flat mica particles.

The effect of mixing different grain shapes is illustrated in Fig. 5. Von Engelhardt presents the mixing rule

$$\phi = m\phi_1 + (1-m)\phi_2 + \beta m(1-m)$$

for a binary mixture, with the mixing coefficient β increasing with the difference in shape. Generally, by mixing particles of comparable size but different shape, porosity is increased.

Intergranular porosity usually decreases during diagenesis due to physical and chemical actions. For clastic sediments, the following stages can be discerned:

a) **packing**, whereby the loosely sedimented particles slide into more stable positions under the pressure of increasing overburden, forming tighter packings of less porosity and more grain-to-grain contacts.

b) **compaction**, further lowering porosity by (mainly mechanical but partially chemical) deformation of grains under still increasing overburden pressure, whereby the point contacts between grains gradually change to flat contacts and finally to concave-convex or sutured contacts (Table 2, Fig. 6).

c) **cementation**, deposition of dissolved material (own or foreign) on free grain surfaces, especially around contacts, thereby still further reducing pore space and also blocking narrow passages.

Besserer Sortierungsgrad steigert im allgemeinen die Porosität. Die größten Porositäten lassen sich für monodisperse Packungen erwarten. Durch Mischung wird die Porosität verringert (Fig. 3). Von Engelhardt [Eng60] beschreibt die Wirkung der Mischung zweier monodisperser Sande durch

worin m und $1-m$ die Anteile und ϕ_1 und ϕ_2 die individuellen Porositäten der Fraktionen eins und zwei sind. α ist ein Mischungskoeffizient, der mit dem Korngrößenverhältnis wächst. Diese Befunde lassen sich sinngemäß auch auf polydisperse Sande erweitern (siehe [Eng60]). Der Einfluß der Form der Korngrößenverteilungskurve auf die Porosität ist qualitativ in Fig. 4 dargestellt.

Der Einfluß der Kornform auf die Porosität läßt sich aus Tab. 1 ersehen. Die kleinsten Porositäten erhält man mit Kugeln und sphärischen, gut gerundeten Körnern. Angulare Partikel ergeben größere Porositäten. Die größten Porositäten werden von stark anisometrischen Teilchen, z.B. flachen Glimmerplättchen, verursacht.

Die Wirkung der Mischung verschiedener Kornformen ist in Fig. 5 demonstriert. Von Engelhardt [Eng60] gibt folgende Mischungsregel für eine binäre Mischung an:

Der Mischungskoeffizient β wächst mit dem Formunterschied. Allgemein gilt, daß durch Mischung von Teilchen vergleichbarer Größe aber unterschiedlicher Form die Porosität vergrößert wird.

Im Zuge der Diagenese verringert sich in der Regel die intergranulare Porosität als Folge physikalischer und chemischer Prozesse. Für klastische Sedimente lassen sich die folgenden Stufen unterscheiden:

a) **Packung**, wobei die lockeren Sedimentpartikel unter dem wachsenden Hangenddruck in stabilere Lagen gleiten und dabei dichtere Packungen mit geringerer Porosität und mehr Korn/Korn-Kontakten bilden.

b) **Kompaktion**, wobei sich die Porosität unter dem weiter anwachsenden Hangenddruck (infolge hauptsächlich mechanischer, aber teilweise auch chemischer Deformierung der Körner) weiter verringert und die ursprünglich punktförmigen Kornkontakte mehr und mehr in ebene und schließlich konkav-konvexe und suturierte Kontakte übergehen (Tab. 2, Fig. 6).

c) **Zementation** durch Ausfällen gelösten Materials (eigener oder fremder Art) und Ablagerung desselben an den freien Kornoberflächen, besonders in der Umgebung von Kornkontakten, wobei der Porenraum weiter verkleinert wird und enge Passagen sogar zugesetzt werden können.

Packing and cementation are the main actions reducing porosity of sands and sandstones, while compaction is the most effective process in clays and shales (Fig. 7). While packing and compaction are mainly mechanical processes under the main influence of lithostatic pressure, cementation is a complex joint effect of lithostatic matrix pressure, hydrostatic pore pressure, temperature, pH-index, dissolved mineral concentration, liquid fluctuation etc. [Sel76, Eng60, Max64, Eng73]. Pressure solution of matrix material at the grain contacts and redeposition at the free grain surfaces has been thought an important process in cementation, however recent investigations by means of cathodoluminescence seem to disprove this assumption [Sel76].

Packing and compaction of grains decreases with decreasing sphericity, roundness, and grain size, for a given pressure or burial depth (Fig. 8).

2.1.2.2 Intragranular porosity — Intragranulare Porosität

Intragranular or intercrystalline porosity can develop during changes in the mineralization by shrinking of crystals. Occasionally, it can eventually develop into a form of microfissure porosity. For example, the porosity of dolomites is generally caused by crystal shrinking during dolomitization of originally massive limestone.

2.1.2.3 Fissure and fracture porosity — Riß- und Kluftporosität

Fissures can develop during shrinking by remineralization, due to temperature changes or by anisotropic pressure release. Effects of dilatation and contraction due to temperature changes might be enhanced by adsorption/desorption effects and capillary forces due to evaporation and condensation of liquids in the fissures [Eng73]. Fissures can widen to fractures due to further mechanical action or chemical dissolution.

2.1.2.4 Vugular porosity — Kavernöse Porosität

Vugular primary porosity can exist in organic sediments (e.g. coral reefs) due to original occupation of space by organic matter and its consecutive deterioration. Vugular porosity can also develop secondarily from fissure or fracture porosity by intense chemical dissolution.

2.1.3 Direct effect of temperature on porosity — Direkter Einfluß der Temperatur auf die Porosität

Little is known about immediate changes of existing porosity with temperature. Theoretically, either an increase, no change or a decrease of porosity with temperature expansion of the matrix is possible (i.e. an over-proportional, a proportional, or an under-proportional expansion of pore space with matrix). However, temperature coefficients of expansion are rather small for all minerals forming sedimentary rocks. Thus, if at all existing, porosity variations with temperature should be expected to be very small, and changes of porosity with depth due to temperature expansion will be negligible against those due to pressure.

However, there is an effect of temperature on cementation, of course, as discussed before.

Wenig ist bisher bekannt über unmittelbare Änderungen vorhandener Porosität mit der Temperatur. Theoretisch ist entweder eine Zunahme, keine Änderung, oder sogar eine Abnahme der Porosität bei Temperaturexpansion der Matrix vorstellbar (d.h. eine überproportionale, proportionale oder unterproportionale Ausdehnung des Porenraums mit der Matrix). Jedoch sind die thermischen Ausdehnungskoeffizienten für alle sedimentbildenden Minerale ziemlich klein, so daß Änderungen der Porosität mit der Temperatur, wenn überhaupt vorhanden, sicher sehr klein sind. Porositätsänderungen mit der Teufe, verursacht durch Temperatur sind vermutlich völlig vernachlässigbar gegenüber solchen, verursacht durch Druck.

Allerdings hat natürlich die Temperatur einen Einfluß auf die Zementation, wie weiter oben bereits diskutiert wurde.

2.1.4 Effect of pressure on porosity — Einfluß des Druckes auf die Porosität

Intergranular porosity has been found to obey an exponential pressure law

Intergranulare Porosität gehorcht, wie sich gezeigt hat, einem exponentiellen Druckgesetz

$$\phi = \phi_0 e^{-p/k}$$

with ϕ_0 the initial porosity at $p=0$ and the parameter k being constant over a wide pressure range [Wag71, Nag65, Nag65a].

The effective pressure causing a compression of the pore space is

mit der Anfangsporosität ϕ_0 bei $p=0$ und einem über einen weiten Druckbereich konstanten Parameter k [Wag71, Nag65, Nag65a].

Der effektive Druck, der eine Kompression des Porenraumes bewirkt (Fig. 9, Tab. 3), ist

$$p = p_{\text{lith}} - v p_{\text{hyd}}$$

with p_{lith} the lithostatic pressure, p_{hyd} the hydrostatic pore pressure, and $v \approx 0.85$ a parameter depending on the compressibility of the solid material (Fig. 5, Table 3).

With

mit dem lithostatischen Druck p_{lith}, dem hydrostatischen Porendruck p_{hyd} und einem Parameter $v \approx 0.85$, der von der Kompressibilität des festen Gesteinmaterials abhängt.

Mit

$$p_{\text{lith}} = \varrho_{\text{mat}} g z \quad \text{and/or} \quad p_{\text{hyd}} = \varrho_{\text{por}} g z$$

ϱ_{mat} being the (average) matrix density and ϱ_{por} the (average) pore fluid density, an exponential law can then be expected to be valid for the dependence of porosity on depth of burial, z, too (Figs. 10, 11).

Although at first sight, such a quasi-elastic exponential law appears reasonable for compaction by reversible framework deformations only, it can even describe plastic compaction, porosity reduction by pressure solution, cementation etc., after a given time, if the rate of change of porosity is considered to be negatively proportional to the porosity itself, and if the factor p of the exponent $-p/k$ exponent p is considered the time integral mean of the pressure. This reasonable assumption is well confirmed by experimental data (Figs. 10···17).

wobei ϱ_{mat} die (durchschnittliche) Matrixdichte und ϱ_{por} die (durchschnittliche) Dichte der Porenfüllung ist, läßt sich auch ein Exponentialgesetz für die Abhängigkeit der Porosität von der Versenkungstiefe z erwarten (Fig. 10, 11).

Obwohl auf den ersten Blick ein solches quasi-elastisches Exponentialgesetz nur für Kompaktion infolge reversibler Korngerüstdeformation sinnvoll erscheint, lassen sich damit auch plastische Kompaktion, Porositätsverringerung durch Drucklösung, Zementation usw. - nach einer gegebenen Zeit - beschreiben, wenn man die zeitliche Änderungsrate der Porosität als negativ proportional der Porosität selbst annimmt und unter dem Faktor p im Exponenten $-p/k$ das zeitliche Integralmittel des Druckes versteht. Diese berechtigte Annahme ist recht gut durch experimentelle Daten bestätigt (Fig. 10···17).

Some authors [Eng60, Eng73, Mat74, Jon75, Füc70] prefer to use a purely empirical relation of the type

$$\phi = A - B \ln z \quad \text{or} \quad \phi = A - B \ln p$$

or of a similar type

$$\frac{\phi}{1-\phi} = A - B \ln z \quad \text{or} \quad \frac{\phi}{1-\phi} = A - B \ln p$$

known as Terzaghi's empirical law.

Although these equations fit the experimental data, too (Figs. 18, 19), there seems to be little theoretical justification for any such law.

In general, the consolidation and porosity reduction depends much on the rate and the type of the acting process. In some cases, quite high porosities can even be found in quite great depths (Table 4, 5). More data are given in Figs. 20···29 and Tables 6···10. Argillaceous rocks are treated in Figs. 21···29. Data on recent marine sediments are shown in Fig. 29 and Tables 7···10. In Table 11, a compilation of more than 900 items of porosity and density data from American, British, German, and Swiss literature are presented [Man63].

Einige Autoren [Eng60, Eng73, Mat74, Jon75, Füc70] bevorzugen eine rein empirische Beziehung der Form

oder eine ähnliche Form von Gleichung

die auch als Terzaghisches Gesetz bekannt ist.

Obwohl die experimentellen Daten diese Gleichung ebenfalls gut erfüllen (Fig. 18, 19), scheint kaum eine theoretische Begründung für Formeln dieser Art vorzuliegen.

Abschließend sei ganz allgemein vermerkt, daß jede Form der Verfestigung und Porositätsverringerung stark von Art und Verlauf des verursachenden Vorganges abhängt. Gelegentlich lassen sich ziemlich große Porositäten auch noch in großen Teufen finden (Tab. 4, 5). Weitere Daten enthalten Fig. 20···29 und Tab. 6···10. Tonige Gesteine sind in Fig. 21···29 behandelt. Daten von rezenten marinen Sedimenten sind in Fig. 29 und Tab. 7···10 aufgeführt. Tab. 11 gibt eine Zusammenstellung von mehr als 900 Porositäts- und Dichteangaben aus amerikanischen, britischen, deutschen und schweizerischen Quellen [Man63].

2.1.5 Relation to density — Verknüpfung mit der Dichte

The bulk density ϱ_{tot} of porous rocks depends on the matrix density ϱ_{mat}, the density ϱ_{por} of the fluid filling the pore space, and the porosity ϕ.

The following equations describe this interrelation

Die Gesteinsdichte (Rohdichte) ϱ_{tot} eines porösen Gesteins hängt ab von der Matrixdichte (Reindichte) ϱ_{mat}, der Dichte ϱ_{por} der Flüssigkeit oder des Gases, das den Porenraum füllt, und der Porosität ϕ.

Die folgenden Gleichungen beschreiben diesen Zusammenhang:

$$\varrho_{tot} = (1-\phi)\varrho_{mat} + \phi \varrho_{por} = \varrho_{mat} - \phi(\varrho_{mat} - \varrho_{por}) \qquad \phi = \frac{\varrho_{mat} - \varrho_{tot}}{\varrho_{mat} - \varrho_{por}}$$

(see also Chapter 1.2. Density of Rocks).

For liquids (fresh water, saltwater, oil), lacking accurate data, generally

(vgl. Kapitel 1.2 Dichte von Gesteinen).

Für Flüssigkeiten (Süßwasser, Salzwasser, Öl) wird in der Geohydrologie und der Erdöl-Erdgas-Technologie in Ermangelung genauerer Daten meist

$$\varrho_{por} \approx 1000 \text{ kg m}^{-3}$$

is assumed in the practice of geohydrology and oil/gas technology. For gases,

gesetzt. Für Gase läßt sich

$$\varrho_{por} \approx 0$$

can be applied.

In Figs. 12···17, depth curves of density and porosity are presented together.

ansetzen.

In Fig. 12···17 sind Teufenabhängigkeiten der Rohdichte und der Porosität gemeinsam dargestellt.

2.1.6 Measurement — Messung

All methods of determination of porosity consist of the measurement of two of the three quantities V_{por}, V_{mat}, V_{tot} and an evaluation according to one of the three equations

Alle Methoden der Porositätsbestimmung beruhen auf der Messung von zweien der drei Größen V_{por}, V_{mat}, V_{tot} und einer Auswertung nach einer der drei Gleichungen

$$\phi = \frac{V_{\text{por}}}{V_{\text{tot}}} = 1 - \frac{V_{\text{mat}}}{V_{\text{tot}}} = \frac{V_{\text{por}}}{V_{\text{por}} + V_{\text{mat}}}$$

Two methods will be discussed here.

Zwei Methoden seien hier behandelt.

2.1.6.1 The Archimedian method — Die Archimedische Methode

The most common method of porosity determination is based on the principle of Archimedes and thus really a density determination. It consists of three weight determinations of the rock sample

Die bekannteste Methode der Porositätsbestimmung basiert auf dem Prinzip des Archimedes und ist somit eigentlich eine Dichtebestimmung. Sie besteht aus drei Gewichtsbestimmungen der Gesteinsprobe

1. in the dry state, i.e. with air filled pore space

1. im trockenen Zustand, d.h. mit luftgefülltem Porenraum

$$m_1 = \varrho_{\text{mat}} V_{\text{mat}}$$

2. in the wet state, i.e. with completely liquid filled pore space

2. im nassen Zustand, d.h. mit vollständig wassergefülltem Porenraum

$$m_2 = \varrho_{\text{mat}} V_{\text{mat}} + \varrho_{\text{por}} V_{\text{por}}$$

3. in the wet state, submerged in the identical liquid (i.e. an apparent mass)

3. im nassen Zustand untergetaucht in identischer Flüssigkeit (d.h. scheinbare Masse)

$$m_3 = (\varrho_{\text{mat}} - \varrho_{\text{por}}) V_{\text{mat}}$$

From those three values, the porosity can be evaluated without knowledge of the liquid density:

Aus diesen drei Werten läßt sich die Porosität ohne Kenntnis der Flüssigkeitsdichte auswerten:

$$\phi = \frac{m_2 - m_1}{m_2 - m_3} = \frac{\varrho_{\text{por}} V_{\text{por}}}{\varrho_{\text{por}} (V_{\text{por}} + V_{\text{mat}})}$$

Care must be taken in saturating the sample to reach full saturation. A low viscosity, well wetting, air free liquid must be used. The sample must be evacuated due time, at least to the order of 1 mbar, before injecting the liquid. Then the sample has to stand covered with liquid for a period of time to allow complete imbibition. The waiting time under vacuum and under liquid depends on the permeability of the sample, and no general rule can be given. Half an hour to two hours under vacuum and half a day up to 14 days under liquid might be necessary for accurate, research type work. The time under liquid can be shortened to hours by applying a pressure of at least 50 bar.

Viel Sorgfalt muß bei der Flüssigkeitssättigung der Probe aufgewendet werden, um wirklich vollständige Sättigung zu erreichen. Eine gut benetzende, luftfreie Flüssigkeit niedriger Viskosität ist hierfür nötig. Vor dem Injizieren der Flüssigkeit muß die Probe genügend lange bei etwa 1 mbar oder weniger evakuiert werden und anschließend noch längere Zeit unter Flüssigkeit stehen bleiben, bis vollständige Imbibition stattgefunden hat. Die Wartezeit unter Vakuum und unter Flüssigkeit hängt von der Permeabilität der Probe ab, und es läßt sich keine allgemeingültige Regel angeben. Eine halbe bis zwei Stunden unter Vakuum und einen halben Tag bis 14 Tage unter Flüssigkeit kann nötig sein für genaue forschungsgemäße Arbeit. Die Zeit unter Flüssigkeit läßt sich auf Stunden verkürzen durch Anlegen eines Druckes von wenigstens 50 bar.

Further care must be taken to use the identical fluid for saturation and submersion, and no matrix material must dissolve in the fluid.

Weiter muß gesichert sein, daß identische Flüssigkeit zum Sättigen und Untertauchen verwendet wird und daß sich kein Matrixmaterial in der Flüssigkeit löst.

Of the three mass values, the second one, m_2, the 'wet mass' is the most critical, error infested one, because it is so hard to decide on the true state of complete saturation: when the sample is taken from the fluid, it is outside covered by an excess fluid film, but when this film is stripped off, some pore fluid might be removed unintentionally, too. Much care is necessary and the whole measurement procedure should be repeated several times.

Von den drei Massenwerten ist der zweite, m_2, die „nasse Masse" am kritischsten und unsichersten, weil eine Entscheidung über den Zustand wirklich vollständiger Sättigung sehr schwierig zu treffen ist: wenn die Probe der Flüssigkeit entnommen wird, ist sie außen mit einem überschüssigen Flüssigkeitsfilm bedeckt; wenn jedoch dieser Film abgestreift wird, kann dabei auch unabsichtlich etwas Flüssigkeit aus dem Porenraum mit entfernt werden. Sehr viel Sorgfalt ist dabei nötig, und die gesamte Prozedur sollte mehrmals wiederholt werden.

2.1.6.2 The Boyle's law method — Die Methode nach dem Boyle-Marietteschen Gesetz

Another usual method is the gas porosimetry using Boyle's law

Eine andere gebräuchliche Methode ist die Gasporosimetrie unter Ausnutzung des Boyle-Marietteschen Gesetzes

$$pV = \text{const}$$

in one way or another. Technical details of the various applications of this law vary widely but the common principle is to determine the gas volume in a chamber with and without a rock sample by monitoring the pressure during compression or expansion. The volume difference then is the matrix volume.

in der einen oder anderen Weise. Die technischen Einzelheiten der verschiedenen Anwendungen dieses Gesetzes weichen stark von einander ab, aber das gemeinsame Prinzip ist die Bestimmung eines Gasvolumens in einer Kammer mit und ohne Gesteinsprobe durch Beobachtung der Druckänderung während einer Kompression oder Expansion. Die Volumendifferenz ergibt dann das Matrixvolumen.

For the calculation of porosity, a separate measurement of the bulk or pore volume is necessary. Geometrically measuring the bulk volume is very inaccurate, even in the case of samples of simple geometrical shape like cylinders, and should not be considered for serious work!

Zur Berechnung der Porosität ist noch eine separate Messung des Totalvolumens nötig. Die geometrische Bestimmung dieses Volumens ist äußerst ungenau, selbst im Falle sehr regelmäßiger, quaderförmiger oder zylindrischer Gestalt, und sollte deshalb für ernsthafte Arbeit gar nicht in Betracht gezogen werden!

Generally, the non-wetting property and high surface tension of mercury is utilized, which practically prevents it from entering the pore space (at ordinary pressure), when a rock sample is submerged into it. The bulk volume thus can be determined by the displaced mercury volume or by the force necessary to submerge the sample.

Üblicherweise nutzt man die Eigenschaft der Nichtbenetzung und die hohe Oberflächenspannung von Quecksilber aus; dasselbe dringt deshalb (unter normalen Druck) nicht in den Porenraum ein, wenn man die Probe darin untertaucht. Das Totalvolumen läßt sich daher aus dem verdrängten Quecksilbervolumen oder aus der Kraft, die zum Untertauchen der Probe nötig ist, bestimmen.

If this force is exerted by the weight of an extra mass m in addition to the dry sample weight m_{mat}, the bulk volume is

Wenn diese Kraft von einer zusätzlichen Masse m, addiert zur Trockenmasse m_{mat} der Probe aufgebracht wird, dann ist das Totalvolumen

$$V_{\text{tot}} = \frac{m + m_{\text{mat}}}{\varrho_{\text{Hg}}}$$

with ϱ_{Hg} being the density of mercury.

worin ϱ_{Hg} die Quecksilberdichte ist.

Errors can arise in the Boyle's law method when not keeping isothermal conditions during gas compression or expansion, or by deviations from the ideal gas behavior, if not compensated in the particular procedure.

Fehler können bei der Methode nach dem Boyle-Marietteschen Gesetz auftreten, wenn man nicht auf isotherme Bedingungen während der Gaskompression oder -expansion achtet. Auch können sich Abweichungen vom idealen Gasverhalten störend auswirken, wenn nicht durch die spezielle Verfahrensweise für eine Kompensation gesorgt wird.

Another source of error rests in the determination of the bulk determination. Erroneously large volumes are measured, when the sample has an irregular shape or a very rough surface, because then the mercury does not attain smoothly to the sample surface, due to its own large surface tension. On the other hand, with extremely coarse grained material with extremely large pores, some mercury might enter the pore space, resulting in an erroneously small bulk volume measured.

For particular realizations of the Boyle's law method, see e.g. [Mül64, Hei61].

Eine andere Fehlerquelle liegt in der Totalvolumenbestimmung. Zu große Volumina werden gemessen, wenn die Probe sehr unregelmäßige Gestalt oder eine sehr rauhe Oberfläche hat, weil dann das Quecksilber auf Grund seiner eigenen großen Oberflächenspannung sich nicht gleichmäßig an die gesamte Probenoberfläche anlegt. Andererseits kann bei Proben aus sehr grobkörnigem Material mit extrem großen Poren etwas Quecksilber auch in den Porenraum eindringen, was zur Messung eines zu kleinen Totalvolumens führt.

Zur Information über spezielle technische Verfahren nach dieser Methode siehe z.B. [Mül64, Hei61].

2.1.7 List of symbols — Symbolliste

Symbol	Description
d [mm]	diameter – Durchmesser
d_{50} [mm]	median diameter – Mediandurchmesser
d_{75}, d_{25} [mm]	quartile diameters – Quartildurchmesser
g	gravitational field strength – Erdbeschleunigung
m [kg]	mass – Masse
p [bar]	pressure – Druck
p_{hyd}	hydrostatic pressure – hydrostatischer Druck
p_{lith}	lithostatic pressure – lithostatischer Druck
V_{mat}	framework (matrix) volume – Feststoffvolumen, $V_{mat} = V_{tot} - V_{por}$
V_{por}	pore volume – Porenvolumen
V_{tot}	total ("bulk") volume – Totalvolumen
z [m]	depth of burial – Bedeckungstiefe
κ [bar^{-1}]	compressibility – Kompressibilität
ϱ_{mat} [kg m^{-3}]	matrix density – Matrixdichte (Reindichte)
ϱ_{por} [kg m^{-3}]	pore fluid density – Dichte der Porenfüllung
ϱ_{tot} [kg m^{-3}]	bulk density – Gesteinsdichte (Rohdichte)
ϕ [%]	porosity – Porosität
ϕ_0	initial porosity at $p=0$ – Anfangsporosität bei $p=0$
ϕ_1, ϕ_2	individual porosities – individuelle Porositäten

2.1.8 Tables and diagrams — Tabellen und Diagramme

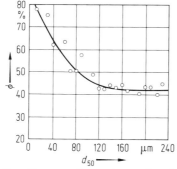

Fig. 1. Median diameter, d_{50}, and porosity, ϕ, of recent North Sea sediments of Wilhelmshaven [Füc60]; cited in [Eng60], p. 15 (80 individual measurements).

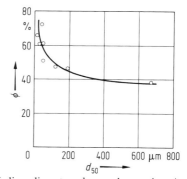

Fig. 2. Median diameter, d_{50}, and porosity, ϕ, of recent shallow sea sediments off the Californian coast, San Diego County [Ham56], cited in [Eng60], p. 15.

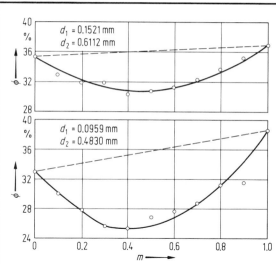

Fig. 3. Porosity, ϕ, of binary mixtures of monodisperse sands [Kin1898], cited in [Eng60], p. 7.

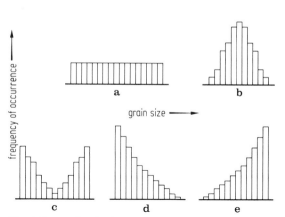

Fig. 4. Types of grain size distributions;
a) uniform, b) symmetrical with maximum, c) symmetrical with minimum, d) asymmetrical, decreasing, e) asymmetrical, increasing [And30], cited in [Eng60], p. 10.

a) For uniform distributions, porosity decreases with increasing variance.

b) Symmetrical maximum distributions show relatively large porosities, increasing with decreasing variance.

c) Symmetrical minimum distributions show very small porosities.

d) Asymmetrical, decreasing distributions show moderately small porosities.

e) Asymmetrical, increasing distributions show moderately small porosities.

Table 1. Influence of grain shape on porosity [Fra35], cited in [Eng60], p. 11.

Material	ϱ_{mat} [g cm^{-3}]	Type packing			
		dry		wet	
		loose	compressed	loose	compressed
Lead Balls	11.21	0.4006	0.3718	0.4240	0.3889
Sulfur Balls	2.024	0.4338	0.3735	0.4414	0.3824
Sea Sand	2.681	0.3852	0.3478	0.4296	0.3504
Shore Sand	2.658	0.4117	0.3655	0.4655	0.3846
Dune Sand	2.681	0.4117	0.3760	0.4493	0.3934
Calcite, ground	2.665	0.5050	0.4076	0.5450	0.4274
Quartz, ground	2.650	0.4813	0.4120	0.5388	0.4396
Rock salt, ground	2.180	0.5205	0.4351		
Mica, ground	2.837	0.9353	0.8662	0.9238	0.8728

Fig. 5. Decrease in tightness of packing for flat circular and triangular discs, depending on the mixing ratio. Model experiments by [Wol59], cited in [Eng60], p. 12.

Table 2. Grain contacts in thin sections of various sandstones from Wyoming [Tay50], cited in [Eng60], p. 12.

	Artifical sand	880 m Mesaverde	1340 m Shannon	2080 m Lower first wall creek	2220 m Frontier	2540 m Morrison
Number of contacts per grain	1.6	2.5	3.5	4.4	4.9	5.2
% grains without contacts	16.6	3				
% tangential contacts	59.4	51.9	21.4	0.9		
% long contacts	40.8	38.1	59.8	51.6	51.5	45.0
% concavo-convex contacts		9.6	19.1	28.5	28.1	23.1
% sutured grains				18.5	19.7	31.8

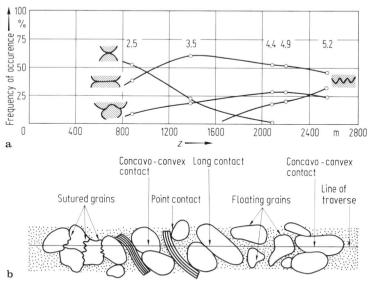

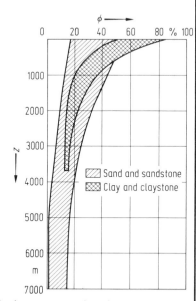

Fig. 6a and b. Frequency of occurrence of point contacts, long contacts, concavo-convex and sutured contacts in North American Cretaceous and Jurassic sandstones in dependence on burial depth. With an illustration of the types of contact (Fig. 6b). The numbers 2.5 to 5.2 are the mean numbers of contacts per grain in thin sections [Tay50], cited in [Pet72] and [Eng60] p. 20.

Fig. 7. Relationship between porosity, ϕ, and depth of burial, z, in non-metamorphic basins. The rapid loss of porosity in clay is due largely to compaction. The gradual loss of porosity in sands is largely due to cementation. The spreads of data are principally due to variations in thermal and pressure gradients and to mineralogical differences [Mea66, Ske70, Max64, Füc67, Pry73], cited in [Sel76], p. 39.

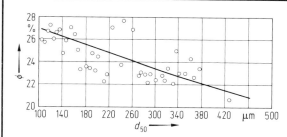

Fig. 8. Porosity, ϕ, of various strata of the Bentheim sandstone (Valendis) from a well in the Scheerhorn field (NW Germany) vs. median grain size d_{50} [Eng60] p. 21.

Table 3. Porosity, ϕ, and compressibility, κ, of 12 porous rocks [Wag71].

	ϕ %	κ 10^{-2} bar^{-1}
Sandstone, medium grained	22,9	1,07
Sandstone, medium grained	21.4	0.10
Sandstone	15.7	0.08
Calcareous sandstone	11.8	0.22
Sandstone	9.9	0.44
Sandstone	7.6	0.10
Sandstone, fine to medium grained	5.9	0.46
Sandstone	4.8	0.19
Sandstone, fine grained	3.8	0.17
Sandstone, fine grained	3.4	0.41
Siltstone	1.5	0.19

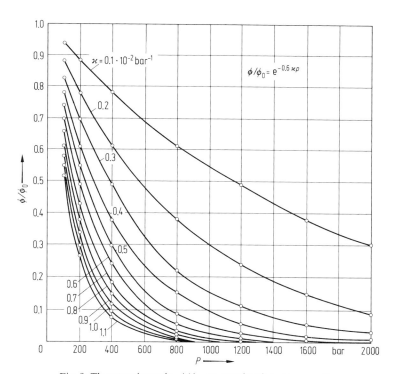

Fig. 9. The porosity ratio, ϕ/ϕ_0, vs. overburden pressure, P, for various values of κ (10^{-2} bar^{-1}) [Wag71].

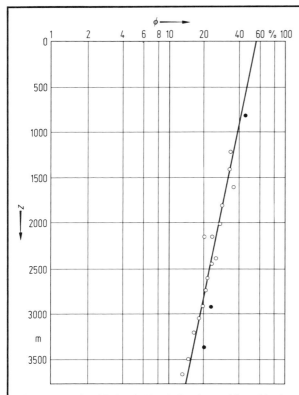

Fig. 10. Porosity (ϕ)/depth (z) relation in semi-logarithmic plot in the well Kambara GS-1, Niigata Prefecture. Open circles: mudstone; full circles, sandstone [Nag65].

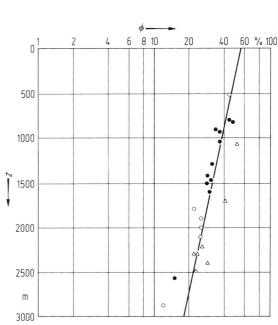

Fig. 11. Porosity (ϕ)/depth (z) relation in semi-logarithmic plot in the well Kasukabe GS-1, Kanto plain. Open circles: mudstone; full circles: sandstone; triangles: tuff [Nag65].

Table 4. Example of deep sandstones of relatively high porosities, cited in [Eng60], p. 16.

Locality	Formation	z m	ϕ %	Ref.
USA				
Rangely, Colorado	Weber, Pennsylvanian	1860	17.6	Win52
Katy, Texas	Cockfield, Eocene	2100	29.8	Win52
University Field, Louisiana	Miocene	2160	28.0	Lev56
Big Medicine Bow, Wyoming	Tennsleep, Pennsylvanian	2280	19.5	Wal41
Davis Lens, Texas	Eocene	2320	27.0	Lev56
Liberty Co., Texas	Eocene	2340	31.5	Win52
Fishers Reef, Texas	Frio, Oligocene	2740	28.2	Win52
Lindsay, Oklahoma	Oil Creek, Ordovician	3260	6.7	Win52
Fillmore, California	Pliocene	4300	20.0	Hen58
Carter Knox Field, Oklahoma	Bromide, Ordovician	4600	5.0	Ste57
Italy				
Cortemaggiore near Piacenza	Lower pliocene	1555	36.0	Agi59
Cortemaggiore near Piacenza	Lower pliocene	1575	23.2	Agi59
Cortemaggiore near Piacenza	Lower pliocene	1595	31.7	Agi59
Cortemaggiore near Piacenza	Lower pliocene	1930	28.0	Agi59
Cortemaggiore near Piacenza	Lower pliocene	1960	27.3	Agi59
Budrio East near Bologna	Upper miocene	2530	30.0	Agi59

Table 5. Very slightly consolidated sandstones from German oil fields [Eng60], p. 16.

Locality	Formation	z m	ϕ %	d_{50} mm	d_{75}/d_{25}	s[1]
Eldingen near Celle	Lias α	1483	28	0.133	1.60	1.12
Eldingen near Celle	Lias α	1463	29	0.105	1.20	1.11
Scheerhorn near Nordhorn	Valendis	1104	23	0.340	1.49	1.04
Scheerhorn near Nordhorn	Valendis	1120	27	0.113	1.60	1.23
Rühlermoor near Meppen	Valendis	842	30	0.270	1.67	1.01
Rühlermoor near Meppen	Valendis	853	33	0.138	1.59	1.10

[1] s = asymmetry of distribution function, $s = d_{50}/\sqrt{d_{10} \cdot d_{90}}$.

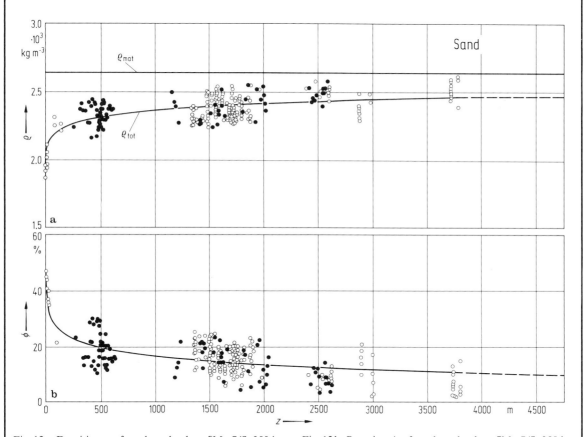

Fig. 12a. Densities, ϱ, of sand vs. depth, z [Mat74]. 355 individual measurements. Open circles: measured in laboratory; full circles: measured in situ.

Fig. 12b. Porosity, ϕ, of sand vs. depth, z [Mat74]. 355 individual measurements. Open circles: measured in laboratory; full circles: measured in situ.

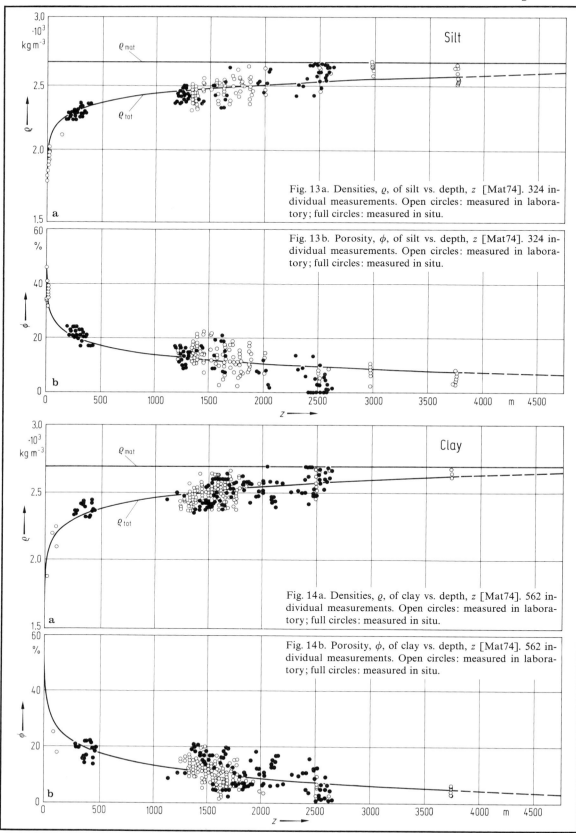

Fig. 13a. Densities, ϱ, of silt vs. depth, z [Mat74]. 324 individual measurements. Open circles: measured in laboratory; full circles: measured in situ.

Fig. 13b. Porosity, ϕ, of silt vs. depth, z [Mat74]. 324 individual measurements. Open circles: measured in laboratory; full circles: measured in situ.

Fig. 14a. Densities, ϱ, of clay vs. depth, z [Mat74]. 562 individual measurements. Open circles: measured in laboratory; full circles: measured in situ.

Fig. 14b. Porosity, ϕ, of clay vs. depth, z [Mat74]. 562 individual measurements. Open circles: measured in laboratory; full circles: measured in situ.

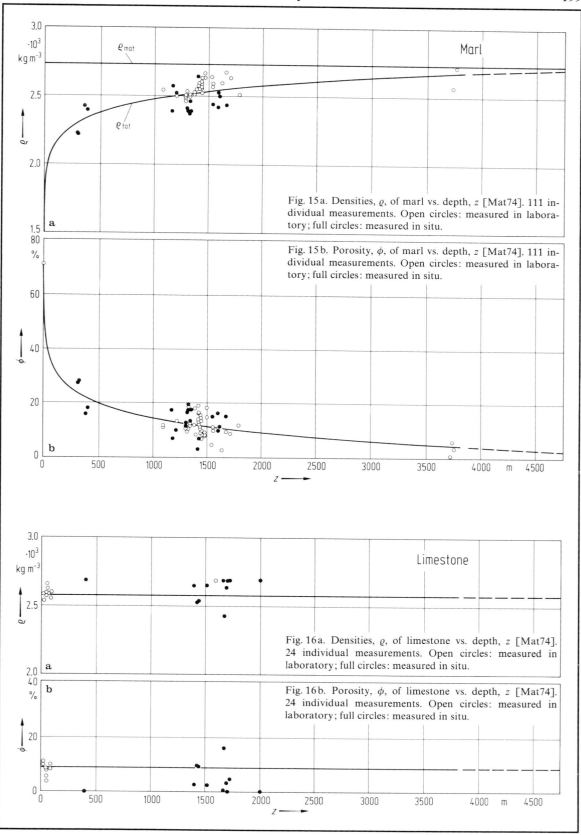

Fig. 15a. Densities, ϱ, of marl vs. depth, z [Mat74]. 111 individual measurements. Open circles: measured in laboratory; full circles: measured in situ.

Fig. 15b. Porosity, ϕ, of marl vs. depth, z [Mat74]. 111 individual measurements. Open circles: measured in laboratory; full circles: measured in situ.

Fig. 16a. Densities, ϱ, of limestone vs. depth, z [Mat74]. 24 individual measurements. Open circles: measured in laboratory; full circles: measured in situ.

Fig. 16b. Porosity, ϕ, of limestone vs. depth, z [Mat74]. 24 individual measurements. Open circles: measured in laboratory; full circles: measured in situ.

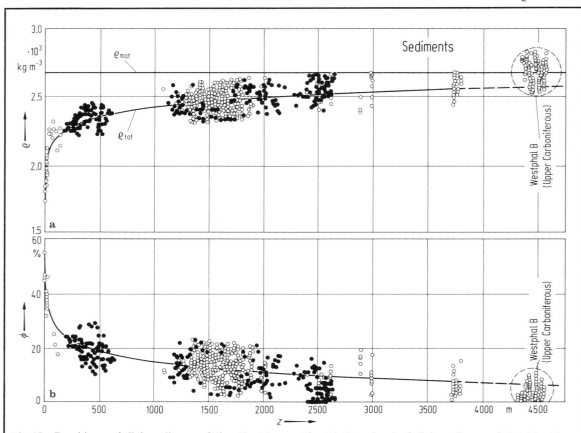

Fig. 17a. Densities, ϱ, of all the sediments of Figs. 12···16 vs. depth, z [Mat74]. Open circles: measured in laboratory; full circles: measured in situ.

Fig. 17b. Porosity, ϕ, of all the sediments of Figs. 12···16 vs. depth, z [Mat74]. Open circles: measured in laboratory; full circles: measured in situ.

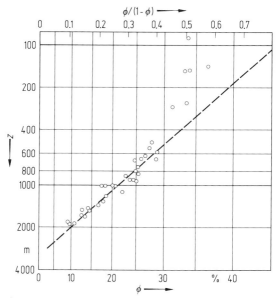

Fig. 18. Porosity, ϕ, and relative pore volume, $\phi/(1-\phi)$, of Tertiary shales from Venezuela vs. depth, z [Hed36], cited in [Eng60], p. 39.

2.1 Porosity of rocks

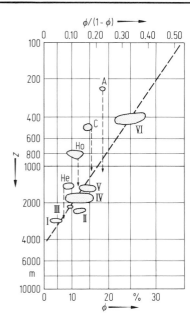

Fig. 19. Porosity, ϕ, and relative pore volume, $\phi/(1-\phi)$, of shaly sediments of Lias α from wells in NW Germany vs. their recent depth, z. After measurements of [Füc55] (number of investigated samples in parentheses), cited in [Eng60] p. 39.
I: Harsebruch 2, SE Celle (20); II: Wettenbostel 2, SW Lüneburg (4); III: Bokel, S Ülzen (10); IV: Oil fields Hohne (6) and Wesendorf, slope (27); furthermore wells Glinde 1, SSE Hamburg (4); Glückstadt 1, WNW Hamburg; Bodenteich 6, S Ülzen (3); Helmerkamp 1, E Celle (7); Unterlüss 1 and 2, NE Celle (9); V: Oil field Eldingen, NE Celle (32); VI: Oil field Wesendorf, top of salt dome (5); He: Gross Hehlen 1, NNW Celle (7); Ho: Oil field Hohenassel, E Hildesheim (3); C: Oil field Calberlah, N Braunschweig (7); A: Oil field Abbensen, ENE Hannover (7). – The points about the straight line, samples I···VI, originate from recent depth equal to maximum depth of burial, while the Lias of He, Ho, C, A has been lifted by salt uplift about the distance marked by arrows, as follows from purely geologic reasoning.

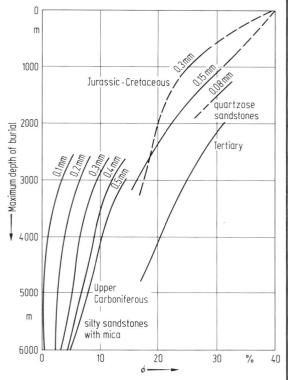

Fig. 20. Decrease of sandstone porosity, ϕ, with maximum depth of burial depending on median diameter (noted along curves). The following facts can be recognized:
1) In fine quartzose sandstones, porosity decreases slower at first, but eventually faster than in coarse ones.
2) Also the time integral of burial depth enters: the Tertiary sandstones buried for shorter time are more porous than the Jurassic-Cretaceous sandstones buried for longer time.
3) In the Upper Carboniferous, porosity of coarse sandstones exceeds that of fine ones, due to the contents of silt and clay increasing with decreasing grain size. Additionally, the phyllosilicates enhance pressure solution here [Füc67] and [Max64], cited in [Füc70] p. 106.

Table 6. Porosity, ϕ, and permeability, K, of some sedimentary rocks from wells in the Federal Republic of Germany. The permeabilities are averages of several samples measured parallel and perpendicular to the bedding plane. [Eng60], cited in [Eng73], p. 214.

	ϕ %	K md*)	d_{50} mm	Karbonate %	$<20\,\mu m$ %	z m
Coarse grained sandstones, unconsolidated						
Ampfing sandstone, lower Oligocene, Ampfing	19.9	4900	0.7	14.6	2.1	1820
Middle Kimmeridge, Ostenwalde	26.2	9900	0.50	0	1.3	1535
Bentheim sandstone, Valendis, Scheerhorn	24.5	5700	0.35	0	1.1	1105
Middle Rhät, Abbensen	18.5	1360	0.31	4	2.3	325
Lower Pechelbronn sandstone, Oligocene, Stockstadt	24.6	3200	0.25	0	2.5	1610
Upper Valendis, Barenburg	25.1	3100	0.25	4	1.5	810
Bentheim sandstone, Valendis, Rühlermoor	29.5	7500	0.25	0	0.8	785
Dogger β, Hankensbüttel	27.8	3250	0.21	1	1.6	1535
Baustein layers, upper Oligocene, Schwabmünchen	28.5	2380	0.2	60	10	1300
Finer grained and diagenetically more consolidated sandstones						
Dogger ε, Kronsberg (calcareous cementation)	24.7	105	0.35	49	6.5	650
Lias α 2, Eldingen	26.9	1570	0.16	0	2	1485
Bentheim sandstone, Valendis, Scheerhorn	27.4	400	0.11	0	9.5	1120
Dogger β, Hankensbüttel	22.8	615	0.09	2	2.6	1605
Dogger β, Meerdorf	19.0	100	0.09	3	3.1	1733
Wealden, Scheerhorn (calcareous cementation)	27.2	180	0.10	26	7.5	1160
Upper Pechelbronn sandstone Oligocene, Stockstadt	10.2	7	0.04	30	15	1550
Lias α 1–2, Eldingen	24.5	35	0,04	2	10	1520
Carbonates						
Coraloolith, Oxford, Hohenassel	19.5	2700				520
Aragonite, Wealden, Lingen	23.6	260				900
Aragonite-Oolith, Portland, Ostenwalde	19.8	65				1495
Main dolomite, Zechstein, Itterbeck	13.0	3				1610

*) $1\,\text{md} = 10^{-3}\,\text{d} \approx 10^{-15}\,\text{m}^2$

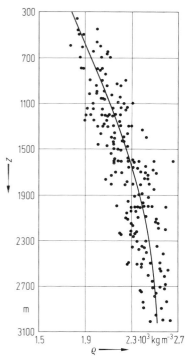

Fig. 21. Density, ϱ, of shales from wells in the Po basin [Sto59].

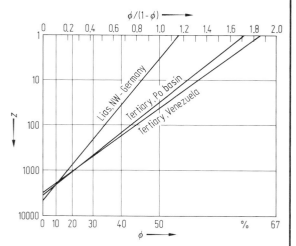

Fig. 22. Schematic of porosity vs. depth for various shales using the empirical formula $\phi/(1-\phi) = E'_1 - b' \lg z$ with the initial pore ratio $E'_1 = \phi_1/(1-\phi_1)$ and the extrapolated zero-porosity depth z_0 with $\lg z_0 = E'_1/b'$ [Eng60], p. 41.

	E'_1	b'	ϕ_1 [%]	z_0 [m]
Tertiary, Venezuela	1.844	0.527	65	3160
Tertiary, Po basin	1.700	0.481	63	3500
Lias, NW Germany	1.160	0.317	54	4570

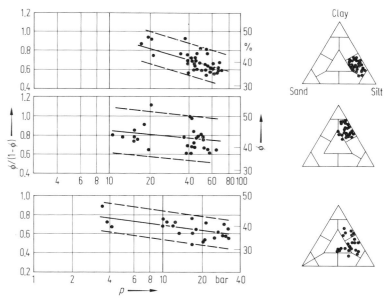

Fig. 23. Porosity, ϕ, of shaly alluvial freshwater sediments from bore holes in San Joaquin and Santa Clara Valleys, California [Mea68], cited in [Eng73], p. 287.

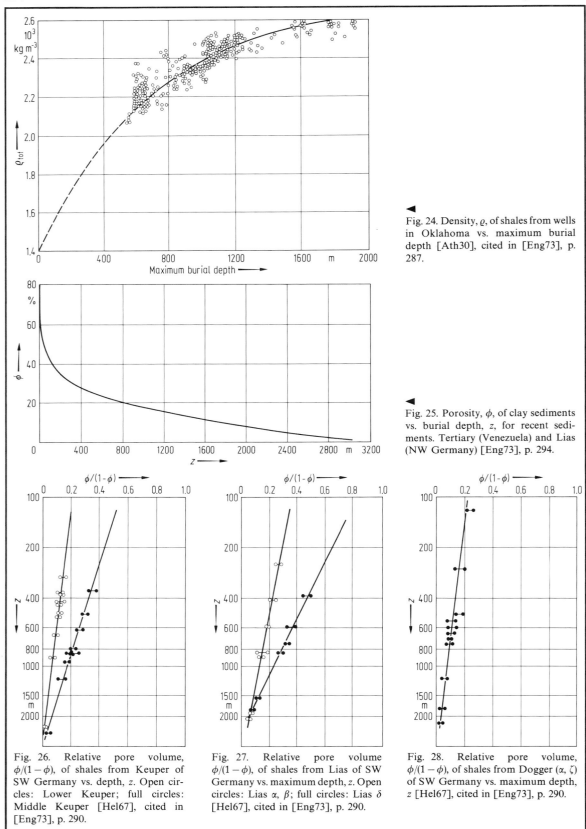

Fig. 24. Density, ϱ, of shales from wells in Oklahoma vs. maximum burial depth [Ath30], cited in [Eng73], p. 287.

Fig. 25. Porosity, ϕ, of clay sediments vs. burial depth, z, for recent sediments. Tertiary (Venezuela) and Lias (NW Germany) [Eng73], p. 294.

Fig. 26. Relative pore volume, $\phi/(1-\phi)$, of shales from Keuper of SW Germany vs. depth, z. Open circles: Lower Keuper; full circles: Middle Keuper [Hel67], cited in [Eng73], p. 290.

Fig. 27. Relative pore volume $\phi/(1-\phi)$, of shales from Lias of SW Germany vs. maximum depth, z. Open circles: Lias α, β; full circles: Lias δ [Hel67], cited in [Eng73], p. 290.

Fig. 28. Relative pore volume, $\phi/(1-\phi)$, of shales from Dogger (α, ζ) of SW Germany vs. maximum depth, z [Hel67], cited in [Eng73], p. 290.

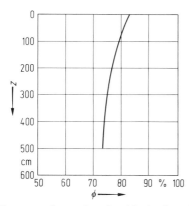

Fig. 29. Decrease of porosity, ϕ, with depth, z, in recent marine sediments off the coast of California (Santa Barbara Basin, water depth 530 m, median diameter $d_{50}=0.0029\cdots 0.0041$ mm) [Eme52], cited in [Eng60], p. 35.

Table 7. Average porosity of clay sediments off the south Californian coast. (25 cm below sea bottom surface; [Eme52], cited in [Eng60], p. 33.)

d_{50} mm	Average porosity %
0.010	73
0.003	80
0.001	89

Table 9. Decrease of porosity, ϕ, with depth, z, in recent clay sediments off the Californian coast [Eme52], cited in [Eng60], p. 36.

z m	ϕ %
0.20	82
0.50	81
1.00	80
2.00	77
3.00	75
4.00	74
5.00	73

Table 8. Porosity, ϕ, of recent tideland sediments of the North Sea off Wilhelmshaven [Füc60], cited in [Eng60], p. 34.

% Clay (<0.02 mm)	ϕ %
52.6	80.7
55.0	75.9
57.8	77.8
59.4	80.1
60.0	76.1
66.8	80.2
67.6	83.4
68.7	82.7
71.4	79.8
73.0	81.0
74.2	82.3
77.0	82.5
77.7	82.0
79.1	86.3
83.0	81.2

Table 10. Decrease of porosity, ϕ, with depth in muds of various sedimentation sites [Füc70], p. 236.

z m	ϕ [%]		
	Black Sea [Saw58]	Santa Barbara basin, California [Eme52]	Zürich lake [Zül56]
0.00			88
0.20		82	78
0.50		81	77
1.00		79	75
2.00	79	77	73
3.00	73	75	71
4.00	72	74	68
5.00	71	73	66
6.00	70		64
7.00	65		62
8.00			60

Table 11. Porosity and bulk density of sedimentary rocks [Man63].

More than 900 items of porosity and bulk density data for sedimentary rocks have been tabulated. Most of the data are from the more accessible American, British, German, and Swiss literature. The number of porosity determinations per item ranges from 1 to 2109. The tabulation reflects the fact that more porosity than bulk density data are available for sedimentary rocks.

The data are tabulated under headings according to rock type and geologic age, and grouped according to geographic locality. To the extent that information is available, the following items are included: The name of the stratigraphic unit, the source of the material or depth below the surface, the number of samples, the average and range of porosity, the average dry and saturated bulk density, the source of the data, and the method of porosity determination.

The methods are described in Table 11f.

see next page

Table 11a. Sandstone, siltstone, quartzite, chert, and conglomerate

Stratigraphic unit	Locality	Source of material or depth [m]	Number of samples	ϕ [%] Minimum	ϕ [%] Maximum	ϕ [%] Average	ϱ_{tot} [gcm^{-3}] Dry	ϱ_{tot} [gcm^{-3}] Water-saturated	Ref.	Method of porosity determination	Remarks
Precambrian											
Goodrich Quartzite	Ishpeming, Michigan	Mine					3.24		Bla56	A-16	
Nonesuch Shale (calcareous sandstone)	White Pine, Michigan					5.3	2.60	2.65	Bla55	A-16	
Cambrian											
Antietam Quartzite	Marticville, Pennsylvania	Outcrop (?)	1			1.7	3.05	3.07	Bla56	A-16	
Chickies Quartzite	Pennsylvania	Outcrop	5	3.8	7.8	5.4			Fan33	T-1	
Mt. Simon sandstone (dolomitic)	Sand Hill well, Wood County, W. Virginia	3964···4013	9	0.2	2.5	0.7	2.69	2.70	Rob62	A-15	
Southern Potsdam sandstone	Wisconsin	Quarry	14	4.8	28.3	11.4	2.30	2.41	Buc1898	A-1	
Northern Potsdam sandstone			16	10.4	22.6	19.4	2.13	2.32	Buc1898	A-1	
Reagan Sandstone	Otis and Penny-Wann fields, Kansas	1051···1123	24	5.5	17.8	11.2			Ral54	A-9	
Sandstone	Conley, Great Britain	Quarry	1			6.1	2.45	2.51	Moo04	A-2	
Upper Cambrian and Lower Ordovician											
Potsdam and Beekmantown Groups (sandstone)	Ontario, Canada	Quarry	6	5.0	12.4	8.0	2.44	2.52	Par12	A-2	5 localities
Ordovician											
Juniata Formation	Sand Hill well, Wood County, W. Virginia	2387	1			1.2	2.67	2.68	Rob62	A-15	
St. Peter Sandstone	Wisconsin	Quarry	2	18.1	20.0	19.1	2.15	2.34	Buc1898	A-1	
	Ozark Plateau, Arkansas	Outcrop	12	3.6	14.1	8.8	2.41	2.50	Bra37	T-2	

2.1 Porosity of rocks

Stratigraphic unit	Locality	Source of material or depth [m]	Number of samples	ϕ [%] Minimum	ϕ [%] Maximum	ϕ [%] Average	ϱ_{tot} [gcm^{-1}] Dry	ϱ_{tot} [gcm^{-1}] Water-saturated	Ref.	Method of porosity determination	Remarks
Crystal Mountain Sandstone	Ouachita Mountains, Arkansas		1			17.5	2.19	2.37	Bra37	T-2	
Simpson Group (sandstone)	Cunningham pool, Kansas	1235…1255	36	5.7	22.7	13.3			Ral54	A-9	
Wilcox sand	Bowlegs field, Oklahoma	1235…1255	2	12.1	12.8	12.5			Bar31	A-6	
	Oklahoma City field, Oklahoma	≈1920	12	16.9	23.3	20.7			Ral54	A-9	
	Ramsay pool, Oklahoma	≈1920	2	8.0	16.9	12.5			Bar31	A-6	
		≈1463				20			Fro40	N-1	Clean, uniform
Seminole field, Oklahoma		≈1310	2	15.6	15.6	15.6			Bar31	A-6	
Second Wilcox sandstone	Arcadia-Coon Creek pool, Oklahoma	≈1829		9	18				Car48a	N-1	
Second Simpson sand	West Edmond field, Oklahoma	2134	2	14.9	15.1	15.0			Ral54	A-9	
First Bromide sand	Fitts pool, Oklahoma	≈1265	1			7.7			Ral54	A-9	
Bromide Formation (sandstone)	Lindsay area, Oklahoma	2896…3444		1	24	14			Swe50	N-1	
	Maysville pool, Oklahoma	2255	1			14.1			Swe50	N-1	
Hammar-Haindl sand	Oklahoma City field, Oklahoma	≈2012	12	7.1	26.5	15.0			Ral54	A-9	
		≈2012	2			14.1			Ber31	A-6	
Johnson sand	Oklahoma City field	≈2012	5	18.2	30.3	24.3			Ral54	A-9	
		≈2012	32	5.3	21.3	14.1			Bar31	A-6	
Mollman sand	Oklahoma City field	≈2012	2	12.5	15.7	14.1			Ral54	A-9	
School Land sand	Oklahoma City field	≈2012	4	11.4	14.8	13.1			Bar31	A-6	
McKee sand	Hare field, New Mexico	2376	2			8.5	2.51	2.60	Hug52	A-5	
Swan Peak quartzite	Afton quadrangle, Wyoming					2.0	2.56	2.58	Nut42	T-2	
Chazy Group	Ontario, Canada	Quarry	1			17.5	2.19	2.37	Pat59	A-2	

(continued)

Table 11a (continued)

Stratigraphic unit	Locality	Source of material or depth [m]	Number of samples	ϕ [%] Minimum	ϕ [%] Maximum	ϕ [%] Average	ϱ_{tot} [gcm^{-1}] Dry	ϱ_{tot} [gcm^{-1}] Water-saturated	Ref.	Method of porosity determination	Remarks
Silurian											
Clinch Sandstone	Lee County, Virginia					9.6			Kru51	N	
Tuscarora Sandstone	Sand Hill well, Wood County, W. Virginia	2337···2366	5	0.9	1.5	1.1	2.59	2.60	Rob62	A-15	
Tuscarora Sandstone, dolomitic		2370···2378	2	0.5	1.1	0.8	2.64	2.65	Rob62	A-15	
Red Mountain Formation	Bessemer, Alabama	Mine, 427				3.1	3.14	3.17	Win49	A-16	Ferruginous sandstones
						2.9	3.26	3.29	Win49	A-16	Red sandstones
						0.8	2.76	2.77	Win49	A-16	Siltstone and shale
Blaylock Sandstone	Ouachita Mountains, Arkansas	Outcrop	4	2.3	17.4	7.3	2.42	2.49	Bra37	T-2	
Medina group	Ontario, Canada	Quarry	2	12.0	14.9	13.5	2.30	2.44	Par12	A-2	2 localities
Devonian											
Oriskany Sandstone	Wayne-Dundee field, New York	≈586	4	7.1	9.0	8.3			Fet38	T	1 well
	State Line field, New York	≈1433	18	7.9	12.2	10.3			Fet38	T	1.7 m of core
Bradford sand	Bradford field, Pennsylvania	≈183···≈701	297	6.0	23.3	15.0	2.25	2.40	Fet34	T-6	Estimated grain density, 2.65 gcm^{-3}
		≈183···≈701	1			13.1			Ral54	A-9	
		≈183···≈701	1			18.6	2.17	2.36	Mel21	T-2	
		≈183···≈701	75	3.8	26.0	12.9			Fan33	A-6	
		≈183···≈701	40	4.5	22.7	12.5			Bar31	A-6	
	Kane field, Pennsylvania	Subsurface	17	2.0	14.4	11.9			Fan33	A-6	
			69	3.0	16.4	10.6			Bar31	A-6	
Chipmunk sand	Cattaraugus County, New York	Subsurface	2	14.5	15.1	14.8			Ral54	A-9	

2.1 Porosity of rocks

Stratigraphic unit	Locality	Source of material or depth [m]	Number of samples	ϕ [%] Minimum	ϕ [%] Maximum	ϕ [%] Average	ϱ_{tot} [gcm^{-3}] Dry	ϱ_{tot} [gcm^{-3}] Water-saturated	Ref.	Method of porosity determination	Remarks
Clarendon sand	Saybrook field, Pennsylvania	Subsurface	2	11.5	13.4	12.5			Bar31	A-6	
	Warren field, Pennsylvania	Subsurface	5	8.8	18.5	14.4			Bar31	A-6	
			13	2.6	25.6	10.6			Fan33	A-6	
Kane sand	Kane field, Pennsylvania	Subsurface	7	14.2	22.2	18.9			Bar31	A-6	
Oriskany Group (sandstone)	Tioga field, Pennsylvania	≈1219	65	2.9	11.8				Fet38	T	19 wells
	Sabinsville pool, Pennsylvania	≈1326	1			9.9			Fet38	T	
	Hebron field, Pennsylvania	≈1564	6	9.1	10.1	9.4			Fet38	T	2 wells
	Hebron gas field, Pennsylvania	1577···1579		9	14				Ree36	N-1	
	Hebron-Ellisville field, Pennsylvania	≈1676				10			Fin49	N	
Sandstone	Butler and Zelienople quadrangle, Pennsylvania	Subsurface	8	4.5	22.2	10.1	2.39	2.49	Mel24	T-2	
	Dorseyville field, Pennsylvania	582	11			8.6	2.42	2.51	Mel21	T-2	
Speechley sand	Oil City field, Pennsylvania	Subsurface	11	3.7	15.7	11.4			Fan33	A-6	
	Butler County, Pennsylvania	619···624	3	4.3	15.4	10.9	2.42	2.53	Mel21	T-2	
Third Bradford sand	Bradford field, Pennsylvania	Subsurface	10	13.1	16.9	14.6	2.25	2.41	Fan33	T-1	
			5	14.9	18.5	16.2			Mel21	T-2	
Third sandstone	Allegheny and Butler Counties, Pennsylvania	518···549	13	4.6	9.1	6.7	2.46	2.53	Mel21	T-2	
Fourth and Top sand	Butler County, Pennsylvania	Subsurface	3	10.6	15.9	12.7	2.31	2.44	Mel21	T-2	
Fifth sandstone	New Kensington quadrangle, Pennsylvania	701	3	6.2	13.2	10.5	2.43	2.54	Mel21	T-2	

(continued)

Table 11a (continued)

Stratigraphic unit	Locality	Source of material or depth [m]	Number of samples	φ [%] Minimum	φ [%] Maximum	φ [%] Average	ϱ_{tot} [gcm^{-3}] Dry	ϱ_{tot} [gcm^{-3}] Water-saturated	Ref.	Method of porosity determination	Remarks
Devonian (continued)											
Thirty-foot sand	Glenshaw field, Pennsylvania	563	7	7.2	9.8	9.0	2.41	2.50	Mel21	T-2	
Oriskany Group (sandstone)	Kanawha County, W. Virginia			6.8	11				Laf38	N-1	
Huntersville chert	Sand Hill well, Wood County, W. Virginia	1053	1			0.7	2.58	2.59	Rob62	A-15	Calcareous
Sandstone	Ravenna, Kentucky	Mine	2	10.4	11.9	11.2	2.55(1)	2.65(1)	Mel21	T-2	
Williamsburg sand	Williamsburg, Kentucky	244(?)	1			12.4			Mel21	T-2	
Hoing sand	Colmar-Plymouth pool, Illinois	126...152	137	5.7	23.9	18.6			Pie40	A-6	
Oriskany sandstone	Ontario, Canada	Quarry	1			6.6	2.48	2.55	Par12	A-2	
Old Red sandstone	Cradley, Great Britain	Outcrop	1			2.5	2.57	2.60	Moo04	A-2	
Graywacke	Germany	Quarry	3	0.5	9.1	3.6			Hir12	A-3	Granular
			3	0.7	7.2	4.3			Hir12	A-3	Slaty
Sandstone	Germany	Quarry	2	7.7	18.7	13.2			Hir12	A-3	
Upper Old Red sandstone	Micheldean, Great Britain	Outcrop	1			16.0	2.18	2.34	Moo04	A-2	Coarse
			1			9.1	2.40	2.49	Moo04	A-2	Fine
Upper Devonian and Lower Mississippian											
Three Forks shale (red sandstone)	Afton quadrangle, Wyoming					6.4	2.62	2.68	Nut42	T-2	
First Venango sand	Franklin Heavy field, Pennsylvania	Subsurface	12	11.2	18.5	13.9			Ral54	A-9	
	Pleasantville field, Pennsylvania	Subsurface	2	12.4	20.7	16.5			Fan33	A-6	
Second Venango sand	Oil City field, Pennsylvania	Subsurface	30	5.8	20.1	15.0			Bar31	A-6	
	Pleasantville field, Pennsylvania	Subsurface	1			17.2			Fan33	A-6	

2.1 Porosity of rocks

Stratigraphic unit	Locality	Source of material or depth [m]	Number of samples	ϕ [%] Minimum	ϕ [%] Maximum	ϕ [%] Average	ϱ_{tot} [gcm^{-3}] Dry	ϱ_{tot} [gcm^{-3}] Water-saturated	Ref.	Method of porosity determination	Remarks
Second sand, Venango oil sand group	Oil City area, Pennsylvania	188···266	115	2.0	28.2	14.6	2.28	2.43	Fet26	T-2	Average grain density, 2.664 gcm^{-3}
Third Venango sand		Subsurface	2	10.2	10.2	10.2			Bar31	A-6	
			17	3.4	22.3	13.7			Fan33	A-6	
Bowlder sand	Allegheny County, Pennsylvania		4	4.4	12.6	7.9	2.45	2.53	Mel21	T-2	
Mississippian											
Berea sandstone	Hancock County, W. Virginia	Subsurface	2	16.7	22.2	19.5			Bar31	A-6	
	Cabin Creek field, West Virginia	≈762				4			Was27	T-2	Quartzite cap
		≈762				16			Was27	T-2	Pay sand
Sandstone, well-sorted	Tucker County, W. Virginia		1			4.4			Rus26	T-1	Carbon ratio, 72.5···77.5%
Berea Sandstone	Harrison County, Ohio	Subsurface	3	15.4	17.2	16.6	2.24	2.40	Mel21	T-2	
	Monroe, Noble, and Belmont Counties, Ohio	439···658	8	4.7	17.1	11.1	2.37	2.48	Mel24	T-2	
	Monroe County, Ohio	Subsurface	1			11.7			Bar31	A-6	
	S. Amherst, West View, and Berea, Ohio	Quarry	6	15.9	17.8	16.6	2.12	2.29	Bow15	A-1	
Berea(?) Sandstone	Berea, Ohio	Outcrop	2	10.2	13.2	11.7	2.06	2.22	Mer08	A-13	Building stone
	Amherst, Ohio	Quarry				16			Win49	A-16	
Keener sand	Monroe County, Ohio	372···478	5	11.3	18.4	14.9	2.25	2.40	Mel21	T-2	Mississippian (?)
Maxville Limestone (siltstone)	Muskingum County, Ohio	≈61	2				2.67		Bla56	A-16	
Sandstones	Monroe County, Ohio	411···457	3	11.3	13.1	12.5	2.35	2.48	Mel24	T-2	Mississippian (?)
Mooretown Formation (sandstone)	Tri-County oil field, Indiana	≈396				5			Esa27	N	Clayey and limy

(continued)

Table 11a (continued)

Stratigraphic unit	Locality	Source of material or depth [m]	Number of samples	ϕ [%] Minimum	ϕ [%] Maximum	ϕ [%] Average	ϱ_{tot} [gcm^{-3}] Dry	ϱ_{tot} [gcm^{-3}] Water-saturated	Ref.	Method of porosity determination	Remarks
Mississippian (continued)											
Sample sand	Francisco pool, Indiana	426				17			Mou29	N	
Waltersburg Sandstone	Lower Wabash area, Illinois and Indiana	579				19.5			Swa51	N-1	Clean, fine sandstone
	Powell's Lake oil field, Kentucky	552···555		14	21	18.5			Ing48	N-1	
Fort Payne Chert	Near Smithville, Tennessee	Subsurface	2	3.8	4.8	4.2	2.65	2.69	Bla56	A-16	
Aux Vases Sandstone	Salem pool, Illinois	536···547	28	7.9	19.2	14.7			Pie40	A-6	
		≈ 549				16.1			Arn39	N-1	
Benoist sand		≈ 526				17.5			Arn39	N-1	
	Hoodville field, Illinois	914···915	6	15.3	17.0	16.1			Rel54	A-6	
Bethel Sandstone	Louden, Centralia, Patoka, and Salem pools, Illinois	415···565	102	10.0	24.9	18.3			Pie40	A-6	
	South-central Illinois	415···525	33	10.0	22.2	18.2			Pye44	N-1	Well sorted
Cypress Sandstone	Bartelso, Carlyle, Lawrence County, Louden, Noble, and Patoka pools, Illinois	308···617	126	5.7	23.8	18.6			Pie40	A-6	
Paint Creek Shale (stray sand)	Fayette County, Illinois	415···525	33	10.0	22.2	18.2			Pye44	N-1	Well sorted
Tar Springs Sandstone	Benton field, Illinois	616···640		13	25	20			How48	N-1	
Carlyle sand	Carlyle, Illinois	313	3	20.6	27.6	23.4	2.02	2.26	Mel21	T-2	
Michigan stray sand	Millbrook field, Illinois	376···391	7	9.7	20.7	19.1			Ral54	A-9	
Sandstones	Ozark Plateau, Arkansas	Outcrop	4	8.6	17.3	13.3	2.30	2.43	Bra37	T-2	
Boone Formation (chert)	Near Picher, Oklahoma	Mine				5.0	2.56	2.61	Bla55	A-16	

Stratigraphic unit	Locality	Source of material or depth [m]	Number of samples	ϕ [%] Minimum	ϕ [%] Maximum	ϕ [%] Average	ϱ_{tot} [gcm^{-3}] Dry	ϱ_{tot} [gcm^{-3}] Water-saturated	Ref.	Method of porosity determination	Remarks
Boone Formation (quartzite)						2.0	2.72	2.74	Bla55	A-16	
Boone Formation (calcareous chert)						8.9	2.39	2.48	Bla55	A-16	
Mississippian or Pennsylvanian											
Deaner sand	Deaner field, Oklahoma	769	1			12.5	2.35	2.47	Mel21	T-2	
Kingwood sand	Deaner field, Oklahoma	810	2	15.7	16.0	15.9	2.27	2.43	Mel21	T-2	
Red sand	Osage County, Oklahoma	Subsurface	5	7.5	18.4	14.0	2.36	2.50	Mel24	T-2	
Pennsylvanian											
Olean Formation (sandstone)	Rock City, New York	Outcrop(?)	2	14.4	15.8	15.1			Bar31	A-6	
Allegheny Formation (sandstone)	Bakerton, Pennsylvania	≈152				1.4	2.70	2.71	Bla56	A-16	
Allegheny Formation (siltstone)		≈152				1.8	2.76	2.78	Bla56	A-16	
	Colver, Pennsylvania	Mine					2.66		Bla56	A-16	
Kanawha Formation (sandstone)	Near Franklin, Pennsylvania	1.5···42.5	3	7.8	12.0	10.3	2.29	2.30	Bla56	A-16	
		0···6					2.46		Bla55	A-16	
		0···6	2			11.0	2.16	2.27	Bla55	A-16	
		0···6					2.15		Bla55	A-16	
Kanawha Formation (sandstone)	Dehue, W. Virginia	Mine					2.70		Bla55	A-16	
Monongahela Formation	Scotts Run, W. Virginia	Mine	3	3.1	4.8	4.1	2.5		Win50	A-16	
Sandstones, well-sorted	West Virginia, Ohio	Outcrop	7	16.5	25.3	21.5			Rus26	T-1	Carbon ratio 52.5···57.5%
			8	6.8	20.1	16.2			Rus26	T-1	Carbon ratio 57.5···62.5%

(continued)

Table 11a (continued)

Stratigraphic unit	Locality	Source of material or depth [m]	Number of samples	φ [%] Minimum	φ [%] Maximum	φ [%] Average	ϱ_tot [gcm⁻³] Dry	ϱ_tot [gcm⁻³] Water-saturated	Ref.	Method of porosity determination	Remarks
Pennsylvanian (continued)											
Sandstone, well-sorted	West Virginia	Outcrop	10	3.0	9.5	5.4			Rus26	T-1	Carbon ratio 72.5···77.5%
			4	1.9	3.9	2.7			Rus26	T-1	Carbon ratio 80···82.5%
Mansfield Formation (sandstone)	Powell's Lake oil field, Kentucky	366		15	23	18.7			Ing48	N-1	
Biehl sand	Allendale pool, Illinois	432···474	93	6.2	20.3	13.3			Pie40	A-6	
Makanda sand	Boskey Dell, Illinois	Outcrop	4	16.2	19.6	18.3			Ral54	A-9	
Robinson sand	Crawford-Main pool, Illinois	276···309	298	3.4	27.1	18.6			Pie40	A-6	
	Flat Rock, Lawrence County, New Hebron, and Parker pools, Illinois	268···314	47	6.3	24.3	16.8			Pie40	A-6	
Upper Partlow sand	North Johnson pool, Illinois	154···171	15	15.5	23.4	20.8			Pie40	A-6	
Bartlesville sand	Anderson County, Kansas					17.5			Kru51	N	Now "Burbank" sand
	Various fields, Kansas	212···361	82	7.0	24.8	20.4			Ral54	A-9	Now "Burbank" sand
Peru sand	Sedan and Cunningham fields, Kansas	60···63	14	14.4	22.4	17.5			Ral54	A-9	
Wayside sand	Montgomery County, Kansas	179···192				9.7	2.47	2.56	Mel21	T-2	
	Jefferson field, Kansas	65···129	16	15.7	23.0	20.0			Ral54	A-9	
Atoka Formation (sandstone)	Ozark Plateau, Arkansas	Outcrop	15	4.7	19.8	11.5	2.32	2.44	Bra37	T-2	
	Arkansas Valley, Arkansas	Outcrop	23	0	20.6	7.6	2.43	2.51	Bra37	T-2	

Stratigraphic unit	Locality	Source of material or depth [m]	Number of samples	ϕ [%] Minimum	ϕ [%] Maximum	ϕ [%] Average	ϱ_{tot} [gcm^{-3}] Dry	ϱ_{tot} [gcm^{-3}] Water-saturated	Ref.	Method of porosity determination	Remarks
Atoka Formation (sandstone)	Ouachita Mountains, Arkansas	Outcrop	13	0	10.4	5.5	2.50	2.55	Bra37	T-2	
Sandstones	Ozark Plateau, Arkansas	Outcrop	2	13.8	18.1	16.0	2.26	2.42	Bra37	T-2	
	Arkansas Valley, Arkansas	Outcrop	12	5.4	10.7	8.2	2.42	2.51	Bra37	T-2	
	Ouachita Mountains, Arkansas	Outcrop	12	0.9	9.6	4.6	2.52	2.57	Bra37	T-2	
Armstrong sand	W. Duncan field, Oklahoma	≈610	63			26.7			Put56	N-1	11 wells
Bartlesville sand	Creek and Osage Counties, Oklahoma	479···817	14	13.5	38.7	20.1	2.17	2.37	Mel21	T-2	
	Cushing field Oklahoma	804···817	2	28.9	32.0	30.5	2.20(1)	2.49(1)	Mel21	T-2	
	Pershing field, Oklahoma	606···650	10	7.6	16.0	13.3	2.31	2.44	Mel21	T-2	
	Various fields, Oklahoma	248···2009	587	3.3	33.7	17.4			Ral54	A-9	
Basal Pennsylvanian sand	Pauls Valley field, Oklahoma	Subsurface	6	5.9	12.3	8.8			Ral54	A-9	
Booch sand	Various fields, Oklahoma	888···(?)	5	17.5	25.4	22.1			Ral54	A-9	
Burbank sand	Burbank field, Oklahoma	890···905	2	15.2	15.7	15.4			Ral54	A-9	
Burgess sand	S. Moore pool, Oklahoma	2375···2376		1.8	35	17			Moo41	N-1	
	Washington County, Oklahoma	≈503	2	16.4	22.0	19.2	2.24(1)	2.40(1)	Moo41	N-1	
Cisco Formation	West Red River field, Oklahoma	≈472	6	22.6	26.1	24.6			Ral54	A-9	
Cleveland sand	Osage County, Oklahoma	Subsurface	1			17.7	2.21	2.39	Mel21	T-2	
Coffeyville Formation (sandstone)	Turley, Oklahoma	Subsurface				16.8	2.14	2.31	Bla56	A-16	

(continued)

Table 11a (continued)

Stratigraphic unit	Locality	Source of material or depth [m]	Number of samples	ϕ [%] Minimum	ϕ [%] Maximum	ϕ [%] Average	ϱ_{tot} [gcm^{-3}] Dry	ϱ_{tot} [gcm^{-3}] Water-saturated	Ref.	Method of porosity determination	Remarks
Pennsylvanian (continued)											
Cromwell sand	Hughes County, Oklahoma		1			12.1			Ral54	A-9	
	Little River field, Oklahoma	≈981	22	16.2	23.2	19.9			Bar31	A-6	
	Kanawa field, Oklahoma	≈701	6	16.6	22.6	19.0			Bar31	A-6	
Deese Formation (sandstone)	Southwest Antioch field, Oklahoma	≈1996	1			15			Res49	N-1	
	Velma oil field, Oklahoma	1015···1317				21			Rut56	N-1	
Dutcher sand	South Depew field, Oklahoma	735	2	14.6	18.2	16.4			Ral54	A-9	
	Slack field, Oklahoma	785···796	5	9.9	14.9	10.8			Mel21	T-2	
Gilcrease sand	Francis field, Oklahoma	≈762	2	27.3	27.5	27.4			Bar31	A-6	
	Holdenville field, Oklahoma	≈975	2	16.8	16.8	16.8			Bar31	A-6	
Glenn sand	Sasakwa, Oklahoma	≈863	2	18.1	18.3	18.2			Bar31	A-6	
	Creek County, Oklahoma	461···485	1			21.4	2.12	2.33	Mel21	T-2	
Healdton sand zone	Healdton field, Oklahoma	259···413	92	2.3	35.8	23.8			Ral54	A-9	
Hewitt sand	Hewitt field, Oklahoma	429···437	3	17.3	21.7	19.5	2.13(1)	2.32(1)	Mel21	T-2	
Hickman sand	Burbank field, Oklahoma	848···916	84	6.1	32.7	20.4	2.15	2.35	Mel21	T-2	
Holdenville Shale (sandstone)	Tulsa, Oklahoma	Subsurface				17.0	2.50	2.67	Bla56	A-16	
Hoover sandstone	Laverne district, Oklahoma	≈1280				18			Pat59	N-1	
Hoxbar Formation	Superior well, Caddo County, Oklahoma	2546···2549	6	3.8	7.8	6.55			Bea50	A	Shaly, limy
		2880···2893	39	4.7	8.7	6.90			Bea50	A	Shaly, limy
		3114···3116	3	9.5	10.9	10.17			Bea50	A	Shaly, limy

Stratigraphic unit	Locality	Source of material or depth [m]	Number of samples	φ [%] Minimum	φ [%] Maximum	φ [%] Average	ϱ_tot [gcm⁻³] Dry	ϱ_tot [gcm⁻³] Water-saturated	Ref.	Method of porosity determination	Remarks
Deese Formation	Superior well, Caddo County, Oklahoma	3988···3991	7	2.1	6.3	4.36			Bea50	A	Shaly, limy
		4595···4598	8	7.2	9.2	8.20			Bea50	A	Shaly, limy
		5167···5168	1			28.2			Bea50	A	Shaly, limy
Humphrey sand	Velma oil field, Oklahoma	Subsurface				17			Rut56	N-1	
Layton gas sand	Pawnee and Creek Counties, Oklahoma	260···465	2	22.1	26.3	24.2	2.02	2.26	Mel21	T-2	
Lower part of Dornick Hills Formation and Springer Formation	Velma pool, Oklahoma	≈1524	3500			21			Dav51	N-1	
		≈1615				19			Dav51	N-1	
		≈1859				16			Dav51	N-1	
		≈2012				14			Dav51	N-1	
		≈2164				12			Dav51	N-1	
Medrano sand	Caddo County, Oklahoma	1401···1775	242	2.2	25.5	17.3			Ral54	A-9	
Middle Rowe Formation	Cement field, Oklahoma	1020···1022	17	3.4	28.9	22.9			Ral54	A-9	
Morrow Series (sandstone)	Laverne district, Oklahoma	Subsurface				14			Pat59	N-1	
Olympic sand	Olympia pool, Oklahoma	≈549		11.9	21.7	18			Til38	N-1	
	Hughes County, Oklahoma					20.5			Kru51	N	
Peru sand	Bartlesville fields, Oklahoma	216···220	17	14.7	20.2	18.6			Ral54	A-9	
Sandstone	Garber field, Oklahoma	544···549	2	10.3	16.4	13.4	2.25(1)	2.41(1)	Mel21	T-2	
	Stone Bluff field, Oklahoma	349···354	1			16.5	2.22	2.38	Mel21	T-2	
Seminole Formation (sandy zone)	Tulsa, Oklahoma	Subsurface				20.5	2.26	2.47	Bla56	A-16	
Sims sand	Velma field, Oklahoma	Subsurface				18···20			Rut56	N-1	
Skinner sand	West Chandler field, Oklahoma	≈1256	1			12.1			Ral54	A-9	
Stray sand	Osage County, Oklahoma	468, or deeper				15.2	2.37	2.47	Mel21	T-2	

(continued)

Table 11a (continued)

Stratigraphic unit	Locality	Source of material or depth [m]	Number of samples	ϕ [%] Minimum	ϕ [%] Maximum	ϕ [%] Average	ϱ_{tot} [gcm^{-3}] Dry	ϱ_{tot} [gcm^{-3}] Water-saturated	Ref.	Method of porosity determination	Remarks
Pennsylvanian (continued)											
Third Deese sand	Southwest Antioch field, Oklahoma	≈1981	400			15.5			Ock51	N-1	
Thomas sandstone	Southwest Randlett field, Oklahoma	≈488		21.8	25.8	22.2			Cip56	N-1	Shaly
Tonkawa sandstone	Laverne district, Oklahoma	Subsurface				18			Pat59	N-1	
Tucker sand	Washington County, Oklahoma	413···418	16	5.3	20.6	16.5			Ral54	A-9	
Upper Rowe Formation	Cement field, Oklahoma	1029···1034	22	15.7	26.2	23.3			Ral54	A-9	
Wayside sand	Bartlesville field, Oklahoma	160···161.5	4	15.7	19.4	18.0			Bar31	A-9	
	Keystone field, Oklahoma	341···376	8	6.7	20.9	15.4	2.3(5)	2.49(5)	Mel21	T-2	
	Muskogee and Pawnee Counties, Oklahoma	340···397(?)	2	15.8	20.1	18.0	2.22	2.40	Mel21	T-2	
Wilson sand	Stephens County, Oklahoma	421	1			20.0	2.16	2.36	Mel21	T-2	
Cisco Group	Archer County oil fields, Texas	274···533		19.4	24.6	22.8			Hub26	T-2	9 wells
Cook sand	Cook Ranch field, Texas	≈396	46	12.5	28.2	24.0			Wil52	N-1	
Hickman(?) sand	Petrolia field, Texas	Subsurface	4	18.5	26.6	22.9	2.08	2.31	Mel24	T-2	
McClesky sand	Eastland County, Texas	955···958	2	6.6	8.5	7.6	2.49(1)	2.55(1)	Mel21	T-2	
Middle Strawn sandstone	Walnut Bend field, Texas	1483···1490	20	8.9	21.7	16.1			Ral54	A-9	
Strawn Group (sandstone)	Cooke County, Texas					22			Kru51	N	
	Colemen County, Texas	594	10	23.6	24.6	24.2			Plu43	A-11	
Strawn Group (sand)	Langston-Kleiner field, Texas	≈1067	Many			15.2			Roa55	N-1	8 wells

Stratigraphic unit	Locality	Source of material or depth [m]	Number of samples	ϕ [%] Minimum	ϕ [%] Maximum	ϕ [%] Average	ϱ_{tot} [gcm^{-3}] Dry	ϱ_{tot} [gcm^{-3}] Water-saturated	Ref.	Method of porosity determination	Remarks
Weber Sandstone	Rangely field, Colorado	1707···1920		6	19				Pic48	N-1	Calcareous
Weber Sandstone (quartzite)		≈1768	10	11.1	13.6	12.2			Ros49	N-1	
Weber Sandstone		1811···1959	36	0.4	15.6	9.8			Gat50	A-9	Fine sandstone, hard
Bell sand	Lance Creek field, Wyoming	1760···1809	2	7.5	10.5	9.0			Wal41	T-2	2 wells
Converse sand	Lance Creek field	1407···1410	1			13.5			Wal41	T-2	
Leo sand	Lance Creek field	1573···1744	7	2.9	16.9	8.6			Wal41	T-2	
Second Leo sand		1598···1608	1			5.9			Wal41	T-2	
Tensleep Sandstone	Big Medicine Bow field, Wyoming	2138···2278	3	9.8	19.5	15.0			Wal41	N-1	
	Elk Basin field, Wyoming	≈1219				10.7			Ste55	N-1	
	Longs Creek, Wyoming	1815···1815.4	2			13.7			Hug52	A-5	
	Salt Creek field, Wyoming	1217···1231	2	10.9	18.7	14.8			Wal41	T-2	
	Steamboat Butte oil fields, Wyoming	2110···2165				12···13			Bar48	N-1	
Pennsylvanian and Permian											
Wells Formation (quartzite)	Afton quadrangle, Wyoming	Outcrop				15.1	2.27	2.42	Nut42	T-2	
Carboniferous											
Lower Coal Measures	Great Britain	Quarry	1			20.5	2.16	2.37	Moo04	A-2	
Millstone grit	Meanwood, Great Britain	Outcrop	2	13.0	18.2	15.6	2.23	2.38	Moo04	A-2	
Sandstone	Germany	Outcrop	3	2.7	13.8	6.9	2.47	2.54	Gar1898	T-3	
		Quarry	9	3.8	23.5	11.7			Hir12	A-3	
	Aktyubin area, USSR	350···1350	9			11.8	2.41	2.53	Nev59	T	
		2320···2594	14			3.9	2.50	2.54	Nev59	T	(continued)

Table 11a (continued)

Stratigraphic unit	Locality	Source of material or depth [m]	Number of samples	φ [%] Minimum	φ [%] Maximum	φ [%] Average	ϱ_{tot} [gcm^{-3}] Dry	ϱ_{tot} [gcm^{-3}] Water-saturated	Ref.	Method of porosity determination	Remarks
Permian											
Washington(?) Formation (sandstone)	Waterford, Ohio	Quarry	3			16	2.17	2.33	Win49	A-16	
First sand	KMA field, Texas	1149···1154	21	7.0	22.3	16.2			Ral54	A-9	Permian
KMA sand	KMA field, Texas	1124···1166	156	5	19.6	7.5			Ral54	A-9	Permian
Spraberry siltstone	Spraberry field, Texas	2073					8		Wil53	N-3	
Sandstone	Great Britain	Outcrop(?)	2	12.1	24.8	18.5	2.22	2.41	Wen48	A-2	
Dyas Sandstone	Germany	Quarry	7	9.1	20.7	15.9			Hir12	A-3	2 localities
Zechstein conglomerate		Outcrop	1			2.8	2.55	2.58	Gar1898	T-3	
Sandstone	Aktyubin area, USSR	Subsurface	45			10.1	2.42	2.52	Nev59	T	
			98			10.5	2.43	2.54	Nev59	T	
Conglomerate		Subsurface	1			10.9	2.59	2.70	Nev59	T	
Permian and Triassic											
Sandstone	Aktyubin area, USSR	Subsurface	3			18.5	2.19	2.38	Nev59	T	
Triassic											
Amherst sandstone	Connecticut Valley	Outcrop	1			23	1.92	2.15	Bor35	A-12	
Stockton Formation (sandstone)	New Jersey		5	1.3	7.9	4.0			Fan33	T-1	
Santa Rosa Sandstone	Guadalupe County, New Mexico	Outcrop(?)	1			18.0	2.17	2.35	Mel21	T-2	
Shublik Formation	Barrow, Alaska	808···843	2	9.3	16.0	12.7			Yus51	A	
Bunter Sandstone	Great Britain	Quarry, outcrop	18	5.8	30.8	20.4	2.09	2.29	Moo04	A-2	
	Caldy Grange, Great Britain	Outcrop	1			14.8	2.22	2.37	Moo04	A-2	7.5 cm from fault
			1			15.5	2.21	2.36	Moo04	A-2	30 cm from fault

Stratigraphic unit	Locality	Source of material or depth [m]	Number of samples	ϕ [%] Minimum	ϕ [%] Maximum	ϕ [%] Average	ϱ_{tot} [gcm^{-3}] Dry	ϱ_{tot} [gcm^{-3}] Water-saturated	Ref.	Method of porosity determination	Remarks
Bunter Sandstone	Caldy Grange, Great Britain		1			22.5	2.02	2.25	Moo04	A-2	60 cm from fault
			1			25.5	1.94	2.20	Moo04	A-2	3.5 m from fault
			1			25.5	1.94	2.20	Moo04	A-2	7.5 m from fault
Keuper Sandstone	Great Britain	Outcrop	16	16.5	28.6	22.6	2.02	2.25	Moo04	A-2	
	Caldy Grange, Great Britain	Outcrop	1			16.5	2.18	2.35	Moo04	A-2	7.5 cm from fault
			1			18.0	2.14	2.32	Moo04	A-2	30 cm from fault
			1			20.1	2.08	2.28	Moo04	A-2	60 cm from fault
			1			22.5	2.02	2.25	Moo04	A-2	3.5 m from fault
			1			22.6	2.01	2.24	Moo04	A-2	7.5 m from fault
Bunter Sandstone	Near Hanover, Germany	In place						2.25	Bir24		Torsion balance
Buntsandstein	Germany	Outcrop	1			20.5	2.09	2.30	Gar1898	T-3	
		Quarry	39	7.7	26.4	18.3			Hir12	A-3	
Keuper Sandstone		Quarry	11	12.1	28.3	19.9			Hir12	A-3	
Sandstone	Cantons of Basel-Land Basel-Stadt, Schwyz and St. Gallen, Switzerland	Quarry	7	3.6	21.1	11.1	2.42	2.53	Gru15	T-4	Dips 0°···20°

Jurassic

Stratigraphic unit	Locality	Source of material or depth [m]	Number of samples	ϕ [%] Minimum	ϕ [%] Maximum	ϕ [%] Average	ϱ_{tot} [gcm^{-3}] Dry	ϱ_{tot} [gcm^{-3}] Water-saturated	Ref.	Method of porosity determination	Remarks
Jones sand	Schuler field, Arkansas	2286···2347		0	35	20.2			Wee42	N-1	
		2306···2326	62	5.4	29.2	20.9			Ell44	N-1	1 well
	Various fields, Arkansas	2295···2336	53	8.1	27.3	15.3			Ral54	A-9	
Morgan sands	Schuler field, Arkansas	1615···1783			26	16			Wee42	N-1	

(continued)

Table 11a (continued)

Stratigraphic unit	Locality	Source of material or depth [m]	Number of samples	φ [%] Minimum	φ [%] Maximum	φ [%] Average	ϱ_{tot} [gcm^{-3}] Dry	ϱ_{tot} [gcm^{-3}] Water-saturated	Ref.	Method of porosity determination	Remarks
Jurassic (continued)											
Morrison Formation (sandstone)	Long Park, Montrose County, Colorado	73···95	62	4.9	24.1	18.6			Cad55	A-9	Well sorted
		655···877	22	5.2	27.8	17.9			Cad55	A-9	Moderately well sorted
		70···81	3	11.7	16.7	14.5			Cod55	A-9	Poorly sorted
Morrison Formation (siltstone)		61···75	22	7.1	27.3	15.6			Cad55	A-9	Poorly sorted
Morrison Formation	Cisco dome, Utah	Outcrop	1	1.9		4.0			Wen46	T-2	Very limy
	Two structures, Wyoming	2516···2560	6		11.1	6.0			Wal41	T-2	2 wells
Nugget Sandstone	Steamboat Butte oil field, Wyoming	≈1585	1?	8.7	24.9				Bar48	N-1	
Nugget Sandstone (quartzite)	Afton quadrangle, Wyoming					8.2	2.46	2.54	Nut42	T-2	
Nugget Sandstone	Fremont County, Wyoming					24.9			Kru51	N	
Sundance Formation	Big Medicine Bow field, Wyoming	1574···1784	24	6.0	23.2	12.9			Ral54	A-9	
		1652	1			12.6			Wal41	T-2	
	Iles dome, Colorado	≈991	1			16.8			Ral54	A-9	
	Iles dome and Lance Creek fields, Colorado and Wyoming	1011···1205	11	2.3	22.1	11.8			Wal41	T-2	5 wells
Sundance Formation, basal part	Lance Creek field, Wyoming	975···1341	Many	20	30	25			Elk49	N-1	
Sundance Formation	Quealy and Lance Creek fields, Wyoming	1235···1311	3	19.1	24.9	22.5			Wal41	T-2	2 wells
	Salt Creek field, Wyoming	828···897	4	4.5	17.6	11.7			Wal41	T-2	2 wells
Preuss Sandstone	Afton quadrangle, Wyoming	Outcrop				3.4	2.59	2.62	Nut42	T-2	

Stratigraphic unit	Locality	Source of material or depth [m]	Number of samples	ϕ [%] Minimum	ϕ [%] Maximum	ϕ [%] Average	ϱ_{tot} [gcm^{-3}] Dry	ϱ_{tot} [gcm^{-3}] Water-saturated	Ref.	Method of porosity determination	Remarks
Stump Sandstone	Afton quadrangle	Outcrop				5.0	2.58	2.63	Nut42	T-2	
Kingak Shale	Barrow, Alaska	530···741	20	8.5	25.0	15.8			Yus51	A	
	Simpson, Alaska	1875···1911	3	17	23	19.0			Yus51	A	
Dogger sandstone	Wesendorf field, Germany	Subsurface(?)				23			Ree46	N	
	Oberg field, Germany	Subsurface(?)		12	22				Ree46	N	
Sandstone	Germany	Quarry	4	5.2	24.6	17.1			Hir12	A-3	
	Luxemburg	Outcrop	1			15.0	2.28	2.43	Gar1898	T-3	Fine sandstone, uniform
Cretaceous											
Raritan Formation (gravel)	New Jersey	56	1			31.9	1.67	1.99	Ste27	T-5	
Raritan Formation (sandstone)	New Jersey	Outcrop and water well	4	41.7	48.4	45.6	1.43	1.89	Ste27	T-5	Medium sand
	New Jersey	Pit	2	39.9	44.8	42.3	1.49	1.91	Ste27	T-5	Fine sand
Raritan(?) Formation (sandstone)	New Jersey	Pit	1			41.5	1.54	1.96	Ste27	T-5	Very coarse sand
			1			35.7	1.75	2.11	Ste27	T-5	Fine gravel
Sands	Dare County, N. Carolina	1977···2006	22	2.5	33.9	26.6			Spa50	N-1	
		2140···2195	15	12.8	32.6	25.5			Spa50	N-1	
Eutaw Formation (sands)		1115···1306	8	15.9	41.2	32.1			Spa50	N-1	
	Gilbertown field, Alabama	1006···1036				30			Cur48	N-1	
Tuscaloosa Group	Brookhaven oil field, Mississippi	3089···3214				24			Wom50	N-1	Pilot zone
						26			Wom50	N-1	Smith zone
						26			Wom50	N-1	Arrington zone
Sturgis zone sand	Fouke field, Arkansas	≈1097		15	36	27.3			Sch48a	N-1	

(continued)

Table 11a (continued)

Stratigraphic unit	Locality	Source of material or depth [m]	Number of samples	φ [%] Minimum	φ [%] Maximum	φ [%] Average	ϱ_{tot} [gcm^{-3}] Dry	ϱ_{tot} [gcm^{-3}] Water-saturated	Ref.	Method of porosity determination	Remarks
Cretaceous (continued)											
Trinity Group	Miller County, Arkansas	Subsurface	2	28.7	29.9	29.3			Ral54	A-9	
Dees sand	Rodessa field, Louisiana		3	13.1	24.9	17.9			Ral54	A-9	
Gas Sand	Monroe gas field, Louisiana	≈655	5	14.3	40.3	27.8	2.04(4)	2.29(4)	Mel21	T-2	
Hill sand	Rodessa field, Louisiana	Subsurface	2	18.0	22.8	20.4			Ral54	A-9	
Tokio sand	Claiborne Parish, Louisiana	Subsurface	2	29.3	30.7	30.0			Ral54	A-9	
	Pine Island field, Louisiana	Subsurface	3	3.2	18.4	13.3			Bar31	A-6	
Trinity Group sand	North Lisbon field, Louisiana	Subsurface	2	20.3	26.4	23.4			Ral54	A-9	
Elledge sand	New Hope field, Texas	≈2499	Many			10.8			Tru50	N-1	
Hill sand	Carthage and New Hope fields, Texas	2241···2296	83	5.4	26.1	15.6			Ral54	A-9	
	New Hope field, Texas	≈2255	Many			16.6			Tru50	N-1	
	Ham Gossett field, Texas	≈1981				20			Wig54	N-1	Calcareous
Lower Pettit sand	Carthage field, Texas	1794···1879	32	3.9	31.0	21.2			Ral54	A-9	
Paluxy Sandstone	Talco field, Texas	≈1219				25			Wen48	N-1	
Pittsburg sand	New Hope field, Texas	2404···2481	315	1.7	20.6	11.5			Ral54	A-9	
		≈2408	Many			13.4			Tru50	N-1	
Travis Peak Formation	Limestone County, Texas	222···223	2	13.5	14.3	13.9	2.08	2.22	Plu43	A-11	
Upper Pettit sand	Carthage field, Texas	1730···1735	16	13.0	20.7	16.7			Ral54	A-9	
Woodbine Formation	Rusk County, Texas	742···1128	10	19.0	32.0	24.7	2.00	2.25	Plu43	A-11	
(sand)	Gregg County, Texas	1071···1132	31	19.8	31.4	26.7			Ral54	A-9	
	Tyler County, Texas					22.1			Kru51	N	

2.1 Porosity of rocks

Stratigraphic unit	Locality	Source of material or depth [m]	Number of samples	ϕ [%] Minimum	ϕ [%] Maximum	ϕ [%] Average	ϱ_{tot} [gcm^{-3}] Dry	ϱ_{tot} [gcm^{-3}] Water-saturated	Ref.	Method of porosity determination	Remarks
Woodbine Formation (sand)	E. Texas field, Texas	≈1113	7	23.8	29.0	26.0			Fan33	A-6	Silty
		1068···1070	1			17.5			Min33	N	Conglomerate
		1074	1			17.6			Min33	N	Chalk conglomerate
		1078···1080	1			5.5			Min33	N	Conglomerate, clayey
		1078···1080	1			6.3			Min33	N	Conglomerate, clayey
		1150···1151	1			11.0			Min33	N	Poor saturation
		1150···1151	1			17.6			Min33	N	Fair saturation
		1150···1151	1			22.2			Min33	N	Excellent saturation
		1151···1152	1			19.7			Min33	N	Good saturation
		1153···1154	1			11.0			Min33	N	Slight saturation
	Ham Gossett field, Texas	≈1113	25			19.0			Min33	N	
		1036···1250		0(?)	27.2	23			Wil53	N-1	10 wells
Lewisville Member	Hawkins field, Texas	≈1478				26.4			Wen46	N-1	
Dexter Member		≈1478				27.4			Wen46	N-1	
Woodbine Formation (sand)	Mexia-Groesbeck field, Texas	≈914	8	10.7	37.7	24.4	2.05	2.29	Mel24	T-2	
	Limestone County, Texas	887···922	100	5.1	29.7	18.6			Ral54	A-9	
Sandstone, well-sorted	Iowa, Nebraska		9	33.1	41.3	37.3			Rus26	T-1	Lignitic
Sandstone, well-sorted, coaly	South Dakota		2	27.7	29.9	28.9			Rus26	T-1	Carbon ratio, 50···55%
Sandstone, cemented	Lincoln County, Kansas	Outcrop	7	1.1	3.4	1.5			Swi47	A-8	Acid solubles, 34···38%
	Ellsworth County, Kansas		1			1.5			Swi47	A-8	Acid solubles, 37%

(continued)

Table 11a (continued)

Stratigraphic unit	Locality	Source of material or depth [m]	Number of samples	ϕ [%] Minimum	ϕ [%] Maximum	ϕ [%] Average	ϱ_{tot} [gcm^{-3}] Dry	ϱ_{tot} [gcm^{-3}] Water-saturated	Ref.	Method of porosity determination	Remarks
Cretaceous (continued)											
Dakota Sandstone (D sand)	Denver Basin, Colorado	Subsurface		8.6	29.5	21.6			Mac55	N-1	
Dakota Sandstone (J sand)	Little Beaver field, Colorado	$\approx$1585		8.9	32.7	19.6			Mac55	N-1	
Dakota Sandstone (D sand)	Little Beaver field, Colorado	$\approx$1585 $\approx$1646				19.7 20.7			Fen55 Fen55	N-1 N-1	Top member Lower member
Dakota Sandstone (J sand)		$\approx$1646				22.8			Fen55	N-1	
Dakota Sandstone	Several structures, Wyoming and Colorado	751…1097	7	13.5	23.4	18.5			Wal41	T-2	4 wells
	Beaver Creek and Nieber domes, Wyoming	2423…2632	6	4.0	7.6	5.4			Wal41	T-2	2 wells
Bearpaw Shale (siltstone)	Rosebud County, Montana	Outcrop	1			41.1	1.57	1.98	Ste27	T-5	
Claggett Formation (siltstone)	Rosebud County, Montana	Outcrop	1			36.3	1.81	2.17	Ste27	T-5	
Cut Bank sand	Glacier County, Montana					15.4			Kru51	N	
First Cat Creek sand	Fergus County, Montana		1			22.6	2.07	2.30	Ste27	T-5	Very fine sandstone
Hell Creek Formation	Rosebud County, Montana	Outcrop	5	24.8	38.3	35.3	1.80	2.15	Ste27	T-5	Fine sandstone
	Yellowstone County, Montana	Outcrop	1			26.8	1.97	2.24	Ste27	T-5	
Judith River Formation	Rosebud County, Montana	Outcrop	2 2	32.3 33.2	44.8 51.2	38.5 42.2	1.66 1.82	2.05 2.24	Ste27 Ste27	T-5 T-5	Very fine sandstone Fine sandstone

Stratigraphic unit	Locality	Source of material or depth [m]	Number of samples	ϕ [%]			ϱ_{tot} [gcm^{-3}]		Ref.	Method of porosity determination	Remarks
				Minimum	Maximum	Average	Dry	Water-saturated			
Lance Formation	Rosebud County, Montana	Outcrop	3	36.7	43.0	39.4	1.66	2.05	Ste27	T-5	Fine sandstone
			1			30.8	1.87	2.18	Ste27	T-5	Fine sandstone
			1			27.4	2.01	2.28	Ste27	T-5	Siltstone
Peay sand	Big Horn Mountains Jack Creek, Montana	Outcrop	1			5.1	2.52	2.57	Mel24	T-2	
		Outcrop	1			5.0	2.58	2.63	Mel24	T-2	
Second Cat Creek sand	Fergus County, Montana	Outcrop	1			23.2	2.06	2.29	Ste27	T-5	Fine sandstone
Third Cat Creek sand			1			25.8	1.96	2.22	Ste27	T-5	Fine sandstone
Virgelle Sandstone, Member of Eagle Sandstone	Yellowstone County, Montana	Outcrop	1			23.6	2.06	2.30	Ste27	T-5	Fine sandstone
	Fergus County, Montana	Outcrop	1			27.1	1.93	2.20	Ste27	T-5	Fine sandstone
Adaville Formation (sandstone)	Sublette County, Wyoming	3879···3881	6	10.3	13.7	12.0			Hay51	A	
		4038···4229	36	7.7	16.5	12.2			Hay51	A	
		4436···4501	11	8.1	11.9	10.1			Hay51	A	
Blair Formation (sandstone)	Sublette County	4805···4811	11	7.8	15.4	11.0			Hay51	A	
Blair(?) Formation (sandstone)		5085···5145	15	4.3	10.4	7.6			Hay51	A	
Frontier Formation (sandstone)	Sublette County	6213···6249	21	4.6	9.5	7.4			Hay51	A	
Bear River Formation	Afton quadrangle, Wyoming		1			7.2	2.48	2.55	Nut42	T-2	
			1			13.8	2.32	2.46	Nut42	T-2	
			1			12.9	2.28	2.41	Nut42	T-2	Quartzite
			1			7.4	2.47	2.54	Nut42	T-2	Conglomerate
Cody Formation	Beaver Creek structure, Wyoming	1136				7.1			Wal41	T-2	Siltstone
Ephraim Conglomerate	Afton quadrangle, Wyoming	Outcrop	1			13.2	2.31	2.44	Nut42	T-2	
Frontier Formation	Grass Creek field, Wyoming	Subsurface				27.0	1.93	2.20	Mel21	T-2	

(continued)

Table 11a (continued)

Stratigraphic unit	Locality	Source of material or depth [m]	Number of samples	ϕ [%] Minimum	ϕ [%] Maximum	ϕ [%] Average	ϱ_{tot} [gcm^{-3}] Dry	ϱ_{tot} [gcm^{-3}] Water-saturated	Ref.	Method of porosity determination	Remarks
Cretaceous (continued)											
Frontier Formation	Afton quadrangle, Wyoming					17.1	2.30	2.47	Nut42	T-2	
Frontier Formation (sandstone)	Wyoming	1301···2102	4	2.7	20.8	14.1			Wal41	T-2	3 wells
		2150···2571	17	0.7	25.9	10.7			Wal41	T-2	3 wells
Dakota Formation (sandstone)	Quealy and Lance Creek structures, Wyoming	1132···1209	4	12.7	23.2	17.6			Wal41	T-2	2 wells
	Nieber dome, Wyoming	2550	1			6.8			Pal41	T-2	
Lower Muddy sand	S. Glen Rock field, Wyoming	1676···2073				20			Cur54	N-1	
Upper Muddy sand		1676···2073				14			Cur54	N-1	
Mesaverde Formation	Beaver Creek structure, Wyoming	617	1			20.2			Wal41	T-2	
Newcastle Sandstone	Lance Creek field, Wyoming	884···888	2	14.9	20.7	17.8	2.29(1)	2.44(1)	Mel21	T-2	
	Osage field, Wyoming	12(?)	12	11.3	26.0	16.7	2.15(5)	2.34(5)	Mel21	T-2	
		431···442	3	21.3	23.6	22.7	2.04(1)	2.28(1)	Mel21	T-2	
	Weston, Crook and Carbon Counties, Wyoming	Outcrop	9	8.8	24.8	21.2	2.17(4)	2.35(4)	Mel21	T-2	
	Weston County, Wyoming	437	2	19.0	19.5	19.3	2.14	2.33	Mel21	T-2	
Torchlight Sandstone Member of Frontier Formation	Big Horn County, Wyoming	Outcrop	2			29.4			Mel24	T-2	
First Wall Creek sand	Carbon County, Wyoming		1			19.9			Mel24	T-2	
Lost Soldier field, Wyoming		88···95	2	17.5	20.1	18.8	2.19 (1)	2.37 (1)	Mel21	T-2	

2.1 Porosity of rocks

Stratigraphic unit	Locality	Source of material or depth [m]	Number of samples	ϕ [%] Minimum	ϕ [%] Maximum	ϕ [%] Average	ϱ_{tot} [gcm^{-3}] Dry	ϱ_{tot} [gcm^{-3}] Water-saturated	Ref.	Method of porosity determination	Remarks
First Wall Creek sand	Salt Creek field, Wyoming	376	1			19.8	2.18	2.38	Mel21	T-2	
	Big Muddy field, Wyoming	927…971	6	17.4	23.4	20.5	2.07	2.29	Mel21	T-2	
Wall Creek Sandstone	Salt Creek field, Wyoming	315…443	4	12.1	20.8	15.3			Wal41	T-2	2 wells
	Natrona County, Wyoming	≈457	1			25.8	1.96	2.22	Mel24	T-2	
		Outcrop	1			7.6	2.47	2.55	Mel24	T-2	Calcareous
Wayan Formation	Afton quadrangle, Wyoming		1			7.7	2.45	2.53	Nut42	T-2	
Ferron Sandstone, Member of Mancos Shale	Wasatch Plateau gas fields, Utah	1402…1554		1	21	12…15			Wal55	N-1	
Dakota Sandstone	North of Chama, N. Mexico	Outcrop	27	3.9	24.5	17.0			Wal46	T-2	
Okpikruak Formation	Oumalik, Alaska	2328…3190	10	2.8	10.0	4.8			Yus51	A	Graywacke
Seabee Formation	Umiat, Alaska	163…176	3	6.1	17	11			Yus51	A	
Topagoruk Formation	Barrow, Alaska	61…584	10	24	29	25.9			Yus51	A	
	Fish Creek, Alaska	891…905	2	25	31	28.0			Yus51	A	
	Oumalik, Alaska	280…840	24	0.4	15.0	8.3			Yus51	A	
	Simpson, Alaska	41…300	22	30	38	35.0			Yus51	A	
	Umiat, Alaska	30…917	97	4	44	13.6			Yus51	A	
Torok Formation	Barrow, Alaska	369…975	44	6.4	24.0	15.0			Yus51	A	Graywacke
Tuktu Formation	Fish Creek, Alaska	1671…1832	8	6.0	10.0	7.8			Yus51	A	Graywacke
	Oumalik, Alaska	989…1145	10	6.0	16.0	10.9			Yus51	A	Graywacke
	Barrow, Alaska	638…639	1			24.0			Yus51	A	Graywacke
Tuluga Member (former usage) of Schrader Bluff Formation	Fish Creek, Alaska	498…500	2	28	33	30.5			Yus51	A	Graywacke
Basal quartz sand	Bellshill Lake field, Alberta, Canada	≈945	25			26.6			Rud59	N-1	

(continued)

Table 11a (continued)

Stratigraphic unit	Locality	Source of material or depth [m]	Number of samples	φ [%] Minimum	φ [%] Maximum	φ [%] Average	ϱ_{tot} [gcm^{-3}] Dry	ϱ_{tot} [gcm^{-3}] Water-saturated	Ref.	Method of porosity determination	Remarks
Cretaceous (continued)											
Ellerlie quartz sands	White Mud oil field, Alberta, Canada	≈1250				25.5			Hun50	N-1	
Viking sandstone	Alberta, Canada	Subsurface	2	17	22	20			Ros49	N-1	
Guasare formation	La Paz field, Venezuela	549…701				22			Car48	N-1	
Eudower sandstone	Germany	Outcrop	2	12.5	15.9	14.2	2.26	2.40	Gar1898	T-3	
Quadersandstein	Germany	Quarry	16	12.2	26.2	21.7			Hir12	A-3	
Sandstone	Germany	Quarry	13	8.8	23.6	18.9			Hir12	A-3	
Valendis sand	Emlichheim field, Germany	Subsurface		25	39				Ree46	N	
Wealden sands	Lingen field, Germany	Subsurface		12	15				Ree46	N	
	Nienhagen and Hanigsen fields, Germany	Subsurface		25	30				Ree46	N	
Paleocene											
Fort Union Formation	North of Buckley, Montana	Outcrop	1			22.6	2.09	2.32	Ste27	T-5	Fine sandstone
Lebo shale Member of Fort Union Formation (sandstone)	Rosebud County, Montana	Outcrop	1			27.7	1.93	2.21	Ste27	T-5	Fine sandstone
Lebo shale Member of Fort Union Formation (siltstone)		Outcrop	1			40.1	1.65	2.05	Ste27	T-5	Siltstone
Tongue River Member of Fort Union Formation	Rosebud County	Outcrop	7	9.4	36.6	27.3	1.96	2.24	Ste27	T-5	Fine stonestone
			3	31.4	53.6	40.0	1.63	2.03	Ste27	T-5	Very fine sandstone
			1			26.2	1.91	2.17	Ste27	T-5	Siltstone
Tullock Member of Fort Union Formation	Rosebud County	Outcrop	2	26.7	36.6	31.7	1.87	2.19	Ste27	T-5	Very fine sandstone
			1			29.8	1.92	2.22	Ste27	T-5	Siltstone

Stratigraphic unit	Locality	Source of material or depth [m]	Number of samples	φ [%] Minimum	Maximum	Average	ϱ_{tot} [gcm^{-3}] Dry	Water-saturated	Ref.	Method of porosity determination	Remarks
Tullock Member of Fort Union Formation	Rosebud County, Montana	Outcrop	1			34.0	1.73	2.07	Ste27	T-5	Medium sandstone
Paleocene and Eocene											
Wasatch Formation	Powder Wash field, Colorado	942···949	4	25.9	30.2	27.7			Ral54	A-9	
Eocene											
Sparta Sand	Gulf coast, USA	109···224	13	33.7	46.5	42.2			Jon51	N-2	
	Gulf coast oil fields USA	≈2743	Many			18···20			Tod40	N-1	
	Nachitoches, Louisiana	169···179	3	41.1	41.8	41.4			Jon51	N-2	
	Ville Platte field, Louisiana	2766				26			Tod40	N-1	
Wilcox Group (sands)	Gulf coast, USA	65···169	10	40.0	43.9	41.6			Jon51	N-2	
	Gulf coast oil fields, USA	Subsurface	Many			20···22			Tod40	N-1	
	Eola field, Louisiana	≈2590	68	9	28	22			Bat41		From core graphs
		2609				24			Tod40	N-1	
Morein (Wilcox) sand	Mamou field, Louisiana	≈3505				19.7			Cre51	N-1	
Cockfield sands	Conroe field, Texas	1600*	200			25···28			Mic36	N-1	
Lower Pawelek (Wilcox) sand	Falls City field, Texas	≈1920	45			24.5			Cru50	N-1	
O'Hern sand	Duval County, Texas					28.4			Kru41	N	
Wilcox Group (sands)	Coastal fields, Louisiana, Texas	2326···3095	Many	7	23	18.4			Cul40	N-1	14 fields
Wilcox Group (massive sand)	Slick-Wilcox field, Texas	≈2438				22			Seb48	N-1	
Yegua Formation (sands)	Katy field, Texas	1905···2271				27			All46	N-1	
Knight Conglomerate	Afton quadrangle, Wyoming		1 (?)			7.4	2.47	2.54	Nut42	T-2	Siltstone

*) Average value.

(continued)

Table 11a (continued)

Stratigraphic unit	Locality	Source of material or depth [m]	Number of samples	ϕ [%] Minimum	ϕ [%] Maximum	ϕ [%] Average	ϱ_{tot} [gcm^{-3}] Dry	ϱ_{tot} [gcm^{-3}] Water-saturated	Ref.	Method of porosity determination	Remarks
Eocene (continued)											
Lyre Formation	Olympic Peninsula, Washington	Outcrop	4	7.4	9.7	8.7			Bro56	A	
Gatchell sand	Pleasant Valley field, California	≈2789	Many			15			Wed51	N-1	
Chorro sands	Infantes field, Columbia	≈671···≈792		15	22				And45	N	
Middle Pauji sandstone	Venezuela	Subsurface				12			Car48	N	
Misoa-Trujillo sandstone						10			Car48	N	
Ramillete sands	La Concepcion field, Venezuela	≈914				21			Car48	N	
Punta Gorda sands		≈1219				18.5			Car48	N	
Upper sands	La Paz field, Venezuela	≈488				26			Car48	N	
Tabla sands	Los Manueles field, Venezuela	1829···2286			20	8···10			Car48	N	
Sandstone	Isle of Wight	Outcrop	1			33.8	1.73	2.07	Mos04	A-2	Banded sandstone
Eocene and Oligocene											
Merecure Formation	Anaco field, Venezuela	≈2134···≈3048	Many			15···20			Fun48	N-1	
Eocene to Miocene											
Sandstone	Wasco field, California	3200···4573				15			Val39	N-1	
Oligocene											
Sands	Saxet field, Texas	≈1768	Many	7	45	31			Poo40	N-1	Medium to coarse
		≈1920	Many	20	31	23			Poo40	N-1	Medium to coarse
		≈2103	Many	21	34	28			Poo40	N-1	Medium to coarse

Stratigraphic unit	Locality	Source of material or depth [m]	Number of samples	ϕ [%] Minimum	ϕ [%] Maximum	ϕ [%] Average	ϱ_{tot} [gcm^{-3}] Dry	ϱ_{tot} [gcm^{-3}] Water-saturated	Ref.	Method of porosity determination	Remarks
B zone sands	La Cara field, Colombia	335···610				22			And45	N	
Sandstone	Cantons of Freiburg, Luzern, Vaud, and Zug, Switzerland	Quarry	6	1.5	13.9	5.8	2.53	2.59	Gru15	T-4	Dip low to 25°, in part calcareous
	Cantons of Unterwalden, Vaud, and Wallis, Switzerland		4	0.8	4.1	2.3	2.64	2.66	Gru15	T-4	Dip 15°···50°, calcareous
Oligocene (?)											
Frio Clay (sand)	Amelia field, Texas	2040···2068				30			Ham39	N-1	
	Anahuac field	2121···2157	22	23.4	37.1	30.4			Ral54	A-9	
Frio Clay (sand No. 1)	South Cotton Lake field, Texas	≈1981				30···35			Wil41	N-1	
Frio Clay (sand No. 2)		≈1981				30···35			Wil41	N-1	
Marginulina sand	South Cotton Lake field, Texas	≈1981				25···30			Wil41	N-1	
Sand in Frio Clay	La Rosa field, Texas	≈1798				32			Fis41	N-1	
Oligocene and Miocene											
Amarillo F sand	Guario dome, Venezuela	≈1981				22···26			Fun48	N-1	
Oficina formation (sands)	Anaco fields, Venezuela	≈1372···2896				18···20			Fun48	N-1	
	Greater Oficina area, Venezuela	914···2347	Many			21···30			Hed47	N-1	Average of averages, 26%
Miocene											
Kirkwood Formation	New Jersey	≈244	5	30.2	44.3	38.0	1.63	2.01	Ste27	T-5	Fine to gravelly
Catahoula Sandstone	Gulf coast, USA	≈71···76	2	40.0	40.9	40.5			Jon51	N-2	Medium to coarse
	Saxet field, Texas	≈1341	Many	29	42	35			Put56	N-1	

(continued)

Table 11a (continued)

Stratigraphic unit	Locality	Source of material or depth [m]	Number of samples	φ [%] Minimum	φ [%] Maximum	φ [%] Average	ϱ_{tot} [gcm^{-3}] Dry	ϱ_{tot} [gcm^{-3}] Water-saturated	Ref.	Method of porosity determination	Remarks
Miocene (continued)											
Fleming Formation of former usage	Gulf coast, USA	105···564	9	31.3	50.1	41.2			Jon51	N-2	
Lombardi sand	San Ardo field, California	≈640		40	45				Bal50	N-1	
Modelo Formation, lower part	Santa Monica Mountains, California	Outcrop	2	22.4	24.0	23.2			Jak37	T-2	Graywacke
Miocene A-2 sand	Wasco field, California	3991···4002		12	24				Val39	N-1	
Salinas Shale (sandstone)	Santa Barbara County, California	Outcrop	1			33.3	1.78	2.11	Mel21	T-2	Dip 55°
Stevens sand zone	Paloma field, California	3050···3102		18	20				Cla40	N-1	
Stevens sand		≈3048	2			21.9	2.08	2.30	Hug52	A-5	
Stevens sand, F-1 section	South Coles Levee field, California	2833···2877	103	3.8	24.9	18.9			Gat50	A-9	Standard deviation, 4.1%
Stevens sand	Ten Section field, California	≈2469	Many	15	30	20			Lie49	N-1	
Temblor Formation (sands)	Kettleman Hills field, California	1905···2844	Many			14			Ges33	N	
Temblor Formation (sandy shales)		1905···2844	Many			7			Ges33	N	
Upper Terminal zone	Wilmington oil field, California	914···1067				25			Bar38	N	91 m of sand
Conglomerate	Cantons of Appenzell and St. Gallen, Switzerland	Quarry	2	1.1	1.1	1.1	2.72	2.73	Gru15	T-4	Dips 16°···17°

2.1 Porosity of rocks

Stratigraphic unit	Locality	Source of material or depth [m]	Number of samples	ϕ [%] Minimum	ϕ [%] Maximum	ϕ [%] Average	ϱ_{tot} [gcm^{-3}] Dry	ϱ_{tot} [gcm^{-3}] Water-saturated	Ref.	Method of porosity determination	Remarks
Sandstone	Cantons of Aargau, Schwyz, Solothurn, St. Gallen, Zug, Appenzell, Basel-Land, Bern, Freiburg, and Luzern, Switzerland		15	13.3	22.1	18.7	2.19	2.37	Gru15	T-4	Dips 7° or less
			53	0.4	17.3	6.5	2.52	2.59	Gru15	T-4	Dips 10° or more
Tipam series sand	Digboi field, Assam	≈1524		3	27	12			Cor49	N-1	
Pliocene											
First grubb pool	San Miguelito field, California	≈1829	800			18			Gle50	N-1	20 wells, poorly sorted sands
First grubb zone		2002···2270				20.6			McC51	N-1	
Second grubb zone		2349···2517				19			McC51	N-1	
Third grubb zone		2582···2724				14.2			McC51	N-1	
Pliocene and Pleistocene											
Sands	Mamou, Louisiana, Gulf coast, USA	155···292	4	36.2	38.7	37.8			Jon51	N-2	
		628···651	3	40.0	40.6	40.3			Jon51	N-2	
Tertiary											
Sandstone	Germany	Quarry	1			29.1			Hir12	A-3	
Age not specified											
Sandstone	Lancashire and Derby, England	Outcrop (?)	2	12.8	13.7	13.3	2.27	2.40	Hol30	T-2	
Sandstone (dolomitic)	Mansfield, England	Outcrop	1			14.1	2.31	2.45	Hol30	T-2	
Quartzite	Nuneaton, England	Outcrop	1			6.4	2.48	2.54	Hol30	T-2	
						0.21			Gei06	A-14	
Conglomerate	St. Marcet anticline, France	≈1524		2	18				Sch48	N	
Pechelbronn sand	Pechelbronn, France	Subsurface	1			17.8			Bar31	A-6	

Table 11b. Limestone, dolomite, chalk, and marble.

Stratigraphic unit	Locality	Source of material or depth [m]	Number of samples	ϕ [%] Minimum	ϕ [%] Maximum	ϕ [%] Average	ϱ_{tot} [gcm^{-1}] Dry	ϱ_{tot} [gcm^{-1}] Water-saturated	Ref.	Method of porosity determination	Remarks
Precambrian											
Grenville Marble	Ontario, Canada	Quarry	9	0.01	1.06	0.35	2.77	2.77	Par12	A-2	
Cambrian											
Bonneterre Dolomite	Near Bonne Terre, Montana	Subsurface				3.3	2.66	2.69	Bla55	A-16	Galena bearing
							3.30		Bla55	A-16	
Gallatin Limestone	Afton quadrangle, Wyoming	Outcrop				8.6	2.61	2.70	Nut42	T-2	
Gros Ventre Formation	Afton quadrangle					11.0	2.40	2.51	Nut42	T-2	
Ophir Formation (limestone)	Ophir, Utah	Mine				0.26	2.78	2.78	Win50	A-16	
Upper Cambrian and Lower Ordovician											
Arbuckle Group (limestone)	Various fields, Kansas	880···1281	26	1.2	19.8	10.3			Ral54	A-9	
	Woodrow field, Oklahoma	≈579				2···8			McB56	N-1	
Lower Ordovician											
Ellenburger Group (limestone)	Riley Mountain, Llano County, Texas	Outcrop	12	0.1	0.7	0.5	2.69	2.70	Gol47	T-2	Sublithographic
Ellenburger Group (dolomite)			23	1.1	12.6	4.3	2.72	2.76	Gol47	T-2	Microgranular
			11	1.3	7.1	3.6	2.73	2.77	Gol47	T-2	Fine grained
			6	1.7	4.3	2.6	2.75	2.78	Gol47	T-2	Medium grained
			2	2.6	5.0	3.8	2.73	2.77	Gol47	T-2	Coarse grained

Stratigraphic unit	Locality	Source of material or depth [m]	Number of samples	ϕ [%] Minimum	ϕ [%] Maximum	ϕ [%] Average	ϱ_{tot} [gcm^{-3}] Dry	ϱ_{tot} [gcm^{-3}] Water-saturated	Ref.	Method of porosity determination	Remarks
Ellenburger Group (cherty limestone)	Riley Mountains, Llano County, Texas	Outcrop	1			0.4	2.69	2.69	Gol47	T-2	
Ellenburger Group (calcitic dolomite)			1			0.8	2.74	2.75	Gol47	T-2	
Ellenburger Group (dolomitic limestone)			1			1.5	2.75	2.77	Gol47	T-2	
	Permain Basin, West Texas	≈2682	63			2.5			Atk48	A-4	Matrix porosity
		2700···2711	36			3.30			Atk48	A-4	All porosity
		2700···2711	36			1.51			Atk48	A-4	Matrix porosity
		2700···2711	36			1.79			Atk48	A-4	Fracture and vug porosity
Ordovician											
Trenton Limestone (medium crystalline)	Rose Hill field, Virginia	Subsurface	3			1.2			Mil48	N-1	Producing, 2 wells
Trenton Limestone (finer crystalline)			7			0.6			Mil48	N-1	Producing, 1 well
Kingsport Formation (dolomite)	Mascot, Tennessee		3			0.7	2.84	2.85	Win49	A-16	
Martinsburg Shale (limestone)	Jefferson City, Tennessee	17···87	7	0.4	2.3	1.3	2.78	2.79	Win49	A-16	
Trenton Limestone (quartzitic)	Sand Hill well, Wood County, W. Virgina	2870	1			0.6	2.71	2.72	Rob62	A-15	
Black River Limestone		2919···2946	6	0.3	2.7	0.9	2.67	2.68	Rob62	A-15	
Beekmantown Group (dolomite)		2984···3204	35	0.1	1.4	0.4	2.70	2.70	Rob62	A-15	
		3210···3641	56	0.1	1.1	0.4	2.80	2.80	Rob62	A-15	
Prairie du Chien Group (dolomite)	Wisconsin	Quarry	4	11.1	13.4	12.4	2.43	2.55	Buc1898	A-1	

(continued)

Table 11b (continued)

Stratigraphic unit	Locality	Source of material or depth [m]	Number of samples	ϕ [%] Minimum	ϕ [%] Maximum	ϕ [%] Average	ϱ_{tot} [gcm^{-3}] Dry	ϱ_{tot} [gcm^{-3}] Water-saturated	Ref.	Method of porosity determination	Remarks
Ordovician (continued)											
Trenton Limestone (dolomite)			2	0.9	1.2	1.0	2.81	2.82	Buc1898	A-1	
Kimmswick Limestone	Dupo pool, Illinois	122···136	44	1.9	17.0	10.8			Pie40	A-6	
		≈183	4	11.6	16.5	15.8			Ral54	A-9	
Viola Limestone	Cunningham field, Kansas	1234···1235	3	2.9	3.1	3.0			Ral54	A-9	
Simpson Group		1235···1255	36	5.7	22.3	13.3			Ral54	A-9	
Bighorn Dolomite	Afton quadrangle, Wyoming	Outcrop				8.6	2.59	2.68	Nut42	T-2	
Beekmantown Group (limestone, dolomite)	Ontario, Canada	Quarry	4	1.3	12.6	4.6	2.66	2.71	Par12	A-2	4 localities
Black River Group (limestone)			11	0.07	1.67	0.46	2.72	2.72	Par12	A-2	10 localities
Silurian											
Brassfield Limestone	Piqua, Ohio	Quarry				1.3	2.8		Win50	A-16	Dolomitic
						2.7	2.6		Win59	A-16	
Niagara Group (dolomite)	Maple Grove, Ohio	Quarry				8.6	2.4		Win50	A-16	
	Gibsonburg, Ohio	Quarry	2	3.4	4.0	3.7	2.6		Win50	A-16	
	Lucky, Ohio	Quarry	2	3.0	8.5	5.8	2.5		Win50	A-16	
Red Mountain Formation (limestone)	Bessemer, Alabama	427				0.9	2.83	2.84	Win49	A-16	
	Wisconsin	Mine				0.6	2.92	2.93	Win49	A-16	Limonitic
Niagara Dolomite	Wisconsin	Quarry	14	0.5	6.7	2.9	2.74	2.77	Buc1898	A-1	
Guelph Dolomite	Ontario, Canada	Quarry	3	14.6	15.9	15.4	2.41	2.56	Par12	A-2	3 localities
Niagara Group (limestone, dolomite)	Ontario, Canada	Quarry	6	4.4	13.4	9.1	2.54	2.63	Par12	A-2	6 localities

Stratigraphic unit	Locality	Source of material or depth [m]	Number of samples	ϕ [%] Minimum	ϕ [%] Maximum	ϕ [%] Average	ϱ_{tot} [gcm^{-3}] Dry	ϱ_{tot} [gcm^{-3}] Water-saturated	Ref.	Method of porosity determination	Remarks
Salina Dolomite	W. Becher pool, Ontario	564···587				10			Rol49	N-1	
Limestone	Various localities, Great Britain		4	1.4	6.3	2.8	2.61	2.64	Moo04	A-2	
Upper Silurian and Lower Devonian											
Hunton Group (limestone)	Hollow pool, Kansas	Subsurface	1			23.6			Ral54	A-9	
Hunton Group, Bois d'Arc Limestone	W. Edmond field, Oklahoma	2117···2224	349	0.10	14				McG46	N-1	
		≈2134	56	0.5	18.2				McG46	N-1	
Hunton Group, Bois d'Arc, Harragan, and Henryhouse Limestone		≈2134	545	1.9	16.7	7.3			Lit48	N-1	8 wells
Devonian											
Columbus Limestone	Barberton, Ohio	Quarry	3			0.7	2.69	2.70	Win49	A-16	
	Columbus, Ohio	Quarry				5.4	2.60	2.65	Bla56	A-16	
Delaware Limestone	Spore, Ohio	Quarry	4	5.2	6.4	5.8	2.5		Win50	A-16	
Devonian Limestone	Crawford-Main pool, Illinois	895···899	14	3.8	11.8	8.1			Pie40	A-6	
Dunudee Limestone (dolomite)	Coldwater field, Michigan	≈1143	16			2.5			Cri54	N-1	Matrix porosity
Rogers City Limestone (dolomite)	Coldwater field	≈1142	39			4.2			Cri54	N-1	Matrix porosity
Ouray Limestone	Rattlesnake field, San Juan County, N. Mexico	2135···2138	6	1.4	2.7	2.0			Hin47	N-1	
		2181···2188	4	7.6	12.9	10.0			Hin47	N-1	Porous zone
Jefferson Dolomite	Afton quadrangle, Wyoming					6.2	2.65	2.71	Nut42	T-2	
D-3 zone dolomite	Leduc field, Alberta, Canada	1478···1646				13			Lay49	N-1	

(continued)

Table 11b (continued)

Stratigraphic unit	Locality	Source of material or depth [m]	Number of samples	φ [%] Minimum	φ [%] Maximum	φ [%] Average	ϱ_{tot} [gcm^{-3}] Dry	ϱ_{tot} [gcm^{-3}] Water-saturated	Ref.	Method of porosity determination	Remarks
Devonian (continued)											
D-3 zone	Leduc field, Alberta,	1570···1615	25	2.5	3.5				War50	N-1	Dense matrix
						6.8			War50	N-1	Entire matrix
Limestone	Various localities, Germany		6	0.6	2.7	1.9			Hir12	A-3	
Mississippian											
McClosky lime	Lawrence County, and Noble pools, Illinois	530···905	60	2.3	25.9	14.2			Pie40	A-6	Oolitic
	Olney and Salem pools, Illinois	579···916	18	5.1	14.1	10.7			Pie40	A-6	Oolitic
	Salem pool, Illinois	≈610				10.3			Arn39	N-1	Oolitic
McClosky limestone	Hitesville Cons. field, Kentucky	770···786				13.2			Byb48	N-1	Oolitic
						13.5			Byb48	N-1	Oolitic
McClosky limestone (D zone)		≈792				15.4			Byb48	N-1	Oolitic
						18.5					
Maxville Limestone	Muskingum County, Ohio	≈61	2	0.9	2.0	1.5	2.77		Bla56	A-16	
		≈61	6				2.67	2.79	Bla56	A-16	
Maxville Limestone (marl)		≈61	3				2.19		Bla56	A-16	
Spergen Limestone	Bedford, Indiana	Quarry	7	11	15.4	13	2.29	2.42	Win49	A-16	
	Ste. Genevieve, Montana	≈30	15				2.59		Win49	A-16	
St. Louis Limestone	Prairie du Rocher, Illinois	76···107	3			0.8	2.68	2.69	Bla55	A-16	
	Ste. Genevieve, Montana	≈15	6				2.62		Bla56	A-16	
Boone Formation, siliceous	Near Picher, Oklahoma	Mine	2			1.2	2.67	2.68	Bla55	A-16	
			2			8.9	2.39	2.48	Bla55	A-16	

2.1 Porosity of rocks

Stratigraphic unit	Locality	Source of material or depth [m]	Number of samples	ϕ [%] Minimum	ϕ [%] Maximum	ϕ [%] Average	ϱ_{tot} [gcm^{-3}] Dry	ϱ_{tot} [gcm^{-3}] Water-saturated	Ref.	Method of porosity determination	Remarks
Brazer Limestone	Afton quadrangle, Wyoming		2			4.0	2.64	2.68	Nut42	T-2	
Madison Limestone	Beaver Lodge and Tioga fields, Montana	≈2560	Many			6			Cox53	N-1	12 wells
	Afton quadrangle, Wyoming		1			3.6	2.61	2.65	Nut42	T-2	
Rundle Formation:											
Upper porous zone	Turner Valley field, Alberta, Canada	≈2103	48	2.0	19.6	10.8			Gal51	N-1	Dolomite, limestone
Lower porous zone		≈2164	84			10.4			Gal51	N-1	Dolomite, limestone
Upper porous zone		1393…2051	1	20	10				Mac40	T-2	Dolomitic
Lower porous zone		1530…2124		1	20	10			Mac40	T-2	Dolomitic
Pennsylvanian											
Lansing and Kansas City Formations	Ellingwood field, Kansas	≈1018	5	6.2	25.3	12.6			Ral54	A-9	
Canyon reef limestone	Scurry field, Texas	≈2073	2109	0	31.6	6.3			Ral54	A-9	
Crinoidal limestone	Todd Deep field, Texas	≈1768				11.7			Imb50	N-1	
Marble Falls Limestone	Erath County, Texas	1025…1055	3	7.3	20.7	15.3			Plu43	A-11	
Carboniferous											
Carboniferous limestone	Midlands, England	Outcrop	24	2.2	14.9	5.7	2.52	2.58	Par22	T-2	30% MgCO$_3$
Limestone	Micheldean, Great Britain		2	9.0	9.4	9.2	2.45	2.53	Moo04	A-2	
	Hilton, Great Britain	Outcrop	1			2.2	2.59	2.61	Moo04	A-2	Under Whinsill
	Germany	Quarry	2	0.6	1.3	1.0			Hir12		2 localities

(continued)

Table 11b (continued)

Stratigraphic unit	Locality	Source of material or depth [m]	Number of samples	ϕ [%] Minimum	ϕ [%] Maximum	ϕ [%] Average	ϱ_{tot} [gcm^{-3}] Dry	ϱ_{tot} [gcm^{-3}] Water-saturated	Ref.	Method of porosity determination	Remarks
Permian											
Brown dolomite	Moore County, Texas	1089···1101	15	3.2	27.1	11.7			Phi60	A-10	Permeable
Limestone	Moore County	1101···1111	8	6.3	12.2	8.5			Phi60	A-10	Impermeable
Permian limestone	Big Springs field, Texas	≈2743	5	11.9	22.2	17.9			Ral54	A-9	Dolomite(?)
Rex chert Member of Phosphoria Formation	Afton quadrangle, Wyoming					17.0	2.33	2.50	Nut42	T-2	
Zechstein	Germany	Outcrop	1			3.1	2.64	2.67	Gar1898	T-3	
	Various localities, Germany	Quarry	3	4.1	21.3	12.3			Hir12	A-3	
Dolomite	Aktyubin area, USSR	Subsurface	4			4.1	2.62	2.68	Nev59	T	
Limestone	Aktyubin area	Subsurface	1			5.2	2.75	2.80	Nev59	T	
Marl	Kungur salt basin, USSR		5			7.6	2.62	2.70	Nev59	T	
	Aktyubin area, USSR		14			8.4	2.53	2.61	Nev59	T	
Paleozoic (?)											
Cockeysville Marble	Cockeysville, Maryland	Quarry				0.6	2.87	2.88	Win49	A-16	Dolomitic
Triassic											
Dinwoody Formation (sandy limestone)	Afton quadrangle, Wyoming					12.9	2.42	2.55	Nut42	T-2	
Thaynes Limestone	Afton quadrangle		1	8.3	8.7						
Ross Fork Limestone	Afton quadrangle					0.7	2.65	2.66	Nut42	T-2	
						7.0	2.52	2.59	Nut42	T-2	
Limestone	Germany	Quarry	2	1.8	2.4	2.1			Nut42	T-2	
	Mutzig, Germany	Outcrop	1			1.0	2.68	2.69	Hir12	A-3	
Muschelkalk	Various localities, Germany	Quarry	19	1.2	36.5	15.5			Gar1898	T-3	
	Galicia, Poland		1			13.4	2.46	2.59	Hir12	A-3	Oolitic
									Gar1898	T-3	

2.1 Porosity of rocks

Stratigraphic unit	Locality	Source of material or depth [m]	Number of samples	φ [%] Minimum	φ [%] Maximum	φ [%] Average	ϱ_{tot} [gcm^{-3}] Dry	ϱ_{tot} [gcm^{-3}] Water-saturated	Ref.	Method of porosity determination	Remarks
Limestone	Cantons of Vaud, Aargau, and Basel-Stadt, Switzerland	≈2377	6	1.1	4.4	2.9	2.67	2.70	Gru15	T-4	Dip 3°···65°; in part dolomitic
	Canton Tessin, Switzerland	2749···2772	7	0.4	4.0	2.0	2.68	2.70	Gru15	T-4	Dip about 25°
Marble	Canton Graubünden, Switzerland		2	2.1	2.9	2.5	2.64	2.67	Gru15	T-4	
Jurassic											
Reynolds oolite	Cairo field, Arkansas	≈2377				17			Goe50	N-1	Porous
Reynolds Oolitic, Member of Smackover Formation	Dorcheat pool, Arkansas	2749···2772		2	20	12			Tra40	N-1	
Reynolds oolite	Schuler field, Arkansas	2332···2362			23	16.7			Wee42	N-1	
Reynolds oolitic limestone	Various fields, Arkansas	2210···2332	4	16.4	20.0	18.0			Ral54	A-9	
Smackover Formation	Various fields, Arkansas	≈2393	150	0	23.9	14.5			Ral54	A-9	
	McKamie-Patton pool, Arkansas	≈2835	1767			14.2			Sch57	N-1	
	McKamie field, Arkansas	2780···2859	14	0	16.4	7.5			Ral54	A-9	
Twin Creek Limestone	Afton quadrangle, Wyoming					0.2	2.75	2.75	Nut42	T-2	
Carmel formation (limestone)	Near Huntington, Utah	1.5···6	6	0.2	4.6	2.0	2.65	2.67	Win50	A-16	
Inferior oolite	Cotswolds, England	Outcrop	3	5.5	24.0	13.4	2.33	2.46	Moo04	A-2	
Oolite	Clove Hill and Lockhampton Hall, England		2	14.3	18.3	16.3	2.25	2.41	Moo04	A-2	
New Red marl	Leamington, England		1	30.0	34.4	32.2			Sor08	A-2	
	England	Deep boring	4	4.8	10.0	7.5			Sor08	A-2	
Portland limestone	Great Britain	Outcrop	1			8.6	2.54	2.63	Hol30	A-2	

(continued)

Table 11b (continued)

Stratigraphic unit	Locality	Source of material or depth [m]	Number of samples	ϕ [%] Minimum	ϕ [%] Maximum	ϕ [%] Average	ϱ_{tot} [gcm^{-3}] Dry	ϱ_{tot} [gcm^{-3}] Water-saturated	Ref.	Method of porosity determination	Remarks
Jurassic (continued)											
White Lias Limestone	Tiverton, England	≈1829	1			9.4	2.44	2.53	Moo04	A-2	
	St. Marcet anticline, France				18···20	10			Sch48	N	
	Germany		1			5.0			Hir12	A-3	
Solnhofen Limestone		Outcrop(?)	23	1.2	5.7	3.9	2.57	2.61	Rob62	A-15	
Limestone	Cantons of Aargau, Basel-Land, Bern, Freiburg, Neuenburg, Solothurn, St. Gallen, Tessin, Vaud, and Zürich, Switzerland	Quarry	114	0.4	25.6	3.6	2.63	2.66	Gru15	T-4	Dip 8° or more or not recorded
	Mostly from Canton Schaffhausen, Switzerland		18	0.9	10.7	5.4	2.57	2.63	Gru15	T-4	Dip 7° or less
No. 3 limestone	Dukhan field, Qatar	1707···2012	Many			16			Dan54	N-1	
No. 4 limestone		1707···2012	Many			21			Dan54	N-1	
Cretaceous											
Dees oolitic limestone	Rodessa field, Louisiana	≈1661	74	8.0	32.0	22.0			Ral54	A-9	
Dees Coquina limestone		≈1661	20	7.0	29.7	20.2			Ral54	A-9	
Kilpatrick zone	Sugar Creek field, Louisiana	≈1378		14	23	18			Cla38	N-1	
Sligo Formation	Haynesville field, Louisiana, Arkansas	≈1615				19.0			Aki51	N-1	
Caddo limestone	Eastland County, Texas	888(?)	2	4.2	4.4	4.3	2.58	2.62	Mel24	T-2	
Glen Rose Limestone	Bell County, Texas	6···10	10	16.0	18.8	16.8	2.21	2.37	Plu43	A-11	
	Ham Gossett field, Texas	1737···1890				18			Wig54	N-1	

Stratigraphic unit	Locality	Source of material or depth [m]	Number of samples	ϕ [%] Minimum	ϕ [%] Maximum	ϕ [%] Average	ϱ_{tot} [gcm^{-3}] Dry	ϱ_{tot} [gcm^{-3}] Water-saturated	Ref.	Method of porosity determination	Remarks
Glen Rose ("Bacon") Limestone	Ham Gossett field Texas	1951···1981				6			Wig54	N-1	
	New Hope field, Texas	2206···2263	61	1.2	23.3	12.9			Ral54	A-9	
		≈2225	Many			18.9			Tru50	N-1	
	Pickton field, Texas	≈2408	380			19			Wel49	N-1	19 wells
Rodessa Formation (limestone)	Ham Gossett field, Texas	2012···2073				16			Wig54	N-1	
Niobrara Formation (chalk)	Pickston, S. Dakota	Subsurface	15				1.63		Bla55	A-16	
Bear River Formation (sandy limestone)	Afton quadrangle, Wyoming					4.6	2.61	2.66	Nut42	T-2	
Peterson Limestone	Afton quadrangle					9.5	2.45	2.54	Nut42	T-2	
Limestone	La Paz field, Venezuela	≈1219···≈2438				1···2			Fic53	N-1	
Chalk	Balmoral, Great Britain	Outcrop	1			9.1	2.44	2.53	Moo04	A-2	
	Various localities, Great Britain	Outcrop	3	17.6	42.8	28.8	1.94	2.23	Moo04	A-2	
	Germany	Quarry	3	2.3	7.0	5.8			Hir12	A-3	
Senonian chalk	Reitbrook field, Germany	Subsurface				25			Ree36	N	
Limestone	Cantons of Bern, Neuenburg, Schwyz, St. Gallen, Unterwalden, Vaud, and Wallis, Switzerland	Quarry	29	0.4	18.3	4.3	2.60	2.65	Gru15	T-4	Folded rocks; dip 10° or more
Globigerina limestone (marly)	Ain Zalah field, Iraq	Subsurface				0···11			Dan54	N-1	
Eocene											
Green River Formation (marlstone)	Rifle, Colorado	Mine	11	0.2	12.0	2.9	2.23	2.26	Win49, Win50	A-16	

(continued)

Table 11b (continued)

Stratigraphic unit	Locality	Source of material or depth [m]	Number of samples	ϕ [%] Minimum	ϕ [%] Maximum	ϕ [%] Average	ϱ_{tot} [gcm^{-3}] Dry	ϱ_{tot} [gcm^{-3}] Water-saturated	Ref.	Method of porosity determination	Remarks
Eocene (continued)											
Green River Formation (limestone)	Rifle, Colorado	Mine	3			1.6	2.10	2.12	Bla55	A-16	
Limestone, porous	Eniwetok Atoll, Marshall Islands	≈1219	4				1.84···1.89		Bla56	A-16	
Limestone, hard		≈1219	2				2.31		Bla56	A-16	
Limestone	Cantons of Bern, Schwyz, St. Gallen, and Vaud, Switzerland	Quarry	4	0.7	2.6	1.7	2.68	2.70	Gru15	T-4	Dip 20°···80°
Miocene											
Limestone, sandy	Eniwetok Atoll, Marshall Island	335···819	2				1.21		Bla56	A-16	
Limestone, porous			2				1.83		Bla56	A-16	
Limestone, semiporous							2.39		Bla56	A-16	
Limestone, hard							2.25···2.51		Bla56	A-16	
Limestone	Canton Schaffhausen, Switzerland	Quarry	1			18.3	2.24	2.42	Gru15	T-4	
Asmari limestone	Masjidi Sulaiman field, Iran	Subsurface	140		22.8	5.6			Lee33	A	Well D
			27	1.1	19.3	6.6			Lee33	A	Well H
					13.0				Lee33	A	Dolomite, 0···10%
			12	1.5	12.0				Lee33	A	Dolomite, 11···25%
			8	2.5	8.9				Lee33	A	Dolomite, 26···50%
			2	13.0	15.8				Lee33	A	Dolomite, 51···75%
			7	4.2	16.1				Lee33	A	Dolomite, 75%

Stratigraphic unit	Locality	Source of material or depth [m]	Number of samples	ϕ [%] Minimum	ϕ [%] Maximum	ϕ [%] Average	ϱ_{tot} [gcm^{-3}] Dry	ϱ_{tot} [gcm^{-3}] Water-saturated	Ref.	Method of porosity determination	Remarks
Quaternary											
Calcareous tufa	Cantons of Freiburg, St. Gallen, and Unterwalden, Switzerland	Outcrop	4	7.0	27.8	19.2	2.03	2.22	Gru15	T-4	
Age not specified											
Marble	Eastern USA (31 localities)	Mostly quarry	100	0.4	0.8	0.6	2.74	2.75	Kes19	T-2	
	Missouri	Quarry	4			2.1	2.66	2.68	Kes19	T-2	
	California	Quarry	4			0.6	2.84	2.85	Kes19	T-2	
	Tokeen, Alaska	Quarry	3			0.5	2.72	2.72	Kes19	T-2	
	Tiree, Great Britain	Outcrop	1			1.0	2.65	2.66	Moo04	A-2	
Oolitic limestone	Great Britain	Outcrop	1			20.3	2.16	2.36	Hol30	T-2	
Dolomite	Micheldean, England	Outcrop	1			8.6	2.54	2.63	Hol30	T-2	
Limestone	Buxton, Darby, England	Outcrop	1			14.1	2.31	2.45	Hol30	T-2	

Table 11c. Shale, claystone, and slate.

Stratigraphic unit	Locality	Source of material or depth [m]	Number of samples	ϕ [%] Minimum	ϕ [%] Maximum	ϕ [%] Average	ϱ_{tot} [gcm^{-3}] Dry	ϱ_{tot} [gcm^{-3}] Water-saturated	Ref.	Method of porosity determination	Remarks
Precambrian											
Goodrich Quartzite (argillite)	Ishpeming, Michigan	Mine	3				2.85		Bla56	A-16	
Negaunee Iron-Formation (white slate)		305	2			0.6	2.93	2.94	Win49	A-16	
Nonesuch Shale (siliceous)	White Pine, Michigan	Mine	6	1.5	1.7	1.6	2.76	2.78	Bla55	A-16	

(continued)

Table 11c (continued)

Stratigraphic unit	Locality	Source of material or depth [m]	Number of samples	ϕ [%] Minimum	ϕ [%] Maximum	ϕ [%] Average	ϱ_{tot} [gcm^{-3}] Dry	ϱ_{tot} [gcm^{-3}] Water-saturated	Ref.	Method of porosity determination	Remarks
Cambrian											
Gros Ventre Formation (shale)	Afton quadrangle, Wyoming	Outcrop				11.1	2.38	2.49	Nut42	T-2	
Ophir Formation (shale)	Ophir, Utah	Subsurface	2			0.9	2.81	2.82	Win50	A-16	
Ophir Formation (silicified shale)			1			0.6	2.80	2.81	Win50	A-16	
Ophir Formation (limestone, shale)			1			0.6	2.92	2.93	Win50	A-16	Mineralized
Ordovician											
Martinsburg Shale	Bangor, Pennsylvania	Quarry				1.0	2.74	2.75	Win49	A-16	Slate
Silurian											
Shale	Various localities, Great Britain	Outcrop	5	2.0	10.1	5.2	2.54	2.59	Moo04	A-2	
Wenlock Shale: Weathered	Malvern, England		1			14.1			Sor08	A-2	
Unweathered			1			5.8			Sor08	A-2	
Devonian											
Slate	Various localities, Germany	Quarry	19	1.7	7.6	3.4			Hir12	A-3	Roofing slate
	Hele, Great Britain	Outcrop	2	1.3	3.5	2.4			Sor08	A-2	
Devonian and Mississippian											
Chattanooga Shale	Irvine field, Kentucky Near Smithville, Tennessee	Subsurface Mine	2	7.4		7.5	2.38 2.53	2.45	Mel24 Bla56	T-2 A-16	
Chattanooga Shale, silty	Near Smithville, Tennessee	Mine	2	1.6	1.7	1.7	2.30	2.32	Bla56	A-16	

Stratigraphic unit	Locality	Source of material or depth [m]	Number of samples	ϕ [%] Min-imum	ϕ [%] Max-imum	ϕ [%] Average	ϱ_{tot} [gcm^{-3}] Dry	ϱ_{tot} [gcm^{-3}] Water-saturated	Ref.	Method of porosity determination	Remarks
Hamilton shale	Hannibal, Montana	Outcrop	1			11.3	2.32	2.43	Hed26	T-2	
Mississippian											
Maxville Limestone (shale)	Muskingum County, Ohio	Subsurface	3				2.56		Bla56	A-16	
Shale overlying Keener sand	Monroe County, Ohio	440···450	2	9.7	11.0	10.4	2.43	2.54	Mel21	T-2	
Ridgetop Shale (silty)	Near Smithville, Tennessee	≈46	6				2.71		Bla56	A-16	
Pennsylvanian											
Allegheny Formation (silty shale)	Colver, Pennsylvania	Mine	3				2.67		Bla56	A-16	
Allegheny Formation (shale, slate)	Bakerton, Pennsylvania	≈152	2			1.7	2.72	2.74	Bla56	A-16	
Kanawha Formation (shale)	Dehue, W. Virginia	Mine	6				2.75		Bla55	A-16	
Monongahela Formation	Scotts Run, W. Virginia	Mine	5			6.1	2.5		Win50	A-16	
Upper Block under clay	New Brazil, Indiana	Trench	1			19.1			Alt59	T-2	
Cherokee Shale	Fulton, Montana	Outcrop	2	17.0	17.2	17.1	2.29	2.46	Hed26	T-2	
Flint clay	Fulton	Outcrop	2	10.1	10.1	10.1	2.37	2.47	Hed26	T-2	
Weston Shale	Bonner Springs, Kansas	Outcrop	2	15.5	16.0	15.8	2.28	2.44	Hed26	T-2	
Chanute Shale	Independence, Kansas	Outcrop	2	14.8	15.0	14.9	2.31	2.46	Hed26	T-2	
Chanute(?) Shale	Montgomery County, Kansas	152···297	4	7.3	10.6	9.1	2.51	2.60	Hed26	T-2	
		396···398	3	7.1	8.5	7.7	2.53	2.61	Hed26	T-2	
Burgess sandstone (shale)	S. Moore pool, Oklahoma	2381···2384				3···4			Moo41	N-1	From graph

(continued)

Table 11c (continued)

Stratigraphic unit	Locality	Source of material or depth [m]	Number of samples	ϕ [%] Minimum	ϕ [%] Maximum	ϕ [%] Average	ϱ_{tot} [gcm^{-3}] Dry	ϱ_{tot} [gcm^{-3}] Water-saturated	Ref.	Method of porosity determination	Remarks
Carboniferous											
Coal Measure clay	Brightside and Darnall, England	Outcrop		12.8	14.3	13.5			Sor08	A-2	
	Near Nottingham, England	155				13.4			Ser08	A-2	
		253				14.6			Sor08	A-2	
		363				10.7			Sor08	A-2	
Middle Coal Measure	St. Helens, Great Britain	Outcrop	2	4.0	7.3	5.7	2.84	2.89	Moo04	A-2	
Shale under basalt	Edinburg Castle, Great Britain		1			1.6	2.58	2.60	Moo04	A-2	
Slate	Various localities, Germany	Quarry	4	1.2	3.8	2.5			Hir12	A-3	Roofing slate
Clay, nonplastic	Near Dabrowa, Poland	Mine, hanging wall	1			0.85	2.44	2.45	Pet26	T-2	
		Sabatzlarer bed. 8	1			1.2	2.48	2.49	Pet26	T-2	
Clay	Aktyubin area, USSR	351···1350	2			15.3	2.30	2.45	Nev59	T	
		2320···2594	1			12.5	2.31	2.44	Nev59	T	
Pennsylvanian and Permian											
Shale	Ponca City and Garber areas, Oklahoma	305				17	2.25	2.42	Ath30	N	From graph
		609				11	2.42	2.53	Ath30	N	From graph
		914				7	2.52	2.59	Ath30	N	From graph
		1219				5	2.57	2.62	Ath30	N	From graph
		1524				4	2.62	2.66	Ath30	N	From graph
Permian											
Wellington Formation (shale)	Selma, Kansas	Outcrop	2	15.3	15.5	15.4	2.40	2.55	Hed26	T-2	

Stratigraphic unit	Locality	Source of material or depth [m]	Number of samples	ϕ [%] Minimum	ϕ [%] Maximum	ϕ [%] Average	ϱ_{tot} [gcm^{-3}] Dry	ϱ_{tot} [gcm^{-3}] Water-saturated	Ref.	Method of porosity determination	Remarks
Permian and Triassic											
Clay	Aktyubin area, USSR	Subsurface	4			16.0	2.33	2.49	Nev59	T	
Triassic											
Ankareh Shale	Afton quadrangle, Wyoming	Outcrop				9.2	2.40	2.49	Nut42	T-2	
Woodside Formation (shale)	Afton quadrangle, Wyoming	Outcrop				16.1	2.24	2.40	Nut42	T-2	
Clay, nonplastic (Bröckelschiefer)	Heidelberg, Germany		1			18.0	2.21	2.39	Pet26	T-2	Wesk folding
Jurassic											
Morrison Formation (claystone)	Long Park, Colorado	62…75	6	8.8	20.3	16.5			Cad55	A-9	
Kimmeridge clay	Oxford and Filey, England	Outcrop	2	19.0	30.7	24.8			Sor08	A-2	
Lias clay	Bath and Robin Hoods Bay, England	Outcrop	3	22.5	27.7	24.4			Sor08	A-2	
Cretaceous											
Middendorf Formation (white clay)	Aiken and Richland Counties, S. Carolina	Outcrop	2	39.0	42.3	40.7	1.55	1.96	Hed26	T-2	
Graneros shale	Hamilton County, Kansas	Outcrop	2	24.6	25.2	24.9	1.99	2.23	Hed26	T-2	Not weathered
Graneros(?) shale	Hamilton County	923…1204	3	9.2	11.6	10.6	2.37	2.47	Hed26	T-2	Ransom well
		1367…1526	3	8.7	9.6	9.1	2.46	2.55	Hed26	T-2	Ransom well
Pennsylvanian(?) shale	Hamilton County	1622…1658	3	7.7	8.4	8.1	2.52	2.61	Hed26	T-2	Ransom well
Mentor Formation (shale)	Falun, Kansas	Outcrop	2	22.9	23.3	23.1	2.06	2.29	Hed26	T-2	
Adaville Formation (mudstone)	Afton quadrangle, Wyoming	Outcrop				23.0	2.07	2.30	Nut42	T-2	

(continued)

Table 11c (continued)

Stratigraphic unit	Locality	Source of material or depth [m]	Number of samples	φ [%] Minimum	φ [%] Maximum	φ [%] Average	ϱ_{tot} [gcm^{-3}] Dry	ϱ_{tot} [gcm^{-3}] Water-saturated	Ref.	Method of porosity determination	Remarks
Cretaceous (continued)											
Adaville Formation (shale)	Afton quadrangle, Wyoming					11.9	2.34	2.45	Nut42	T-2	
	Sublette County, Wyoming	4151	1			7.8			Hay51	A	Permeability, 0.1 md
		4174	1			5.7			Hay51	A	Permeability, <0.1 md
Hilliard Formation (shale)	Afton quadrangle, Wyoming	Outcrop(?)		13.9	26.8		1.98…2.28	2.25…2.42	Nut42	T-2	
Shale	Black Hills, Wyoming, Montana		3	32.5	37.6	34.5	1.80	2.14	Rub30	T-2	Dip 1°…5°
			3	25.4	26.0	25.6	1.99	2.24	Rub30	T-2	Dip 5°…10°
			1			23.8	2.00	2.24	Rub30	T-2	Dip 33°
			1			35.8	1.56	1.92	Rub30	T-2	Dip 45°
			1			25.3	1.99	2.24	Rub30	T-2	Dip 50°
Wayan Formation (clay)	Afton quadrangle, Wyoming					25.3	1.80	2.05	Nut42	T-2	
Wayan Formation (mudstone)						28.6	1.90	2.19	Nut42	T-2	
Gault clay	Aylesford and Folkestone, England	Outcrop	3	18.9	28.1	24.0			Sor08	A-2	
Specton clay	England		1			13.6			Sor08	A-2	Weathered
			1			8			Sor08	A-2	Unweathered
Gosauschichten	Austria(?)	Outcrop(?)	3	0.8	4.7	2.2	2.61	2.63	Pet26	T-2	Folded
Paleocene											
Fort Union Formation: Lebo Shale Member	Rosebud County, Montana	Outcrop(?)	1			21.2	2.07	2.28	Ste27	T-5	
Tongue River Member	Rosebud County	Outcrop	2	23.5	36.9	30.2	1.87	2.17	Ste27	T-5	

Stratigraphic unit	Locality	Source of material or depth [m]	Number of samples	ϕ [%] Minimum	ϕ [%] Maximum	ϕ [%] Average	ϱ_{tot} [gcm^{-3}] Dry	ϱ_{tot} [gcm^{-3}] Water-saturated	Ref.	Method of porosity determination	Remarks
Eocene and Oligocene											
Shales, sandy	Los Manueles field, Venezuela	914···1524				20			Car48	N	
Oligocene											
Clay (plastic)	Baden, Germany	Pit	1			26.0	1.90	2.16	Pet26	T-2	
Oligocene and Miocene											
Shale, undisturbed and nearly horizontal	Eastern Venezuela	89···281	6	31.3	35.8	33.5	1.73	2.06	Hed36	T-2	Wells AB, CD, EF
		499···585	3	27.4	28.7	28.0	1.93	2.21	Hed36	T-2	Wells AB, EF
		619···913	9	22.9	28.9	25.4	2.00	2.25	Hed36	T-2	Well AB
		919···1211	9	17.8	25.6	21.1	2.14	2.35	Hed36	T-2	Well AB
		1322···1478	3	14.2	17.8	16.3	2.22	2.38	Hed36	T-2	Well AB
		1526···1677	4	12.8	14.6	13.5	2.32	2.46	Hed36	T-2	Well AB
		1833···1882	3	9.1	10.6	9.6	2.42	2.52	Hed36	T-2	Well AB
		1988	1			12.1			Hed36	T-2	Well GH
		2362···2437	2	10.3	10.4	10.4			Hed36	T-2	Well GH
Miocene											
Kirkwood Formation	Yorktown, N.Jersey	1.2	1			51.9	1.30	1.86	Hed26	T-2	Yellow clay
		3.4	1			40.3	1.62	2.02	Hed26	T-2	Yellowish clay
		4	1			38.6	1.70	2.08	Hed26	T-2	Blue clay
Shale in Stevens sand	S. Coles Levee field, California	2876···2877	1			8.0			Gat50	A-9	
Temblor Formation	Kettleman Hills, California	1905···2844	Many			$\lesssim 3$			Ges33	N	
Clay, nonplastic	Austria	Mine	1			7.0	2.69	2.76	Pet26	T-2	Slight folding
		Subsurface	1			22.5	2.01	2.24	Pet26	T-2	Above coal

(continued)

Table 11c (continued)

Stratigraphic unit	Locality	Source of material or depth [m]	Number of samples	φ [%] Minimum	Maximum	Average	ϱ_{tot} [gcm^{-3}] Dry	Water-saturated	Ref.	Method of porosity determination	Remarks
Miocene (continued)											
Clay (plastic)	Austria	Pit	1(?)			24.0	1.90	2.14	Pet26	T-2	Miocene(?)
	Near Egger, Austria		2	42.0	45.0	43.5	1.43	1.87	Pet26	T-2	Miocene(?)
Miocene(?) and Pliocene(?)											
Cohansey sand (clay)	Crossley, N. Jersey	Outcrop	2	36.6	37.4	37.0	1.67	2.04	Hed36	T-2	
Pliocene											
Clay (plastic in Congeria beds)	Austria	Pit	2	26.0	26.1	26.1	1.80	2.06	Pet26	T-2	Impure
Tertiary											
Clays	Missouri	Pit or bank						1.96…2.13	Whe1896		Natural state(?)
Clay	Various localities, England	Outcrop	3	28.3	29.8	28.8			Sor08	A-2	
Age not specified											
Clay	49 localities, New Jersey	Outcrop	49				1.83		Coo1878		Range 1.53…2.17 gcm^{-3} air dried
Kaolins	Missouri	Pit or bank						1.90	Whe1896		Natural state(?)
Loess clays	Missouri	Pit or bank						2.05	Whe1896		Natural state(?)
Gumbo clays	Missouri	Pit or bank						2.01	Whe1896		Natural state(?)
Fire clay	Missouri	Subsurface						2.40	Whe1896		Natural state(?)
Shales	Missouri	Pit or bank						2.38	Whe1896		Natural state(?)

2.1 Porosity of rocks

Stratigraphic unit	Locality	Source of material or depth [m]	Number of samples	ϕ [%] Minimum	ϕ [%] Maximum	ϕ [%] Average	ϱ_{tot} [gcm^{-3}] Dry	ϱ_{tot} [gcm^{-3}] Water-saturated	Ref.	Method of porosity determination	Remarks
Shale	Phillips well, Russell County, Kansas	427···453	4	22.0	23.3	22.5	2.15	2.37	Hed26	T-2	
		785	2	16.5	17.6	17.1	2.31	2.48	Hed26	T-2	
Shale, weathered	Near Ponca City, Oklahoma			37	48			1.52···1.85	Ath30	N-2	Moist
Clay	Cornwall, Great Britain	Outcrop	1			53			Gei06	A-14	
Slate	Moffat, Great Britain	Outcrop		3.8	5.2	5.9			Sor08	A-2	
Slate, black						4.5			Sor08	A-2	
Slate, red						3.4			Sor08	A-2	
Slate, green						2.1			Sor08	A-2	
Slate, black	Westmoreland, Great Britain	Outcrop	3	0.40	0.50	0.49			Sor08	A-2	
Slate, purple	Penrhyn, Great Britain	Outcrop				0.24			Sor08	A-2	
Slate	Great Britain Settle, Great Britain	Outcrop	6	1.3	5.2	3.2			Sor08	A-2	
			1			0.9	2.73	2.74	Sor08	A-2	
Slate, ordinary	Great Britain		2	0.6	0.6	0.6	2.83	2.84	Moo04	A-2	
Slate	Cornwall, Great Britain		2	4.0	6.0	5.0	2.60	2.65	Moo04	A-2	
Schist	Great Britain		2	1.2	2.3	1.8	2.73	2.75	Moo04	A-2	
	Tasmania		1			6.0	2.64	2.70	Moo04	A-2	

Table 11d. Unconsolidated materials.

Sand, clay, gravel, and alluvium of Quaternary age

Stratigraphic unit	Locality	Source of material or depth [m]	Number of samples	ϕ [%] Minimum	ϕ [%] Maximum	ϕ [%] Average	ϱ_{tot} [gcm^{-3}] Dry	ϱ_{tot} [gcm^{-3}] Water-saturated	Ref.	Method of porosity determination	Remarks
Beach sand, well-sorted	Revere Beach, Massachusetts	Surface	3	39.2	39.4	39.3			Fra35	T-7	1 mm in diameter
			3	40.1	40.2	40.1			Fra35	T-7	0.5 mm in diameter

(continued)

Table 11 d (continued)

Stratigraphic unit	Locality	Source of material or depth [m]	Number of samples	φ [%] Minimum	Maximum	Average	ϱ_{tot} [gcm^{-3}] Dry	Water-saturated	Ref.	Method of porosity determination	Remarks
Sand, clay, gravel, and alluvium of Quaternary age (continued)											
Beach sand	Revere Beach, Massachusetts	Surface	9	38.7	44.8	42.9			Fra35	T-7	Adjacent to boulders
	Lynn Beach, Massachusetts		4	41.1	43.6	42.9			Fra35	T-7	
	Marblehead Beach, Massachusetts		1			42.1			Fra35	T-7	Fine, wet
			1			42.9			Fra35	T-7	Fine, dry
			1			39.0			Fra35	T-7	Coarse, damp
			1			34.0			Fra35	T-7	Coarser, damp
Sand, artificially packed			1			37.9			Fra35	T-7	Wet
			1			37.0			Fra35	T-7	Dry
Cape May Formation (sand)	Runyon, N. Jersey	Mostly pits	8	30.8	45.3	40.1	1.50	1.90	Ste27	T-5	Fine to medium grained
	Pleasantville, N. Jersey	≈0.3	4	25.5	30.0	27.7	1.74	2.02	Ste27	T-5	Poorly sorted
	Absecon, N. Jersey	0···1	4	30.8	39.9	36.6	1.63	2.00	Ste27	T-5	Medium grained
Cape May Formation (gravel)	Absecon	0···0.6	2	23.4	27.4	25.4	1.83	2.08	Ste27	T-5	Loose
Silt	Princeton, N. Jersey	0.3	1			53.2	1.25	1.78	Ste27	T-5	
Sand	Old Bridge, N. Jersey	0.6···2	4	43.6	46.6	45.0	1.47	1.92	Ste27	T-5	
Sand and gravel	Pine Island, Louisiana	37···211	19	24.9	40.1	32.4			Jon51	N-2	Recent(?)
Sands	Gulf coast, USA	19···26	2	36.4	40.8	38.6			Jon51	N-2	Recent(?)
Terrace gravel	Various localities, Montana	Surface	4	23.6	27.1	25.0	2.03	2.28	Ste27	T-5	
Gravel	Yellowstone River, Rosebud County, Montana	Surface	1			20.2	2.19	2.39	Ste27	T-5	

2.1 Porosity of rocks

Stratigraphic unit	Locality	Source of material or depth [m]	Number of samples	ϕ [%] Minimum	ϕ [%] Maximum	ϕ [%] Average	ϱ_{tot} [gcm^{-3}] Dry	ϱ_{tot} [gcm^{-3}] Water-saturated	Ref.	Method of porosity determination	Remarks
Gravel (clinkered shale and sandstone)	Tongue River, Montana	Surface	1			29.3	1.36	1.65	Ste27	T-5	
Sand, very fine	Tongue River	Surface	1			49.9	1.36	1.86	Ste27	T-5	
Gravel	Fergus County, Montana	Surface	1			25.1	1.89	2.14	Ste27	T-5	
Clay	Clark County, Idaho	17	1			42.4	1.51	1.93	Ste27	T-2	
	Jefferson County, Idaho	0.8	1			62.9	1.00	1.63	Ste27	T-2	
Surface material	Rogers Dry Lake, California	Surface	2	37.8	38.3	38.1			Mel21	T-2	
Upper clay member	Mojave River, California	Outcrop	1			43.1	1.55	1.98	Mel21	T-2	Above gravel bank
Lower clay member			1			35.7	1.73	2.09	Mel21	T-2	Below gravel bank
Coarse sand	San Diego County, California	Outcrop		39	41				Lee19	T-5	Valley fill
Medium sand				41	48				Lee19	T-5	Valley fill
Fine sand				44	49				Lee19	T-5	Valley fill
Fine sandy loam				50	54				Lee19	T-5	Valley fill
Clay	Old lakebed, Sewerby, England	Outcrop	1			49.5			Sor08	A-2	
Boulder clay	Bridlington, England	Outcrop	3	23.4	25.5	24.8			Sor08	A-2	Pleistocene
	Balby, England	Outcrop	2	23.9	24.1	24.0			Sor08	A-2	Pleistocene
Alluvial clay	Orgreave, England	Outcrop	1			30.2			Sor08	A-2	Pleistocene
Glacial clay	Mecklenburg, Germany	Railroad cut	4	37	51	45			Pfe28	A	Porosity by water content
Glacial fine sand	Pomerania, Germany	3.7···4	1			42	1.52	1.94	Pfe28	T-2	
		5.5···5.8	1			45	1.46	1.91	Pfe28	T-2	
		8	1			39	1.63	2.02	Pfe28	T-2	
		13···13.4	1			41	1.58	1.99	Pfe28	T-2	

(continued)

Table 11 d (continued)

Stratigraphic unit	Locality	Source of material or depth [m]	Number of samples	φ [%] Minimum	φ [%] Maximum	φ [%] Average	ϱ_{tot} [gcm^{-3}] Dry	ϱ_{tot} [gcm^{-3}] Water-saturated	Ref.	Method of porosity determination	Remarks
Soils of Recent age											
Black Gumbo	Carbon County, Idaho	Surface	1			54.1	1.19	1.73	Ste27	T-5	
Gravelly loam	Fremont County, Idaho	Surface	1			53.7	1.19	1.73	Ste27	T-5	
Loess soil	Caribou, Fremont, and Jerome Counties, Idaho		3	53.2	69.4	61.2	1.00	1.61	Ste27	T-5	
Soils	USA			45	65	55			Fil06	N	Common range
Arable soil	Near Hamburg, Germany	0···0.6	13					1.73	Koc25	N	Natural state(?)
Loamy sand	Near Hamburg	0.2···2.7	16				1.88		Koc25	N	Natural state
Marshy loam	Near Hamburg	Surface(?)				84.0			Sch23	T	82% organic material
Sandy loam	Near Hamburg, Germany	0.6···3	3				2.00		Koc25	N	Natural state (?)
Wet sandy loam	Near Hamburg	1.5···2.5	2				2.08		Koc25	N	Natural state
Subaqueous materials of Recent age											
Coarse sand	San Diego, California	Sea floor sediments from 0···2.5 cm below depositional surface	3			38.3		2.08	Ham56	A-7	
Medium sand			3			40.9		2.00	Ham56	A-7	
Fine sand			54			46.2		1.93	Ham56	A-7	
Very fine sand			15			47.7		1.92	Ham56	A-7	
Sandy coarse silt			7			51.2		1.86	Ham56	A-7	
Silty very fine sand			7			61.3		1.68	Ham56	A-7	

Stratigraphic unit	Locality	Source of material or depth [m]	Number of samples	ϕ [%] Minimum	ϕ [%] Maximum	ϕ [%] Average	ϱ_{tot} [gcm^{-3}] Dry	ϱ_{tot} [gcm^{-3}] Water-saturated	Ref.	Method of porosity determination	Remarks
Medium silt			2			60.9		1.69	Ham56	A-7	
Clayey fine silt			4			65.6		1.60	Ham56	A-7	
Sand, silt, and clay			3			74.7		1.44	Ham56	A-7	
Fine sand	Channel Islands region, California	Sea bottom				29.5			Tra31	N	
Fine silt						66			Tra31	N	
Mud	Hudson River, near Canal Street, New York City, N. York	River bottom		77.2	88.4				Lew24	A-7	
Mud	Hudson River at Jersey City, N. Jersey	Mud on a submerged crate				88.2			Lew24	A-7	
Silt	Hudson River	15 m below riverbed				55			Hew22	A-7	83% through No. 200 sieve
Newly deposited material	Mississippi River Delta					80…90			Sha23	A	
Mud	Seacoasts, USA		9	40	>90				Sha15	A	
Soft mud	Clyde Sea, Great Britain	0…2.5 cm in mud	9	80	87	82			Moo31	A	
		22.5…25 cm in mud	9	72	80	75			Moo31	A	

Table 11e. Other rock types.

Stratigraphic unit	Locality	Source of material or depth [m]	Number of samples	ϕ [%] Minimum	ϕ [%] Maximum	ϕ [%] Average	ϱ_{tot} [gcm^{-3}] Dry	ϱ_{tot} [gcm^{-3}] Water-saturated	Ref.	Method of porosity determination	Remarks
Phosphoria Formation	Idaho, Utah, and Wyoming		9				2.91		Man27		Permian

(continued)

Table 11e (continued)

Stratigraphic unit	Locality	Source of material or depth [m]	Number of samples	φ [%] Minimum	φ [%] Maximum	φ [%] Average	ϱ_tot [gcm⁻³] Dry	ϱ_tot [gcm⁻³] Water-saturated	Ref.	Method of porosity determination	Remarks
Negaunee Iron-Formation (ore)	Ishpeming, Michigan	Mine	12				4.53		Bla56	A-16	Precambrian
Negaunee Iron-Formation (hematitic)			3				4.07		Bla56	A-16	Precambrian
Wabana series (ore)	Wabana, Newfoundland	91 m sea water plus 305 m rock	3				3.46		Bla56	A-16	Ordovician
			3				4.26		Bla56	A-16	Ordovician
Red Mountain Formation (ore)	Bessemer, Alabama	≈427	2	2.3	3.0	2.7	3.73	3.76	Win49	A-16	Silurian
Biwabik Iron-Formation	Virginia, Minnesota	Subsurface	1			0.3	2.75	2.75	Bla55	A-16	Precambrian
"Granite"	Sand Hill well, Wood County, W. Virginia	4058···4059	2	0.2	0.4	0.3	2.78	2.78	Rob62	A-15	
Younger rock salt	Near Hanover, Germany	In place						2.1	Bir24		Torsion balance
Older rock salt	Near Hanover	In place						2.1	Bir24		Torsion balance
Anhydrite	Near Hanover	In place						2.9	Bir24		Torsion balance
Potash bed	Near Hanover	In place						1.6	Bir24		Torsion balance
Gypsum and anhydrite		In place						2.6	Bir24		Torsion balance
Gypsum		In place						2.2	Bir24		Torsion balance
Borax, Ricardo Formation (Pliocene)	Boron, California	91···305					2.14		Bla56	A-16	33% ore
		91···305					1.74		Bla56	A-16	80% ore
		91···305					1.72		Bla56	A-16	92% ore
Serpentine	Hilbig oil field, Texas	≈762		2.2	25.7	22			Bla35	N	Productive
Altered basalt	Oil field, Texas	Subsurface		21.7	35.6				Sel32	N	

Unit conversion to SI-Unit of density: 1 gcm⁻³ = 10³ kgm⁻³

Table 11f. Methods of porosity determinations.

Total porosity

T   Total porosity is determined but the method is not stated.

T-1   Russell's method. Bulk volume (V_{tot}) and grain volume are measured by the displacement of an organic liquid in a volumeter (Russell tube).

T-2   Bulk density-grain density method. Bulk density (ϱ_{tot}) is obtained from the dry weight and the loss of weight in water. Grain density (ϱ_{mat}) is obtained by pycnometry. Percent porosity is $100\,(1-\varrho_{tot}/\varrho_{mat})$.

T-3   Bulk density-grain density method. ϱ_{tot} is obtained from the dry weight and micrometer measurement for volume. ϱ_{mat} is obtained by pycnometry.

T-4   Bulk volume-grain volume method. V_{tot} is determined by loss of weight in water. V_{mat} is determined by the displacement of liquid as in method T-1.

T-5   Bulk density-grain density method. ϱ_{tot} is determined from the dry weight and the volume of unconsolidated rock in a sampling tube. ϱ_{mat} is determined by pycnometry.

T-6   Bulk density-grain density method. ϱ_{tot} is determined from the dry weight and by liquid displacement. ϱ_{mat} is assumed to be constant at 2.65 gcm^{-3}.

T-7   Bulk volume-grain volume method. V_{tot} is determined by loss of weight in water. V_{mat} is determined by pycnometry.

Apparent or effective porosity

A   Apparent porosity is determined but the method is not stated. The following methods are all based on the determination of pore volume (V_{por}) and bulk volume (V_{tot}). Percent porosity is $100\,V_{por}/V_{tot}$.

A-1 V_{por} is determined by water absorption by immersion in hot or boiling water, followed by evacuation to the vapor pressure of water at room temperature for about 1 day. V_{tot} is determined by the loss of weight in water.

A-2 V_{por} is determined by water absorption by immersion in water at room temperature while evacuated at the vapor pressure of water at room temperature. V_{tot} is determined from pore volume and grain density.

A-3 V_{por} is determined as in method A-2, but water absorption is followed by the application of pressure from 50 to 100 bar. V_{tot} is determined by the displacement of water.

A-4 V_{por} is determined by water absorption under high vacuum. V_{tot} is determined by the displacement of water. Large openings are covered with rubber sheeting to include fracture and vuggy porosity with matrix porosity.

A-5 V_{por} is determined by water absorption under high vacuum (3 µbar), followed by application of pressure of 1000 bars. Method of determining V_{tot} is not stated.

A-6   Barnes' method. V_{por} is determined by the volume of organic liquid absorbed under vacuum. V_{tot} is determined by the volume of organic liquid displaced in a volumeter.

A-7 V_{por} is determined as the volume of natural-state water. V_{tot} is determined as the volume of a sampling chamber or by loss of weight in water. Where A-7 is enclosed in parentheses, method of determining bulk volume is not stated.

A-8   A.S.T.M. method C127-42. V_{por} is determined by water absorption during boiling for 2 hours. V_{tot} is determined by water displacement.

A-9   U.S. Bureau of mines method. V_{por} is determined by the pressure and volume relationships of a gas system with and without a rock specimen. V_{tot} is determined by mercury displacement.

A-10 V_{por} is determined by the volume of mercury injected at 67 bar. V_{tot} is determined by mercury displacement.

A-11 V_{por} is determined by the volume of liquid injected by fluid flow. V_{tot} is determined by the loss of weight in liquid.

A-12 V_{por} is determined as the volume of water absorbed after evacuation. V_{tot} is determined from the bulk density of the dry specimen.

A-13 V_{por} is determined as the volume of water imbibed at room temperature and pressure. V_{tot} is determined by the loss of weight in water.

A-14 V_{por} is obtained by Full's conversion of Geikie's values of water absorbed at atmospheric pressure. V_{tot} is obtained from an assumed natural-state bulk density of 2.65 (?) gcm^{-3}.

A-15 V_{por} is determined by the volume of water absorbed after evacuation to 0.15 mbar at 80 °C. V_{tot} is determined from the loss of weight in water.

(continued)

Table 11f (continued)

A-16 V_{por} is determined by the volume of water imbibed for one week at room temperature and pressure by a specimen previously oven dried. V_{tot} is determined by dimensional measurement.

Porosity method not certain

N Method is not specified.

N-1 Very likely apparent porosity.

N-2 Probably apparent porosity. V_{por} probably is determined by water displacement. V_{tot} is determined by the volume of a containing cell.

2.1.9 References for 2.1 — Literatur zu 2.1

Agi59	AGIP Mineraria: I Giacimenti Gassiferi dell' Europa Occidentale. Vol. II. Acad. Nazl. dei Lincei. Roma **1959**.
Aki51	Akins, D.W.: Am. Inst. Mining Metall. Engineers Trans. **192** (1951) 239.
All46	Allison, A.P., Beckelhymer, R.L., Benson, D.G., Hutchins, Jr., R.M., Lake, C.L., Lewis, R.C., O'Bannon, P.H., Self, S.R., Warner, C.A.: Am. Assoc. Petroleum Geologists Bull. **30** (1946) 157.
Alt59	Altschaeffel, A.G., Harrison, W.: Jour. Sed. Petrology **29** (1959) 178.
And30	Andreasen, A.H.M.: Kolloid Z. **50** (1930) 217.
And45	Anderson, J.L.: Am. Assoc. Petroleum Geologists Bull. **29** (1945) 1065.
Arn39	Arnold, Jr., H.H.: Am. Assoc. Petroleum Geologists Bull. **23** (1939) 1352.
Ath30	Athy, L.F.: Am. Assoc. Petroleum Geologists Bull. **14** (1930) 1.
Atk48	Atkinson, Burton, Johnston, D.: Am. Inst. Mining Metall. Engineers Trans. **179** (1948) 128.
Bal50	Baldwin, T.A.: Am. Assoc. Petroleum Geologists Bull. **34** (1950) 1981.
Bar31	Barb, C.F., Branson, E.R.: International Petroleum Technology **8** (1931) 325. Also quoted in: Francher, G.H., Lewis, J.A., Barnes, K.B.: Third Pennsylvania Mineral Industries Conf., petroleum and natural gas sec., Proc.: Pennsylvania State Coll. Mineral Industries Expt. Sta. Bull. 12 (1933).
Bar14	Barrell, J.: Jour Geology **22** (1914) 214.
Bar48	Barton, H.E.: in: Howell, J.V. (Ed.), Structure of typical American oil fields, vol. 3: Am. Assoc. Petroleum Geologists **1948**, 480.
Bar38	Bartosh, E.J.: Am. Assoc. Petroleum Geologists Bull. **22** (1938) 1048.
Bat41	Bates, F.W.: Am. Assoc. Petroleum Geoligists Bull. **25** (1941) 1363.
Bea50	Beaudry, D.A.: Unpublished dissertation, Univ. Cincinati, 1950.
Bir24	Birnbaum, A.: Kali **18** (1924) 144.
Bla35	Blackburn, W.C.: Am. Assoc. Petroleum Geologists Bull. **19** (1935) 1023.
Bla55	Blair, B.E.: Physical properties of mine rock, Part III: U.S. Bur. Mines Rept. Inv. 5130, 69 p. (1955).
Bla56	Blair, B.E.: Physical properties of mine rock, Part IV: U.S. Bur. Mines Rept. Inv. 5244, 69 p. (1956).
Bor35	Born, W.T., Owen, J.E.: Am. Assoc. Petroleum Geologists Bull. **19** (1935) 9.
Bow15	Bownocker, J.A.: Geological Survey of Ohio, 4th ser., Bull. **18**, 160 p. (1915).
Bra37	Branner, G.C.: Am. Assoc. Petroleum Geologists Bull. **21** (1937) 67.
Bro56	Brown, Jr., R.D., Snavely, Jr., P.D., Gower, H.D.: Am. Assoc. Petroleum Geologists Bull. **40** (1956) 94.
Buc1898	Buckley, E.R.: Wisconsin Geol. and Nat. History Survey Bull. **4** (1898) 401.
Byb48	Bybee, H.H.: Am. Assoc. Petroleum Geologists Bull. **32** (1948) 2063.
Cad55	Cadigan, R.A., Caraway, W.H., Gates, G.L., Morris, F.C.: Private communication. **1955**.
Car48	Caribbean Petroleum Company, staff: Am. Assoc. Petroleum Geologists Bull. **32** (1948) 517.
Car48a	Carver, G.E., in: Howell, J.V., ed., Structure of typical American oil fields, vol. 3: Am. Assoc. Petroleum Geologists **1948**, 319.
Cip56	Cipriani, Jr., D., in: Petroleum Geology of Southern Oklahoma, vol. 1: Am. Assoc. Petroleum Geologists **1956**, 311.
Cla38	Clark, C.C.: Am. Assoc. Petroleum Geologists Bull. **22** (1938) 1504.
Cla40	Clark, R.W.: Am. Assoc. Petroleum Geologists Bull. **24** (1940) 742.
Coo1878	Cook, G.H.: Geol. Survey of New Jersey, p. 284–285 (1878).

Cor49	Corps, E.V.: Am. Assoc. Petroleum Geologists Bull. **33** (1949) 1.
Cox53	Cox, H.M.: Am. Assoc. Petroleum Geologists Bull. **37** (1953) 2294.
Cre51	Crego, W.O., Hanagan, J.M.: Am. Inst. Mining Metall. Engineers Trans. **192** (1951) 263.
Cri54	Criss, C.R., McCormick, R.L.: Am. Inst. Mining Metall. Engineers Trans. **201** (1954) 23.
Cru50	Crutchfield, J.W., Bowers, E.F.: Am. Inst. Mining Metall. Engineers Trans. **189** (1950) 335.
Cul40	Culbertson, J.A.: Am. Assoc. Petroleum Geologists Bull. **24** (1948) 1891.
Cur48	Current, A.M., in: Howell, J.V., ed., Structure of typical American oil fields, vol. 3: Am. Assoc. Petroleum Geologists (1948) p. 1.
Cur54	Curry, Jr., W.H., Curry, III, W.H.: Am. Assoc. Petroleum Geologists Bull. **38** (1954) 2119.
Dan54	Daniel, E.J.: Am. Assoc. Petroleum Geologists Bull. **38** (1954) 774.
Dav51	Davis, W.B.: Am. Inst. Mining Metall. Engineers Trans. **192** (1951) 29.
Elk49	Elkins, L.F., French, R.W., Glenn, W.E.: Am. Inst. Mining Metall. Engineers Trans. **179** (1949) 222.
Ell44	Elliott, G.R.: Am. Assoc. Petroleum Geologists Bull. **28** (1944) 217.
Eme52	Emery, K.O., Rittenberg, S.C.: Am. Assoc. Petroleum Geologists Bull. **36** (1952) 735.
Eng60	von Engelhardt, W.: Der Porenraum der Sedimente, Berlin-Göttingen-Heidelberg: Springer-Verlag **1960**.
Eng73	von Engelhardt, W.: Die Bildung von Sedimenten und Sedimentgesteinen, Stuttgart: Schweizerbart'sche Verlagsbuchhandlung **1973**.
Esa27	Esarey, R.E.: Am. Assoc. Petroleum Geologists Bull. **11** (1927) 601.
Fan33	Fancher, G.H., Lewis, J.A., Barnes, K.B., in: Third Pennsylvania Mineral Industries Conference, petroleum and natural gas section, Proc: Pennsylvania State Coll. Mineral Industries Expt. Sta. Bull. **12**, p. 65–171 (1933).
Fen55	Fentress, G.H.: Am. Assoc. Petroleum Geologists Bull. **39** (1955) 155.
Fet26	Fettke, C.R., in: Petroleum Development and Technology (1926) Am. Inst. Mining Metall. Engineers, Petroleum Div., p. 219–234.
Fet34	Fettke, C.R.: Am. Assoc. Petroleum Geologists Bull. **18** (1934) 191.
Fet38	Fettke, C.K.: Am. Assoc. Petroleum Geologists Bull. **22** (1938) 241.
Fic53	Fichter, H.J., Renz, H.H., in: Mencher, E., and others, 1953, Geology of Venezuela and its oil fields: Am. Assoc. Petroleum Geologists Bull. **37** (1953) 690.
Fin49	Finn, F.H.: Am. Assoc. Petroleum Geologists Bull. **33** (1949) 303.
Fis41	Fisher, Barney: Am. Assoc. Petroleum Geologists Bull. **25** (1941) 300.
Fra35	Fraser, H.J.: J. Geology **43** (1935) 910.
Fro40	Frost, V.L.: Am. Assoc. Petroleum Geologists Bull. **24** (1940) 1995.
Füc55	Füchtbauer, H.: Erdoel und Kohle **8** (1955) 616.
Füc60	Füchtbauer, H., Reineck, H.: Unpublished results, cited in [Eng 60].
Füc67	Füchtbauer, H.: 7th World Petrol. Congr. Mexico, Panel Disc. 3, **1967**.
Füc70	Füchtbauer, H., Müller, G.: Sedimente und Sedimentgesteine, Stuttgart: Schweizerbart'sche Verlagsbuchhandlung **1970**.
Ful06	Fuller, M.L.: U.S. Geological Survey Water-Supply Paper 160, p. 59–72 (1906).
Fun48	Funkhouser, H.J., Sass, L.C., Hedberg, H.D.: Am. Assoc. Petroleum Geologists Bull. **32** (1948) 1851.
Gal51	Gallup, W.B.: Am. Assoc. Petroleum Geologists Bull. **35** (11951) 797.
Gar1898	Gary, N.: Königliche technische Versuchanstalten zu Berlin, Mitt., **5** (1898) 243.
Gat50	Gates, G.L., Morris, F.C., Caraway, W.H.: U.S. Bur. Mines Rept. Inv. 4716 (1950).
Gei06	Geikie, A., quoted in Fuller, M.L.: U.S. Geol. Survey Water-Supply Paper 160, p. 59–72 (1906).
Ges33	Gester, G.C., Galloway, J.: Am. Assoc. Petroleum Geologists Bull. **17** (1933) 1161.
Gle50	Glenn, W.E.: Am. Inst. Mining Metall. Engineers Trans. **189** (1950) 243.
Goe50	Goebel, L.A.: Am. Assoc. Petroleum Geologists Bull. **34** (1950) 1954.
Gol47	Goldich, S.S., Parmelee, E.B.: Am. Assoc. Petroleum Geologists Bull. **31** (1947) 1982.
Gru15	Grubenmann, U., Niggli, P., and others: Beitr. Geologie Schweiz, Geotechnische Serie, **5** (1915) 423 p.
Ham39	Hamner, E.J.: Am. Assoc. Petroleum Geologists Bull. **23** (1939) 1635.
Ham56	Hamilton, E.L., Menard, H.W.: Am. Assoc. Petroleum Geologists Bull. **40** (1956) 754.
Har26	Hartmann, F.: Berichte der Fachausschüsse des Vereins deutscher Eisenhüttenleute, Ber. Nr. 82 (1926) 9 p.
Hay51	Hays, F.R.: Unpublished dissertation, Univ. Cincinati, 1951.
Hed26	Hedberg, H.D.: Am. Assoc. Petroleum Geologists Bull. **10** (1926) 1035.

Hed36	Hedberg, H.D.: Am. J. Sci., **31** (1936) 241.
Hed47	Hedberg, H.D., Sass, L.C., Funkhouser, H.J.: Am. Assoc. Petroleum Geologists Bull. **31** (1947) 2089.
Hei61	Heim, A.H.: J. Petrol. Tech. **13** (1961) 87.
Hel67	Heling, D.: Contr. Miner. Petrol. **15** (1967) 224.
Hen58	Henriksen, D.A.: Am. Assoc. Petrol. Geol., Ann. Meeting Los Angeles **1958**.
Hew22	Hewett, B.H.M., Johannesson, S.: Shield and compressed air tunneling: New York, McGraw-Hill, p. 291–295, **1922**.
Hin47	Hinson, H.H., Am. Assoc. Petroleum Geologists Bull. **31** (1947) 731.
Hir12	Hirschwald, J.: Handbuch der bautechnischen Gesteinprüfung: Berlin, Borntraeger, 923 p., **1912**.
Hol30	Holmes, A.: Petrographic methods and calculation (revised ed.) London, Murby & Co., p. 52, **1930**.
How48	Howell, J.V.: Am. Assoc. Petroleum Geologists Bull. **32** (1948) 745.
Hub26	Hubbard, W.E., Thompson, W.C.: Am. Assoc. Petroleum Geologists Bull. **10** (1926) 457.
Hug52	Hughes, D.S., Kelly, J.L.: Geophys. **17** (1952) 739.
Hun50	Hunt, W.C.: Am. Assoc. Petroleum Geologists Bull. **34** (1950) 1795.
Imb50	Imbt, R.F., McCollum, S.V.: Am. Assoc. Petroleum Geologists Bull. **34** (1950) 239.
Ing 48	Ingham, W.I.: Am. Assoc. Petroleum Geologists Bull. **32** (1948) 34.
Jak37	Jakosky, J.J., Hooper, R.H.: Geophys. **2** (1937) 33.
Jon51	Jones, P.H., Buford, T.B.: Geophys. **16** (1951) 115.
Jon75	Jones jr., F.O.: J. Petrol. Tech. **27** (1975) 21.
Kes19	Kessler, D.W.: U.S. Bur. Standards Tech. Paper no. 123, 54 p. (1919).
Kin1898	King (1898), cited in J. Sedimentary Petrol. **23** (1953) 180.
Koc25	Koch, E., in: Königsberger, J., Über die heute mit der Drehwage von Eötvös bei Feldmessungen erreichbare Genauigkeit und über den Einfluß der geologischen Beschaffenheit des Terrains hierauf: Zeitschr. prakt. Geologie 33 (1925) 169.
Kru51	Krumbein, W.C., Sloss, L.L.: Stratigraphy and sedimentation: San Francisco, Freeman & Co., p. 91, **1951**.
Laf38	Lafferty, R.L.: Am. Assoc. Petroleum Geologists Bull. **22** (1938) 175.
Lay49	Layer, D.B., and Members of Staff, Imperial Oil Limited. Am. Assoc. Petroleum Geologists Bull. **33** (1949) 572.
Lee19	Lee, C.H.: Water in the major valleys, in: Ellis, A.J., Lee, C.H., Geology and ground waters of the western part of San Diego County, California: U.S. Geol. Survey Water-Supply Paper 446, p. 121, **1919**.
Lee33	Lees, G.M.: Am. Assoc. Petroleum Geologists Bull. **17** (1933) 229.
Lev56	Levorsen, A.J.: Geology of petroleum. San Francisco **1956**.
Lew24	Lewis, J.V.: Geol. Soc. America Bull. **35** (1924) 557.
Lie49	Lietz, W.T.: Am. Inst. Mining Metall. Engineers Trans. **186** (1949) 251.
Lit48	Littlefield, M., Gray, L.L., Godbold, A.C.: Am. Inst. Mining Metall. Engineers Trans. **174** (1948) 131.
Mac40	MacKenzie, W.D.C.: Am. Assoc. Petroleum Geologists Bull. **24** (1940) 1620.
Mac55	MacQuown, W.C., Jr., Millikan, W.E.: Am. Assoc. Petroleum Geologists Bull. **39** (1955) 630.
Man27	Mansfield, G.R.: U.S. Geol. Survey Prof. Paper 152, p. 210, **1927**.
Man63	Manger, G.E.: Porosity and Bulk Density of Sedimentary Rocks, Geol. Surv. Bull. 1144-E, Washington D.C., **1963**.
Mat74	Matthesius, G.: Diss. Braunschweig, Germany **1974**.
Max64	Maxwell, J.C.: Am. Assoc. Petrol. Geol. Bull. **48** (1964) 697.
McB56	McBee, W., Jr., Vaughn, L.G., in: Petroleum geology of Southern Oklahoma: Am. Assoc. Petroleum Geologists vol. 1 (1956) 355.
McC51	McClellan, H.W., Haines, R.B.: Am. Assoc. Petroleum Geoligists Bull. **35** (1951) 2542.
McG46	McGree, D.A., Jenkins, H.D.: Am. Assoc. Petroleum Geologists Bull. **30** (1946) 1797.
Mea66	Meade, R.H.: J. Sed. Petrol. **36** (1966) 1085.
Mea68	Meade, R.H.: U.S. Geol. Surv. Prof. Paper 497-D, p. 1, **1968**.
Mel21	Melcher, A.F.: Am. Inst. Mining Metall. Engineers Trans. **65** (1921) 469.
Mel24	Melcher, A.F.: Am. Assoc. Petroleum Geologists Bull. **8** (1924) 716.
Mer08	Merrill, G.P.: Stones for building and decoration: New York, John Wiley & Sons, 551 p., **1908**.
Mic36	Michaux, F.W., Buch, E.O.: Am. Assoc. Petroleum Geologists Bull. **20** (1936) 736.
Mil48	Miller, R.L., in: Howell, J.V., Ed., Structure of typical American oil fields: Am. Assoc. Petroleum Geologists vol. 3 (1948) 452.

Min33	Minor, H.E., Hanna, M.A.: Am. Assoc. Petroleum Geologists Bull. **17** (1933) 757.
Moo41	Moore, C.A.: World Oil **126**, June 16 (1941) 38.
Moo04	Moore, C.C.: Geol. Soc. Liverpool Proc. **9** (1900–1904) p. 129–162.
Moo31	Moore, H.B.: Marine Biol. Assoc. [Great Britain] Jour. **17**, new series (1931) 325.
Mou29	Moulton, G.F., Bell, A.H., in: Howell, J.V., Ed., Structure of typical American oil fields: Am. Assoc. Petroleum Geologists vol. 2 (1929) 115.
Mül64	Müller, G.: Methoden der Sedimentuntersuchung, Stuttgart: Schweizerbart'sche Verlagsbuchhandlung **1964**.
Nag65	Nagumo, S.: Bull. Earthq. Res. Inst. Tokyo **43** (1965) 317.
Nag65a	Nagumo, S.: Bull. Earthq. Res. Inst. Tokyo **43** (1965) 339.
Nev59	Nevolin, N.V., Galakfionov, A.B., Serova, A.D.: Prikladnaya Geofizika **22** (1959) 129.
Nut30	Nutting, P.G.: Am. Assoc. Petroleum Geologists Bull. **14** (1930) 1337.
Nut42	Nutting, P.G., quoted in: Birch, F., Schairer, J.F., and Spicer, H.C., eds., Handbook of Physical Constants: Geol. Soc. America Spec. Paper 36, p. 25–26, **1942**.
Ock51	Ockershauser, T.E.: Am. Inst. Mining Metall. Engineers Trans. **192** (1951) 199.
Par12	Parks, W.A.: Report on the building and ornamental stones of Canada, vol. 1: Canada Dept. Mines, 376 p., **1912**.
Par22	Parsons, L.M.: Geol. Mag. [Great Britain] **59** (1922) 51.
Pat59	Pate, J.D.: Am. Assoc. Petroleum Geologists Bull. **43** (1959) 39.
Pet26	Petrascheck, W., Wilser, B.: Berg und Hüttenm. Jahrb. **74**, pt. 2 (1926) 57.
Pet72	Pettijohn, F.J., Potter, P.E., Siever, R.: Sand and Sandstone, Berlin-Heidelberg-New York: Springer-Verlag **1972**.
Pfe28	Pfeiffer, H., Dienemann, W.: Preuß. Geol. Landesanstalt Jahrb. **49** (1928) 304.
Phi60	Philips Petroleum Co.: private communication, 1960.
Pic48	Pickering, W.Y., Dorn, C.L., in: Howell, J.V., Ed., Structure of typical American oil fields: Am. Assoc. Petroleum Geologists vol. 3 (1948) 132.
Pie40	Piersol, R.J., Workman, L.E., Watson, M.C.: Illinois Geol. Survey, Rept. Inv. no. 67, 72 p. **1940**.
Plu43	Plummer, F.B., Tapp, P.F.: Am. Assoc. Petroleum Geologists Bull. **27** (1943) 64.
Poo40	Poole, J.C.: Am. Assoc. Petroleum Geologists Bull. **24** (1940) 1805.
Pry73	Pryor, W.A.: Am. Assoc. Petrol. Geol. Bull. **57** (1973) 162.
Put56	Putman, D.M., in: Petroleum Geology of Southern Oklahoma: Am. Assoc. Petroleum Geologist vol. 1 (1956) 319.
Pye44	Pye, W.D.: Am. Assoc. Petroleum Geologists Bull. **28** (1944) 63.
Ral54	Rall, C.G., Hamontre, H.C., Taliaferro, D.B.: U.S. Bur. Mines Rept. Inv. 5025, 24 p., **1954**.
Ree36	Reeves, J.R.: Am. Assoc. Petroleum Geologists Bull. **20** (1936) 1019.
Ree46	Reeves, F.: Am. Assoc. Petroleum Geologists Bull. **30** (1946) 1546.
Roa55	Roark, G.E., Lindner, J.D.: Am. Inst. Mining and Metall. Engineers Trans. **204** (1955) 16.
Rob62	Robertson, E.C.: private communication, 1962.
Rol49	Roliff, W.A.: Am. Assoc. Petroleum Geologists Bull. **33** (1949) 153.
Ros49	Rose, W., Bruce, W.A.: Am. Inst. Mining Metall. Engineers Trans. **186** (1949) 127.
Rub30	Rubey, W.W.: U.S. Geol. Survey Prof. Paper 165-A, p. 1–54, 1930.
Ruc58	Ruchin, L.B.: Grundzüge der Lithologie. Lehre von den Sedimentgesteinen. – Akademie Verlag, Berlin **1958**, 806 pp. (Translation from Russian: A. Schüller).
Rud59	Rudolph, J.C.: Am. Assoc. Petroleum Geologists Bull. **43** (1959) 880.
Rus26	Russell, W.L.: Am. Assoc. Petroleum Geologists Bull. **10** (1926) 939.
Rut56	Rutledge, R.B., in: Petroleum Geology of Southern Oklahoma: Am. Assoc. Petroleum Geologists vol. 1 (1956) 260.
Saw58	Sawelejew: Cited in [Ruc58].
Sch57	Schauer, P.E., Jr.: Am. Inst. Mining Metall. Petroleum Engineers Trans. **210** (1957) 108.
Sch48	Schneegans, D.: Am. Assoc. Petroleum Geologists Bull. **32** (1948) 198.
Sch48a	Schwartz, C.B., in: Howell, J.V., Ed., Structure of typical American oil fields: Am. Assoc. Petroleum Geologists vol 3. (1948) 5.
Sch23	Schwarz, T., quoted in: Prinz, E.: Handbuch der Hydrologie 2d ed.: Berlin, Julius Springer, p. 133, **1923**.
Seb48	Sebring, L., Jr.: Am. Assoc. Petroleum Geologists Bull. **32** (1948) 228.
Sel32	Sellards, E.H.: Am. Assoc. Petroleum Geologists Bull. **16** (1932) 741.
Sel76	Selley, R.G.: An Introduction to sedimentology, London-New York-San Francisco: Academic Press **1976**.

Sha15	Shaw, E.W.: Am. Inst. Mining Metall. Engineers Bull. **103** (1915) 1449.
Sha23	Shaw, E.W.: U.S. Geol. Survey Water-Supply Paper 489, p. 8, **1923**.
Ske70	Skempton, A.W.: Q.J. Geol. Soc. Lond. **125** (1970) 373.
Sor08	Sorby, H.C.: Geol. Soc. London Quart. J. **64** (1908) 171.
Spa50	Spangler, W.B.: Am. Assoc. Petroleum Geologists Bull. **34** (1950) 100.
Ste24	Steinhoff, E., Mell, M.: Berichte der Fachausschüsse des Vereins deutscher Eisenhüttenleute, Ber. Nr. 44, 6 p., **1924.**
Ste27	Stearns, N.D.: U.S. Geol. Survey Water-Supply Paper 596, p. 121–176, **1927**.
Ste55	Stewart, F.M., Garthwaite, D.L., Krebill, F.K.: Am. Inst. Mining Metall. Engineers Trans. **204** (1955) 49.
Ste57	Stearns, G.M.: Petroleum Engr. B-61, **1957**.
Sto59	Storer, I.: Giacementi Gassiferi dell' Europa Occidentale, Vol. II Acad. Nazl. dei Lincei, Roma 1959.
Swa51	Swann, D.H.: Am. Assoc. Petroleum Geologists Bull. **35** (1951) 2561.
Swe49	Swendenborg, E.A.: Am. Inst. Mining Metall. Engineers Trans. **186** (1949) 163.
Swe50	Swesnik, R.M.: Am. Assoc. Petroleum Geologists Bull. **34** (1950) 386.
Swi47	Swineford, A.: Kansas Geol. Survey Bull. 70, p. 53–104, **1947**.
Tay50	Taylor, J.M.: Am. Assoc. Petrol. Geol. Bull. **34** (1950) 701.
Til38	Tillotson, A.W.: Am. Assoc. Petroleum Geologists Bull. **22** (1938) 1579.
Tod40	Todd, J.D., Roper, F.C.: Am. Assoc. Petroleum Geologists Bull. **24** (1940) 701.
Tra40	Trager, H.H.: Am. Assoc. Petroleum Geologists Bull. **24** (1940) 738.
Tra31	Trask, P.D.: Am. Assoc. Petroleum Geologists Bull. **15** (1931) 271.
Tru50	Trube, A.S., Jr., DeWitt, S.W.: Am. Inst. Mining Metall. Engineers Trans. **189** (1950) 325.
Val 39	Vallat, E.H.: Am. Assoc. Petroleum Geologists Bull. **23** (1939) 1564.
Wag71	Wagner, C., Voigt, M.: Z. Angew. Geol. **17** (1971) 413.
Wal41	Waldschmidt, W.A.: Am. Assoc. Petroleum Geologists Bull. **25** (1941) 1834.
Wal46	Waldschmidt, W.A.: Am. Assoc. Petroleum Geologists Bull. **30** (1946) 561.
Wal55	Walton, P.T.: Am. Assoc. Petroleum Geologists Bull. **39** (1955) 385.
War50	Waring, W.W., Layer, D.R.: Am. Assoc. Petroleum Geologists Bull. **34** (1950) 295.
Was27	Wasson, Theron, Wasson, I.B.: Am. Assoc. Petroleum Geologists Bull. **11** (1927) 705.
Wed51	Weddle, H.W.: Am. Assoc. Petroleum Geologists Bull. **35** (1951) 619.
Wee42	Weeks, W.B., Alexander, C.W.: Am. Assoc. Petroleum Geologists Bull. **26** (1942) 1467.
Wel49	Welsh, J.R., Simpson, R.E., Smith, J.W., Yust, C.E.: Am. Inst. Mining Metall. Engineers Trans. **186** (1949) 55.
Wen48	Wendlandt, E.A., Shelby, T.H., Jr., in: Howell, J.V., Ed., Structure of typical American oil fields: Am. Assoc. Petroleum Geologists vol. 3 (1948) 432.
Wen46	Wendlandt, E.A., Shelby, T.H., Bell, J.S.: Am. Assoc. Petroleum Geologists Bull. **30** (1946) 1830.
Whe1896	Wheeler, H.A.: Missouri Geological Survey **11** (1896) 90.
Wig54	Wiggins, P.N., III: Am. Assoc. Petroleum Geologists Bull. **38** (1954) 306.
Wil53	Wilkinson, W.M.: Am. Assoc. Petroleum Geologists Bull. **37** (1953) 250.
Wil41	Wilson, J.M.: Am. Assoc. Petroleum Geoligsts Bull. **25** (1941) 1898.
Wil52	Wilson, W.W.: Am. Inst. Mining Metall. Engineers Trans. **195** (1952) 77.
Win49	Windes, S.L.: U.S. Bur. Mines Rept. Inv. 4459, 79 p., **1949**.
Win50	Windes, S.L.: U.S. Bur. Mines Rept. Inv. 4727, 37 p., **1950**.
Win52	Winsauer, W.O., Shearin, H.M., Masson, P.H., Williams, M.: Am. Assoc. Petrol. Geol. Bull. **36** (1952) 253.
Wol59	Wolf, K.L.: Physik und Chemie der Grenzflächen, Band II, Berlin-Göttingen-Heidelberg: Springer 1959.
Wom50	Womack, R., Jr.: Am. Assoc. Petroleum Geoligsts Bull. **34** (1950) 1517.
Yus51	Yuster, S.T.: U.S. Geol. Survey Oil and Gas Inv. Map OM 126, Sheet No. 2, **1951**.
Zül56	Züllig, H.: Schweiz. Z. Hydrol. **18** (1956) 5.

2.2 Specific internal surface and capillarity of rocks — Spezifische innere Oberfläche und Kapillarität der Gesteine

2.2.1 Introduction — Einleitung

2.2.1.1 General remarks — Allgemeine Bemerkungen

Rocks can have a very large internal surface. This surface constitutes an interface between rock matrix and pore fluid. Interactions between these two components can occur at this interface. Thus the internal surface may contribute strongly to the physical properties (hydraulic, electrical or even elastic) of rocks. The great importance of internal-surface effects has been widely overlooked in the past but current research is showing more and more that the omnipresent surface effects are equivalent to, or sometimes by far exceed the volume effects (see e.g. 2.3 Permeability; 5.3 Conductivity).

Gesteine können eine sehr große innere Oberfläche haben. Diese Oberfläche bildet eine Grenzfläche zwischen Gesteinsmatrix und Porenfüllung. Wechselwirkungen zwischen diesen beiden Komponenten können an dieser Grenzfläche vorsichgehen. Dadurch kann die innere Oberfläche einen starken Einfluß auf die physikalischen (hydraulischen, elektrischen und sogar elastischen) Gesteinseigenschaften ausüben. Die große Bedeutung der Effekte an der inneren Oberfläche wurde in der Vergangenheit oft übersehen, aber die gegenwärtige Forschung zeigt mehr und mehr, daß die allgegenwärtigen Oberflächeneffekte den Volumeneffekten gleichwertig oder gelegentlich sogar bei weitem vorrangig sind (vgl. z.B. 2.3 Permeabilität; 5.3 Leitfähigkeit).

If more than one phase, wetting the rock material differently, is present in the pore space, the resulting differences in surface energy lead to capillarity forces causing an imbibition of the wetting phase and a displacement of the non-wetting one. Usually but not exclusively, water constitutes the wetting phase in an oil/water or gas/water system. In a gas/oil system without water, oil is usually the wetting phase. The magnitude of these forces, expressed by the material-independent term "capillarity" is governed by the magnitude of the specific internal surface. Hence these quantities are discussed together here.

Wenn mehr als eine Phase mit unterschiedlicher Benetzung des Gesteinsmaterials im Porenraum anwesend ist, führen die resultierenden Unterschiede in der Oberflächenenergie zu Kapillarkräften, die eine Imbibition der benetzenden und eine Verdrängung der nichtbenetzenden Phase bewirken. In einem Öl/Wasser- oder Gas/Wasser-System bildet Wasser gewöhnlich, aber nicht ausschließlich, die benetzende Phase. In einem Gas/Öl-System ohne Wasser ist normalerweise Öl die benetzende Phase. Der Betrag dieser Kräfte wird beschrieben durch den materialunabhängigen Ausdruck „Kapillarität" und hängt vom Betrag der spezifischen inneren Oberfläche ab. Daher werden diese beiden Größen hier gemeinsam diskutiert.

2.2.1.2 Definitions — Definitionen

By the term "specific internal surface" various normalized quantities of the internal surface are meant. It can be normalized to the dry rock material mass (S_m), to the total rock volume (bulk volume) (S_{tot}), to the rock matrix volume (S_{mat}) or to the pore volume (S_{por}). Between these quantities, the following equations apply:

Unter dem Ausdruck „spezifische innere Oberfläche" werden verschiedene normierte Größen der inneren Oberfläche verstanden. Diese läßt sich auf die Masse des trockenen Gesteins beziehen (S_m), auf das gesamte Gesteinsvolumen (S_{tot}), auf das Volumen des festen Gesteinsmaterials (S_{mat}) oder auf das Porenvolumen (S_{por}). Zwischen diesen Größen gelten die folgenden Gleichungen:

$$S_m = S_{mat}/\varrho_{mat} \qquad (1)$$

$$S_{tot} = \phi S_{por} = (1-\phi) S_{mat} \qquad (2), (3)$$

with ϱ_{mat} being the matrix density and ϕ the porosity. S_{tot} is also called surface density.

mit der Matrixdichte ϱ_{mat} und der Porosität ϕ. S_{tot} wird auch Oberflächendichte genannt.

Internal surfaces of natural materials are not smooth, but wavy with many superimposed frequencies; i.e. they contain many superimposed levels of fine structures. Every method of measuring the internal surface invariably applies a certain high frequency choke to the measurement, i.e. a somehow smoothed surface only is measured. Thus, the measured value always depends on the power of resolution of the method. It can

Die innere Oberfläche natürlicher Materialien ist nicht glatt, sondern rauh mit einer Welligkeit aus vielen überlagerten Frequenzen; d.h. sie enthält viele überlagerte Ordnungen von Feinstrukturen. Jede Meßmethode für die innere Oberfläche bewirkt zwangsläufig eine gewisse Tiefpaßfilterung in der Messung, d.h. es wird nur eine teilweise geglättete Oberfläche gemessen. Daher hängt der gemessene Oberflächenwert stets vom

differ by orders of magnitude within the conventional methods and does not even converge toward a limiting value but theoretically goes beyond all limits with increasing resolution. (This problem is called the "coast line of Britain problem" due to the fact that the length of a coast line determined from a map depends on the scale). A practical limit is imposed by the atomistic structure of matter which ultimately causes the term "internal surface" to loose its meaning completely at the molecular level. (As the term "coast line" does at the scale of the sand grains on the beach).

Capillarity C is defined to be the capillary pressure p_c normalized to the product of the interfacial tension γ and the cosine of the contact angle ω:

$$C = p_c / \gamma \cos \omega \qquad (4)$$

It is a purely geometrical attribute of the acting capillary system, i.e. in rocks the pore system. Its physical dimension is an inverse length and can be considered an incremental pore-volume-specific surface. Interesting information can be derived from the capillarity curve $C(\Sigma)$ which represents the capillarity as a function of the saturation Σ of a wetting or non-wetting phase. Its inverse differential quotient $d\Sigma/dC$ describes the increment $d\Sigma$ of this phase occupying an increment of the pore space that has a capillarity between C and $C + dC$. Thus the inverse function $\Sigma(C)$ can be considered a cumulative frequency distribution of capillarity.

The mean capillarity

$$\bar{C} = \int_0^1 C \, d\Sigma \equiv S_{\text{por}} \qquad (5)$$

is identical with the specific pore space surface S_{por}.

The capillarity distribution can be interpreted in terms of an apparent pore size distribution by modelling the porous medium by a bundle of cylindrical tubes of varying circular cross-section. In this case, it would be

$$C = \frac{c}{a} = \frac{2}{r} \qquad (6), (7)$$

if c is the circumference, a the area and r the radius of the circular cross-sections of the assumed individual tubes. $4/C$ can thus be considered an apparent pore diameter. Hence, the quantity $1/C$ is often called "hydraulic radius". However, it should be noted that

$$\bar{C} = \left(\overline{\frac{2}{r}}\right) \neq \frac{2}{\bar{r}} \qquad (8), (9)$$

and that the mean capillarity does not inform about the mean apparent pore size.

Auflösungsvermögen des Meßverfahrens ab. Er kann sich bei den üblichen Methoden um Größenordnungen unterscheiden und konvergiert nicht einmal gegen einen Grenzwert, sondern wächst mit beliebig wachsendem Auflösungsvermögen theoretisch über alle Grenzen. (Dieses Problem ist bekannt unter dem Namen „Coast-Line-of-Britain-Problem", weil in gleicher Weise die aus einer Karte bestimmte Länge einer Küstenlinie vom Maßstab der Karte abhängt). Eine praktische Grenze ist lediglich durch die atomistische Struktur der Materie gegeben, durch die der Begriff „innere Oberfläche" im molekularen Bereich völlig seinen Sinn verliert. (Wie es auch dem Begriff „Küstenlinie" im Größenbereich einzelner Sandkörner am Strand ergeht).

Die Kapillarität C ist definiert als Kapillardruck p_c, normiert auf das Produkt aus Grenzflächenspannung γ und Kosinus des Randwinkels ω.

Sie ist eine rein geometrische Größe des wirksamen Kapillarsystems, d.h. in Gesteinen, des Porensystems. Ihre physikalische Dimension ist eine reziproke Länge und läßt sich auch als eine inkrementale porenraumspezifische Oberfläche auffassen. Wichtige Information läßt sich aus der Kapillaritätskurve $C(\Sigma)$ ableiten, einer graphischen Darstellung der Kapillarität C als Funktion des Sättigungsgrades Σ einer benetzenden oder nichtbenetzenden Phase. Der inverse Differentialquotient $d\Sigma/dC$ beschreibt das Inkrement $d\Sigma$ dieser Phase, das ein Inkrement des Porenraumes mit einer Kapillarität zwischen C und $C + dC$ besetzt. Daher läßt sich die inverse Funktion $\Sigma(C)$ als kumulative Häufigkeitsverteilung der Kapillarität betrachten.

Die mittlere Kapillarität

ist identisch mit der spezifischen Porenraumoberfläche S_{por}.

Die Kapillaritätsverteilung läßt sich als scheinbare Porengrößenverteilung interpretieren unter der Modellannahme eines Bündels kreiszylindrischer Röhren unterschiedlichen Querschnitts für den Porenraum. In diesem Fall wäre

wobei c der Umfang, a die Querschnittsfläche und r der Radius der Einzelröhren sei. $4/C$ läßt sich somit als scheinbarer Porendurchmesser ansehen. Daher läuft die Größe $1/C$ auch oft unter dem Namen „hydraulischer Radius". Man sollte aber beachten, daß

so daß also die mittlere Kapillarität noch keine Aussagen über die mittlere scheinbare Porengröße macht.

Moreover, such an apparent pore size can be no more than a purely mathematical quantity without direct physical meaning for a real porous medium because of the latter's complex geometry. Eq. (6) indeed still applies for prismatic tubes of non-circular cross-section if c and a are still perimeter and area, respectively. Nevertheless, all the other deviations of the real pore system from a simple tube bundle model have a disturbing influence on the function $C(\Sigma)$ [Fat56].

As has been experimentally confirmed, the capillarity curve shows a hysteresis, which reflects the size difference between pore bulbs and pore necks, during an imbibition/drainage cycle. For the imbibition curve which gives the smaller capillarity values, the pore bulbs are determinative; for the drainage curve which gives the larger capillarity values, the pore necks are determinative. Here imbibition means increasing the saturation of the wetting phase or decreasing the non-wetting phase saturation; drainage is the opposite.

Darüberhinaus kann eine solche scheinbare Porengröße nicht mehr als eine reine Rechengröße sein, ohne direkte physikalische Bedeutung für das wirkliche poröse Medium, wegen dessen komplexer Geometrie. Tatsächlich gilt Gl. (6) auch noch für prismatische Röhren nicht-kreisförmigen Querschnitts, wenn c und a weiterhin Umfang und Flächeninhalt bleiben. Dennoch dürfen auch die anderen Abweichungen des wahren Porensystems von einem einfachen Kapillarenbündelmodell, die einen Störeinfluß auf die Funktion $C(\Sigma)$ haben [Fat56], nicht außer Acht bleiben.

Mit experimenteller Bestätigung zeigen die Kapillaritätskurven innerhalb eines Imbibitions-Drainage-Zyklus eine Hysterese, die den Größenunterschied zwischen Porenbäuchen und Porenhälsen widerspiegelt. Für die Imbibitionskurve, die die kleineren Kapillaritätswerte liefert, sind die Porenbäuche bestimmend, für die Drainagekurve, die die größeren Werte liefert, die Porenhälse. Hier bedeutet Imbibition Steigerung der Sättigung einer benetzenden Phase oder Senkung der Sättigung einer nicht-benetzenden Phase, Drainage das Umgekehrte.

2.2.1.3 Petrographical aspects — Petrographische Aspekte

The specific matrix surface S_{mat} of loose sediments is primarily related to mean grain size and grain shape, secondarily to grain size distribution (Table 1). Packing leaves S_{mat} practically unchanged, while it obviously increases S_{tot} and S_{por}. Compaction, cementation and other diagenetical processes can change the surface density in either direction. Clay deposit, encrustation of iron hydroxide and similar actions can drastically increase the fine structured surface density.

Grain size distribution and pore size distribution are connected very loosely only.

The petrographical influence on the capillarity curve can be seen in Figs. 2 and 4.

Die spezifische Matrixoberfläche S_{mat} von Lockersedimenten ist in erster Linie verknüpft mit der mittleren Korngröße und Kornform, in zweiter Linie mit der Korngrößenverteilung (Tab. 1). Bei der Packung bleibt S_{mat} praktisch unverändert, während natürlich S_{tot} und S_{por} wachsen. Kompaktion, Zementation und andere diagenetische Prozesse können die Oberflächendichte sowohl vergrößern als auch verkleinern. Tonablagerungen, Verkrustungen mit Eisenhydroxid und ähnliche Vorgänge können die Feinstruktur-Oberflächendichte bedeutend erhöhen.

Korngrößenverteilung und Porengrößenverteilung sind nur sehr lose miteinander verknüpft.

Der petrographische Einfluß auf die Kapillaritätskurve läßt sich in Fig. 2 und 4 erkennen.

2.2.1.4 Methods of measuring internal surface area — Meßmethoden für die innere Oberfläche

Of the many methods for measuring the internal surface area two groups are most important: physical methods based on surface-proportional adsorption of gases, liquids, dyes etc., and stereological methods to determine surface density or component-specific surfaces from section images. In the first case, the resolution of fine surface structure depends on the effective size of the adsorbed particles, molecules or the like. In the second case, the resolution of the optical measurement is determinative. Two samples, one from either group, will be discussed here.

Von den vielen Methoden zur Messung der inneren Oberfläche sind zwei Gruppen besonders wichtig: einerseits physikalische Methoden auf der Basis einer oberflächenproportionalen Adsorption von Gasen, Flüssigkeiten, Farbstoffen usw., anderseits stereologische Methoden der Bestimmung der Oberflächendichte oder komponentenspezifischen Oberfläche aus Schnittbildern (Anschliffen). Im ersten Falle hängt das Auflösungsvermögen von der wirksamen Größe der adsorbierten Partikel, Moleküle oder dergleichen ab. Im zweiten Fall ist die Auflösung der optischen Messung bestimmend. Zwei Beispiele, eines aus jeder Gruppe, werden im folgenden diskutiert.

2.2.1.4.1 The BET-method — Die BET-Methode

This method, developed by Brunauer, Emmet and Teller [Bru38] and simplified – among others – by Haul and Dümbgen [Hau63], is based on the monomolecular condensation of nitrogen gas on the internal surface at temperatures near to the boiling point of liquid nitrogen.

From the space required by the adsorbed nitrogen molecules and the amount of adsorbed gas, the adsorbing surface area can be determined, provided only monomolecular adsorption is occurring. Monomolecular adsorption occurs in a certain range of the adsorption isotherm. It can be checked graphically by the so-called BET-plot which is to yield a straight line [Bru38]. If, by appropriately setting the initial gas pressure, it can be guaranteed that the working point is safely within that range, one point of the adsorption isotherm is sufficient for the surface area determination [Hau63].

The mass-specific surface is calculated with the measured absolute area and the known weight of the sample. The other conventional quantities (Eqs. (1), (2), (3)) can be calculated with the rock density and porosity. Using nitrogen, the minimum fine structure resolved is about 10^{-10} m.

Prior to any adsorption measurement, the removal of foreign adsorbed matter is always necessary. For the BET-method, the rock samples have to undergo a thorough drying and de-gassing by alternate heating in vacuum and in a nitrogen atmosphere. The necessary pre-treatment duration depends on the rock matter and must be determined experimentally. After insufficient pre-treatment the area measured is too small.

Diese Methode wurde von Brunauer, Emmet und Teller [Bru38] entwickelt und unter anderen von Haul und Dümbgen [Hau63] vereinfacht. Sie basiert auf der monomolekularen Kondensation von Stickstoffgas an der inneren Oberfläche in der Nähe der Siedetemperatur des flüssigen Stickstoffs.

Aus dem Platzbedarf der adsorbierten Stickstoffmoleküle und der Menge des adsorbierten Gases läßt sich die adsorbierende Oberfläche bestimmen, vorausgesetzt, es findet monomolekulare Adsorption statt. Letzteres ist jedoch in einem gewissen Bereich der Adsorptionsisotherme der Fall. Es läßt sich graphisch überprüfen mittels des sogenannten BET-Plots, der eine Gerade ergeben muß [Bru38]. Wenn durch geeignete Wahl des Anfangsdruckes gewährleistet ist, daß der Arbeitspunkt sicher im linearen Bereich liegt, genügt ein einziger Punkt der Adsorptionsisotherme zur Bestimmung der inneren Oberfläche [Hau63].

Aus der gemessenen absoluten Oberfläche und der Masse der Probe folgt die massenspezifische Oberfläche. Mit Gesteinsdichte und Porosität ergeben sich dann die anderen Größen nach Gl. (1), (2), (3). Bei Stickstoff ist die noch aufgelöste minimale Feinstruktur etwa 10^{-10} m.

Vor jeder Adsorptionsmessung ist die Entfernung von adsorbierten Fremdstoffen nötig. Für die BET-Messung müssen die Proben einer intensiven Trocknung und Entgasung unterzogen werden durch wiederholtes abwechselndes Ausheizen im Vakuum und in Stickstoffatmosphäre. Die notwendige Dauer der Vorbehandlung hängt vom Gesteinsmaterial ab und muß experimentell festgestellt werden. Bei ungenügender Vorbehandlung wird ein zu niedriger Oberflächenwert gemessen.

2.2.1.4.2 Stereological methods — Stereologische Methoden

Stereology provides an estimate of quantitative geometrical properties of a higher dimensional space by measurements in a lower dimensional section or projection of that space.

In this way, the volume ratio of components in a compound sample can, for example, be estimated from the area ratio in a random section through that sample by using the principle of Delesse [Del1847, 1848]; by the line ratio on a random line intersecting that sample using the principle of Rosival [Ros1898]; or by the number ratio of random points within that sample using the principle of Glagoleff [Gla33] (point counting method).

In a similar way, the surface density S_{tot} can be estimated by the length L of the traces of the surface in a section plane, divided by the total investigated area A_{tot} of that plane:

Die Stereologie ermöglicht, Schätzwerte für statistische quantitative geometrische Eigenschaften eines höherdimensionalen Raumes aus Meßwerten von niedrigerdimensionalen Schnitten oder Projektionen dieses Raumes zu gewinnen.

So läßt sich das Volumenverhältnis von Komponenten in einer zusammengesetzten Probe aus dem Flächenverhältnis in Zufallsschnitten durch die Probe abschätzen (Prinzip von Delesse [Del1847, 1848]). In gleicher Weise läßt sich das kumulative Streckenverhältnis entlang einer zufälligen Schnittlinie durch die Probe nutzen (Prinzip von Rosival [Ros1898]) oder auch das Zahlenverhältnis von Zufallspunkten innerhalb der Probe (Punktzählmethode, Prinzip von Glagoleff [Gla33]).

In ähnlicher Weise ergibt sich eine Abschätzung der Oberflächendichte S_{tot} aus der Länge L der Schnittspuren in einer Schnittebene pro gesamte untersuchte Fläche A_{tot}:

$$S_{\text{tot}} = \frac{4}{\pi} L/A_{\text{tot}} \qquad (10\text{a})$$

This formula can be further reduced to

Diese Formel reduziert sich weiter zu

$$S_{\text{tot}} = 2 N/L_{\text{tot}} \qquad (10\text{b})$$

with N the number of intersections of a random line of total length L_{tot} with the surface in space, or the number of intersections of the trace line in the section plane. The volume specific surface of particular phases or components can also be estimated from sections by combining Eqs. (10) with the aforesaid principles of determining volume ratios. For the specific surfaces S_{por} and S_{mat} of the pore space and the rock matrix, we thus obtain

wenn N die Anzahl der Durchstoßpunkte einer Zufallsstrecke der Länge L_{tot} im Raum durch die Oberfläche – bzw. in einer Schnittebene durch die Spuren – ist. Auch die volumenspezifischen Oberflächen einzelner Phasen oder Komponenten lassen sich so aus Schnitten gewinnen, indem die Gln. (10) mit den vorgenannten Prinzipien der Bestimmung von Volumenverhältnissen kombiniert werden. Für die spezifischen Oberflächen S_{por} und S_{mat} des Porenraumes und der Gesteinsmatrix ergibt sich demzufolge

$$S_{\text{por}} = S_{\text{tot}}/\phi = \frac{4}{\pi} L/A_{\text{por}} = 2 N/L_{\text{por}} \qquad (11\text{a})$$

$$S_{\text{mat}} = S_{\text{tot}}/(1-\phi) = \frac{4}{\pi} L/A_{\text{mat}} = 2 N/L_{\text{mat}} \qquad (11\text{b})$$

if A_{por}, A_{mat}, L_{por}, L_{mat} are the area and line sections of pore space and rock matrix, respectively.

The quality of the estimate depends on the randomness of the structure and the statistical representativity of the section. The measurements can be done visually or by automatic image analysis.

For more detailed information, the reader is referred to the special literature on stereology (cf. [Eli67, Wei67, Und70]).

worin A_{por}, A_{mat}, L_{por}, L_{mat} die Flächen- und Streckenanteile im Porenbereich und Matrixbereich sind.

Die Güte der Abschätzung hängt von der Zufälligkeit der Struktur und der statistischen Repräsentativität des Schnittes ab. Die Messung kann visuell oder mittels automatischer Bildanalyse durchgeführt werden.

Für detailliertere Information sei der Leser auf die spezielle Literatur über Stereologie verwiesen (s. [Eli67, Wei67, Und70]).

2.2.1.5 Methods of measuring capillarity — Meßmethoden für die Kapillarität

Capillarity is always measured during variation of wetting phase and non-wetting phase saturation usually while the wetting phase is displaced by the non-wetting phase ("drainage"). Sometimes the operation is run in reverse direction ("imbibition") or a full drainage/imbibition cycle is run.

The methods common in practice differ in the media used for the wetting and non-wetting phase, in the way the pressure is applied and according to whether the operation is run manually or automatically.

Die Kapillarität wird stets gemessen, indem die Sättigung der benetzenden oder nicht-benetzenden Phasen verändert wird. Gewöhnlich wird dabei die benetzende durch die nicht-benetzende Phase verdrängt (Drainage); gelegentlich mißt man in umgekehrter Richtung (Imbibition), oder es wird ein voller Drainage/Imbibitions-Zyklus durchlaufen.

Die in der Praxis üblichen Methoden unterscheiden sich in den Substanzen, die als benetzende und nichtbenetzende Phase verwendet werden, in der Art und Weise, wie der Druck erzeugt wird, und in manuellem oder automatischem Betrieb.

2.2.1.5.1 The restored-state method — Die "Restored-State"-Methode

In the so-called restored-state method, water is displaced by air or oil. The sample is placed with one face against a disc of very finely porous material inserted in the wall of a pressure chamber. Initially, sample and porous disc are completely water-saturated after prior evacuation. Then the chamber is filled, i.e. the sample is surrounded on its free sides, by air or oil; pressure in the non-wetting phase is raised in steps, and the water forced from the sample through the disc after each pressure rise is volumetrically measured. The

Bei der sogenannten Restored-State-Methode wird Wasser durch Luft oder Öl verdrängt. Die Probe wird mit einer Stirnfläche auf eine Platte aus einem sehr feinporigen Material gesetzt, die in die Wandung einer Druckkammer eingesetzt ist. Am Anfang sind Probe und poröse Platte vollständig wassergesättigt, nach vorheriger Evakuierung. Dann wird die Kammer mit Luft oder Öl gefüllt, so daß die Probe an den freien Seiten davon umgeben ist. Schrittweise wird nun der Druck in der nicht-benetzenden Phase erhöht und das

water-saturated disc prevents the non-wetting phase from passing through it until its own capillary breakthrough pressure is reached. This pressure limits the measurable capillarity range. An apparent pore diameter of 1 µm corresponds to about 0.8 bar for an oil/water system and about 2.8 bar for an air/water system.

For the displacing phase, oil is better than air because air could dissolve in the water under pressure, diffuse through the porous disc, free itself under atmospheric pressure outside the disc and disturb the volume measurement there. The restored-state method is very time-consuming because of the slow equilibrating after each pressure rise. 3 weeks to 1 month may be necessary for a complete capillarity curve under routine conditions.

For further references, see the literature on reservoir engineering (e.g. [Mon75, Cal60, Cla60]).

2.2.1.5.2 The mercury-injection method — Die Quecksilber-Injektionsmethode

In this method [Pur49] mercury is used for the non-wetting phase and injected under pressure into the evacuated rock sample in a special pycnometer pressure chamber. The volume of mercury injected in dependence on the applied pressure is determined. The method is quick and easy and equipment for manual or automatic operation is commercially available.

A disadvantage is the fact that residual mercury will be trapped in the pore space after complete removal of pressure, rendering the samples useless for further investigation as well as leading to an extensive mercury loss and causing problems with the disposal of the mercury contaminated samples. Recovery of the mercury by destillation is possible, but the samples generally cannot be recovered unaltered.

Regularly shaped samples with smooth outer surfaces should be used, since the mercury does not cling tightly to concave surfaces at low pressures and thus an error is introduced into the capillarity curve [Den64, Rie62].

A surface tension γ and contact angle ω of

$$\gamma = 0.480\ N \cdot m^{-1};\ \omega = 140°$$

are usually assumed, although there is some doubt in the constancy of the contact angle. With these values, an apparent pore diameter of 1 µm corresponds to a pressure of 14.7 bar.

bei jedem Schritt aus der Probe durch die Platte hindurch verdrängte Wasser volumetrisch gemessen. Dabei hindert die wassergesättigte poröse Platte bis zu ihrem eigenen Durchbruchs-Kapillardruck die nichtbenetzende Phase am Austritt. Dieser Druck begrenzt damit den meßbaren Kapillaritätsbereich. Ein scheinbarer Porendurchmesser von 1 µm entspricht dabei einem Druck von etwa 0.8 bar in einem Öl/Wasser-System und etwa 2.8 bar in einem Luft/Wasser-System.

Öl als verdrängende Phase ist geeigneter als Luft, weil sich Luft unter Druck im Wasser lösen, durch die poröse Platte hindurch diffundieren, außerhalb der Platte unter Atmosphärendruck wieder frei werden und dort die Volumenmessung stören kann. Die Restored-State-Methode ist sehr zeitaufwendig wegen der langsamen Einstellung des Gleichgewichtszustandes nach jeder Druckerhöhung. 3 Wochen bis 1 Monat können für eine vollständige Kapillaritätskurve unter Routinebedingungen nötig sein.

Für weitere Information sei auf die Literatur der Erdöl-Erdgas-Lagerstätten-Physik und -Technik (Reservoir Engineering) verwiesen (z.B. [Mon75, Cal60, Cla60]).

Bei dieser Methode [Pur49] wird Quecksilber als nichtbenetzende Phase benutzt und unter Druck in die evakuierte Gesteinsprobe injiziert, die sich in einer speziellen Pyknometerkammer befindet. Das injizierte Quecksilbervolumen wird in Abhängigkeit vom angelegten Druck bestimmt. Die Methode ist leicht und schnell durchführbar, und Apparaturen für manuellen oder automatischen Betrieb sind im Handel.

Ein Nachteil ist, daß am Ende der Messung auch nach vollständiger Druckentlastung, restliches Quecksilber im Porenraum eingefangen bleibt. Hierdurch werden die Proben unbrauchbar für weitere Messungen, es entsteht ein kostspieliger Verlust von Quecksilber und die gefahrlose Beseitigung der kontaminierten Proben macht Schwierigkeiten. Rückgewinnung durch Destillation ist möglich; aber man erhält die Probe dabei nicht unverändert zurück.

Regelmäßig geformte Proben mit glatter äußerer Oberfläche sollten verwendet werden; denn bei konkaven Oberflächen legt sich das Quecksilber bei niedrigen Drucken nicht dicht an die Oberfläche an, so daß ein Fehler in der Kapillaritätskurve entsteht [Den64, Rie62].

Eine Oberflächenspannung γ und ein Randwinkel ω von

werden gewöhnlich zugrundegelegt, obwohl Zweifel an der Konstanz des Randwinkels bestehen. Nach diesen Werten entspricht ein scheinbarer Porendurchmesser von 1 µm einem Druck von 14.7 bar.

2.2.1.5.3 The Centrifuge method — Die Zentrifugen-Methode

In this method the water-saturated sample is placed in a centrifuge and the wetting phase is drained under centrifugal force. The drained water can be collected in a small graduated vessel revolving with the sample and watched under stroboscopic illumination, while the corresponding capillarity is calculated from the angular velocity.

The method is quick. Water can be used as saturating fluid. The samples can be reused. However, the accuracy of the measurement is not so good.

Bei dieser Methode befindet sich die wassergesättigte Probe in einer Zentrifuge. Durch Zentrifugalkraft wird die benetzende Phase ausgetrieben. Das ausfließende Wasser läßt sich in einem kleinen Meßglas sammeln, das mit der Probe rotiert und im stroboskopischen Licht beobachtet werden kann. Die zugehörige Kapillarität ergibt sich aus der jeweiligen Winkelgeschwindigkeit.

Die Methode ist schnell durchführbar. Man kann Wasser zur Sättigung verwenden. Die Proben lassen sich weiterverwenden. Allerdings ist die Genauigkeit der Messung nicht besonders gut.

2.2.1.6 Relations to other petrophysical quantities — Verknüpfungen mit anderen petrophysikalischen Größen

Boundary layer effects at the internal surface, which are a function of surface density or even capillarity and lead to interactions between pore fluid and solid phase, have an influence on almost every petrophysical quantity (especially electrical, hydraulic and elastomechanical). The importance of these surface effects has been quite overlooked in the past, but recent research has shown that they must be seriously considered beside the volume effects if serious errors are to be avoided (cf. 2.3 Permeability, 5.3 Conductivity).

Grenzflächeneffekte an der inneren Oberfläche, die eine Funktion der Grenzflächendichte oder auch der Kapillarität sind und zu Wechselwirkungen zwischen Porenflüssigkeit und fester Phase führen, haben einen Einfluß auf fast alle petrophysikalischen Größen (besonders elektrische, hydraulische und elastomechanische). Die Wichtigkeit dieser Grenzflächeneffekte wurde in der Vergangenheit vielfach übersehen, aber jüngste Forschungen haben gezeigt, daß merkliche Fehler entstehen, wenn diese Erscheinungen nicht neben den Volumeneffekten beachtet werden (vgl. 2.3 Permeabilität, 5.3 Leitfähigkeit).

2.2.1.7 List of symbols — Symbolliste

a	area — Querschnittfläche
A_{tot}	total area — Gesamtfläche
C	capillarity — Kapillarität
c	circumference — Umfang
d_c [m]	capillary diameter — Kapillardurchmesser
K [d]	permeability — Permeabilität
L [m]	length — Länge
L_{tot} [m]	total length — Gesamtlänge
N	number of intersections — Anzahl der Durchstoßpunkte
p_c [bar]	capillary pressure — Kapillardruck
r [m]	radius — Radius
S_m [m² kg⁻³]	specific internal surface normalized to the dry rock material mass — spezifische innere Oberfläche bezogen auf die Masse des trockenen Gesteins
S_{mat} [m⁻¹]	specific internal surface normalized to the rock matrix volume — spezifische innere Oberfläche bezogen auf das Matrixvolumen
S_{por} [m⁻¹]	specific internal surface normalized to the pore volume — spezifische innere Oberfläche bezogen auf das Porenvolumen
S_{tot} [m⁻¹]	specific internal surface normalized to the total rock volume (bulk volume) — spezifische innere Oberfläche bezogen auf das ganze Gesteinsvolumen
V_{por}	pore volume — Porenvolumen
ϱ_{mat}	matrix density — Matrixdichte
Σ	saturation — Sättigung
Σ_w	water saturation — Sättigung mit Wasser
ϕ [%]	porosity — Porosität
ω [deg]	contact angle — Randwinkel

2.2.2 Tables and figures — Tabellen und Abbildungen
2.2.2.1 Specific surface — Spezifische Oberfläche

Since the specific surface heavily depends on grain size and grain size distribution, superimposed by cementation and other diagenetical effects, and furthermore the measured values depend so much on the applied method, there is not much sense in listing specific surface values for classes of sedimentary rocks. The general tendency of variation of the pore-space-specific surface S_{por} with diagenetical stage can be recognized from Fig. 1. The general trend of the variation of the matrix-specific surface S_{mat} with grain parameters is given in Table 1.

Da die spezifische Oberfläche stark von der Korngröße und der Korngrößenverteilung abhängt, denen Einflüsse von Zementations- und anderen diagenetischen Effekten überlagert sind, und im übrigen die Meßwerte so stark durch die angewandte Meßmethode bestimmt werden, hat es wenig Sinn, Werte der spezifischen Oberfläche für Klassen von Sedimentgesteinen zusammenzustellen. Der generelle Trend der Veränderung der porenraumspezifischen Oberfläche S_{por} in den verschiedenen diagenetischen Zuständen läßt sich aus Fig. 1 erkennen. Den Verlauf der Änderung der matrixspezifischen Oberfläche S_{mat} gibt Tab. 1 an.

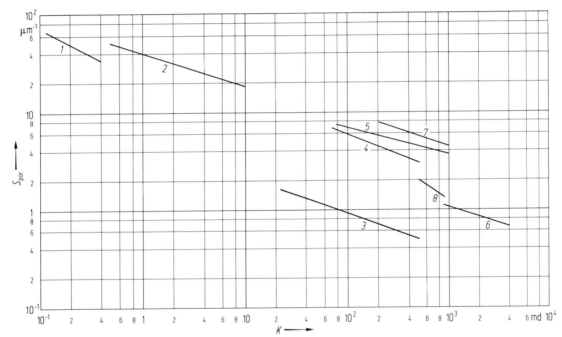

Fig. 1. Pore-space-specific surface S_{por} vs. permeability K from wells in Northern Germany: *1*, Hoya Z1, Upper Carboniferous; *2*, Norderney Z1, Redlaying; *3*, Hardesse 30, Rhaet; *4*, Hohne 4, Lias-α; *5*, Plön-E 75, Dogger-β; *6*, Hankensbüttel-S 41, Dogger-β, Upper zone; *7*, Leiferde 21, Valendis; *8*, Hankensbüttel-S 56a, Dogger-β, Lower zone. Recalculated and redrawn from [Gai73].

Table 1. Variation of matrix-specific surface with mean grain size, mean grain shape and grain size variance for loose sediments.

S_{mat} increases with	decreasing mean grain size
	decreasing grain sphericity
	decreasing grain roundness
	decreasing size variance

2.2.2.2 Capillarity curves — Kapillaritätskurven

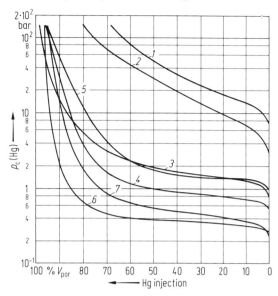

Fig. 2. Capillary pressure curves (mercury injection) for sandstones from wells in Northern Germany (for numbering of the curves, see Fig. 1) [Gai73].

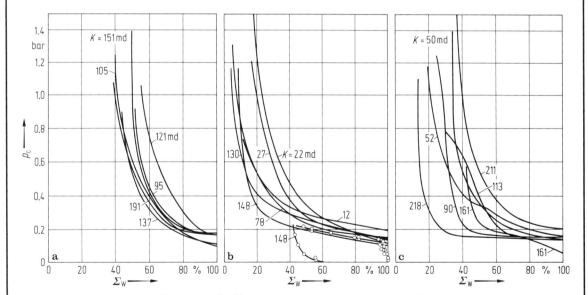

Fig. 3a···c. Capillary pressure p_c vs. water saturation Σ_w (restored state) [Mus49] after [Has44]. Numbers are permeabilities K in [md]. a) For shaly sandstones of a Californian field; b) For dolomites of a West Texas field; c) For sandstones of a Mid Continent field.

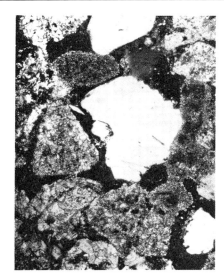 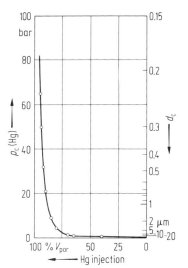

Fig. 4a. Wedge pores.
Left: Very sandy dolomite arenite with calcareous cement. Tertiary Molassis of the Alp forelands (Baustein layers, well Schwabmünchen 1). 35.6% sand (white) (median diameter 0.19 mm), 12% calcite, 52.4% dolomite. 26.1% porosity, 2500 md permeability. (Width of image: 0.9 mm, crossed nicols; pores black).

Right: Capillary pressure curves of the same sample. The right hand scale shows capillary radii according to the formula $r = 2\gamma \cos \omega / p$ (r = capillary radius, γ = surface tension, here 0.480 $N \cdot m^{-1}$, ω = contact angle, here 140°, p = applied pressure). It can be read: 12% of the capillary diameters are below 1 µm, 11% between 1 and 5 µm, 27% between 5 and 20 µm, 50% between 20 and 50 µm (according to friendly information by Dr. Miessner, Hannover). [Füc70].

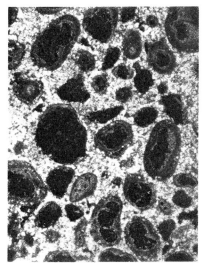 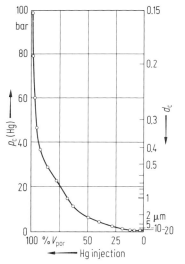

Fig. 4b. Hollow pores.
Left: Dolomite oncolith; oncoides mostly dissolved. Upper Permian 2-Dolomite, W Nienburg (Weser). Oncoides cryptocrystalline; crystal size about 35 µm, porosity 30.1%, air permeability 11 md. (Width of image 2.3 mm, crossed nicols; pores black).

Right: Capillary pressure curve of the same sample (for explanation, see Fig. 4a). Capillary radii: 32% < 1 µm, 34% 1···5 µm, 17% 5···20 µm, 17% 20···50 µm.

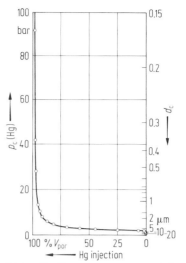

Fig. 4c. Intercrystalline pores.
Left: Pill Dolomite, idiomorphic. Paleocene, Libya. Mean crystal size 20···25 µm. Porosity 36.0%, air permeability 170 md. (Width of image 0.4 mm, crossed nicols; pores black).

Right: Capillary pressure curve of the same sample (for explanation, see Fig. 4a). Capillary radii: 4% < 1 µm, 32% 1···5 µm, 62% 5···7.5 µm, 2% > 7.5 µm.

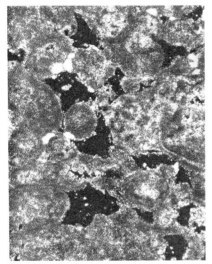
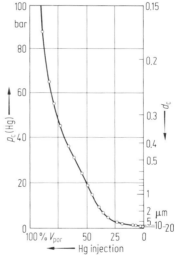

Fig. 4d. Cavernous pores.
Left: Dolomite oncolith, basic material (dolomite, calcite) partially dissolved. Upper Permian 2-Dolomite, W Nienburg (Weser). Oncoides 0.05···0.5 mm, cryptocrystalline. Porosity 23.9%, air permeability 160 md. (Width of image 0.9 mm, crossed nicols; pores black).

Right: Capillary pressure curves of the same sample (for explanation, see Fig. 4a). Capillary radii: 54% < 1 µm, 19% 1···5 µm, 27% 5···20 µm.

2.2.3 References for 2.2.1 and 2.2.2 — Literatur zu 2.2.1 und 2.2.2

Bru38	Brunauer, S., Emmet, P.H., Teller, E.: J. Am. Chem. Soc. **60** (1938) 309.
Cal60	Calhoun jr., J.C.: Fundamentals of Reservoir Engineering, Norman: University of Oklahoma Press, **1960**.
Cla60	Clarc, N.J.: Elements of Petroleum Reservoirs, Dallas: E.J. Storm Printing Co., **1960**.
Del1847	Delesse, M.A.: C.R. Acad. Sci. Paris **25** (1847) 544–545.
Den64	Denecke, A.: Erdöl-Z. **80** (1964) 171–184, 215–220.
Eli67	Elias, H. (ed.): Stereology, Berlin-Heidelberg-New York: Springer, **1967**.

Fat56	Fatt, I.: Trans. Am. Inst. Min. Metall. Petrol. Engrs. **207** (1956) 144–160.
Füc70	Füchtbauer, H., Müller, G.: Sedimente und Sedimentgesteine, Stuttgart: Schweizerbarth, **1970**.
Gai73	Gaida, K. H., Rühl, W., Zimmerle, W.: Erdöl Erdgas Z. **89** (1973) 336–343.
Gla33	Glagoleff, A. A.: Trans. Inst. Econ. Mineral. Moskov **59** (1933).
Has44	Hassler, G. L., Brunner, E., Deahl, T.J.: Trans. Am. Inst. Min. Metall. Petrol. Engrs. **155** (1944) 155.
Hau63	Haul, R., Dümbgen, G.: Chem.-Ing.-Tech. **32** (1960) 5; **35** (1963) 8.
Mon75	Monicard, R.: Caracteristiques des roches reservoirs — Analyse des carottes (Cours de production, Tome I), Paris: Edition Technip, **1975**.
Mus49	Muskat, M.: Physical principles of oil production, New York: McGraw-Hill **1949**.
Pur49	Purcell, W. R.: Trans. Am. Inst. Min. Metall. Petrol. Engrs. **189** (1949) 39.
Rie62	Rieckmann, M., Becker, J.: Erdöl-Z. **78** (1962) 629–637.
Und70	Underwood, E. E.: Quantitative Stereology, Reading, Mass.: Addison-Wesley, **1970**.
Wei67	Weibel, E. R., Elias, H. (ed.): Quantitative Methoden der Morphologie, Berlin-Heidelberg-New York: Springer, **1967**.

2.3 Permeability of rocks — Permeabilität der Gesteine

Permeability is a measure for the ability of porous rocks to permit fluid flow through its pore space. It generally increases with porosity, but it also depends on other structural parameters of the rocks, e.g. the specific surface of the pore space. It can vary over an extensive range of about 8 orders of magnitude in loose sediments and sedimentary rocks. Traces of permeability can even be found in metamorphic and igneous rocks.

Die Permeabilität ist ein Maß für die Fähigkeit poröser Gesteine, Flüssigkeitsströmung durch ihren Porenraum zu erlauben. Sie wächst im allgemeinen mit der Porosität, hängt aber zusätzlich noch von anderen Strukturparametern des Gesteins, wie z.B. der spezifischen Oberfläche des Porenraumes, ab. In Sedimenten und Sedimentgesteinen kann sie in einem sehr weiten Bereich von etwa 8 Größenordnungen variieren, und Spuren von Permeabilität lassen sich sogar in metamorphen und vulkanischen Gesteinen finden.

2.3.1 Definitions — Definitionen

The permeability K is a physical parameter characteristic of a given material. It has the dimension of an area and is defined by d'Arcy's law, a linear relation between flow rate and pressure proved in 1856 by Henri d'Arcy.

In this law

Die (hydraulische) Permeabilität K ist eine physikalische Materialgröße mit der Dimension einer Fläche und ist definiert durch das d'Arcysche Gesetz, eine lineare Beziehung zwischen Flußrate und Druck, das 1856 von Henri d'Arcy gefunden wurde.

In diesem Gesetz

$$\boldsymbol{u} = \frac{K}{\eta}(\varrho \boldsymbol{g} - \operatorname{grad} p) \tag{1}$$

ϱ is the fluid density, $\boldsymbol{g}$ the gravity field strength, $\boldsymbol{u}$ the volume flow density (flow rate per unit area of cross-section), p the applied pressure, η the dynamic viscosity of the streaming fluid, and K the permeability. $\boldsymbol{u}$ has the dimension of a velocity and is sometimes also called "approach velocity". It is connected with the so-called "advance velocity" v (the true mean velocity of fluid volume elements within the pore space in the direction of the macroscopic flow) by the law of Dupuis and Forchheimer

ist ϱ die Flüssigkeitsdichte, $\boldsymbol{g}$ die Schwerefeldstärke, $\boldsymbol{u}$ die Volumenflußdichte (Flußrate durch die Flächeneinheit des Querschnitts), p der angelegte Druck, η die dynamische Zähigkeit der strömenden Flüssigkeit und K die Permeabilität. $\boldsymbol{u}$ hat die Dimension einer Geschwindigkeit und ist mit der mittleren wahren Geschwindigkeit v der Volumenelemente der Flüssigkeit innerhalb des Porenraumes in der makroskopischen Strömungsrichtung verknüpft durch das Gesetz von Dupuis und Forchheimer

$$\boldsymbol{v} = \boldsymbol{u}/\phi, \tag{2}$$

where ϕ is the porosity.

For homogeneous horizontal flow, d'Arcy's law can be written

worin ϕ die Porosität ist.

Für homogenen horizontalen Fluß läßt sich das d'Arcysche Gesetz folgendermaßen schreiben:

$$J = \frac{A}{L}\frac{K}{\eta}\Delta p \tag{1a}$$

where J is the volume flow rate, A the available cross-section of porous material and L the length and Δp the pressure difference between intake and outlet.

In the form (1a), d'Arcy's law properly applies to incompressible fluids only, but it can be written in the same way for compressible fluids, if the flow volume is understood to be normalized to the mean between the intake and outlet pressure.

If the outlet pressure is arbitrary, and the flow measurement is made at atmospheric pressure, as usual, the law can be written in the following ways:

$$J_{atm} = \frac{A}{L} \frac{K}{\eta} \frac{\Delta(p^2)}{2 p_{atm}} = \frac{A}{L} \frac{K}{\eta} \frac{p_{in}^2 - p_{out}^2}{2 p_{atm}}$$
$$= \frac{A}{L} \frac{K}{\eta} \frac{(p_{in} - p_{out})(p_{in} + p_{out})}{2 p_{atm}} \quad (3)$$
$$= \frac{A}{L} \frac{K}{\eta} \Delta p \frac{p_{in} + p_{out}}{2 p_{atm}}$$

J_{atm} = volume flow rate at atmospheric pressure, p_{atm} = atmospheric pressure, p_{in} = intake pressure, p_{out} = outlet pressure.

The permeability K is a scalar quantity for isotropic media. However, if the pore space has an anisotropic structure, the flow resistance depends on the flow direction and then K is a symmetric tensor.

Fluid flow through porous media is described by d'Arcy's law as long as the driving force is counterbalanced by viscous forces only. At higher flow velocities, inertial forces might become significant and then lead to a square flow law

$$|\text{grad } p| = a_0 + a_1 u + a_2 u^2 \quad (4)$$

where the constant a_1 contains the viscosity and a_2 the density of the fluid; in addition both constants depend on the porous structure. a_2 might even show a very slight dependence on u itself.

Law (4) is valid above a certain critical flow density u_{crit}, whereas up to there, the true linear d'Arcy law holds. There is a smooth transition between those ranges at u_{crit}.

A dimensionless number, formally similar to a Reynolds' number, can be given for u_{crit}:

$$\text{Re}_{crit} = \frac{\varrho}{\eta} \frac{\sqrt{K}}{\phi} u_{crit} \quad (5)$$

i.e. the critical velocity depends on the permeability itself [Hof74].

For the "effective length" $\sqrt{K}/\phi$ in this equation, the mean grain diameter $\bar{d}$ is sometimes introduced:

$$\mathrm{Re}_{\mathrm{crit}} = \frac{\varrho}{\eta} \bar{d}\, u_{\mathrm{crit}} \qquad (6)$$

leading to a slightly different value of the critical Reynolds' number.

However, this is not a fixed value, but a range must be considered: $\approx 1 \cdots 10$ in Eq. (6), for most reservoir rocks.

At very low liquid flow velocities, viscous forces can be overridden by always present interactive forces between fluids and solids. These are caused by reactions at the internal surface of the rock matrix and the resulting structural viscosity in the liquid.

This applies especially to polar liquids, and to electrolytes in semipermeable – i.e. fine porous – materials. Flow of water, especially fresh water, is significantly retarded by such effects in most porous rocks [Far54, Eng54]. Those effects increase with the specific internal surface of the pore space.

At polyphase flow, capillary forces due to interfacial tensions between the phases can significantly control the flow. A d'Arcy type law then can be written for each individual phase. However the effective phase permeabilities K_{eff} then depend on the phase saturations and generally do not add up to the absolute (single-phase) permeability K. Sometimes, so-called relative permeabilities

$$k_{\mathrm{rel}} = K_{\mathrm{eff}}/K \qquad (7)$$

are used instead.

Molecular slip can contribute to the flow of gases, when the pore dimensions approach the mean free path of the gas molecules; the apparent permeability K_{a} becomes dependent on the mean absolute gas pressure $\bar{p}$:

$$K_{\mathrm{a}} = K\left(1 + \frac{\alpha}{\bar{p}}\right) \qquad (8)$$

The constant α, called Klinkenberg constant, depends on rock structure, molecular weight of the streaming gas and absolute temperature [Schei60]. This effect is known in physics as "Kundt-Warburg Effect" and in petroleum engineering as "Klinkenberg Effect". In engineering and geoscience practice, the Klinkenberg constant α is often expressed as a function of the true permeability K. Several empirical formulas are in use, e.g.

$$\alpha = 0.5\, K^{-0.37} \qquad (9)$$

with α in [bar] and K in [md].

The Klinkenberg effect might even lead to a "gas chromatographic" separation of gases of different molecular weight, in porous rocks.

In geohydrology, a simplified version of d'Arcy's law is commonly used for horizontal flow:

$$\mathbf{u} = -k \, \text{grad} \, h \tag{10}$$

Here h is the hydrostatic head. The so-called permeability coefficient k with the dimension of a velocity is related to the permeability proper according to

$$k = K \frac{\varrho}{\eta} g \tag{11}$$

Usually, this equivalence (11) is based on the density and viscosity of pure water at 10 °C.

The SI unit for the permeability K of appropriate size would be μm^2. However, in practice the unit "Darcy" (d) is commonly used, which originally was defined as

$$1 \, d = 1 \, cm \, s^{-1} \cdot 1 \, cP/1 \, atm \, cm^{-1} \tag{12}*$$

resulting in the equivalence

$$1 \, d = 0.9869 \, \mu m^2 \tag{13}$$

Later on, in continental European countries, the "physical atmosphere" (atm) has often been replaced by the "technical atmosphere" (at or kp/cm² or kg (force)/cm²), resulting in the equivalence [Eng60]

$$1 \, d = 1.0197 \, \mu m^2 \tag{14}*$$

One should always be aware of this slight disagreement in the definitions. However, within practical measuring accuracy, the Darcy can be assumed to be equal to the corresponding SI unit, leading to a third, metric, definition

$$1 \, d = 1 \, \mu m^2 = 1 \, cm \, s^{-1} \cdot 1 \, cP/1 \, bar \, cm^{-1}, \tag{15}*$$

which lies between the two former definitions. Further units in use are:

$$1 \, md = 10^{-3} \, d \quad \text{and} \quad 1 \, \mu d = 10^{-6} \, d \tag{16}$$

A quantity related to permeability is the transmissivity, which is the product of permeability and the bed thickness of permeable strata. This quantity is a measure of the flow rate from the formation stratum toward a borehole of a given diameter at a given pressure drop.

---

*) $1 \, cP = 10^{-3} \, Pa \, s$
$1 \, atm = 101\,325 \, Pa$
$1 \, at = 98\,066.5 \, Pa$
$1 \, bar = 10^5 \, Pa$.

2.3.2 Origin and different types of permeability — Entstehungsursache und unterschiedliche Arten der Permeabilität

The origin of permeability is rather the same as for porosity. Thus, we can classify permeability in a similar way (see 2.1.2):

– Intergranular permeability
– Intragranular permeability
– Fissure and fracture permeability
– Vugular permeability

Sometimes, one can discern between rock permeability and formation permeability. Rock permeability means the intergranular and intragranular permeability which can be measured even with small rock samples, while formation permeability means the mean value of a larger formation volume, that might either contain inhomogeneous rock of varying inter- or intragranular permeability, or massive rock penetrated by fractures, fissures, or vugs.

Like porosity, permeability is influenced by packing, compaction and cementation. However, it is much more sensitive to such influences than the porosity. Even small changes in pore structure that hardly affect pore volume at all, can cause tremendous changes in permeability. Accordingly, permeability is very sensitive also to grain size, grain shape and sorting [Kru42, Hsü77].

Generally, permeability decreases with progressing diagenesis. However such a trend can be concealed considerably by facial changes in grain size, grain shape, sorting, clay content, chemical effects, etc. Thus, prediction of permeability is very hard, and no good correlation can be observed, neither with geological conditions nor with other single physical parameters.

Hinsichtlich der Entstehung gilt für die Permeabilität das gleiche wie für die Porosität. Daher läßt sich die Permeabilität auch in ähnlicher Weise klassifizieren (vgl. 2.1.2):

– Intergranulare Permeabilität
– Intragranulare Permeabilität
– Riß- und Kluftpermeabilität
– kavernöse Permeabilität

Gelegentlich wird noch zwischen Gesteinspermeabilität und Gebirgspermeabilität unterschieden. Hierbei bedeutet Gesteinspermeabilität die intergranulare und intragranulare Permeabilität, sie sich auch an kleinen Gesteinsproben messen läßt, während unter Gebirgspermeabilität ein Mittelwert über ein größeres Gebirgsvolumen zu verstehen ist, das entweder inhomogenes Gestein veränderlicher inter- oder intragranularer Permeabilität enthält oder von Rissen, Klüften oder Kavernen durchzogenes massives Gestein umfaßt.

Wie die Porosität wird auch die Permeabilität durch Packung, Kompaktion und Zementation beeinflußt. Jedoch ist sie wesentlich empfindlicher gegenüber solchen Einflüssen als die Porosität; denn bereits kleinste Änderungen der Porenraumstruktur, die noch gar keine Wirkung auf das Porenvolumen haben, können erhebliche Änderungen der Permeabilität verursachen. Dementsprechend ist die Permeabilität auch sehr empfindlich gegenüber Korngröße, Kornform und Sortierungsgrad [Kru42, Hsü77].

Allgemein ist ein Trend der Abnahme der Permeabilität mit fortschreitender Diagenese zu beobachten. Jedoch kann ein solcher Trend stark maskiert sein durch fazielle Änderungen von Korngröße, Kornform, Sortierungsgrad, Tongehalt, chemischen Bedingungen usw. Daher ist eine Vorhersage der Permeabilität sehr schwierig, und man beobachtet meistens keine gute Korrelation, weder mit geologischen Bedingungen noch mit anderen physikalischen Einzelparametern.

2.3.3 Relations to other petrophysical quantities — Verknüpfungen mit anderen petrophysikalischen Größen

Permeability is related to porosity but never uniquely. Many other independent structural quantities enter this relation. Very important is pore size and pore size distribution, as well as pore shape and pore shape distribution.

Pore size and shape also control surface density or specific internal surface, respectively. Thus, a relation between permeability and surface density can be expected. Such a relation is given by the empirical Kozeny-Carman equation [Koz27, Car56]

Die Permeabilität ist verknüpft mit der Porosität, jedoch nicht eindeutig. Viele andere unabhängige Strukturgrößen gehen in diesen Zusammenhang ein. Sehr bedeutend ist Porengröße und Porengrößenverteilung, ferner Porenform und Porenformverteilung.

Porengröße und Porenform bestimmen auch die Oberflächendichte bzw. spezifische innere Oberfläche. Daher ist ein Zusammenhang zwischen Permeabilität und Oberflächendichte zu erwarten. Eine solche Verknüpfung wird durch die empirische Kozeny-Carman-Gleichung

$$K = \frac{\phi^3}{k_{\text{Koz}} S_{\text{tot}}^2} = \frac{\phi}{k_{\text{Koz}} S_{\text{por}}^2} = \frac{\phi^3}{k_{\text{Koz}}(1-\phi)^2 S_{\text{mat}}^2}, \tag{17}$$

where the specific surfaces S_{tot}, S_{por}, S_{mat} are referred to the total rock volume, the pore volume and the matrix volume, respectively (cf. 2.2.1.2, p. 267).

Taking theoretical models into consideration, the empirical "Kozeny constant" k_{Koz} can be written

$$k_{Koz} = 2T, \qquad (18)$$

where T is the hydraulic tortuosity, a quantity representing the impeding influence of curvatures, constrictions, etc. of the flow path in porous rocks, and thus a rock structural constant.

If the hydraulic tortuosity T is assumed to be approximately equal to the electrical tortuosity X (see 5.3.2.2, Eq. (8)), then the Kozeny-Carman equation can also be written

$$FK = \frac{1}{2} \frac{\phi^2}{S_{tot}^2} \qquad (19)$$

relating permeability K to porosity ϕ, to surface density S_{tot} and to the so-called formation resistivity factor F (cf. 5.3.2), which can be determined by electrical measurements [Scho66, 67].

In all of the above equations, the effects of pore size and shape distribution are neglected. Using capillarity C (cf. 2.2.1), which can be determined from capillary pressure curves, for a measure of pore size and shape, a modified Kozeny-Carman equation can be established:

$$K = \left| \frac{1}{2} \frac{\phi}{T} \int_0^1 \frac{1}{C^2} d\Sigma \right| \qquad (20)$$

or

$$FK = \left| \frac{1}{2} \int_0^1 \frac{1}{C^2} d\Sigma \right|, \qquad (21)$$

where Σ is the saturation of the porous formation by a wetting or non-wetting fluid [Eng60, Pur49].

However, even in this latter equation, many vital structural properties of porous rocks – like e.g. the topology of the pore network [Fat56, Scho66] or pore constrictions – are neglected possibly leading to quite erroneous permeability values.

Generally, much care should be taken in the practical application of all such relations between permeability and other petrophysical quantities, because they are inevitably based on model assumptions that are too simple. Even the use of empirical correction factors is of no great help, because of the strong sensitivity of permeability to structural changes, and the great variability of pore structure in natural rocks.

[Koz27, Car56] dargestellt, worin die spezifischen Oberflächen S_{tot}, S_{por}, S_{mat} jeweils auf das totale Gesteinsvolumen, das Porenvolumen und das Matrixvolumen bezogen sind (vgl. 2.2.1.2, S. 267).

Auf Grund theoretischer Modellüberlegungen läßt sich die empirische „Kozeny-Konstante" k_{Koz} auch ausdrücken als

worin T die hydraulische Tortuosität ist, eine Strukturgröße des Gesteins, die den strömungsbehindernden Einfluß von Krümmungen, Verengungen u.ä. der Strömungspfade im porösen Gestein erfaßt.

Nimmt man die hydraulische Tortuosität T als annähernd gleich der elektrischen Tortuosität X an (vgl. 5.3.2.2, Gl. 8), dann läßt sich die Kozeny-Carman-Gleichung auch folgendermaßen schreiben:

Diese Gleichung verbindet die Permeabilität K mit der Porosität ϕ, der Oberflächendichte S_{tot} und dem sogenannten Formationsfaktor F (vgl. 5.3.2), der sich aus elektrischen Messungen bestimmen läßt, [Scho66, 67].

In allen bisherigen Gleichungen ist der Einfluß der Porengrößen- und Porenformverteilung vernachlässigt. Zieht man die Kapillarität C (vgl. 2.2.1), die sich mittels Kapillardruckkurven ermitteln läßt, als ein Maß für die Porengröße und Porenform heran, dann läßt sich eine erweiterte Kozeny-Carman-Gleichung

aufstellen, worin Σ der Sättigungsgrad des Porenraumes mit einer benetzenden oder nicht benetzenden Flüssigkeit ist [Eng60, Pur49].

Aber auch in diesen letzten Gleichungen bleiben immer noch wesentliche Struktureigenschaften des porösen Gesteins – wie z.B. die Topologie des Porennetzwerks [Fat56, Scho66] oder Porenkanalverengungen – unberücksichtigt, was mitunter zu beträchtlich fehlerhaften Permeabilitätswerten führen kann.

Allgemein sollte man äußerste Vorsicht bei der praktischen Anwendung solcher Beziehungen zwischen Permeabilität und anderen petrophysikalischen Größen walten lassen, weil sie unvermeidlich auf zu einfachen Modellvorstellungen beruhen. Auch empirische Korrekturfaktoren bieten keine große Hilfe, wegen der großen Empfindlichkeit der Permeabilität gegen Strukturänderungen und der starken Variabilität der Porenstruktur in natürlichen Gesteinen.

2.3.4 Measurement — Messung

The measurement of permeability consists in measuring the volume flow per unit time of a fluid of given viscosity under a given pressure, while a simple flow field geometry – homogeneous or (two-dimensionally) radial – is maintained. Liquids or gases might be used for the flowing medium, either one with its particular advantages and disadvantages. Flow velocity must always be kept in the d'Arcy range.

Die Messung der Permeabilität besteht in einer Bestimmung des Volumendurchflußes pro Zeiteinheit eines Fluids gegebener Viskosität unter gegebenem Druck, bei Einhaltung einer einfachen Strömungsgeometrie, wie homogener Strömung oder (zweidimensional) radialer Strömung. Als Strömungsmedium lassen sich Flüssigkeiten oder Gase verwenden, jedes mit speziellen Vor- und Nachteilen. Die Strömungsgeschwindigkeit muß stets im d'Arcy-Bereich bleiben.

2.3.4.1 Liquid flow measurement — Messung mit Flüssigkeiten

The measurement using a liquid is easy, because liquids can be considered incompressible, and volumes can be easily measured by calibrated vessels, burettes and the like, and because the outlet pressure can always be atmospheric. The disadvantage of liquids in comparison to gases is their higher viscosity, limiting low permeability measurements, but mainly their not truly inert, not truly Newtonian behavior under all circumstances.

Care must be taken in selecting the liquids. Electrolytes and polar liquids are not well suited. Also solvents of rock components are obviously inappropriate. Apparently, organic liquids with long straight-chain molecules are less recomendable than those of a side-chain type; iso-octane has been often used with good success. Special "core test fluids" are also on the market, consisting of higher order hydrocarbons of a particular molecular structure. – However, in low-permeability high-internal-surface rock samples, every liquid seems to develop strange structural properties, that lead to an apparently much higher viscosity, which even increases steadily with flow time.

Die Messung unter Verwendung von Flüssigkeiten ist einfach, weil Flüssigkeiten als inkompressibel angenommen werden können, weil sich die Volumina leicht messen lassen, mittels Büretten o.ä., und weil am Ausgang stets Atmosphärendruck anliegen kann. Der Nachteil von Flüssigkeiten gegenüber Gasen ist ihre größere Viskosität, durch die die Messung zu niedrigen Permeabilitäten hin begrenzt wird, aber vor allem ihr nicht völlig inertes, nicht unter allen Umständen echt Newtonsches Verhalten.

Sorgfältige Auswahl der Flüssigkeiten ist nötig. Elektrolyte und polare Flüssigkeiten sind ungeeignet, natürlich auch alle Flüssigkeiten, die Gesteinskomponenten lösen. Ferner sind organische Flüssigkeiten mit langen, geraden Kettenmolekülen offenbar weniger geeignet als solche mit verzweigten Kettenmolekülen. Iso-Oktan hat sich vielfach gut bewährt. Im übrigen sind spezielle "Core Test Fluids" auf dem Markt, die aus Kohlenwasserstoffen höherer Ordnung mit spezieller Molekülstruktur bestehen. – Jedoch scheint in Gesteinen sehr kleiner Permeabilität und sehr großer innerer Oberfläche mehr oder weniger jede Flüssigkeit ungewöhnliche Struktureigenschaften zu entwickeln, die scheinbar zu einer sehr viel größeren Viskosität führen, welche noch dazu oft monoton mit der Durchflußzeit ansteigt.

2.3.4.2 Gas flow measurement — Messung mit Gasen

For permeability measurements gases have the advantage of inertness and low viscosity. However, their compressibility must be taken into account, and the Klinkenberg effect must be corrected for.

The measurement and its evaluation can be based on equation (3) in one of its forms. The flow rate should be measured at atmospheric pressure, which can be assumed to be practically constant at its normal value. Then two pressure measurements are necessary, either p_{in} and p_{out} or one of each and the pressure difference Δp (by a differential gage).

Gase haben für Permeabilitätsmessungen den Vorteil geringer Viskosität und inerten Verhaltens. Dagegen muß ihre Kompressibilität bei der Auswertung berücksichtigt werden, und es ist eine Klinkenberg-Korrektur nötig.

Die Messung und Auswertung kann auf Grund der Gleichung (3) in einer ihrer Formen vorgenommen werden. Die Durchflußrate sollte bei Atmosphärendruck gemessen werden, der sich praktisch als Konstante mit seinem Normalwert einsetzen läßt. Dann sind nur zwei Druckmessungen nötig: entweder für p_{in} und p_{out} oder für einen dieser beiden Drucke und die Druckdifferenz Δp (mit einem Differenzdruckmesser).

From the obtained pressure values the mean pressure for the Klinkenberg correction can also be derived. If one measurement only is made at only one mean pressure value, an approximate correction is possible according to one of the empirical formulas like (9).

However, a true correction can be made according to Eq. (8) by repeating the measurement at various mean pressures $\bar{p}$ and plotting K_a vs. $1/\bar{p}$. A straight line can then be drawn through the measurement points, with a slope $K\alpha$ and intercepting the ordinate at K. – While varying the mean pressure, the pressure difference should be kept rather constant, to avoid non-d'Arcy effects.

Aus den so erhaltenen Druckwerten läßt sich auch der Mitteldruck für die Klinkenberg-Korrektur gewinnen. Falls nur eine einzelne Messung bei einem einzigen Mitteldruck durchführbar ist, kann eine Näherungskorrektur nach einer der empirischen Formeln wie (9) angebracht werden.

Jedoch läßt sich nach Gl. (8) auch eine strenge Korrektur durchführen, indem man die Messung bei mehreren verschiedenen Mitteldrucken $\bar{p}$ wiederholt und graphisch K_a gegen $1/\bar{p}$ aufträgt. Dann läßt sich durch die Meßpunkte eine Gerade mit der Steigung $K\alpha$ und dem Ordinatenabschnitt K legen. – Im Zuge der Variation des Mitteldruckes sollte die Druckdifferenz Δp möglichst konstant bleiben, damit Fehler durch Abweichungen vom d'Arcyschen Gesetz vermieden werden.

2.3.5 List of symbols — Symbolliste

d [cm, mm, µm]	diameter – Durchmesser
$\bar{d}$	arithmetic mean diameter – arithmetisch gemittelter Durchmesser
$\tilde{d}$	geometric mean diameter – geometrisch gemittelter Durchmesser
d_{50}	median diameter – Mediandurchmesser
F	formation factor – Formationsfaktor
G	granular parameter – Granularparameter
i_{crit}	critical hydraulic gradient – kritischer hydraulischer Gradient
J [cm^3 s^{-1}]	volume flow rate – Volumenflußrate
K [d]	permeability – Permeabilität
k [%]	relative permeability – relative Permeabilität
L [cm]	length – Länge
p [Pa, bar]	pressure – Druck
$\bar{p}$	mean pressure – mittlerer Druck
p_{hyd} [bar]	hydrostatic pressure – hydrostatischer Druck
p_{lith} [bar]	lithostatic pressure – lithostatischer Druck
S [µm^{-1}]	specific internal surface, surface density – spezifische innere Oberfläche, Oberflächendichte
S_{mat}	specific internal surface referred to matrix volume – spezifische innere Oberfläche bezogen auf das Matrixvolumen
S_{por}	specific internal surface referred to pore volume – spezifische innere Oberfläche bezogen auf das Porenvolumen
S_{tot}	specific internal surface referred to the total rock volume – spezifische innere Oberfläche bezogen auf das gesamte Gesteinsvolumen
T [°C]	temperature – Temperatur
T_F [°C]	formation temperature – Formationstemperatur
T	hydraulic tortuosity – hydraulische Tortuosität
u [m s^{-1}]	volume flow density – Volumenflußdichte
u_{crit} [m s^{-1}]	critical flow density – kritische Flußdichte
X	electrical tortuosity – elektrische Tortuosität
z [m]	depth – Tiefe
α [bar]	Klinkenberg constant – Klinkenbergkonstante
η [cP, Pa·s]	dynamic viscosity of streaming fluid – dynamische Zähigkeit der strömenden Flüssigkeit
ϱ [g cm^{-3}, kg m^{-3}]	fluid density – Flüssigkeitsdichte
Σ	saturation of the porous formation – Sättigungsgrad des Porenraumes
σ_ϕ	logarithmic standard deviation of grain size distribution – logarithmische Standardabweichung der Korngrößenverteilung
ϕ [%]	porosity – Porosität
Φ	$-\log_2 d$ = negative binary logarithm of diamter – negativer binärer Logarithmus des Durchmessers

2.3.6 Tables and figures — Tabellen und Abbildungen

Table 1. Rock properties and flow response [Pet72].

Rock Property	Effects on permeability and porosity
Texture	
Grain Size	Permeability decreases with grain size; porosity unchanged.
Sorting	Permeability and porosity decrease as sorting becomes poorer.
Packing	Although little data is available, tighter packing favors both lesser permeability and porosity.
Fabric	In absence of lamination, controls anisotropy of permeability; permeability is a maximum parallel to mean shape fabric.
Cement	The more cement, the less permeability and porosity.
Sedimentary structures	
Parting lineation	Maximum permeability most probably parallels fabric in plane of bedding.
Crossbedding	Scant available data suggests that horizontal permeability parallels direction of inclination and that the steeper the dip of the foreset, the weaker the horizontal vector of permeability.
Ripple mark	Little data, but fine grain and more laminations combine to cause low permeability and hence ripple zones are commonly barriers to flow.
Grooves and flutes	As judged by fabric, permeability should parallel long dimension.
Slump structures	No data, but probably always greatly reduce horizontal permeability.
Biogenic structures	Destroy depositional fabric and bedding and thus drastically reduce permeability and cause minimal, if any, horizontal anisotropy of permeability. Effect on porosity is unknown, but may be negligible.
Lithology	
Sandstone	Thicker beds tend to be coarser grained and thus more permeable, if cement is not a factor. If weakly or uncemented, ratio of maximum to minimum permeability is perhaps less than 5 to 1; if cement controlled, ratio may reach 100 to 1 or more.
Shale	The prime barrier to flow that outshadows all others by far. Thus it is the *arrangement* of sand and shale much more than permeability variation *within* the sand that controls flow in most reservoirs.

For Table 2, see next page.

Table 3. Magnitude of permeability, K, of gravel, sand, silt, and clay [Pet72]

	← Permeability K			
	10^5 10^4 10^3 10^2 10	1	10^{-1} 10^{-2} 10^{-3}	10^{-4} 10^{-5} d
Material	Clean gravel	Clean sands; mixtures of clean sands and gravels	Very fine sands; silts; mixtures of sand, silt, and clay; glacial till; stratified clays; etc.	Unweathered clays
Flow characteristics	Good aquifers		Poor aquifers	Impervious

Table 2. Hierarchical sequence of primary controls on permeability [Pet72].

Control	Remarks
Texture and fabric Defined by grain size, sorting, packing, and shape orientation of framework grains. Scale: 1 to a few cm^3.	Fundamental "building blocks" that define the primary pore system. Depositional fabric may be completely destroyed by burrowing organisms.
Sedimentary structures Crossbedding, ripple mark, and parting lineation are most common and nearly always have preferred orientation and anisotropic fabrics. Scale: 1 to 10^2 m^3.	Directional structures consist of anisotropic fabrics so that individual structures should behave as "flow packets."
Bedding facies Defined by bed thickness, types and abundances of sedimentary structures and frequency of shale beds. Scale: 10^2 to 10^5 m^3.	Probably the most important primary control on permeability distribution in a sandstone body. Shale beds act as impermeable barriers to flow and are one of the more continuous lithologies.
Composite sand bodies Superposition of one "cycle" of sand upon another, cycles commonly separated by unconformities. Scale: 10^6 to 10^{10} m^3.	Charactteristic of many alluvial and deltaic sands. Multilateral as well as multistory bodies possible.

For Table 3, see previous page.

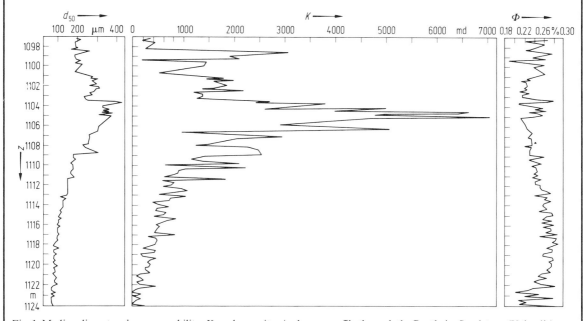

Fig. 1. Median diameter, d_{50}, permeability, K, and porosity, ϕ, along a profile through the Bentheim Sandstone (Valendis), Scheerhorn oil field near Lingen, NW Germany [Eng60]. z = depth.

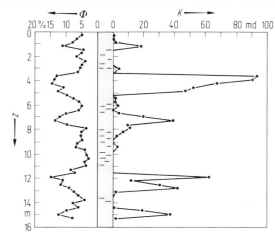

Fig. 2. Vertical section of a sandstone sequence to show the variations in porosity, ϕ, and permeability, K. The observed relationship between increasing porosity and permeability is not uncommon [Sel76].

Table 4. Calculation of permeability, K, of loose sands (after v. Engelhardt and Pitter, 1951) [Eng73].

No.		S_{mat}[2]) [cm^{-1}]	K_{calc}[3]) [d]	K[4]) [d]
Monodisperse Sands[1])				
	d [cm]			
1	0.0200	175	118	72
2	0.0138	255	55	55
3	0.0113	311	37	37
4	0.00875	399	23	26
5	0.00625	560	11	10
6	0.00437	802	5.6	5.4
7	0.00337	1040	3.3	2.5
Mixtures of sands[1])				
8	50% 1, 50% 2	215	78	60
9	50% 5, 50% 6	679	7.8	10
10	Je 25% 1, 2, 5, 6	448	18	20
11	75% 1, 25% 6	331	33	42

[1]) Sieve separates of a diluvial sand and mixtures of it.
[2]) Calculated from grain size.
[3]) Calculated from S_{mat} using Eq. (17).
[4]) Measured at $\phi = 40\%$.

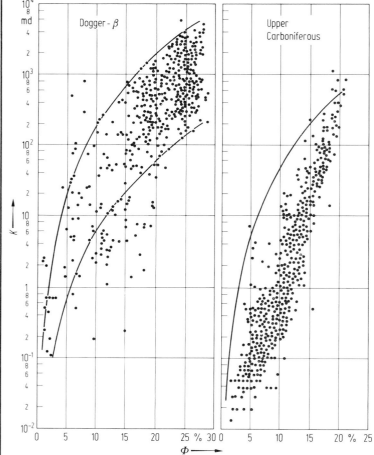

Fig. 3. Logarithmic plots of permeability, K, against porosity, ϕ. Left, Dogger-β (Jurassic) and right, Upper Carboniferous sandstone. Note correlation (Füchtbauer, 1967) [Pet72].

Table 5. Petrographical character, porosity ϕ, and permeability K of some reservoir rocks from oil and gas fields of Germany [Eng60]. d_{50} = median diameter; $d < 20$ µm: grain content with $d < 20$ µm; z = depth.

Location	Formation	z m	d_{50} mm	$d < 20$ µm %	Carbonate %	ϕ	K md
Sandstones							
Schwabmünchen near Augsburg	Baustein layers, Chattian sands[1])	1300	0.2	10	60	0.285	2380
Ampfing near Mühldorf/Inn	Ampfing sandstone, Lattorf[1])	1820	0.7	2.1	14.6	0.199	4900
Stockstadt near Darmstadt	Upper Pechelbronn layers, Oligocene	1550	0.04	15	30	0.102	7
Stockstadt near Darmstadt	Lower Pechelbronn layers, Oligocene[1])	1610	0.25	2.5	0	0.246	3200
Barenburg near Nienburg/Weser	Upper Valendis[1])	810	0.25	1.5	4	0.251	3100
Wietingsmoor near Nienburg/Weser	Middle Valendis[1])	850	0.4	6	6	0.236	510
Rühlermoor near Meppen/Ems	Bentheim sandstone, Valendis[1])	785	0.25	0.8	0	0.295	7500
Scheerhorn near Lingen/Ems	Bentheim sandstone, Valendis[1])	1105	0.35	1.1	0	0.245	5700
Scheerhorn near Lingen/Ems	Bentheim sandstone, Valendis, Base	1120	0.11	9.5	0	0.274	400
Scheerhorn near Lingen/Ems	Wealden[1])	1160	0.10	7.5	26	0.272	180
Ostenwalde near Meppen/Ems	Middle Kimmeridge	1535	0.50	1.3	0	0.262	9900
Kronsberg near Hannover	Dogger-ε[1])	650	0.35	6.5	49	0.247	105
Hankensbüttel, S of Uelzen	Dogger-β, upper sandstone[1])	1535	0.21	1.6	1	0.278	3250
Hankensbüttel, S of Uelzen	Dogger-β, lower sandstone	1605	0.09	2.6	2	0.228	615
Meerdorf near Braunschweig	Dogger-β[1])	1733	0.09	3.1	3	0.190	100
Eldingen near Celle	Lias-α 2[1])	1485	0.16	2	0	0.269	1570
Eldingen near Celle	Lias-α 1, 2 [1])	1520	0.04	10	2	0.245	35
Abbensen near Peine	Middle Rhaetian	325	0.31	2.3	4	0.185	1360
Carbonate rocks							
Lingen/Ems	Wealden, Schalenkalk[1])	900				0.236	260
Ostenwalde near Meppen/Ems	Portland Schalenkalk, Oolite[1])	1495				0.198	65
Hohenassel near Braunschweig	Oxford Coral Oolite[1])	520				0.195	2700
Itterbeck near Nordhorn, Emsland	Zechstein, Main Dolomite[1])	1610				0.130	3

[1]) Especially permeable layers.

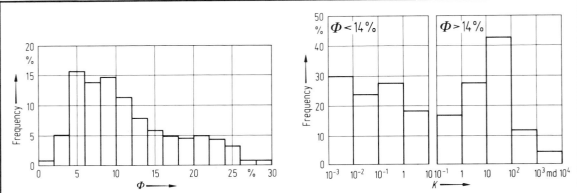

Fig. 4a. Porosity, ϕ, and permeability, K, distribution in the Lithothamnia Limestone of Wolfersberg [Sta74].

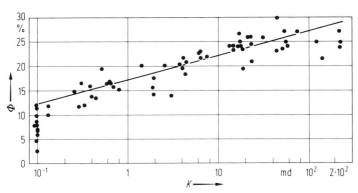

Fig. 4b. Crossplot of porosity, ϕ, and permeability, K, for the Lithothamnia Limestone of Wolfersberg (cut-off 14% porosity) [Sta74].

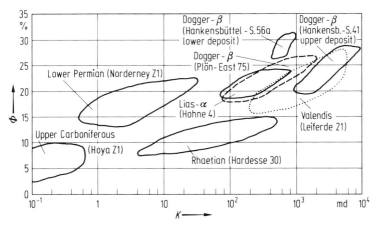

Fig. 5a. Relations between porosity, ϕ, and permeability, K, of sandstones of NW Germany [Gai73].

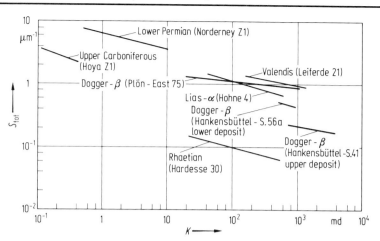

Fig. 5b. Relations between specific internal surface, S_{tot} in [cm^2/cm^3 rock volume], and permeability, K, of sandstones of NW Germany [Gai73].

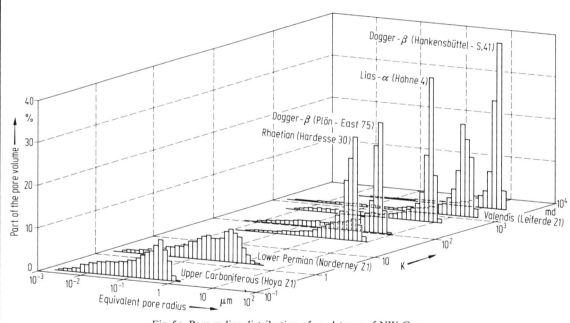

Fig. 5c. Pore radius distribution of sandstones of NW Germany vs. permeability, K [Gai73].

Table 6. Approximate ranges of porosity, ϕ, and permeability, K, in the more important porous formations of Germany with brief comments [Scho75].

Formation	ϕ %	K md	Comment
Miocene, Burdigal	15⋯30	400⋯1200	potentially good sandy reservoirs
Oligocene, Chattian sands	25⋯30	200⋯1000	very porous, little consolidated sands of fair to good permeability
Eocene, Lithothamnia limestone	5⋯30	0⋯200	bioclastic limestone, calcareous-dolomitic-marly cemented; very differentiated: from little consolidated to very cemented; potential reservoirs of poor quality
Lower Cretaceous, Valendis, Bentheim sandstone	18⋯32	100⋯1100	reservoirs of fair to excellent porosity and permeability, N German basin
Dogger-β, sandstones	17⋯31	20⋯9000	sandstones of poor to excellent porosity and permeability, N German region
Lias-α, sandstones	18⋯31	40⋯1000	sandstones of poor to good porosity and permeability, Weser-Elbe region
Keuper, Rhaetian sandstones	5⋯15	0⋯5000	mainly poorly consolidated sandstones of fair permeability; probably very good reservoirs, SE periphery of Lüneburg Heath
Bunter sandstone (Lower Triassic)	5⋯25	0⋯1000	varying and somewhat complex petrophysical properties; partly major content of clay; potential poor to good reservoirs
Lower Permian, Cornberg sandstone, Walkenried sands, Grey Layers	0⋯15	0⋯100	generally, fairly to very strongly consolidated, and mostly very tight; NW German seaboard, Emsland, Weser-Elbe region, Schleswig-Holstein; possibly poor reservoirs
Upper Carboniferous, Carboniferous sands	0⋯10	0⋯50	mainly, strongly consolidated and very tight; mainly of very little porosity and permeability; poor reservoirs, Münsterland, Emsland, Weser-Ems region

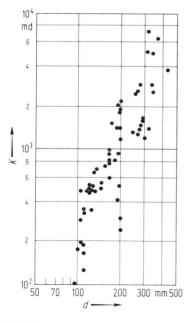

Fig. 6. Logarithmic plot of permeability, K, vs. grain size, d, in the Bentheim Sandstone of the Scheerhorn oil field near Lingen, Germany. Regression equation is $\log_{10} K = -2.1007 + 2.221 \log_{10} d$, where K is permeability in [md] and d is the grain size in [μm]. Scatter diagram based on random selection of data; from von Engelhardt, 1960 [Pet72].

Table 7. Typical values of porosity, ϕ, and permeability, K, listed according to geographical areas, with depths, z, and estimates of formation temperatures, T_F [Scho75, Sta74, Gai73].

Formation or reservoir	Region, locality, or well	z m	ϕ %	K md	T_F °C
Burdigal	Ostmolasse	1121	30	800···900	50
		1124	15···18	3	50
		1126	30	1200	50
		1127···1131	28···30	400···600	50
Upper Chattian sands	Ostmolasse	1080	30	300···1000	50
Lithothamnia limestone	Wolfersberg	2928···2934	13···25	0.5···150	100
		2934···2944	5···17	0···1	100
		2944···2964	5···25	0···50	100
Bentheim sandstone	Rühlermoor	760···800	32	2100	35···40
		785	30	1100	35
		785	30	8700	35
		1097	24	300	
		1099	25	2000···3000	
		1097···1103	20···28	200···3000	
		1103···1107	20···28	1000···7000	
		1107···1112	22···26	600···3000	
		1112···1122	24···28	100···1000	
Valendis	Leiferde Z1		16···28	150···6000	
		817	18	1210	40
		818	19	337	40
Dogger	Plön-E 75		17···27	20···2000	
		2715	26	960	105
		2717	20	101	105
Dogger	Hankensbüttel-S 41		19···29	700···9000	
		1566	23	3550	65
	Hankensbüttel-S 56a		26···31	400···1000	
		1492	25	592	60
Lias	Eldingen	1483	28	420	75
		1490	31	950	75
Lias	Hohne 4		18···24	40···800	
		1763	24	547	85
Rhaetian	Hardesse 30		7···15	4···500	
		2578	13	435	130
Bunter sandstone	Emsland	1801		580	70···80
		1805	21	130	
		1809	23	310	
		1814		770	
		1821	25	810	
		1829	24	1130	
		1830	25	2600	
		1831	18	580	
		1832	25	470	
		1834	27	940	
		1838	23	290	
		1839	26	660	
		1840	23	270	
		1843	22	300	

(continued)

Table 7 (continued)

Formation or reservoir	Region, locality, or well	z m	ϕ %	K md	T_F °C
Bunter sandstone (continued)	Emsland	1845	22	390	
		1846	24		
		1847		1030	
		1848	22	370	
		1854	25	430	75···85
		1855	25	850	
		1856	23	190	
		1857		403	
		1858	22	260	
		1863	23	280	
		1865	20	740	
		1866	24	500	
		1867	21	410	
		1870	22	800	
		1871	23	410	
		1875	21	1060	
		1877	22	310	
		1878	22	250	80···90
Lower Permian	Norderney Z1		13···23	0.5···30	
		4754	20	1.5	100
Upper Carboniferous	Rehden	2103.8···2111.4	16	12···25	105
		2103.8···2111.4	15	5···11	105
		2103.8···2111.4	16	30···40	105
		2111.4···2120.4	8	1	105
		2139.4···2155.4	5···8	1	110
		2155.4···2161.5	14···15	1	110
		2295.6···2305.8	14	1	115
		2305.8···2323.8	14···15	1	115
		2305.8···2323.8	15	1	115
		2470.5···2488.5	8	1	125
		2470.5···2488.5	12	1	125
		2518.5···2529.0	12	1	125
		2518.5···2529.0	13	1	125
		2580.7···2599.0			125
	Hoya Z1		2···10	0.7	
		4486	8.5	0.1	100

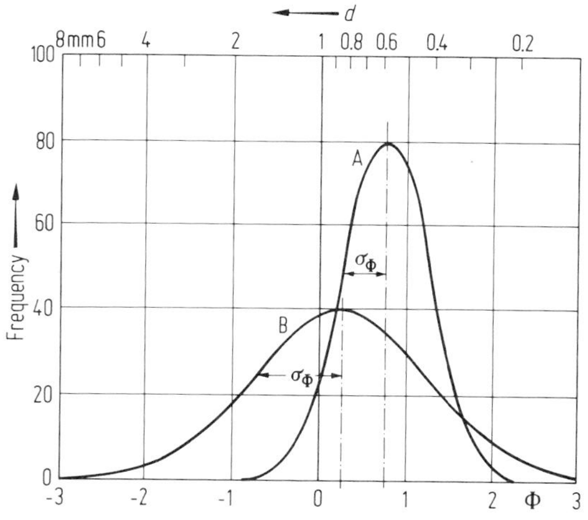

Fig. 7. Definition of the Φ scale: $\Phi = -\log_2 d$, the negative binary logarithm of grain size d in [mm]; and of the logarithmic standard deviation, σ_Φ [Kru42].
Curve A: $\sigma_\phi = 0.04$; $\Phi = 0.75$
Curve B: $\sigma_\phi = 0.21$; $\Phi = 0.25$

Table 8. List of data to Fig. 8a···c [Kru42] $\Phi = -\log_2 \tilde{d}$; $\sigma_\phi = \Phi$ standard deviation; $\tilde{d}$ = geometric mean diameter; K = permeability (see also Figs. 7 and 8 for definition).

Sample No.	Φ	σ_ϕ	$\tilde{d}$ mm	K d
a	1.88	0.04	0.273	57.0
b	1.63	0.04	0.324	60.5
c	1.38	0.04	0.38	89.3
d	1.13	0.04	0.46	139
e	0.88	0.04	0.55	207
f	0.63	0.04	0.65	298
g	0.38	0.04	0.77	403
h	0.13	0.04	0.92	615
i	−0.13	0.04	1.09	812
j	−0.38	0.04	1.30	1195
1	1.25	0.21	0.42	106
2	1.00	0.21	0.50	148
3	0.75	0.21	0.59	213
4	0.50	0.21	0.71	275
5	0.25	0.21	0.84	401
6	0.00	0.21	1.00	590
7	−0.25	0.21	1.19	810
8	−0.50	0.21	1.41	1120
9	−0.75	0.21	1.69	1555
10	0.00	0.15	1.00	618
11	0.00	0.21	1.00	590
12	0.00	0.28	1.00	520
13	0.00	0.34	1.00	470
14	0.00	0.40	1.00	465
15	0.00	0.47	1.00	430
16	0.00	0.53	1.00	361
17	0.00	0.61	1.00	355
18	0.00	0.66	1.00	311
19	0.00	0.74	1.00	294
20	0.00	0.80	1.00	257

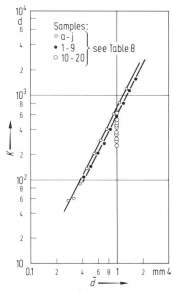

Fig. 8a. Permeability, K, vs. geometric mean grain size, $\tilde{d}$. $K \propto \tilde{d}^2$ [Kru42].

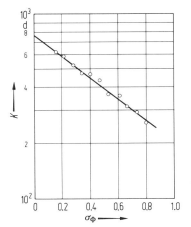

Fig. 8b. Permeability, K, vs. the logarithmic standard deviation σ_ϕ of the grain size distribution [Kru42].

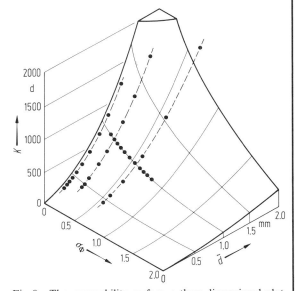

Fig. 8c. The permeability surface, a three-dimensional plot of K vs. $\tilde{d}$ and σ_ϕ. $K \propto \tilde{d}^2 \cdot 2^{-1.89\sigma_\phi}$ [Kru42].

Table 9. Permeability, K, of sandstones from German oil fields to air and various NaCl solutions at a constant pressure gradient $dp/dx = 0.02$ bar m^{-1} at a temperature $T = 20\,°C$.
V marks samples of Bentheim sandstone of Valendis, Lower Cretaceous; mineral content: above 20 μm grain size, 70···90% quartz, the rest alkali feldspar; below 20 μm grain size, 15···25% quartz, ≈15% Illite, and 30···40% Kaolinite.
L marks samples of a Lias-α sandstone; mineral content: above 20 μm grain size, more than 90% quartz, the rest alkali feldspar; below 20 μm grain size, 15% quartz, 10···20% Illite, and 50···70% Kaolinite [Far54, Eng54].
$d < 20$ μm = grain content with $d < 20$ μm, d_{50} = median grain size, ϕ = porosity, η = viscosity, J = volume flow rate.

Sample No.	Field, well	Formation	$d < 20$ μm %	d_{50} mm	ϕ %	Air K md	NaCl J cm^{3}s^{-1}	sol. %	η mPa·s	K md
1 V	Georgsdorf 212, depth 960···970 m	Valendis (Bentheim sandstone)	2	0.10	28.3	530	0.000880	10	1.19	520
							0.000950	5	1.07	510
							0.000880	1	1.01	450
							0.000880	0	1.00	440
2 V	Georgsdorf 212, depth 960···970 m	Valendis (Bentheim sandstone)	2	0.10	26.7	560	0.000922	10	1.19	550
							0.001036	5	1.07	550
							0.000998	1	1.01	500
							0.000964	0	1.00	480
3 V	Georgsdorf 212, depth 960···970 m	Valendis (Bentheim sandstone)	3	0.10	28.3	410	0.000628	10	1.19	370
							0.000636	5	1.07	340
							0.000544	1	1.01	280
							0.000442	0	1.00	220
4 V	Georgsdorf 77, depth 1012.2···1023 m	Valendis (Bentheim sandstone)	≈1	0.17	25.4	1300	0.002218	10	1.19	1320
							0.002104	5	1.07	1130
							0.001828	1	1.01	920
							0.001661	0	1.00	830
5 V	Georgsdorf 94a, depth 1134.7···1141.5 m	Valendis (Bentheim sandstone)	≈5	0.10	26.1	189	0.000300	10	1.19	180
							0.000310	5	1.07	170
							0.000310	1	1.01	160
							0.000296	0	1.00	150
6 V	Georgsdorf 94a, depth 1134.7···1141.5 m	Valendis (Bentheim sandstone)	≈5	0.10	25.6	134	0.000194	10	1.19	120
							0.000206	5	1.07	110
							0.000208	1	1.01	110
							0.000190	0	1.00	100
7 V	Georgsdorf 94a, depth 1134.7···1141.5 m	Valendis (Bentheim sandstone)	≈4	0.10	26.5	270	0.000368	10	1.19	220
							0.000384	5	1.07	210
							0.000374	1	1.01	190
							0.000362	0	1.00	180
8 L	Eldingen 4, depth 1575···1580 m	Lias (sandstone)	4	0.10	25.7	500	0.000620	17	1.37	430
							0.000180	0	1.00	90
9 L	Eldingen 13b, depth 1588.8···1592.3 m	Lias (sandstone)	2	0.10	25.0	200	0.000234	17	1.37	160
							0.000202	10	1.19	120
							0.000168	5	1.07	90
							0.000044	0	1.00	20

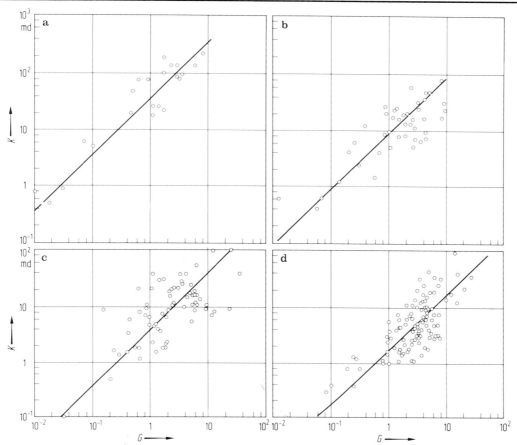

Fig. 9 a···d. Influence of grain size, sorting, and compaction on permeability. Data [Hsü77] for compacted but little cemented pliocene sands of Ventura oil field, California.

The granular parameter G [Kru42] is defined as $G = 10^3 d_{50}^2 \, e^{-1.31\sigma_\phi}$, where d_{50} is the median diameter and σ_ϕ the logarithmic standard deviation of the grain size distribution.

a) upper pliocene, 950···1300 m;   b) lower pliocene, 1525···1740 m;
c) lower pliocene, 2000···2205 m;   d) lower pliocene, 2195···2460 m.

Data show a proportionality between permeability and granular parameter and an exponential compaction law $K = K_0 \, e^{-\alpha z}$, with the uncompacted sand permeability K_0, the depth of burial, z, and here $\alpha = 1.8$ km^{-1}.

Table 10. Permeabilities, K, of Lias sandstones from German wells; for various fluids [Eng60]. $d < 20$ µm: grain content with $d < 20$ µm.

Sample No.	$d < 20$ µm %	ϕ	K in [d] to				
			Air	CCl$_4$	C$_6$H$_{14}$	C$_6$H$_6$	H$_2$O
10 L	5	0.253	0.67	0.70	0.75	0.62	0.27
11 L	5	0.263	0.63	0.61	0.62	0.59	0.13
12 L	5	0.276	0.64	0.66	0.63	0.63	0.07
13 L	4	0.273	1.10	1.11	1.08	1.06	0.26

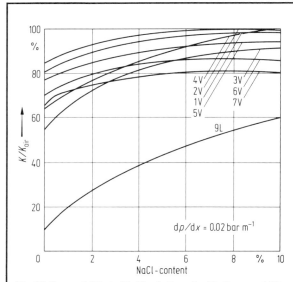

Fig. 10. Permeability to NaCl solutions (in % of permeability to air) of various sandstones [Far54]. For sample numbers, see Table 9.

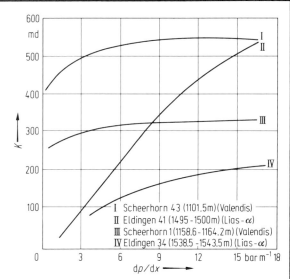

Fig. 11. Permeability, K, of various sandstones to destillated water [Far54].

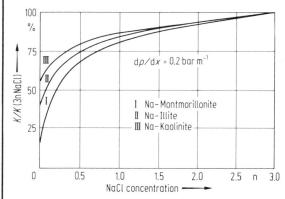

Fig. 12. Permeability of mixtures of quartz sand with 4% Na clay, to NaCl solutions (in % of the permeability to 3n NaCl solution) [Far54].

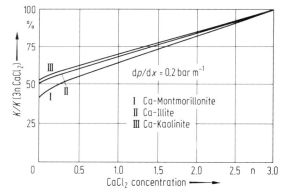

Fig. 13. Permeability of mixtures of quartz sand with 4% Ca clay, to $CaCl_2$ solutions (in % of the permeability to 3n $CaCl_2$ solution) [Far54].

Table 11. Summary of permeability data for air, salt water, and fresh water [Joh45].

Number of wells	Number of tests	Permeability in [md] to			Permeability ratios		
		Air	Salt water	Fresh water	Air/Salt water %	Fresh water/Salt water	
						Average %	Range %
3	20	53	17	1.1	310	6.5	2⋯12
		384	212	185	139	87	68⋯94
2	109	289	211	4.3	137	5.0	2.8⋯8.2
		905	603	231	150	38	5⋯100
3	71	1965	975	8.3	200	0.85	0.7⋯1.0
		1207	637	110	189	17	4⋯27
2	13	1580	1040	144	152	14	1.3⋯19
		1100	820	516	134	63	48⋯100
1	8	450	286	233	157	81	54⋯100
1	8	34	17	12	200	71	55⋯92
1	1	18	0.39	0.25	4600	64	64
3	30	982	304	1.2	323	0.40	0.2⋯0.7
		513	139	49	370	35	9⋯75
2	11	1233	134	0.0	920	0.0	0
		13350	4610	418	290	9.0	0.0⋯22
2	5	8670	5070	15	171	3.0	0.5⋯21
		1590	1280	429	124	34	34
2	26	2280	550	0.05	415	0.01	0.0⋯0.02
		2450	1115	17.7	220	1.55	0.5⋯10
1	5	3240	2035	1245	159	61	2⋯91
1	6	136	104	86	131	82	69⋯100
1	2	1925	124	73	155	59	58⋯60
1	5	221	18.5	12.5	1600	68	56⋯100
1	8	161	22	13	732	59	30⋯100
1	4	1908	334	209	572	63	44⋯94
2	12	50	2.7	0.8	1850	30	27⋯80
		864	375	198	230	53	24⋯73
2	54	94	27	5.8	350	21	1.3⋯90
		553	350	218	158	62	37⋯100
10	294	6444	1628	0.0	395	0.0	0
		4890	3280	2380	149	88	35⋯100
1	5	1356	443	2.4	306	0.5	0.0⋯3.2
1	5	3750	2760	628	136	23	7⋯63
5	42	1830	542	225	338	41	5.3⋯70
		2685	1645	1540	163	94	77⋯100
1	25	2100	1383	479	152	35	2⋯76
2	52	2100	1990	1500	106	75	49⋯100
		3100	3000	2800	103	93	76⋯100
1	13	3460	1580	2.9	220	0.2	0.006⋯1.1
1	15	4410	1549	1049	285	68	24⋯91
1	12	45	13	6.8	345	52	0.0⋯94
4	29	52	10	0.004	520	0.04	0.0⋯1.5
		339	108	23	315	21	9⋯97
1	2	801	2.1	0.8	38000	38	0.0⋯56
4	15	607	378	1.5	160	0.4	0.2⋯0.5
		108	47	33	230	71	30⋯75
1	5	600	21	4.3	2850	20	2.9⋯56
3	78	3190	1630	869	195	53	35⋯63

(continued)

Table 11 (continued)

Number of wells	Number of tests	Permeability in [md] to			Permeability ratios		
		Air	Salt water	Fresh water	Air/Salt water %	Fresh water/Salt water	
						Average %	Range %
16	78	4630	1840	5.0	250	0.3	0.0···2.0
		2674	1343	1318	198	98	88···100
5	67	2030	1495	188	136	13	1.5···95
		2770	1920	1040	144	54	24···70
1	3	402	84	16	480	19	10···24
3	22	216	54	39	400	72	24···100
		3010	2450	1920	123	78	73···91
6	20	275	180	71	153	39	39
		54	39	32	138	82	82
2	12	4.3	3.2	1.4	134	43	30···82
		18	10	8	180	80	56···82
2	14	48	5.4	2.1	890	39	22···60
		28	13	8.9	215	68	31···100
2	9	86	3.8	1.8	2260	47	36···52
		32	17	16	188	94	78···100
1	13	70	68	44	103	65	44···100
1	5	9	6.1	1.9	148	31	6···58

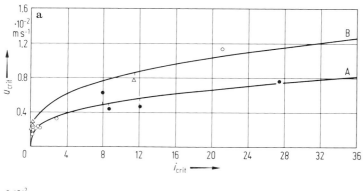

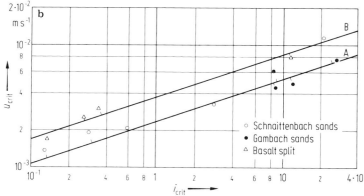

Fig. 14a, b. Critical flow density, u_{crit}, vs. the critical hydraulic gradient, i_{crit} ($i = |\text{grad } p|/\varrho g$); a) linear plot; b) double logarithmic plot. Curve A: quartz sand; curve B: basalt split. A relation $u_{\text{crit}} \propto \sqrt[3]{i_{\text{crit}}}$ can be seen from the curves; thus also follows $u_{\text{crit}} \propto 1/\sqrt{K}$ [Hof74].

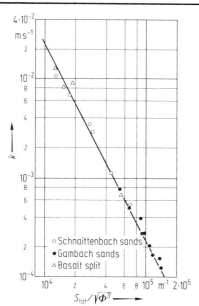

Fig. 15. Permeability coefficient $k = K g \varrho/\eta$ vs. $S_{tot}/\sqrt{\phi^3}$ with S in [μm^{-1}] [Hof74]. See Eq. 17 in 2.3.3.

Table 12. Critical pressure gradients, (grad p)$_{crit}$, and flow densities, u_{crit}, for the validity of d'Arcy's law in rocks (for water) [Eng60]. d = grain size, K = permeability

d m	u_{crit} m·s^{-1}	(grad p)$_{crit}$ [bar m^{-1}]	
		$K = 0.1$ d	$K = 1$ d
10^{-3}	10^{-3}	100	10
10^{-4}	10^{-2}	1000	100
10^{-5}	10^{-1}	10000	1000

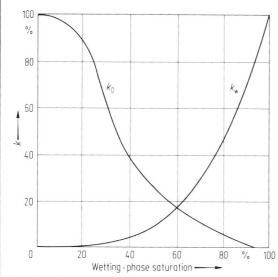

Fig. 17. Relative permeabilities, k, for a wetting (k_w) and a non-wetting phase (k_0) vs. the wetting-phase saturation [Eng73].

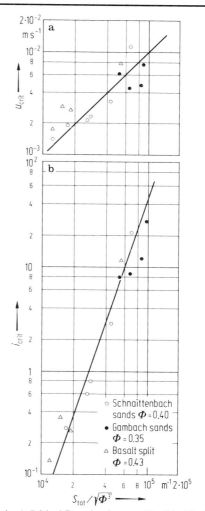

Fig. 16a, b. a) Critical flow density u_{crit}, b) critical hydraulic gradient i_{crit} ($i = |\text{grad } p|/\varrho g$) vs. $S_{tot}/\sqrt{\phi^3}$ with S in [μm^{-1}] [Hof74].

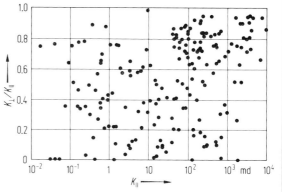

Fig. 18. Dependence of the ratio of the permeabilities normal ($K_\perp$) and parallel ($K_\parallel$) to the bedding plane on $K_\parallel$; for Lias, Dogger, and Valendis sandstones of the Gifhorn Trough, NW Germany. With increasing $K_\parallel$, the average of $K_\perp/K_\parallel$ approaches one. After Rühl and Schmid, 1957 [Eng60].

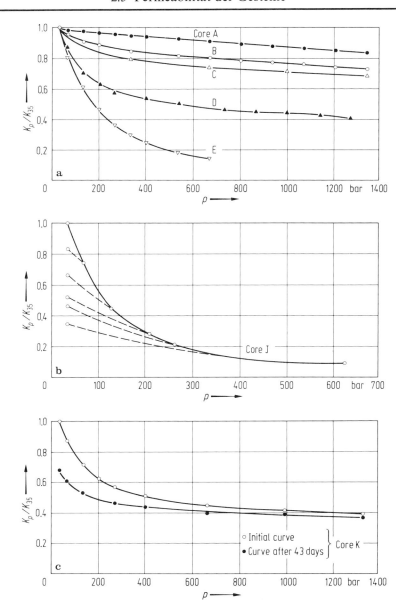

Fig. 19a⋯c. Effect of confining pressure, p, on permeability, K_p. K_{35} permeability at $p \approx 35$ bar, the initial confining pressure, [Vai71].
Core A: Chanute sandstone, $K_{35} = 191$ md; core B: San Andres carbonate, $K_{35} = 15$ md; core C: Springer sandstone, $K_{35} = 1.7$ md; core D: Springer sandstone, $K_{35} = 0.186$ md; core E: Frio sandstone, $K_{35} = 0.04$ md; core J: Morrow sandstone, $K_{35} = 0.377$ md; core K: devonian dolomite, $K_{35} = 0.872$ md.
a) Permeability ratio vs. net confining pressure.
b) Hysteresis. Left final points of dashed lines: recovery one hour after quick removal of various stresses.
c) Reproducibility of stress curve after 43 days without stress.

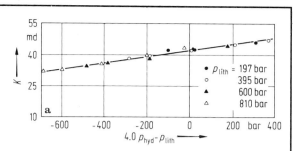

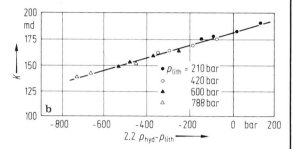

Fig. 20a, b. Dependence of permeability on the effective stress, $p_{eff}=p_{lith}-\zeta p_{hyd}$, with lithostatic pressure p_{lith} on the framework and hydrostatic pressure p_{hyd} in the pores. Experimental results for Berea sandstone, a) for flow normal to the bedding plane, $\zeta=4$; $dK/dp_{eff}=14.3$ μd/bar; b) for flow in the bedding plane, $\zeta=2.2$; $dK/dp_{eff}=59.8$ μd/bar, [Zob75]. Data show a linear approximation of a more general exponential stress law $K/K_0 = e^{-\alpha p_{eff}} \approx 1-\alpha\, p_{eff}$. Measurement with light lubricating oil.

2.3.7 References for 2.3.1···2.3.6 — Literatur zu 2.3.1···2.3.6

Car56	Carman, P.C.: Flow of Gases through Porous Media. London, **1956**.
Eng51	von Engelhardt, W., Pitter, H.: Heidelbg. Beitr. Min. Petrogr. **2** (1951) 477.
Eng54	von Engelhardt, W., Tunn, W.: Heidelbg. Beitr. Min. Petrogr. **4** (1954) 12.
Eng60	von Engelhardt, W.: Der Porenraum der Sedimente. – Berlin-Göttingen-Heidelberg: Springer **1960**.
Eng73	von Engelhardt, W.: Die Bildung von Sedimenten und Sedimentgesteinen. Stuttgart: Schweizerbart **1973**.
Far54	Farahmand, F.: Dissertation, Technische Universität Clausthal **1954**.
Fat58	Fatt, I.: Bull. Amer. Assoc. Petrol. Geol. **42** (1958) 1914.
Fat52	Fatt, I., Davis, D.H.: J. Petrol. Tech. **4** (1952) 329.
Fat56	Fatt, I.: Trans. Amer. Instn. Mining, Metall., Petrol. Engrs. **207** (1956) 144 and 160.
Gai73	Gaida, K.H., Rühl, W., Zimmerle, W.: Erdöl Erdgas Z. **89** (1973) 336.
Hof74	Hofedank, R.H.: Dissertation, Universität Giessen **1974**.
Hsü77	Hsü, K.H.: Bull. Amer. Assoc. Petrol. Geol. **61** (1977) 169.
Joh45	Johnston, N., Beeson, C.M.: Trans. Amer. Instn. Mining, Metall., Petrol. Engrs. **160** (1945) 43 and 55.
Koz27	Kozeny, J.: Sitz. Ber. Akad. Wiss. Wien, Math. Nat. (Abt. IIa) **136a** (1927) 271.
Kru42	Krumbein, W.C., Monk, G.D.: Trans. Amer. Instn. Mining, Metall., Petrol. Engrs. **151** (1942) 153.
Pet72	Pettijohn, F.J., Potter, P.E., Siever, R.: Sand and Sandstone. Berlin-Heidelberg-New York: Springer **1972**.
Pur49	Purcell, W.R.: Trans. Amer. Instn. Mining, Metall., Petrol. Engrs. **186** (1949) 39.
Schei60	Scheidegger, A.E.: The Physics of Flow through Porous Media. Univ. of Toronto Press, Toronto **1960**.
Scho66	Schopper, J.R.: Geophys. Prosp. **14** (1966) 301.
Scho67	Schopper, J.R.: Geophys. Prosp. **15** (1967) 262.
Scho75	Schopper, J.R.: Intergranulare Porositäten und Permeabilitäten im Untergrund der Bundesrepublik Deutschland. Unveröffentlichte Zusammenstellung, Clausthal **1975**.
Sel76	Selley, R.C.: An Introduction to Sedimentology. London-New York-San Francisco: Academic Press **1976**.
Sta74	Stanciu, M.: Erdöl Erdgas Z. **90** (1974) 333.
Tod60	Todd, D.K.: Ground Water Hydrology. New York: Wiley **1960**.
Vai71	Vairogs, J., Hearn, C.L., Dareing, D.W., Rhoades, V.W.: J. Petrol. Tech. **23** (1971) 1161.
Zob75	Zoback, M.D., Byerlee, J.D.: Bull. Amer. Assoc. Petrol. Geol. **59** (1975) 154.

3 Elasticity and inelasticity — Elastizität und Inelastizität

3.1 Elastic wave velocities and constants of elasticity of rocks and rock-forming minerals — Geschwindigkeiten elastischer Wellen und Elastizitäts-Konstanten von Gesteinen und gesteinsbildenden Mineralen

3.1.1 Introduction — Einleitung
(H. Gebrande)

3.1.2 Elastic wave velocities and constants of elasticity at normal conditions — Geschwindigkeiten elastischer Wellen und Elastizitäts-Konstanten bei Normalbedingungen
(H. Gebrande)

3.1.3 Elastic wave velocities and constants of elasticity of rocks at normal temperature and pressures up to 1 GPa — Geschwindigkeiten elastischer Wellen und Elastizitäts-Konstanten von Gesteinen bei Zimmertemperatur und Drucken bis 1 GPa
(H. Gebrande)

3.1.4 Elastic wave velocities and constants of elasticities of rocks at elevated pressures and temperatures — Geschwindigkeiten elastischer Wellen und Elastizitäts-Konstanten bei erhöhten Drucken und Temperaturen
(H. Kern)

3.2 Fracture and flow of rocks and minerals — Bruch und Inelastizität von Gesteinen und Mineralen
(F. Rummel)

See Subvolume V/1b, page 1 ff

4 Thermal properties — Thermische Eigenschaften

4.1 Thermal conductivity and specific heat of minerals and rocks — Wärmeleitfähigkeit und Wärmekapazität der Minerale und Gesteine

4.1.1 Introductory remarks — Einführung

4.1.1.1 Definitions — Definitionen

The process of temperature equalization is connected to the flow of heat from places at higher temperature to those at lower temperature; this flow of heat, q, is proportional to the existing temperature difference and to the thermal conductivity, k, of the medium in question:

Temperaturausgleich ist mit Wärmefluß verbunden, wobei die Wärme von Orten höherer Temperatur zu jenen mit niedriger Temperatur fließt; der Wärmefluß, q, ist proportional zur Temperaturdifferenz und zur Wärmeleitfähigkeit, k, des betreffenden Mediums:

$$q = k \frac{dT}{dx}, \qquad (1)$$

where dT/dx is the temperature gradient.

wo dT/dx den Temperaturgradienten bezeichnet.

The transfer of heat is realized by four main physical processes: (a) *conduction*: heat passes due to lattice interaction, i.e. through the substance of the body itself (k_l); (b) *convection*: heat is transferred by relative motions within the heated body (k_c); (c) *radiation*: heat is transferred by electromagnetic waves (k_r); (d) *excitation* (k_{ex}).

Wärmetransport kann durch vier verschiedene physikalische Vorgänge erfolgen: (a) *Leitung*: die Wärme wird durch Gitterschwingungen d.h. durch den Körper selbst weitergegeben (k_l); (b) *Konvektion*: die Wärme wird durch Relativbewegungen innerhalb des erwärmten Körpers transportiert (k_c); (c) *Strahlung*: die Wärmeübertragung erfolgt mittels elektromagnetischen Wellen (k_r) und (d) *Excitonen* (k_{ex}).

The effective thermal conductivity in the earth's interior is thus the sum of these components,

Die effektive Wärmeleitfähigkeit im Erdinneren kommt durch die Überlagerung der erwähnten Komponenten zustande:

$$k_{eff} = k_l + k_c + k_r + k_{ex}. \qquad (2)$$

In a solid body at moderate temperature and pressure, convection is absent and radiation as well as excitation are negligible. For geophysical studies of the earth's lithosphere, only conduction is of importance and the most substantial thermal property of rocks is their *thermal conductivity*. Even within the same rock type thermal conductivity k can vary over a considerable range (Fig. 1).

In Festkörpern bei mäßiger Temperatur und Druck tritt keine Konvektion auf; Strahlungs- sowie Excitonen-Wärmeleitung können ebenfalls vernachlässigt werden. Für geophysikalische Betrachtungen der Lithosphäre ist deshalb nur die Wärmeleitung im engeren Sinn von Belang und die maßgebende Größe ist die konduktive *Wärmeleitfähigkeit* der Gesteine. Innerhalb desselben Gesteinstyps kann die Wärmeleitfähigkeit (k) über einen weiten Bereich variieren (Fig. 1).

The differential equation which describes the conduction of heat in an isotropic medium with no internal heat sources is

Die Differentialgleichung der Wärmeleitung in einem isotropen Medium ohne Wärmequellen lautet

$$\frac{\partial T}{\partial t} = \kappa \Delta T \qquad (3)$$

where κ is the *thermal diffusivity*, Δ the Laplace operator and $\partial T/\partial t$ the time derivative of the temperature. The thermal diffusivity is related to the thermal conductivity k according to

wobei κ die *Temperaturleitfähigkeit* (Diffusivität), Δ den Laplace-Operator und $\partial T/\partial t$ die Ableitung der Temperatur nach der Zeit bedeuten. Die Temperaturleitfähigkeit hängt mit der Wärmeleitfähigkeit k wie folgt zusammen

$$\kappa = \frac{k}{c_p \varrho} \qquad (4)$$

where c_p is the *specific heat* (= heat capacity per weight unit) at constant pressure and ϱ is the density.

wobei c_p die *spezifische Wärme* bei konstantem Druck und ϱ die Dichte bedeuten.

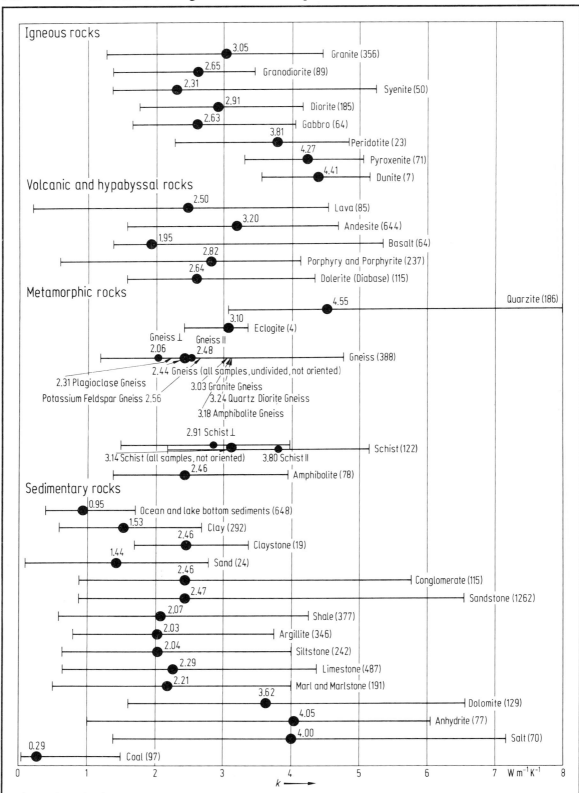

Fig. 1. Thermal conductivities, k, of various rocks (range of measured values and mean value, data from Table 7).

4.1.1.2 Units — Einheiten

The units in the SI system, the cgs units and the conversion factors are given in Table 1.

Die SI-Einheiten, die cgs-Einheiten und die Umrechnungsfaktoren sind in Tabelle 1 aufgeführt.

Table 1. SI and cgs units of thermal properties.

Parameter	Symbol	SI unit	cgs unit	Conversion
Heat flow — Wärmefluß	q	$W\,m^{-2}$	$cal\,cm^{-2}\,s^{-1}$	$1\,W\,m^{-2} = 2.388 \cdot 10^{-5}\,cal\,cm^{-2}\,s^{-1}$ $1\,\mu cal\,cm^{-2}\,s^{-1} = 41.87\,mW\,m^{-2}$
Thermal conductivity — Wärmeleitfähigkeit	k	$W\,m^{-1}\,K^{-1}$	$cal\,cm^{-1}\,s^{-1}\,°C^{-1}$	$1\,W\,m^{-1}\,K^{-1}$ $= 2.388 \cdot 10^{-3}\,cal\,cm^{-1}\,s^{-1}\,°C^{-1}$ $1\,mcal\,cm^{-1}\,s^{-1}\,°C^{-1}$ $= 0.4187\,W\,m^{-1}\,K^{-1}$
Thermal diffusivity — Temperaturleitfähigkeit	κ	$m^2\,s^{-1}$	$cm^2\,s^{-1}$	$1\,m^2\,s^{-1} = 1 \cdot 10^4\,cm^2\,s^{-1}$ $1\,cm^2\,s^{-1} = 1 \cdot 10^{-4}\,m^2\,s^{-1}$
Specific heat — spezifische Wärme	c_p	$J\,kg^{-1}\,K^{-1}$	$cal\,g^{-1}\,°C^{-1}$	$1\,J\,kg^{-1}\,K^{-1}$ $= 2.388 \cdot 10^{-4}\,cal\,g^{-1}\,°C^{-1}$ $1\,cal\,g^{-1}\,°C^{-1} = 4.187\,kJ\,kg^{-1}\,K^{-1}$

4.1.1.3 Anisotropy — Anisotropie

Anisotropy of thermal conductivity is most pronounced in minerals. As rocks are polycrystalline aggregates of individual minerals, their thermal conductivities are determined by the conductivities of the mineral constituents.

Rock forming minerals are predominantly of low symmetry. Thus the relationship (3), valid for isotropic material, is not applicable for small elements of rocks. Microanisotropy of thermal conductivity is related to the arrangement of mineral particles while macroanisotropy occurs in large rock volumes due to bedding, schistosity and also to fracturing and tectonic disturbances. Usually the conductivities in two main directions are important: (a) parallel and normal to the optical axis of minerals (see Table 5) and (b) parallel and normal to the plane of bedding or schistosity in rocks (see Table 12). The ratio of parallel thermal conductivity to normal thermal conductivity is called the *anisotropy factor*.

Am ausgeprägtesten manifestiert sich die Richtungsabhängigkeit der Wärmeleitfähigkeit an Mineralen. Da Gesteine aus polykristallinen Mineralaggregaten bestehen, werden ihre Wärmeleitfähigkeiten durch die Leitfähigkeiten der Mineralbestandteile sowie deren Orientierung bestimmt.

Gesteinsbildende Minerale besitzen meist niedrige Symmetrie. Die Beziehung (3), gültig für isotropes Material, ist für sehr kleine Gesteinselemente nicht anwendbar. Mikroanisotropie der Wärmeleitfähigkeit hängt von der räumlichen Anordnung von Mineralkörnern im Kleinbereich ab, während für die Makroanistropie großer Gesteinsvolumina die Schichtung, Schieferung sowie Klüftung und eventuell auch tektonische Störungen maßgebend sind. Meist genügt die Angabe der Wärmeleitfähigkeit in zwei Hauptrichtungen: (a) parallel und senkrecht zur optischen Achse bei Mineralen (siehe Tab. 5) und (b) parallel und senkrecht zur Schichtung oder Schieferung bei Gesteinen (siehe Tab. 12). Das Verhältnis der parallelen zur senkrechten Wärmeleitfähigkeit wird als *Anisotropie-Faktor* bezeichnet.

4.1.1.4 Temperature and pressure dependence — Temperatur- und Druckabhängigkeit

For studies of the thermal state of the earth, it is very important to know how the thermal conductivity, k, varies with increasing depth, i.e. the dependence of conductivity on temperature, T, and pressure, p. In first approximation the following simple expression can be used:

Für das Verständnis thermischer Vorgänge im Erdinneren ist die Tiefenabhängigkeit der Wärmeleitfähigkeit, k, von ausschlaggebender Bedeutung, d.h. die Abhängigkeit von Druck, p, und Temperatur, T, welche in erster Näherung durch den folgenden einfachen Ausdruck beschrieben wird

$$k = \frac{b\,f(p)}{T} \qquad (5)$$

where $b/T = k_0$ is the conductivity under surface conditons, $f(p) = 1$ at the earth's surface and $f(p) = 23$ at the core-mantle boundary [Cla57a].

wobei $b/T = k_0$ die Wärmeleitfähigkeit bei Bedingungen an der Erdoberfläche ist, ferner $f(p) = 1$ an der Oberfläche und $f(p) = 23$ an der Kern-Mantel-Grenze [Cla57a].

Only limited knowledge on the behaviour of thermal conductivity with increasing pressure is available. At low pressure ($p \leq 100$ MPa) the conductivity may increase by up to 10% due to the closing of pores and microcracks and to better thermal contact of the individual mineral particles. At higher pressure which influences also the elastic properties, the change of conductivity, k, with pressure will be given by

$$k = k_0 (1 + \alpha p) \qquad (6)$$

where α is a constant. The thermal diffusivity, κ, shows a similar behaviour. Within the earth's crust the pressure dependence is negligible compared with the much more pronounced temperature effect.

The *temperature dependence* of thermal conductivity in the crust and upper mantle (i.e. in the temperature range of $0 \cdots 1600$ °C) is generally determined by two mechanisms, lattice conductivity and radiative conductivity. While the former decreases with increasing temperature approximately with $1/T$, the radiative component increases with T^3.

The temperature dependence of the thermal conductivity in the broad range of $0 \cdots 1200$ °C is given in [Hae73] by the expression

$$k \, [\text{Wm}^{-1}\text{K}^{-1}] = 3.6 - 0.49 \cdot 10^{-2} T + 0.61 \cdot 10^{-5} T^2 - 2.58 \cdot 10^{-9} T^3 \qquad (7)$$

The temperature dependence for most rocks in the upper part of the crust (at temperatures $0 \cdots 600$ °C) can be expressed by the simple relation

$$k_T = \frac{1}{A + BT} \qquad (8)$$

where A and B are constants; after rearrangement one obtains an expression for the ratio of the thermal conductivity at surface conditions (k_0) to the conductivity at temperature T:

$$\frac{k_0}{k_T} = 1 + CT \qquad (9)$$

where $C = A/B$.

The thermal conductivity of some rocks with high feldspar content is relatively independent of temperature or even may increase with increasing temperature. Similar phenomena are observed for some glasses.

Quartz is very frequently used as standard (reference) material in experimental determinations of thermal conductivity. Table 2 summarizes the data available on the temperature dependence of thermal conductivity of fused quartz, k_f, and of quartz single crystals perpendicular to the optical axis, $k_\perp$.

Specific heat, c_p, usually increases with increasing temperature. For rocks, the following expression is given in [Eng78]

$$c_p = 0.754 (1 + 6.14 \cdot 10^{-4} T - 1.928 \cdot 10^4 / T^2) \qquad (10)$$

where T is in [K] and c_p in [kJ kg^{-1} K^{-1}].

4.1.1.5 Effects of density/porosity and water content — Einfluß von Dichte/Porosität und Wassergehalt

The thermal conductivity is related to the density, ϱ, and within the interval 2000···3000 kg m^{-3} the density dependence for various rock types can be expressed by individual linear functions (Fig. 2)

Zwischen Wärmeleitfähigkeit und Dichte (ϱ) besteht eine Korrelation, und im Bereich von 2000···3000 kg m^{-3} kann dieser Zusammenhang für verschiedene Gesteinstypen durch individuelle lineare Beziehungen (Fig. 2) angegeben werden in der Form

$$k = A + B\varrho \qquad (11)$$

Porosity and moisture content can affect the thermal conductivity or rocks to an important degree. This applies especially to sedimentary rocks, for which the laboratory and *in-situ* value may differ considerably.

Porosität und Wassergehalt können die Wärmeleitfähigkeit von Gesteinen wesentlich beeinflussen. Dies gilt insbesondere für Sedimentgesteine, bei welchen die im Labor gemessenen Werte unter Umständen stark von den *in situ*-Werten abweichen können.

Fig. 3 shows the ratio of "dry" and "wet" conductivity of sedimentary rocks as a function of the water content. There is an increase of conductivity with increasing water content which reaches a maximum at a porosity of about 20···30%. An empirical formula was proposed to express "wet" conductivity, k_w, as a function of the porosity:

Fig. 3 zeigt das Verhältnis der „trockenen" zur „feuchten" Leitfähigkeit von Sedimentgesteinen in Abhängigkeit vom Wassergehalt. Die Wärmeleitfähigkeit nimmt mit zunehmendem Wassergehalt zunächst zu und erreicht ein Maximum im Porositätsbereich 20···30%. Eine empirische Beziehung zwischen „feuchter" Leitfähigkeit (k_w) und der Porosität (ϕ) lautet

$$k_w = k_d \exp(0.024\,\phi) \qquad (12)$$

where k_d is the conductivity measured for dry samples and ϕ is the porosity in vol%.

wobei k_d die „trocken" gemessenen Wärmeleitfähigkeit bezeichnet; die Porosität ist in vol%.

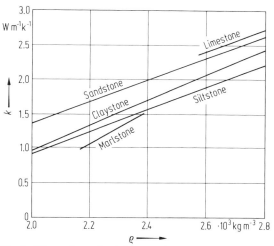

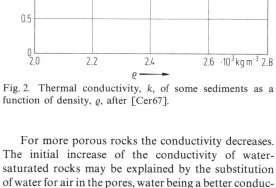

Fig. 2. Thermal conductivity, k, of some sediments as a function of density, ϱ, after [Cer67].

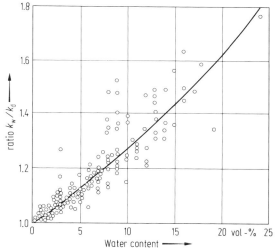

Fig. 3. Ratio of k_w/k_d ("wet" conductivity to "dry" conductivity) as a function of volume water content of sedimentary rocks (after [Cer67]).

For more porous rocks the conductivity decreases. The initial increase of the conductivity of water-saturated rocks may be explained by the substitution of water for air in the pores, water being a better conductor. During further increase of water content, however, the solid particles are gradually replaced by water and the bulk conductivity decreases (Fig. 4). This fact is quite important for sea bottom sediments, with porosities in excess of 50%.

Bei Gesteinen mit noch höherer Porosität nimmt die Wärmeleitfähigkeit wieder ab. Die anfängliche Leitfähigkeitszunahme in wassergesättigten Gesteinen rührt vom Ersatz der Porenluft durch das besser leitende Wasser her. Bei weiterer Zunahme des Wassergehaltes werden jedoch zunehmend auch feste Partikel durch Wasser ersetzt, was zu einer Abnahme der Gesamt-Wärmeleitfähigkeit führt (Fig. 4). Dieser Effekt ist besonders ausgeprägt bei Seeboden-Sedimenten, welche oft Porositäten über 50% aufweisen.

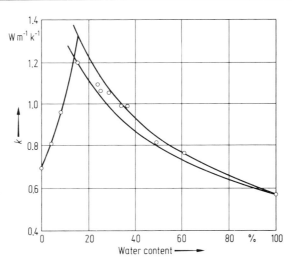

Fig. 4. Relation between thermal conductivity, k, of powdered rocks and water content, after [Hor60].

4.1.2 Tables — Tabellen
4.1.2.1 Minerals — Minerale

Table 2. Thermal conductivity of fused and crystalline quartz.

Temperature range [°C]	Formula $k = f(T)$ k in [W m^{-1} K^{-1}], T in [K]	Ref.
Fused quartz		
60···240	$k_f = 1.336 + 0.000775\,T$	Kay26
−38···952	$k_f = 0.682 + 0.00160\,(T + 273)$	See28
−50···100	$k_f = 1.340 + 0.000766\,T$	Ben39
38···121	$k_f = 1.348 + 0.00374\,T$	Col52
−150···50	$k_f = 1.323 + 0.00193\,T$ $\quad - 0.0000067\,T^2$	Rat59
Crystalline quartz (perpendicular to the optical axis)		
0···100	$k_\perp = (0.1428 + 0.000619\,T)^{-1}$	Kay26
−50···100	$k_\perp = 6.574 - 0.01633\,T$	Ben39
0···400	$k_\perp = (0.1476 + 0.000551\,T$ $\quad - 0.000000291\,T^2)^{-1}$	Bir40
0···120	$k_\perp = (0.1450 + 0.000578\,T)^{-1}$	Rat59

Table 3. Thermal conductivity, k, and specific heat, c_p, of rock-forming minerals (monomineralic, polycrystalline aggregates). k at room temperature [Hor71], c_p at 0 °C [Gor42].

Mineral	Formula	k W m^{-1} K^{-1}	c_p kJ kg^{-1} K^{-1}
Halide			
Fluorite	CaF_2	9.50	0.85
Phosphates			
Fluor-apatite	$Ca_5(PO_4)_3F$	1.37	
Chlor-apatite	$Ca_5(PO_4)_3Cl$	1.38	
Monazite	$CaLa(PO_4)$	1.10	
Carbonates			
Calcite	$CaCO_3$	3.57	0.793
Aragonite	$CaCO_3$	2.23	0.78
Magnesite	$MgCO_3$	5.83	0.864
Dolomite	$CaMg(CO_3)_2$	5.50	0.93 (at 60 °C)
Siderite	$FeCO_3$	3.00	0.683
Sulfates			
Anhydrite	$CaSO_4$	4.76	0.52
Barite	$BaSO_4$	1.33	0.45
Sulfides			
Pyrite	FeS_2	19.2	0.500
Chalcopyrite	$CuFeS_2$	8.19	0.54 (at 50 °C)
Sphalerite	ZnS	12.7	0.45
Galena	PbS	2.28	0.207
Hydroxides			
Gibbsite	$Al(OH)_3$	2.60	
Goethite	α-$FeO \cdot OH$	2.91	
Oxides			
Haematite	α-Fe_2O_3	11.3	0.61
Rutile	TiO_2	5.12	0.70
Spinel	$MgAl_2O_4$	9.48	
Magnetite	$FeFe_2O_4$	5.10	0.60
Chromite	$FeCr_2O_4$	2.52	
Silica minerals			
Quartz	α-SiO_2	7.69	0.698
Chert	SiO_2	4.53	
Flint	SiO_2	3.71	
Vitrous silica	SiO_2	1.36	0.70
Ortho- and Ring-silicates			
Olivines			
$Fo_{98}Fa_2$	$Fo = Mg_2SiO_4$ (Forsterite)	5.06	
$Fo_{54}Fa_{46}$	$Fa = Fe_2SiO_4$ (Fayalite)	3.44	0.79 (at 36 °C)
Fo_4Fa_{96}		3.16	0.55
Zircon	$ZrSiO_4$	4.54	0.61 (at 60 °C)
Sphene	$CaTiSiO_4(OH)$	2.33	
Garnets			
Pyrope	$Mg_3Al_2Si_3O_{12}$	3.18	0.74 (at 58 °C)
Almandine	$Fe_3Al_2Si_3O_{12}$	3.31	
Grossularite	$Ca_3Al_2Si_3O_{12}$	5.48	

(continued)

Table 3 (continued)

Mineral	Formula	k W m^{-1}K^{-1}	c_p kJ kg^{-1}K^{-1}
Sillimanite	} Al_2SiO_5	9.09	0.743
Kyanite		14.2	0.70
Andalusite		7.57	0.77
Staurolite	$Fe_3MgAl_{18}O_{12}(SiO_4)_8(OH)_4$	3.46	
Zoisite	$Ca_2Al \cdot Al_2O \cdot OH \cdot Si_3O_{11}$	2.15	
Epidote	$Ca_2Fe \cdot Al_2O \cdot OH \cdot Si_3O_{11}$	2.82	
Chain silicates			
Pyroxenes			
Enstatite $En_{98}Fs_2$	$En = MgSiO_3$ (Enstatite)	4.34	0.80 (at 60 °C)
Bronzite $En_{78}Fs_{22}$	$Fs = FeSiO_3$ (Ferrosilite)	4.16	0.752
Diopside	$CaMgSi_2O_6$	5.02	0.69
Augite	$CaMgFeSi_3O_9$	3.82	
Wollastonite	$CaSiO_3$	4.03	0.67
Amphiboles			
Antophyllite	$Mg_7Si_8O_{22}(OH)_2$	3.96	0.740
Tremolite	$Ca_2MgSi_8O_{22}(OH)_2$	4.08	
Hornblende	$Ca_2Mg_3FeAlSi_7AlO_{22}(OH)_2$	2.85	
Glaucophane	$Na_2Mg_3Al_2Si_8O_{22}(OH)_2$	2.17	
Sheet silicates			
Micas			
Muscovite	$K_3Al_4Si_6Al_2O_{20}(OH)_4$	2.32	
Biotite	$K_2Mg_3Fe_2^{2+}Al_{1/2}Fe_{1/2}^{3+} \cdot Si_6Al_2O_{20}(OH)_3$	1.17	
Chlorite	$Mg_5AlSi_3O_{10}(OH)_8$	5.14	
Talc	$Mg_6Si_8O_{20}(OH)_4$	6.10	0.87 (at 58 °C)
Framework silicates			
Alkali feldspar			
Microcline		2.49	0.680
Orthoclase	$KAlSi_3O_8$	2.31	0.61
Sanidine		1.65	
Plagioclase			
$Ab_{99}An_1$	$Ab = NaAlSi_3O_8$ (Albite)	2.31	0.709
$Ab_{46}An_{54}$		1.53	0.70
Ab_4An_{96}	$An = CaAl_2Si_2O_8$ (Anorthite)	1.68	0.70
Nepheline	$Na_3KAl_4Si_4O_{16}$	1.73	
Leucite	$KAlSi_2O_6$	1.15	0.745 (at 60 °C)
Natrolite (Zeolite)	$Na_2Al_2Si_3O_{10} \cdot 2H_2O$	2.00	

Table 4. Temperature dependence of thermal conductivity, k, specific heat, c_p, and thermal diffusivity, κ, of minerals.

Mineral	T [°C] 0	50	100	200	300	400	500	600	700	800	900	1000	1100	1200	Ref.
Quartz, parallel															
k [W m^{-1} K^{-1}]		12.0	9.16	6.88	5.22	4.34	3.56	4.05	4.67	4.96					Kan68
c_p [kJ kg^{-1} K^{-1}]	0.698	0.766	0.833	0.969	1.06	1.13	1.14	1.15	1.16	1.17				1.33	Gor42
κ [·10^{-7} m^2 s^{-1}]		59.2	41.5	26.8	18.6	14.5	11.8	13.3	15.2	16.0					Kan68
Quartz, perpendicular															
k [W m^{-1} K^{-1}]		5.96	4.99	4.05	3.40	2.98	2.69	2.93	3.41	3.90					Kan68
κ [·10^{-7} m^2 s^{-1}]		29.4	22.6	15.8	12.1	9.95	8.90	9.60	11.1	12.6					Kan68
Fused silica															
k [W m^{-1} K^{-1}]	0.700	1.55	1.14	1.39	1.56	1.85	1.99	2.26	2.92	3.35	4.07	5.17	6.26	7.98	Bir42
c_p [kJ kg^{-1} K^{-1}]		0.749	0.821	0.938	1.02	1.08	1.13	1.16	1.19	1.21				1.34	Gor42
κ [·10^{-7} m^2 s^{-1}]		7.23	7.18	7.08	6.98	7.11	7.33	7.82	8.59						Kan68
Olivine (Fo$_{18}$Fa$_{82}$)[1]															
k [W m^{-1} K^{-1}]		4.98	4.81	4.35	3.98	3.87	3.95	4.16	4.60	5.27					Kan68
c_p [kJ kg^{-1} K^{-1}]		0.816	0.883	0.988	1.03	1.08	1.10	1.13	1.15	1.17					Ver55
κ [·10^{-7} m^2 s^{-1}]		17.7	15.8	12.9	11.1	10.3	10.3	10.7	11.6	13.0					Kan68
Periclase (MgO)															
k [W m^{-1} K^{-1}]			50.2	37.0	28.9	24.4	20.5	17.7	16.0	14.6					Kan68
c_p [kJ kg^{-1} K^{-1}]	0.870			1.09	1.13	1.16	1.18	1.19	1.21	1.24				1.30	Gor42
κ [·10^{-7} m^2 s^{-1}]				96.0	71.4	57.7	48.7	41.7	36.7	33.2					Kan68
Calcite, parallel															
k [W m^{-1} K^{-1}]	4.00		2.99	2.55	2.29	2.13									Bir42
c_p [kJ kg^{-1} K^{-1}]	0.793			1.00		1.13									Gor42
Calcite, perpendicular															
k [W m^{-1} K^{-1}]	3.48		2.72	2.37	2.16	2.06									Bir42
Halite (NaCl)															
k [W m^{-1} K^{-1}]	6.10		4.20	3.12	2.49	2.08									Bir42
c_p [kJ kg^{-1} K^{-1}]	0.855			0.915		0.975				1.09					Gor42
Sanidine (Or$_{61}$Ab$_{34}$An$_5$)[2]															
κ [·10^{-7} m^2 s^{-1}]		7.06	6.77	6.51	6.62	6.90	7.19	7.65	8.12	8.79				1.27	Kan68
c_p [kJ kg^{-1} K^{-1}]	0.700			0.950		1.05				1.17					Gor42

(continued)

Table 4 (continued)

Mineral	T [°C]													Ref.
	0	50	100	200	300	400	500	600	700	800	900	1000	1100	1200	
Garnet ($Py_{50}Al_{50}$)[3]															
κ [$\cdot 10^{-7}$ m^{2}s^{-1}]		10.8	10.2	9.25	8.65	8.15	8.05	8.00	8.05	8.15					Kan68
c_p [kJ kg^{-1} K^{-1}]		0.740													Gor42
Spinel (MgAl$_2$O$_4$)															
κ [$\cdot 10^{-7}$ m^{2}s^{-1}]				36.5	32.2	29.4	26.5	24.5	23.1	21.3					Kan68
Jadeite (NaAlSi$_2$O$_6$)															
κ [$\cdot 10^{-7}$ m^{2}s^{-1}]		14.8	13.5	11.6	10.1	9.99	8.45	8.30		9.35					Kan68
Corundum (Al$_2$O$_3$)															
κ [$\cdot 10^{-7}$ m^{2}s^{-1}]		66.0		59.2	37.2	29.6	26.0	22.5	19.1	16.8					Kan68

[1] Fo = Forsterite, Fa = Fayalite   [2] Or = Orthoclase, Ab = Albite, An = Anorthite   [3] Py = Pyrope, Al = Almandine

Table 5. Thermal conductivity, k, of anisotropic rock-forming minerals at room temperature and atmospheric pressure.

Mineral	Formula	Symmetry	k_{11} W m^{-1} K^{-1}	k_{33} W m^{-1} K^{-1}	Ref.
Gypsum	CaSO$_4 \cdot$ 2H$_2$O	monoclinic, C$_{2h}$	2.6	3.7	Dre74
Muscovite	K$_3$Al$_4$Si$_6$.Al$_2$O$_{20}$(OH)$_4$	monoclinic, C$_{2h}$	0.84	5.1	Cla66
Orthoclase	KAlSi$_3$O$_8$	monoclinic, C$_{2h}$	2.9	4.6	Dre74
Anhydrite	CaSO$_4$	orthorhombic, V$_h$	5.6	5.9	Dre74
Calcite	CaCO$_3$	trigonal, D$_{3d}$	3.2	3.7	Bir42
Dolomite	CaCO$_3 \cdot$ MgCO$_3$	trigonal, D$_{3d}$	4.7	4.3	Dre74
Haematite	Fe$_2$O$_3$	trigonal, D$_{3d}$	14.7	12.1	Dre74
Quartz	SiO$_2$	trigonal, D$_{3d}$	6.5	11.3	Dre74
Graphite	C	hexagonal, D$_{6h}$	355.0	89.4	Dre74
Beryll	3BeO $\cdot$ Al$_2$O$_3 \cdot$ 6SiO$_2$	hexagonal, D$_{6h}$	7.8	9.4	Dre74
Ice (0 °C)	H$_2$O	hexagonal, D$_{6h}$	1.9	2.3	Dre74
Rutile	TiO$_2$	tetragonal, D$_{4h}$	9.3	12.9	Dre74
Zircon	ZrO$_2 \cdot$ SiO$_2$	tetragonal, D$_{4h}$	3.9	4.8	Dre74

$\bar{k} = \frac{1}{3}(2k_{11} + k_{33}) = \frac{1}{3}(2k_\perp + k_\parallel)$ $k_\perp$: perpendicular to optical c-axis
k_{11}, k_{33}: components of thermal conductivity tensor $k_\parallel$: parallel to optical axis
$\bar{k}$: average conductivity

Table 6. Pressure dependence of diffusivity, $\kappa(p) = \kappa(0) + \alpha p$ for selected minerals at 40 °C.

Mineral	$\kappa(0)$ 10^{-6} m^2 $\cdot$ s^{-1}	Pressure range GPa	Pressure derivative, α 10^{-6} m^2 $\cdot$ s^{-1} GPA^{-1}	Ref.
Quartz, κ_{11}	6.00	0.1···3.0	0.31	Yuk74
Quartz, κ_{33}	3.10	0.1···3.0	0.53	Kie76
Fused silica	0.68	0.1···3.6	−0.0067	Kie76
Halite (NaCl)	3.10	0.1···1.8	0.95	Kie76

4.1.2.2 Rocks — Gesteine

Table 7. Thermal properties of rocks: thermal conductivity, k; thermal diffusivity, κ; specific heat, c_p
n = number of determination, s.d. = standard deviation

Rock	Locality	Ref.	T [°C]	k W m^{-1} K^{-1}			κ 10^{-7} m^2 s^{-1}			c_p kJ kg^{-1} K^{-1}		
				n	range	mean	n	range	mean	n	range	mean
Igneous rocks												
Alkali Granite	Lake Baikal shore, USSR	Lub75	RT	2	2.1···2.4	2.27	2	9.4···9.8	9.6	2	0.84···0.92	0.88
Granite	Norway	Swa74	25	34		3.38						
	Loetschberg Tunnel, Switzerland	Cla50	RT	12	2.6···3.8	3.25						
	Switzerland (Alpine rocks)	Wen69	20	8	1.3···2.5	2.19	8	6.8···12.8	11.1	8	0.67···0.80	0.75
	Germany	Hae76	50	16	s.d. ±0.45	3.09	16	s.d. ±2.12	13.8	16	s.d. ±0.06	0.86
	Germany	Kap74	50	12	2.1···3.1	2.96	8	10.3···14.3	14.2	8	0.79···0.98	0.82
	Germany	Hae71	RT	11	2.5···4.0	3.10	11	11.3···17.6	13.3	11	0.81···0.96	0.88
	Oberpfalz, Germany	Cre64	RT			2.76						
	Sweden	Eri79	RT	33	s.d. ±0.67	4.43						
	Bohemian Massif, Czechoslovakia	Cer68	25	3	2.4···2.6	2.52						
	Kola peninsula, USSR	Sta73	RT	22	2.4···3.3	2.88	22	8.1···13.4	11.3	22	0.80···1.09	0.96
	Ukrainian shield, USSR	Lub64a	RT	8	2.2···3.2	2.95						
	USSR, shield areas	Lub66	RT	13	1.8···4.0	2.78	13	5.6···10.4	8.4	13	0.92···1.55	1.22
	USSR	Tim70	RT	46	1.4···3.6	2.51						
	Lake Baikal shore, USSR	Lub75	RT	7	2.2···2.9	2.53	7	8.0···12.8	10.0	7	0.88···1.05	0.94
	Southern Siberia, USSR	Duc74	RT	45	1.7···3.4	2.47						
	Godavari valley, India	Rac70	RT	5	2.7···3.5	2.98						
	Korea	Miz70	RT	1		3.73						
	North-East USA	Hor72	23	5	2.7···3.3	2.92						
	Transvaal, South Africa	Bul39	25	4	2.7···3.1	2.85						
	Transvaal, South Africa	Car55	40	2	2.0···2.9	2.47						
	New South Wales, Australia	Bec65	RT	48	2.7···4.5	3.57						
		Raz69	20		2.9···4.1							
		Bir40	25	4		2.62						
		Win52	50	2		2.61						
		Moi66	50			2.16						
		Moi68	20	13	1.6···2.8	2.36	13	5.0···15.1	8.9	13	0.88···1.38	1.13

(continued)

Table 7 (continued)

Rock	Locality	Ref.	T [°C]	k W m⁻¹ K⁻¹			κ 10⁻⁷ m² s⁻¹			c_p kJ kg⁻¹ K⁻¹		
				n	range	mean	n	range	mean	n	range	mean
Plagiogranite	Lake Baikal shore, USSR	Lub75	RT	2	2.1···2.4	2.27	2	9.4···9.8	9.6	2	0.84···0.91	0.88
Granite and Quartz Monzonite	Adams Tunnel, Colorado, USA	Bir50	30	59	2.8···3.6	3.30						
	USA	Hor72	23	4	2.7···2.9	2.80						
Granodiorite	Ukrainian shield, USSR	Lub64	RT	10	1.9···3.3	2.63						
	USSR, shield areas	Lub66	RT	2	2.5···3.3	2.91	2	7.8···11.8	9.8	2	1.05···1.17	1.11
	USSR	Tim70	RT	34	1.4···3.2	2.22						
	New Hampshire, USA	Hor72	23	1		3.49						
	Nevada, USA	Bir66	RT	5	2.6···2.9	2.78						
	California, USA	Cla57	27	14	2.9···3.5	3.19						
	West Australia	Sas64	RT		3.0···3.3	3.22						
		Moi68	20	9	1.6···2.3	2.03	9	5.0···9.1	7.2	9	0.84···1.26	1.09
		Ner67	28	14	2.6···3.5	3.19						
Quartz Diorite	Southern Siberia, USSR	Duc74	RT	108	1.7···4.0	2.72						
	Ontario and Quebec, Canada	Mis 51	20	2	2.1···3.3	2.70						
Alkali Syenite (Alaskite)	Saskatchewan, Canada	Lew69	RT	6		3.76						
Granosyenite	Southern Siberia, USSR	Duc74	RT	83	1.8···3.4	2.34	15	4.5···8.5	6.0			
Syenite	Norway	Swa74	25	15		2.02						
	USSR	Tim70	RT	4	1.4···5.3	3.35						
	Southern Siberia, USSR	Duc74	RT	24	1.7···3.4	2.22	1		6.3			
	Ontario and Quebec, Canada	Mos51	20	3	2.8···3.2	3.04						
	Arkansas, USA	Hor72	23	1		1.63						
	Transvaal, South Africa	Car55	40	1		2.6						
		Bir40	50	1		2.2						
		Mis55	16	1		3.16						
Nepheline Syenite	South Greenland	Sas72	25	83	1.5···3.8	2.38						
Syenite and Syenite Porphyry	Ontario, Canada	Mos51	20	37	2.6···4.0	3.21						
	Minnesota, USA	Hor72	23	1		2.34						
Monzonite		Bir40	25	1		3.06						
Tonalite		Bir40	25	1		2.63						

Rock	Locality	Ref.	T [°C]	k $\text{W m}^{-1}\text{K}^{-1}$			κ $10^{-7}\text{m}^2\text{s}^{-1}$			c_p $\text{kJ kg}^{-1}\text{K}^{-1}$		
				n	range	mean	n	range	mean	n	range	mean
Diorite	Finland	Jär79	RT			2.47						
	USSR, shield areas	Lub66	RT	3	2.2···2.8	2.46	3	7.3···8.6	7.8	3	1.13···1.17	1.14
	Southern Siberia, USSR	Duc74	RT	46	1.8···2.9	2.26	8	5.4···8.5	6.4			
	Korea	Miz70	RT	1		2.39						
	Ontario and Quebec, Canada	Mis51	20	2	2.3···3.0	2.63						
	Quebec, Canada	Lew77	RT	129		3.19						
	Mid-Atlantic Ridge	Hyn71	25	1		2.08						
		Kaw64	25	1		2.04						
		Moi68	20	1		1.91	1		9.5			
Quartz Gabbro	Ontario and Quebec, Canada	Mis51	20	1		6.53						
Gabbro	Kola peninsula, USSR	Sta73	RT	9	2.5···4.1	2.99	9	9.3···12.2	9.7	9	0.88···1.13	1.00
	Baltic shield, USSR	Lub72	RT	38	2.3···2.9	2.61						
	Ontario and Quebec, Canada	Mis51	20	1		2.77						
	Maine, USA	Hor72	23	1		1.82						
	West Australia	Sas64	RT		2.6···3.0	2.76						
		Bir40	25	1		2.29						
		Bir40	50	1		1.93						
		Bir40	25	1		1.99						
		Win52	25	2		2.14						
		Win52	30	2		2.51						
Olivine Gabbro	New Hampshire, USA	Hor72	23	2	1.7···2.0	1.87						
Gabbro and Pyroxenite	USSR	Tim70	RT	5	2.3···4.0	3.15						
Anorthosite	Norway	Swa74	25	9		1.71						
	Quebec, Canada	Hor72	23	1		1.75						
	Transvaal, South Africa	Bir66	RT	1		2.09						
		Bir40	25	3	1.6···1.9	1.76						
Norite	Ontario, Canada	Mis51	20	5	2.3···3.1	2.69						
Peridotite	Finland	Jär79	RT			2.78						
	Kola peninsula, USSR	Sta73	RT	14	3.8···4.9	4.37	14	11.9···14.1	13.3	14	0.92···1.09	1.00
	USSR, shield areas	Lub66	RT	1		2.77						
	Ontario and Quebec, Canada	Mis51	20	5	2.3···2.9	2.65						
		Kaw64	25			3.98						
		Kaw64	50			3.60						

(continued)

Table 7 (continued)

Rock	Locality	Ref.	T [°C]	k $Wm^{-1}K^{-1}$ n	range	mean	κ $10^{-7} m^2 s^{-1}$ n	range	mean	c_p $kJkg^{-1}K^{-1}$ n	range	mean
Hyperstenite		Bir40	25			4.40						
Bronzitite		Bir40	25			4.25						
Pyroxenite	Kola peninsula, USSR	Sta73	RT	64	3.4···5.1	4.31	64	9.4···14.9	12.8	64	0.88···1.21	1.00
	Alaska, USA	Bir66	RT	3		4.06						
	($\varrho = 3.31$ g cm^{-3})	Bir66	RT	3		3.63						
	($\varrho = 3.25$ g cm^{-3})	Bir66	RT	1		4.94						
	Transvaal, South Africa											
(Pyroxene)	Lake Baikal shore, USSR	Lub75	RT	1		1.74	1		7.3	1		0.81
(Diopside)		Bir66	RT	1		4.27						
Dunite	Washington, USA	Hor72	23	1		4.23						
	Transvaal, South Africa	Bir66	RT	1		3.64						
		Kaw64	25	1		3.98						
		Win52	25	1		4.79						
		Bir40	25	3	4.4···5.2	4.74						
Olivinite	Mid-Atlantic Ridge	Hyn71	25			3.50						
Lherzolite		Kaw66	50			3.56						
Hornblendite	Alaska, USA	Bir66	RT	4		2.83						
Cumberlandite	Rhode Island, USA	Hor72	23	1		2.75						
Volcanic rocks												
Lava	Cyprus (Basaltic pillow lava)	Mor75	RT	6	1.3···4.0	1.85						
	USSR	Moi68	20	5	0.2···0.7	0.49						
	Michigan, USA (Dense flow)	Bir54	RT	27	1.7···2.8	2.10						
	(Amygdaloidal tops)	Bir54	RT	10	2.3···3.8	2.68						
	Transvaal, South Africa	Mos51	RT	15	2.7···3.3	3.01						
	Orange Free State, South Africa	Bul39	25	9	2.6···3.6	3.10						
	South Africa	Car55	40	13	2.6···4.6	3.25						
Rhyolite	Ontario and Quebec, Canada	Mos51	20	5	3.1···4.1	3.52	5	3.0···5.4	3.77	5	0.67···1.38	1.08
	Quebec, Canada	Lew77	RT	152		4.20						
Altered Rhyolite	Ontario, Canada	Mos51	20	6	3.1···3.7	3.45						
Liparite	Sweden	Eri79	RT	26	s.d. ±0.51	3.26						

4.1 Thermal conductivity and specific heat of rocks

Rock	Locality	Ref.	T [°C]	k W m^{-1} K^{-1} n	range	mean	κ 10^{-7} m^2 s^{-1} n	range	mean	c_p kJ kg^{-1} K^{-1} n	range	mean
Trachydolerite	Hungary	Bol64	RT	3	2.1···2.8	2.38						
Andesite	Hungary	Bol63	RT	6	1.6···2.5	1.97						
	Siberia and Kamchatka, USSR	Duc74	RT	2	2.3···2.7	2.51						
	Quebec, Canada	Lew77	RT	636		3.21						
Basalt	Germany	Hae71	50	3	1.6···1.8	1.68	3	6.3···6.9	6.58	3	0.88···0.89	0.88
	Cyprus	Mor75	RT	1		1.56						
	USSR	Tim70	RT	9	1.4···3.6	2.60						
	Siberia and Kamchatka, USSR	Duc74	RT	5	1.6···2.3	1.76						
	Texas, USA	Hor72	23	1		2.29						
	Hawai, USA	Wat76	RT	9	1.4···2.1	1.73						
	West Australia	Sas64	RT		2.5···2.9	2.81						
	West Australia (Amphibolite Basalts)	Sas64	RT		3.2···5.0	4.06						
	West Australia (Chlorite Basalts)	Sas64	RT		3.8···5.4	4.35						
	Mid-Atlantic Ridge (dredge samples)	Hyn71	20	11	1.4···2.7	1.76						
	Mid-Atlantic Ridge (core samples)	Hyn71	20	16	1.4···1.7	1.62						
		Win52	19			1.38						
		Web67	0			1.33						
		Bir66	20			2.81						
		Bir40	25			2.28						
		Bri24	30			1.69						
		Kaw64	25			2.29						
		D'An49	25	2	1.4···1.7	1.55						
Obsidian	Armenia, USSR	Sak73	16	3	1.3···1.6	1.45						
Tuff	Eastern Carpathians, USSR	Kut70	20	6	1.4···2.1	1.82	3	5.4···5.9	5.31	3	2.34···2.84	2.58
	Kola peninsula, USSR	Sta73	RT	3	2.7···4.4	3.45	3	9.4···13.3	11.1	3	1.00···1.17	1.09
		Raz69	50			0.63						
Tuff Breccia	USSR	Tim70	RT	32	1.0···3.6	2.03						

(continued)

Table 7 (continued)

Rock	Locality	Ref.	T [°C]	k W m^{-1} K^{-1} n	range	mean	κ 10^{-7} m^2 s^{-1} n	range	mean	c_p kJ kg^{-1} K^{-1} n	range	mean
Hypabyssal rocks												
Quartz Porphyry	Germany	Bac68	RT	5	2.1···3.0	2.47						
	Switzerland	Cla56	RT	4	s.d. ±0.4	3.54						
	New Mexico, USA	Hor72	23	1		2.27						
		Raz69	20			2.33 ∥						
		Raz69	20			1.40 ⊥						
Quartz Feldspar Porphyry	Orange Free State, South Africa	Bul39	25	5	3.2···3.6	3.35						
Syenite Porphyry	Ontario and Quebec, Canada	Mis51	20	35	2.6···4.0	3.23						
	Minnesota, USA	Hor72	23	1		2.26						
Rhyolite Porphyry	Oklahoma, USA	Hor72	23	1		3.64						
Porphyry	Sweden	Eri79	RT	54	s.d. ±0.46	3.80						
	WestAustralia	Sas64	RT		4.1···4.2	4.19						
Porphyrite	Kola peninsula, USSR	Sta73	RT	8	2.4···3.2	2.83	8	9.0···10.7	9.86	8	0.92···0.96	0.93
	Southern Siberia, USSR	Duc74	RT	67	1.7···3.2	2.39						
	USSR	Tim70	RT	48	0.6···2.7	2.01						
	Korea	Miz70	RT	5	1.8···3.2	2.46						
Porphyrite and Diabase	California, USA	Cla57	27	21	2.6···3.4	2.99						
Dolerite (=Diabase)	Norway	Swa74	25	5		2.16						
	Kola peninsula, USSR	Sta73	RT	20	2.3···3.1	2.74	20	8.5···11.2	9.78	20	0.75···1.00	0.91
	USSR, shield areas	Lub66	RT	2	2.4···2.5	2.46	2	9.0···10.2	9.59	2	0.80···0.92	0.86
	Southern Siberia, USSR	Duc74	RT	10	2.4···3.0	2.55						
	Virginia, USA	Hor72	23	1		2.29						
	Orange Free State, South Africa	Mos51	35	9	1.6···2.3	2.01						
	West Australia	Sas64	RT		2.6···4.4	3.47			8.41			0.70
		Moi68	25			2.14						
		Win52	0			2.20						
Quartz Dolerite	West Australia	Sas64	RT		3.4···4.4	3.85						
Gabbro Diabase	Kola peninsula, USSR	Smi79	RT	41	2.2···3.2	2.68	41	8.3···14.3	10.7	41	0.63···1.17	0.84
	Moscow syneclise, USSR	Smi79	RT	2	2.3···2.6	2.45	1		8.8			
Meta Gabbro Diabase	Kola peninsula, USSR	Smi79	RT	29	2.4···3.2	2.89	29	9.8···12.4	10.9	29	0.67···1.09	0.84

4.1 Thermal conductivity and specific heat of rocks

Rock	Locality	Ref.	T [°C]	k W m^{-1} K^{-1} n	range	mean	κ 10^{-7} m^2 s^{-1} n	range	mean	c_p kJ kg^{-1} K^{-1} n	range	mean
Pegmatite	Switzerland (Alpine rocks)	Wen69	20	3	1.5···3.5	2.51	3	7.8···17.8	12.8	3	0.67···0.84	0.75
	USSR, shield areas	Lub66	RT	2	3.4···3.7	3.56	2	9.2···9.4	9.32	2		1.42
Lamprophyre	Ontario, Canada	Mis51	20	1		3.44						
	New Hampshire, USA	Hor72	23	1		1.96						
Carbonatite	Finland	Jär79	RT			2.58						
Metamorphic rocks												
Marble	Switzerland (Alpine rocks)	Wen69	20	9	2.0···3.4	2.41	9	9.2···15.3	10.9	10	0.67···0.92	0.80
	Southern Siberia, USSR	Duc74	RT	22	2.0···5.5	2.76	10	5.5···11.5	7.7			
		Raz69	20		2.1···3.5							
		Win52	0			3.05						0.80
		Bir40	25			2.86			10.7			0.75
		D'An49	25			2.79						
		D'An49	45			2.43						
Quarzite	Germany	Kap74	50	1		6.18	1		29.5	1		0.79
	Switzerland (Alpine rocks)	Wen69	20	2	3.7···4.4	4.06	2	17.6···20.8	19.2	2	0.71···0.92	0.81
	Sweden	Eri79	RT	92	s.d. ±0.74	3.36						
	Thuringia Basin, Germany	Mei67	RT	1		2.92						
	East European platform	Lub64b	RT	5	5.1···7.6	6.43	5	14.8···20.9	18.4	5	1.00···1.34	1.14
	USSR	Tim70	RT	8	3.5···7.8	5.23						
	South Dakota, USA	Bir66	RT	6	5.9···7.4	6.72						
	Transvaal, South Africa	Bul39	25	17	3.6···8.0	5.99						
	Transvaal, South Africa	Mos51	RT	28	3.1···7.9	5.61						
	Transvaal, South Africa	Car55	40	13	4.1···6.9	5.56						
		Win52	50	10		5.82						0.77
		Sas71	25			5.72						
		Raz69	20			6.05						0.80
		D'An49	25			6.05						
Serpentinite	Finland	Jär79	RT			2.35						
	Switzerland (Alpine rocks)	Wen69	20	1		2.51 ∥	1		10.5 ∥	1		0.88
	Switzerland (Alpine rocks)	Wen69	20	1		1.72 ⊥	1		7.2 ⊥			
	Kola peninsula, USSR	Sta73	RT	5	2.3···2.9	2.63	5	8.4···9.8	8.92	5	0.96···1.13	1.00
	Mid-Atlantic Ridge	Hyn71	20	2	1.4···2.2	1.80						
(Serpentine)		Bir54	RT	2	1.8···2.5	2.14						

(continued)

Table 7 (continued)

Rock	Locality	Ref.	T [°C]	k W m^{-1} K^{-1} n	range	mean	κ 10^{-7} m^2 s^{-1} n	range	mean	c_p kJ kg^{-1} K^{-1} n	range	mean
(Talc = Soapstone)	Switzerland (Alpine rocks)	Wen69	20	1		5.28	1		17.9	1		1.00
		Bir54	RT			2.97						
		Bir66	RT			5.0						
Serpentinized Peridotite	Quebec, Canada	Mis51	20	5	2.4···2.9	2.65						
Hornfels	East European platform, USSR	Lub64b	RT	1		6.07	1		14.6	1		1.47
	Southern Siberia, USSR	Duc74	RT	3	3.0···3.4	3.22						
	USSR	Tim70	RT	2	4.4···4.8	4.61						
Skarn	Southern Siberia, USSR	Duc74	RT	17	1.8···3.1	2.43						
	Korea	Miz70	RT	1		2.72						
Eclogite	USSR	Moi66	25			3.35						
	USSR	Moi68	50			3.10						
	Mid-Atlantic Ridge	Hyn71	25			3.40						
		Kaw64	25			2.46						
Gneiss	Finland	Jär79	RT			3.35						
	Sweden	Eri79	RT	47	s.d. ±0.46	2.37						
	Gotthard Tunnel, Switzerland	Cla56	RT	15	2.1···3.3	2.80						
	Simplon Tunnel, Switzerland	Cla56	RT	8	2.5···4.8	3.73 ∥						
	Simplon Tunnel, Switzerland	Cla56	RT	22	1.9···3.2	2.65 ⊥						
	Switzerland (Alpine rocks)	Wen69	20	55	1.2···3.1	2.12 ∥	55	6.0···15.7	10.6 ∥	55	0.46···0.92	0.75
	Switzerland (Alpine rocks)	Wen69	20	55	1.2···2.6	1.74 ⊥	55	6.2···12.8	8.7 ⊥			
	Thuringia Basin, Germany	Mei67	RT	1		2.07						
	Germany	Hae71	RT	2	2.5···2.6	2.58	2	11.5···12.0	11.76	2	0.79···0.80	0.80
	Germany	Kap74	50	4	2.6···2.9	2.70	4	11.3···14.1	12.24	4	0.77···0.87	0.81
	Kola peninsula, USSR	Lub72	RT			2.49						
	Moscow Syneclise, USSR	Smi79a	RT	2	2.6···2.7	2.62	2	9.9···11.1	10.5	1		0.84
	Lake Baikal shore, USSR	Lub75	RT	6	1.7···2.6	2.13	6	5.9···9.8	8.47	6	0.80···0.98	0.90
	Korea	Miz70	RT	10	1.7···3.7	2.52						
	Chester, Vermont, USA	Cla66	RT	9	2.6···4.4	3.49 ∥						
	Chester, Vermont, USA	Cla66	RT	9	2.1···3.6	2.61 ⊥						
	South Greenland	Sas72	25	59	2.1···4.0	2.90						
		Bir40	50			2.93 ∥						
		Bir40	50			2.09 ⊥						

Rock	Locality	Ref.	T [°C]	k $W m^{-1} K^{-1}$			κ $10^{-7} m^2 s^{-1}$			c_p $kJ kg^{-1} K^{-1}$		
				n	range	mean	n	range	mean	n	range	mean
Granite Gneiss	Norway	Swa74	25	6		3.14						
	Moscow Syneclise, USSR	Smi79a	RT	2	2.5···2.9	2.68	2	10.9···11.9	11.4			
Quartz Diorite Gneiss	Adams Tunnel, Colorado, USA	Bir50	30	17	2.8···3.6	3.24						
Orthogneiss	Lake Baikal shore, USSR	Lub75	RT	1		2.08	1		5.7	1		1.35
Potassium feldspar Gneiss	Bohemian Massif, Czechoslovakia	Cer68	25	2	2.6···2.8	2.72						
	Kola peninsula, USSR	Sta73	RT	5	2.7···3.2	2.97	5	8.7···13.0	10.2	5	0.92···1.38	1.13
	USSR, shield areas	Lub66	RT	1		1.34	1		6.0	1		1.00
	Southern Siberia, USSR	Duc74	RT	15	1.6···4.0	2.51	6	3.5···6.5	4.8			
	Lake Baikal shore, USSR	Lub75	RT	1		2.08	1		5.7	1		1.34
Plagioclase Gneiss	Bohemian Massif, Czechoslovakia	Cer67a	25	2	2.5···2.6	2.55						
	Kola peninsula, USSR	Sta73	RT	9	2.2···2.7	2.44	9	6.3···8.3	7.32	9	0.96···1.26	1.13
	USSR, shield areas	Lub66	RT	4	1.4···2.6	2.17	4	7.1···8.5	7.5	4	0.80···1.17	1.05
	Southern Siberia, USSR	Duc74	RT	7	1.8···2.6	2.34						
	Lake Baikal shore, USSR	Lub75	RT	6	1.7···2.6	2.13	6	5.9···9.8	8.47	6	0.80···1.00	0.91
Amphibolite Gneiss	Norway	Swa74	25	7		3.18						
Injection Gneiss and Schist	Adams Tunnel, Colorado, USA	Bir50	30	41	1.7···4.6	3.24						
Migmatite	Lake Baikal shore, USSR	Lub75	RT	3	2.0···2.4	2.17	3	9.2···9.8	9.5	3	all 0.84	0.84
Agmatite	Lake Baikal shore, USSR	Lub75	RT	2	2.0···2.4	2.18	2	9.4···9.8	9.6	2	0.82···0.84	0.83
Albitite		Bir40				2.03						
Leptite	Sweden	Eri79	RT	255	s.d. ±0.76	3.36						
Schist	Switzerland (Alpine rocks)	Wen69	20	18	1.7···4.1	2.88 ∥	18	7.8···18.3	13.1 ∥	18	0.67···1.05	0.80
	Switzerland (Alpine rocks)	Wen69	20	18	1.0···3.1	2.05 ⊥	18	4.5···14.1	9.3 ⊥			
	Scotland	Ric76	RT	39	2.5···5.2	3.84 ∥						
	Scotland	Ric76	RT	39	1.8···3.9	2.86 ⊥						
	Southern Siberia, USSR	Duc74	RT	28	1.6···3.4	2.51						
	Godavari Valley, India	Rao70	RT	4	2.2···4.6	3.39 ∥						
	Godavari Valley, India	Rao70	RT	4	2.8···4.0	3.35 ⊥						
	Ontario and Quebec, Canada	Mis51	20	8	2.5···4.8	3.14						
		Raz69	20		2.3···3.7 ∥							
		Raz69	20		1.5···2.0 ⊥							

(continued)

Table 7 (continued)

Rock	Locality	Ref.	T [°C]	k W m^{-1} K^{-1}			κ 10^{-7} m^2 s^{-1}			c_p kJ kg^{-1} K^{-1}		
				n	range	mean	n	range	mean	n	range	mean
Quartz Mica Schist	Norway	Swa74	25	36		2.68						
Mica Schist	Finland	Jär79	RT			2.92						
	Bohemian Massif, Czechoslovakia	Cer68	25	1		2.80						
	Kola peninsula, USSR	Sta73	RT	5	2.5···2.8	2.66	5	6.4···8.8	7.73	5	1.05···1.21	1.13
	Korea	Miz70	RT	1		2.68						
Slate	Germany	Hae76	50	11	s.d. ±0.74	2.63	11	s.d. ±3.32	11.17	11	s.d. ±0.07	0.91
	Korea	Miz70	RT	1		2.39						
		Kaw64	25			2.60						
		Bir40	25			1.89						
Clay Slate	Germany	Kap74	50	5	1.4···3.7	2.15	5	6.4···15.2	9.26	5	all 0.85	0.86
Salt Slate	Germany	Cre65	50	7	1.2···4.2	2.76	7	6.4···21.7	13.9			
Phyllite	Thuringia Basin, Germany	Mei67	RT	1		1.93						
	Kola peninsula, USSR	Lub72	RT	21	2.1···3.8	2.95						
	Kola peninsula, USSR	Sta73	RT	7	2.6···4.4	3.81	7	10.8···14.5	12.4	7	1.00···1.17	1.09
	USSR, shield areas	Lub66	RT	2	3.1···3.3	3.2	2	9.3···11.0	10.2	2	0.96···1.17	1.07
Calcareous Mica Phyllite	South Dakota, USA	Bir66	RT	9	4.0···5.9	4.95 ∥						
(Chlorite)		Bir66	RT	7	2.7···3.8	3.30 ⊥						
		Bir54	RT			5.23						
Amphibolite	Norway	Swa74	25	6		3.35						
	Switzerland (Alpine rocks)	Wen69	20	8	1.3···1.7	1.54 ∥	8	5.9···7.7	7.1 ∥	8	0.67···0.88	0.75
	Switzerland (Alpine rocks)	Wen69	20	8	1.0···1.7	1.28 ⊥	8	4.6···7.7	5.9 ⊥			
	Bohemian Massif, Czechoslovakia	Cer68	25	4	2.8···3.5	3.19						
	Kola peninsula, USSR	Sta73	RT	8	2.1···3.1	2.53	8	5.7···8.2	7.48	8	1.05···1.26	1.13
	Kola peninsula, USSR	Smi79	RT	16	1.4···2.8	2.26						
	USSR, shield areas	Lub66	RT	1		2.50	1		8.1	1		1.13
	Quebec, Canada	Mis51	20	1		2.71						
	Saskatchewan, Canada	Lew69	RT	2		3.51						
	South Dakota, USA	Bir66	RT	6	2.5···3.8	2.90						
	Godavari valley, India	Rao70	RT	9	2.4···4.0	3.19 ∥						
	Godavari valley, India	Rao70	RT	9	1.9···2.8	2.40 ⊥						

4.1 Thermal conductivity and specific heat of rocks

Rock	Locality	Ref.	T [°C]	k W m^{-1} K^{-1} n	range	mean	κ 10^{-7} m^2 s^{-1} n	range	mean	c_p kJ kg^{-1} K^{-1} n	range	mean
Mylonite	Saskatchewan, Canada	Lew69	RT	10		5.08						
Greenstone	Norway	Swa74	25	234		2.92						
Sedimentary rocks												
Ocean and	Arctic Ocean	Lub76	RT	38	0.75···1.13	0.89						
Sea Sediments	Norwegian and Greenland Sea	Hae79	RT	62	0.42···1.58	0.97						
	Atlantic Ocean	Hae79	RT	120	0.50···1.26	0.92						
	Mediterranean Sea	Hae79	RT	87	0.75···1.34	0.98						
	Black Sea	Hae79	RT	76	0.70···1.28	0.85						
	Caspian Sea	Lub76	RT	23	0.65···1.02	0.92						
	West of Oregon Coast, Pacific Ocean	Hut68	RT	12	0.79···1.44	1.06						
Lake Bottom Sediments	Lake Baikal, USSR	Gup77	RT	230	0.6···1.7	0.99						
Clay	Germany	Kap74	50	3	2.1···2.3	2.22	3	8.5···10.2	9.50	3	0.89···1.00	0.93
	Germany	Hae76	50	19	s.d. ±0.64	2.78	19	s.d. ±3.15	10.84	19	s.d. ±0.04	0.88
	Thuringia Basin, Germany	Mei67	RT	12	1.1···2.0	1.67						
	Pannonian Basin, Hungary	Bol66	RT	15		1.68						
	Poland	Ple66	RT			1.22			8.2			0.92
	Eastern Carpathians, USSR	Kut70	20	21	1.4···2.4	1.75						
	Crimea, USSR	Leb69	RT	14	1.2···1.7	1.48						
	East European platform, USSR	Lub64	RT	12	1.4···2.2	1.79	12	2.5···4.9	3.2	12	2.14···3.56	2.89
	Daghestan, USSR	Dja69	RT	62	0.6···2.0	1.23						
	USSR	Tim70	RT	128	0.6···2.7	1.36						
Marly Clay	Germany	Kap74	50	2	1.7···2.0	1.89	2	7.2···10.7	8.94	2	0.77···0.99	0.88
Schistose Clay	Germany	Kap74	50	3	1.9···2.3	2.15	3	8.1···10.2	9.37	3	0.91···0.93	0.92
Claystone	Germany	Kap74	50	15	1.7···3.4	2.38	9	8.2···15.8	12.18	9	0.82···0.93	0.88
	Saarland, Germany	Hüc66	RT	4	2.3···3.3	2.74	4	12.0···16.0	14.0			
Sand	East-European platform, USSR	Lub64	RT	8	1.3···2.8	1.97	8	3.5···6.0	4.7	8	1.97···3.18	2.30
	USSR	Tim70	RT	16	0.1···2.3	1.17						
Gravel		Raz69	20									1.84

(continued)

Table 7 (continued)

Rock	Locality	Ref.	T [°C]	k W m^{-1} K^{-1} n	range	mean	κ 10^{-7} m^2 s^{-1} n	range	mean	c_p kJ kg^{-1} K^{-1} n	range	mean
Conglomerate	Saarland, Germany	Hüc66	RT	1		4.73						
	Germany	Hae71	RT	1		3.9	1		14.6	1		1.15
	East-European platform, USSR	Lub64b	RT	1		5.74	1		17.4	1		0.96
	Crimea, USSR	Lyu73	RT	4	2.3⋯3.9	2.96						
	Southern Siberia, USSR	Duc74	RT	77	1.5⋯3.8	2.30	10	6.8⋯9.8	7.5			
	Michigan, USA	Bir54	RT	31	0.9⋯3.3	2.09						
Tuff-conglomerate	Siberia and Kamchatka, USSR	Duc74	RT	15	1.1⋯2.6	2.01						
Breccia	Southern Siberia, USSR	Duc74	RT	31	1.5⋯2.9	2.30	10	3.5⋯8.5	5.8			
	Ontario, Canada	Mis51	20	2	3.1⋯3.3	3.19						
Quartz Breccia	Ontario, Canada	Mis51	20	1		6.80						
Sandstone	Finland	Jär79	RT			3.35						
	Nottinghamshire, England	Bul51	RT	6	2.5⋯3.2	2.77						
	Saarland, Germany	Hüc66	RT	14	2.9⋯5.1	3.74	14	14.0⋯23.0	17.0			0.82
	Thuringia Basin, Germany	Mei67	RT	10	1.3⋯2.4	1.84						
	Germany	Bec68	RT	26	1.8⋯3.1	2.41						
	Germany	Hae71	RT	17	2.3⋯3.9	3.03	17	10.6⋯21.1	13.2	17	0.76⋯1.08	0.96
	Bohemian Massif, Czechoslovakia	Cer67a	25	3	2.2⋯2.4	2.3 dry						
	Bohemian Massif, Czechoslovakia	Cer67a	25	3	2.3⋯2.5	2.4 wet						
	Czechoslovakia	Cer67	25	140	1.0⋯3.1	2.19 dry						
	Czechoslovakia	Cer67	25	140	1.6⋯3.1	2.42 wet						
	Poland	Ple66	RT			2.64			12.0			0.84
	Pannonian Basin, Hungary	Bol64	RT	18	2.0⋯5.5	3.64						
	Eastern Carpathian, USSR	Kut70	20	12	2.2⋯4.1	3.02						
	East-European platform, USSR	Lub64b	RT	15	1.8⋯3.5	2.89	15	2.5⋯10.5	7.2	15	1.05⋯3.35	1.47
	Krivoy Rog, USSR	Lub64	RT	10	2.4⋯3.5	3.00						
	Moscow, Syneclise, USSR	Smi79a	RT	14	1.7⋯3.0	2.30	9	5.5⋯12.0	9.5	9	0.75⋯1.38	1.00
	Crimea, USSR	Leb69	RT	4	1.8⋯2.6	2.18						
	Crimea, USSR	Lyu73	RT	5	1.3⋯2.8	2.14						
	Daghestan, USSR	Dja69	RT	34	1.4⋯3.8	2.20						

(continued)

4.1 Thermal conductivity and specific heat of rocks

Rock	Locality	Ref.	T [°C]	k W m⁻¹ K⁻¹ n	range	mean	κ 10⁻⁷ m² s⁻¹ n	range	mean	c_p kJ kg⁻¹ K⁻¹ n	range	mean
Sandstone (continued)	Siberia and Kamchatka, USSR	Duc74	RT	102	1.1···3.4	2.00	22	3.5···9.5	5.9			
	USSR	Tim70	RT	360	0.2···4.0	2.17						
	Cyprus	Mor75	RT	1		2.12						
	Camby Basin, India	Gup70	RT	4		1.74 dry						
	Camby Basin, India	Gup70	RT	4		2.70 wet						
	Godavari Valley, India	Rao70	RT	89	1.3···3.7	2.43 dry						
	Godavari Valley, India	Rao70	RT	94	2.1···4.5	3.14 wet						
	Korea	Miz70	RT	1		2.09						
	Michigan, USA	Bir54	RT	8	2.1···4.3	2.84						
	USA	Hut68	RT	28	1.8···6.6	3.42 wet						
	Orange Free State, South Africa	Mos51	35	7	1.4···3.2	1.97						
		Kap74	50	54	2.2···5.1	3.24	31	10.9···23.6	16.45	31	0.76···1.07	0.82
		Zie56	RT	37	1.5···5.8	3.41 wet						
Limy Sandstone		Zie56	RT	3	3.1···3.3	3.16 wet						
Quartz Sandstone	Poland	Ple66	RT			5.54						
	USA	Woo61	30	6	0.5···6.5	2.93 dry						
	USA	Woo61	30	6	2.0···7.4	5.03 wet						
Tuffite	Kola peninsula, USSR	Sta73	RT	9	2.6···3.3	3.02	9	9.4···11.5	10.5	9	0.92···1.17	1.00
Shale	England	Bul51	RT	11	1.2···1.8	1.36						
	Saarland, Germany	Hüc66	RT	3	1.8···2.3	1.97	3	8.0···10.0	8.7			0.86
	Saarland, Germany	Hüc66	RT	4	1.4···2.2	1.71 ⊥	4	6.0···10.0	7.5			
	Saarland, Germany	Hüc66	RT	5	3.1···3.7	3.35 ∥	5	14.0···16.0	15.0			
	Poland	Ple66	RT			2.42			11.0			0.84
	East-European platform, USSR	Lub64b	RT	17	2.3···4.0	2.98	17	7.0···12.7	9.16	17	0.88···1.42	1.18
	Crimea, USSR	Leb69	RT	5	2.2···2.9	2.65						
	Krivoy Rog, USSR	Lub64	RT	4	2.3···3.0	2.75						
	Siberia and Kamchatka, USSR	Dja74	RT	63	1.6···4.0	2.57	11	5.3···9.4	7.9			
	USSR	Tim70	RT	165	0.6···3.6	2.01						
	Cambay Basin, India	Gup70	RT	18		1.03 dry						
	Cambay Basin, India	Gup70	RT	18		1.5 wet						

(continued)

Table 7 (continued)

Rock	Locality	Ref.	T [°C]	k W m⁻¹ K⁻¹			κ 10⁻⁷ m² s⁻¹			c_p kJ kg⁻¹ K⁻¹		
				n	range	mean	n	range	mean	n	range	mean
Shale (continued)	Godavari Valley, India	Rao70	RT	8	1.3···1.8	1.57 dry						
	Godavari Valley, India	Rao70	RT	14	1.5···2.3	1.91 wet						
	Korea	Miz70	RT	4	3.9···4.3	4.06						
	California, USA	Ben47	RT	31	1.1···2.3	1.64						
	Orange Free State, South Africa	Mos51	35	6	1.9···2.9	2.39						
Dolomitic Shale	Queensland, Australia	Hyn66	25	36		4.06						
Pyritic Shale	Queensland, Australia	Hyn66	25	6		6.57						
Carbonaceous Shale	Queensland, Australia	Hyn66	25	10		3.39						
Limestone	Germany	Hae71	RT	2	2.3···2.5	2.40	2	10.2···11.1	10.7	2		0.89
	Germany	Bec68	RT	3	1.9···3.1	2.60						
	Germany	Kap74	50	11	1.7···2.7	2.21	11	8.2···12.2	10.54	11	0.82···0.95	0.85
	Switzerland (Alpine rocks)	Wen69	20	2		2.09	2		10.1	2		0.80
	Czechoslovakia	Cer67	25	5	2.5···2.8	2.62						
	Poland	Ple66	RT			2.00			7.9			1.00
	Pannonian Basin, Hungary	Bol66	RT	5		2.47						
	Eastern Carpathian, USSR	Kut70	20	12	1.9···3.1	2.67						
	East-European platform, USSR	Lub64	RT	8	2.3···4.4	3.07	8	7.4···11.8	9.7	8	0.96···1.72	1.18
	Ukrainian Shield, USSR	Lub64a	RT	18	2.5···3.1	2.86						
	Crimea, USSR	Leb69	RT	12	2.0···2.6	2.37						
	Crimea, USSR	Lyu73	RT	30	0.9···2.6	1.81						
	Moscow Syneclise, USSR	Smi79a	RT	11	1.3···3.3	2.26	8	5.0···10.8	8.5	8	0.84···1.26	1.00
	Daghestan, USSR	Dja69	RT	13	1.2···2.8	1.96						
	Southern Siberia, USSR	Duc74	RT	52	1.6···3.8	2.47	14	6.0···9.7	7.8			
	USSR	Tim70	RT	247	0.6···3.2	2.17						
	Cyprus	Mor75	RT	1		3.05						
	Iran	Cos47	RT	21	s.d. ±0.4	2.18						
	Cambay Basin, India	Gup70	RT	3		1.35 dry						
	Cambay Basin, India	Gup70	RT	3		2.2 wet						
	Korea	Miz70	RT	3	3.2···3.6	3.39						
	Ontario, Canada	Mis51	20	6	1.9···3.0	2.56						
	Resolute Bay, Canada	Mis55	RT	5	2.7···3.4	3.06						

(continued)

4.1 Thermal conductivity and specific heat of rocks

Rock	Locality	Ref.	T [°C]	k W m⁻¹ K⁻¹			κ 10⁻⁷ m² s⁻¹			c_p kJ kg⁻¹ K⁻¹		
				n	range	mean	n	range	mean	n	range	mean
Limestone (continued)	Resolute Bay, Canada	Raz69	20			2.21						
		Bir40	25			2.79						
		Sas71	25	7		2.78 wet						
		Zie56	RT	9	1.6···3.8	2.22 wet						
Compact Limestone	Germany	Kap74	50	6	2.3···3.5	2.83	6	10.8···15.2	12.18	6	0.82···0.92	0.88
Marl	England	Bul51	RT	5	0.9···2.8	1.83						
	Germany	Hae76	50	19	s.d. ±0.39	2.13	19	s.d. ±1.48	9.34	19	s.d. ±0.08	0.90
	Cyprus	Mor75	50	5	2.1···2.2	2.10						
		Kap74	50	3	2.3···3.2	2.70	3	9.9···13.8	11.18	3	0.91···0.93	0.92
Clay Marl	Germany	Kap74	50	7	1.4···2.6	2.04	7	8.0···11.7	9.34	7	0.78···0.98	0.86
Sandy Marl	Cyprus	Mor75	RT	5	all 2.33	2.33						
Lime Marl	Germany	Kap74	50	2	1.8···2.4	2.12	2	9.0···9.6	9.34	2	0.84···0.95	0.90
	Germany	Hae71	RT	6	2.3···3.5	2.69	6	9.6···15.2	11.4	6	0.86···0.95	0.90
Marlstone	Germany	Bec68	RT	7	1.1···2.0	1.59						
	Germany	Hae71	RT	2	2.3···2.6	2.43	2	9.8···10.3	10.1	2	0.91···0.92	0.91
	Czechoslovakia	Cer67	25	6	1.0···1.6	1.38 dry						
	Czechoslovakia	Cer67	25	6	1.6···2.3	1.87 wet						
	Pannonian Basin, Hungary	Bol64	RT	50	1.7···4.0	2.73						
	East-European platform, USSR	Lub64b	RT	9	1.4···2.8	1.79	9	3.1···8.3	4.7	9	1.09···3.10	2.23
	Ukrainian shield, USSR	Lub64a	RT	10	2.1···2.8	2.52						
	Crimea, USSR	Leb69	RT	2		2.29						
	Crimea, USSR	Lyu73	RT	9	0.5···1.7	1.23						
	USSR	Tim70	RT	38	1.0···2.7	2.07						
Dolomite	Germany	Bec68	RT	2	2.7···3.0	2.86						
	Germany	Hae71	RT	4	2.5···3.5	2.87	4	10.7···13.0	12.3	4	0.86···1.00	0.92
	Germany	Hae76	50	9	s.d. ±0.44	3.11	9	s.d. ±1.34	12.7	9	s.d. ±0.04	0.93
	Thuringia Basin, Germany	Mei67	RT	31	3.0···4.5	3.70						
	Poland	Ple66	RT			2.64			10.3			0.92
	East-European platform, USSR	Lub64b	RT	3	2.9···3.8	3.35	3	8.7···11.9	10.2	3	1.21···1.26	1.24
	Moscow Syneclise, USSR	Smi79a	RT	8	2.0···3.9	2.81	6	5.4···11.9	9.8	5	0.84···1.55	1.06
	Daghestan, USSR	Dja69	RT	3	1.6···2.8	2.10						
	Southern Siberia, USSR	Duc74	RT	33	2.5···4.2	3.45	11	7.2···11.2	8.7			
(continued)											(continued)	

Table 7 (continued)

Rock	Locality	Ref.	T [°C]	k W m⁻¹ K⁻¹			κ 10⁻⁷ m² s⁻¹			c_p kJ kg⁻¹ K⁻¹		
				n	range	mean	n	range	mean	n	range	mean
Dolomite (continued)	Transvaal, South Africa	Car55	40	8	4.3···6.6	5.19						
	Transvaal, South Africa	Ben39	25	7	4.0···5.0	4.56						
	Queensland, Australia	Hyn66	25	12		4.15						
		Raz69	20									
		Bir40	25			4.65						
		Win52	50			4.44				8.50		
		Kap74	50	6	2.5···3.8	3.34	6	10.7···15.0	11.17	6	0.92···1.00	0.95
Dolomite and Anhydrite	Switzerland	Cla56	RT	7	3.7···5.8	5.00						
Anhydrite	Germany	Cre65	50	7	4.1···6.1	5.28	7	17.0···25.7	22.41			
	Germany	Bec68	RT	5	2.9···3.6	3.33						
	Germany	Hae71	RT	2	5.1···6.1	5.64	2	21.5···22.5	22.0	2	0.81···0.94	0.88
	Thuringia Basin, Germany	Mei57	RT	49	2.4···4.2	3.82						
	Poland	Ple66	RT			5.54			19.8			1.05
	Switzerland	Euc11	RT	3		5.61						
	East-European platform, USSR	Lub64b	RT	5	1.0···4.5	2.25						
	Iran	Cos47	RT	3		4.90						
	Louisiana, USA	Her56	RT	1		5.74						
	New Mexico, USA	Her56	RT	1		5.40						
Flint	Sweden	Eri79	RT	59	s.d. ±0.57	4.24						
Chert	Korea	Miz70	RT	9	1.4···2.8	1.83						
Rock Salt	England	Ben39	RT			7.20						
	Germany	Cre65	50	14	4.5···5.7	5.52	14	25.2···33.8	30.60			
	Thuringia Basin, Germany	Mei67	RT	45	1.9···4.0	3.15						
	Poland	Ple66	RT			3.63			19.8			0.84
	Iran	Cos47	RT			6.78						
	N.Mexico and Oklahoma, USA	Her56	RT	4		5.34						
	Colorado and Michigan, USA	Len66	RT	4		5.56						
		Raz69	20									
Sylvinite	Germany	Cre65	50	10	4.6···5.8	5.29	10	26.4···32.9	29.90			0.92

Rock	Locality	Ref.	T [°C]	k W m⁻¹ K⁻¹			κ 10⁻⁷ m² s⁻¹			c_p kJ kg⁻¹ K⁻¹		
				n	range	mean	n	range	mean	n	range	mean
Gypsum	Poland	Ple66	RT			1.28						1.13
	Iran	Cos47	RT			1.3			5.3			
Hematite		Bir54	RT			10.5						
Coal	Saarland, Germany	Hüc66	RT	1		0.32 ∥						
	Southern Siberia, USSR	Duc74	RT	6	0.08···0.29	0.17						
	USSR	Tim70	RT	90	0.04···1.5	0.30						

Table 8. Temperature effect on the thermal conductivity, k, of rocks.

Rock	Ref.	ϱ 10³ kg m⁻³	k [W m⁻¹ K⁻¹]							κ 10⁻⁷ m² s⁻¹								c_p kJ kg⁻¹ K⁻¹					
			T [°C] 0	50	100	200	300	400	500	600	700	800	900	1000	1100	1200	1300	1400					

Igneous rocks

Rock	Ref.	ϱ	0	50	100	200	300	400	500	600	700	800	900	1000	1100	1200	1300	1400
Alkali Granite	Sak73	2.81	2.80	2.80	2.80	2.60	2.36	2.22	2.04									
Granite	Sak73	2.62	3.52	3.50	3.24	2.74	2.50	2.31	2.20									
	Bir40	2.61	3.52	3.27	3.01	2.70	2.45											
	Bir40	2.61	3.80	3.48	3.21	2.85												
	Bir40	2.65	2.79	2.61	2.47	2.30												
	Bir40	2.64	2.43	2.34	2.27	2.14												
	Moi68			2.00	1.70	1.37	1.19	1.08	0.99	0.92	0.89	0.88	0.84	0.84	0.88	0.89	0.97	1.05
	Moi66					1.23		1.05		0.90		0.84		0.82		0.89	0.98	
	Hyn71			2.17	1.93	1.54	1.35	1.23	1.14	1.08	1.02	0.99	0.98	0.98	1.00	1.07	1.16	
Quartz Diorite	Sak73	2.72	2.52	2.68	2.60	2.38	2.20	2.09	2.02									
	Kaw64	2.64	3.56	3.37	3.24	3.06	2.93	2.85	2.83	2.81								
Syenite	Bir40	2.80		2.20	2.13	2.09												
Monzonite	Bir40	2.64	3.17	2.92	2.74	2.47												
Tonalite	Bir40	2.74	2.69	2.58	2.47	2.31												
Diorite	Kaw64	2.92	2.07	2.01	1.97	1.84	1.76	1.67	1.59	1.57	1.55	1.06	1.01	1.01	1.05	1.15		
	Hyn71			1.98	1.87	1.60	1.48	1.39	1.30	1.20	1.11							
Gabbro	Bir40	3.03	2.32	2.25	2.20	2.15												
	Bir40	2.86		1.93	1.95	1.99												
	Bir40	2.88	1.99	1.99	1.99	1.99	2.00	2.01										

(continued)

Table 8 (continued)

Rock	Ref.	ϱ [10³ kg m⁻³]	k [W m⁻¹ K⁻¹]															
			T [°C] 0	50	100	200	300	400	500	600	700	800	900	1000	1100	1200	1300	1400
Gabbro (Pyroxene)	Moi66	3.06						1.41		1.49		1.38						
Hornblende Gabbro	Kaw64	2.71	3.24	3.16	3.08	2.91	2.76	2.68	2.58	2.47	2.43			1.12	1.10			
Anorthosite	Bir40	2.70	1.85	1.88	1.90	1.96												
	Bir40	2.74	1.73	1.75	1.76	1.82	1.88											
	Bir40		1.68	1.69	1.71	1.79												
Peridotite	Kaw64	3.05	4.10	3.60	3.35	2.99	2.76	2.60	2.43	2.30								
	Kaw64	2.95	4.61	3.85	3.64	3.29	3.04	2.81	2.68	2.51								
Hyperstenite	Bir40	3.26	4.65	4.19	3.90	3.64												
Bronzitite	Bir40	3.26	5.19	3.85	3.56	3.28	3.06											
Dunite	Win52		4.19	4.40	3.94	3.39												
	Kaw64	3.21	4.94	3.77	3.48	3.01	2.70	2.43	2.22	2.05	1.93							
	Bir40	3.27	5.99	4.19	3.69	3.15												
	Bir40	3.25	4.69	4.77	4.23	3.67												
	Bir40	3.25		4.23	3.89	3.38												
	Kaw66				3.73	4.06	3.54	3.52	3.55	3.58	3.60	3.66	3.50	3.50	3.43	3.27	2.95	
Olivinite	Hut68			3.16	2.80	2.08	1.87	1.89	1.83	1.74	1.66	1.60	1.58	1.57	1.55	1.51	1.49	1.47
	Hyn71			3.33	2.99	2.46	2.30	2.18	2.09	2.02	1.95	1.89	1.88	1.84	1.81	1.78	1.77	1.76
Lherzolite	Kaw66			3.56	3.18	3.79	3.22	3.10	3.10	3.01	2.83	2.74	2.74	2.69	2.64			

Volcanic and hypabyssal rocks

Rock	Ref.	ϱ	T [°C] 0	50	100	200	300	400	500	600	700	800	900	1000	1100	1200	1300	1400
Basalt	Kaw64	2.58	2.37	2.22	2.09	1.82	1.59	1.47	1.38	1.36	1.34							
	Sak73	2.62	2.09	2.16	2.20	2.12	1.98	1.89	1.83									
	Sak73	2.96	2.20	2.39	2.51	2.51	2.36	2.23	2.18									
	Sak73	2.80	2.22	2.28	2.32	2.31	2.20	2.06	1.99									
	Sak73	2.82	2.40	2.51	2.60	2.58	2.42	2.34	2.32									
	Sak73	2.84	2.36	2.45	2.48	2.40	2.20	2.00	1.95									
Basalt Glass	Sak73	2.72	2.61	2.62	2.63	2.72	2.78	2.85	2.96									
	Sak73	2.74	2.72	2.76	2.80	2.88	2.96	3.00	3.00									
	Sak73	2.90	2.68	2.70	2.72	2.80	2.86	2.93	2.94									
Obsidian	Sak73	2.38	1.40	1.45	1.50	1.56	1.62	1.65	1.67									
	Sak73	2.38	1.36	1.41	1.47	1.55	1.62	1.68	1.71									
	Sak73	2.40	1.48	1.53	1.58	1.65	1.70	1.72	1.72									
	Bir40		1.34	1.40	1.46	1.57	1.67	1.78	1.89									

(continued)

4.1 Thermal conductivity and specific heat of rocks

Rock	Ref.	ϱ 10^3 kg m^{-3}	k [W m^{-1} K^{-1}] T [°C] 0	50	100	200	300	400	500	600	700	800	900	1000	1100	1200	1300	1400
Obsidian (continued)				0.35	0.50	0.71	0.75	0.78	0.85	0.12	1.04	1.16	1.24	0.87				
					0.55	0.73	0.78	0.86	0.95	1.08	1.21	1.35						
Tuff	Sak73	2.64	4.04	3.65	3.68	3.28	2.98	2.79	2.70									
	Sak73	2.65	2.82	2.80	2.78	2.46	2.20	2.10	2.05									
	Sak73	2.44	2.86	2.80	2.76	2.51	2.14	1.85	1.72									
Quartz Porphyry	Sak73	2.56	2.29	2.27	2.22	2.11	1.85	1.74	1.72									
	Sak73	2.46	1.81	1.92	1.98	1.85	1.69	1.66	1.65									
Porphyrite	Sak73	2.72	1.91	1.82	1.85	1.68	1.70	1.71	1.70									
	Sak73	2.92	3.86	3.65	3.47	3.12	2.78	2.50	2.31									
	Sak73	2.32	1.65	1.72	1.76	1.77	1.70	1.70	1.70									
	Sak73	2.48	1.72	1.87	1.93	2.00	2.00	1.94	1.93									
	Sak73	2.36	1.91	1.96	2.03	2.03	1.97	1.94	1.90									
	Sak73	2.65	1.55	1.99	2.20	2.18	2.09	2.10	2.12									
Diabase, Dolerite	Win52		2.20	2.18	2.16	2.15												
	Bir40	3.01	2.35	2.28	2.24	2.25												
	Bir40	2.96	2.19	2.16	2.14	2.11	2.09											
	Bir40	2.96	2.11	2.10	2.10	2.10	2.11	2.12										
	Bir40		1.15	1.20	1.26	1.37	1.48											
	Moi66					1.26		1.27		1.29	1.35	1.43	1.37	1.36	1.39			
	Moi68	2.82		1.60	1.49	1.26	1.27	1.28	1.28	1.29		1.40		1.33	1.31			
Metamorphic rocks																		
Marble																		
$\parallel$	Bir40	2.69	3.08	2.70	2.49	2.17												
$\perp$	Bir40	2.69	3.01	2.65	2.39	2.11												
(not oriented)	Win52		3.05		2.45	2.16												
Quarzite	Win52		6.24		5.23													
Serpentinized Peridotite	Kaw64	2.74	4.61	4.19	3.87	3.37	3.08	2.76	2.58	2.41	2.26							
Eclogite	Hyn71			3.10	2.52	1.77	1.58	1.49	1.42	1.33	1.36	1.14	1.06	0.98	0.90	0.82		
	Kaw64	3.50	2.51	2.41	2.30	2.11	1.99	1.84	1.76	1.74								
	Moi68			3.10	2.60	1.80	1.58	1.52	1.23	1.21	1.15	1.07	0.95	0.89	0.82			
	Moi66					1.62		1.53	1.37	1.24		1.20		0.96		0.82		

(continued)

Table 8 (continued)

Rock	Ref.	ϱ 10^3 kg m^{-3}	k [W m^{-1} K^{-1}]														
			T [°C] 0	50	100	200	300	400	500	600	700	800	900	1000	1100	1200	1300	1400
Gneiss																		
$\parallel$	Bir40	2.64		2.93	2.87													
$\perp$	Bir40	2.64		2.09	2.02													
Albitite	Bir40	2.61	2.16	2.03	2.01	1.97	1.91											
Slate																		
(not oriented)	Win52	2.70	2.05	2.01	1.88	1.63	1.47											
$\parallel$	Kaw64	2.66		2.53	2.45	2.30	2.18	2.07	2.01	2.01								
$\perp$	Bir40	2.76	1.94	1.84	1.77	1.71												
Sedimentary rocks																		
Quartz Sandstone																		
$\parallel$	Bir40	2.64	5.69	4.94	4.44	3.77												
$\perp$	Bir40	2.64	5.48	4.77	4.31	3.62												
Sandstone	Sak73	2.62	2.42	2.33	2.45	1.85	1.66	1.60										
	Sak73	2.60	2.57	2.41	2.37	2.00	1.70	1.58	1.57									
Shale	Win52			0.91	0.94	1.00	1.06	1.12	1.18	1.25	1.28	1.32	1.36	1.39				
Dolomite	Bir40	2.83	4.98	4.31	3.89	3.33												
	Win52		4.98		3.89	3.35												
Limestone																		
(not oriented)	Bir40	2.60	3.01	2.57	2.32	2.00												
$\parallel$	Bir40	2.60	3.45	3.16	2.95	2.74												
$\perp$	Bir40	2.69	2.55	2.38	2.27													
(not oriented)	Win52		3.01		2.30	2.01	1.40											
(not oriented)	Sak73	2.70	3.10	2.85	2.60	2.09	1.67	1.39	1.27									
(not oriented)	Sak73	2.75	3.00	2.91	2.75	2.21	1.75	1.53	1.43									
(not oriented)	Sak73	2.60	2.40	2.24	2.18	1.77	1.52	1.40	1.37									

Table 9. Temperature effect on the thermal diffusivity, κ, of rocks, after [Sak73].

Rock	ϱ 10^3 kg m^{-3}	κ [10^{-7} m^2 s^{-1}]						
		T [°C] 0	50	100	200	300	400	500
Alkali Granite	2.81	10.2	9.3	8.5	7.3	6.3	5.6	4.9
Granite	2.62	11.8	10.0	8.7	7.2	6.2	5.4	4.9
Quartz Diorite	2.72	8.0	7.3	6.5	5.8	5.3	4.9	4.7
Basalt	2.62	5.2	4.9	4.7	4.5	4.3	4.2	4.2
	2.96	7.0	6.5	6.0	5.6	5.4	5.0	4.9
	2.80	7.0	6.4	6.1	5.7	5.3	5.0	4.9
	2.82	6.4	5.8	5.4	4.8	4.4	4.2	4.2
	2.84	6.1	5.8	5.7	5.4	4.8	4.2	4.2
Basalt Glass	2.72	4.4	4.1	3.9	3.8	3.8	3.8	3.7
	2.74	4.3	4.1	3.9	4.0	4.0	3.9	3.9
	2.90	5.7	5.4	5.2	4.8	4.8	4.8	4.7
Obsidian	2.38	5.8	5.4	5.1	4.9	4.8	4.7	4.7
	2.38	6.0	5.7	5.5	5.2	5.1	5.0	5.0
	2.40	5.5	5.3	5.2	5.0	4.9	4.8	4.8
Tuff	2.64	11.1	9.4	8.2	6.9	6.1	5.6	5.4
	2.65	7.2	6.5	5.9	4.9	4.3	3.9	3.8
	2.44	10.5	9.3	8.2	6.6	5.5	4.8	4.7
Quartz Porphyry	2.56	7.5	6.6	6.1	5.5	5.4	4.7	4.0
	2.46	6.7	6.4	6.0	5.3	4.9	4.7	4.6
Porphyrite	2.72	6.7	6.2	5.8	5.2	4.9	4.9	4.9
	2.92	10.4	9.6	8.9	7.7	6.6	5.8	5.3
	2.32	6.0	5.5	5.2	4.8	4.6	4.6	4.6
	2.48	5.0	4.9	4.8	4.6	4.5	4.5	4.4
	2.36	6.0	5.8	5.6	5.4	5.3	5.1	4.9
	2.65	5.7	5.6	5.4	5.2	5.1	4.9	4.7
Sandstone	2.62	7.7	6.8	6.0	4.8	4.2	3.9	3.7
	2.60	7.7	6.8	6.0	4.8	4.2	3.9	3.8
Limestone	2.70	11.0	9.7	8.7	6.8	5.3	4.4	4.0
	2.75	10.0	9.6	8.9	6.9	5.4	4.6	4.2
	2.60	7.6	7.0	6.4	5.2	4.3	3.8	3.6

Table 10. Temperature effect on the specific heat, c_p, of rocks.

Rock	Ref.	c_p [kJ kg^{-1} K^{-1}]									
		T [°C] 0	100	200	300	400	500	600	700	800	1200
Granite	Leo67		0.846	0.938	1.005	1.059	1.097	1.130	1.164	1.189	
	Win52	0.800	0.875	0.950	1.017	1.089	1.118	1.235	1.311	1.390	
	Che72	0.69		0.96							
Granodiorite	Che72	0.71		0.96							
Diorite	Che72	0.63		1.00							
Gabbro	Che72	0.68		1.0							
Peridotite	Leo67		1.005	1.101	1.160	1.214	1.256				
Basalt	Leo67		0.871	0.955	1.017	1.051	1.084	1.118	1.147	1.177	
	Win52	0.858	0.963	1.038	1.097	1.143	1.189	1.239	1.281	1.319	1.491
	Che72	0.84		1.1							

(continued)

Table 10 (continued)

Rock	Ref.	c_p [kJ kg^{-1} K^{-1}]									
		T [°C] 0	100	200	300	400	500	600	700	800	1200
Diabase	Win52	0.699	0.795	0.871	0.934	0.988	1.034	1.084	1.097	1.189	1.361
	Che72	0.74		0.88							
Marble	Win52	0.749	0.896	1.001	1.072	1.130	1.172				
	Che72	0.80		1.00							
Quartzite	Win52	0.699	0.850	0.971	1.068	1.130	1.151	1.151	1.160	1.172	
	Che72	0.71		0.96							
Serpentinite	Leo67		1.093	1.193	1.264	1.327	1.386				
Gneiss	Win52	0.741	0.875	1.013							
	Che72	0.79		1.0							
Schist	Che72	0.71		1.0							
Slate	Win52	0.708	0.904	1.001	1.065	1.101	1.130	1.156	1.172	1.202	
Clay	Win52	0.800	0.871	0.938	1.013	1.080	1.168	1.281		1.784	1.784
	Win52	0.749	0.846	0.938	1.034	1.130	1.223	1.323	1.415	1.512	
	Che72	0.79		0.92							

For Table 11, see p. 338

Table 12. Anisotropy of the thermal conductivity, k, of rocks; n = number of determinations; ∥, ⊥: parallel, normal to the plane of bedding or schistosity, respectively.

Rock	Location	Ref.	ϱ 10^3 kg m^{-3}	k [W m^{-1} K^{-1}]				$A = \dfrac{k_\parallel}{k_\perp}$
				n	$k_\parallel$	n	$k_\perp$	
Marble	Switzerland	Wen69	2.76	10	2.46	10	2.31	1.06
	Vermont, USA	Bir40	2.69	1	3.08	1	3.01	1.02
Serpentinite	Switzerland	Wen69	2.72	1	2.51	1	1.72	1.46
Granite and Granite Gneiss	Austria	Cla61	2.61	13	3.71	9	2.87	1.29
Granite Gneiss	Massachusetts, USA	Bir40	2.64	1	3.10	1	2.16	1.44
Quartz Diorite Gneiss	Colarado, USA	Bir50		11	3.36	4	3.01	1.11

(continued)

Table 12 (continued)

Rock	Location	Ref.	ϱ 10^3 kg m^{-3}	k [W m^{-1} K^{-1}]				$A = \dfrac{k_\parallel}{k_\perp}$
				n	$k_\parallel$	n	$k_\perp$	
Gneiss	Switzerland	Cla56	2.71	8	3.73	22	2.65	1.41
	Switzerland	Wen69	2.67	55	2.12	55	1.74	1.22
	Vermont, USA	Cla66		9	3.49	9	2.61	1.34
Gneiss and Mica Schist	Austria	Cla61		7	4.50	8	3.03	1.49
	Switzerland	Cla56	2.70	6	3.90	12	2.63	1.49
	Washington, D.C., USA	Dim64		34	3.61	35	2.77	1.30
	Carolina, USA	Dim65	2.71	10	2.94	10	2.31	1.27
Injection schist and Gneiss	Colorado, USA	Bir50		17	3.58	15	2.91	1.23
Quartz rocks and Gneiss	Australia	Sas63		4	4.94	4	2.93	1.69
Pegmatite	Switzerland	Wen69	2.61	3	2.55	3	2.43	1.05
Schist	Switzerland	Wen69	2.76	18	2.88	18	2.05	1.40
	Scotland	Ric76	2.77	39	3.84	39	2.86	1.34
	Bihar, India	Ver66	2.70	12	3.47	12	3.02	1.15
	Godavari valley, India	Rao70		4	3.35	4	3.39	1.03
Schistes Lustrées	Switzerland	Cla56	2.71	7	3.14	8	2.40	1.31
Calcareous Mica Phyllite	South Dakota, USA	Bir66		9	4.95	7	3.30	1.50
Amphibolite	Switzerland	Wen69	2.89	8	1.54	8	1.28	1.20
	Godavari valley, India	Rao70		9	3.19	9	2.40	1.36
Quartz Sandstone	USA	Bir40	2.64	1	5.69	1	5.48	1.04
Quartz rocks and Sandstone	Germany	Hur65		17	3.24	17	2.99	1.08
Sandstone	Rheinland, Germany	Müc62	2.57	6	3.76	6	3.36	1.12
	Saarland, Germany	Hüc66	2.73	3	3.66	3	3.45	1.06
	Thüringia, Germany	Wei66		19	2.29	28	1.92	1.19
Claystone and Schist	Thüringia, Germany	Wei66		2	2.52	2	1.52	1.67
Clay Slate	Rheinland, Germany	Müc62	2.66	4	3.29	4	1.71	1.92
	Saarland, Germany	Hüc66	2.64	6	2.87	6	1.91	1.50
Shale	Saarland, Germany	Hüc66	2.64	5	3.35	6	1.75	1.92
Limestone	Switzerland	Wen69	2.60	1	2.04	1	2.20	0.93
	Pennsylvania, USA	Bir40	2.69	1	3.45	1	2.55	1.35
Dolomite	Thüringia, Germany	Wei66		61	3.98	58	3.91	1.02
Anhydrite	Thüringia, Germany	Wei66		12	3.65	13	3.58	1.02
Sylvinite	Germany	Cre65	2.10	1	5.28	1	5.15	1.02
Salt	Saarland, Germany	Cre65	2.17	1	5.69	1	5.74	0.99

Table 11. Temperature effect on the thermal conductivity, k, of soils (Thermal conductivity values derived from measured effective mean thermal conductivities as a function of hot-side temperature; cold-side temperature kept at 25 °C), after [Fly69].

Type of soil	ϱ 10^3 kg m^{-3}	k [W m^{-1} K^{-1}]												
		T [°C] 100	200	400	600	800	1000	1200	1300	1400	1500	1600	1650
Calcareous (Natural weathered limestone)	2.00	0.78	0.74	0.68	0.63	0.58	0.54	0.54	0.62	0.82	1.32	2.4	3.4
Granitic detrital (Weathered, decomposed granite soil)	1.92	0.88	0.83	0.77	0.74	0.75	0.82	1.05	1.44	2.5	5.2[1]		
Dune Sand (Windblown sand)	1.57	0.30	0.34	0.40	0.44	0.51	0.70	1.25	1.77	2.5	3.7	5.2[1]	
Magnesian (Magnesium-aluminium silicate)	1.79	0.66	0.65	0.64	0.65	0.67	0.73	0.93	1.17	1.6			
Podzol (Leached organic timberland soil)	1.75	0.52	0.53	0.51	0.52	0.64	0.98	1.63	2.1	2.7	3.4	4.2[1]	
Coastal plains clay (Coastal flood plain soil)	1.335	0.28	0.30	0.34	0.40	0.47	0.56	0.75	0.96	1.38	2.2	3.7	4.8
Laterite (Tropical rainforest soil)	1.49	0.32	0.27	0.19	0.13	0.13	0.30	0.85	1.38	2.2	3.3	4.9	5.9
Estacia Playa (Highly saline playa soil)	1.53	0.38	0.40	0.42	0.41	0.35	0.28	0.38	0.76	1.8	4.4		
Ottawa sand (Silica artificial soil)	1.76	0.43	0.51	0.68	0.85	1.04	1.24	1.45	1.56	1.68	1.80		
Ottawa sand (Silica artificial soil)	1.57	0.30	0.34	0.45	0.57	0.75	0.98	1.29	1.48	1.70			

[1]) Extrapolated value.

For Table 12, see p. 336

Table 13. Pressure effect on the thermal conductivity, k, of rocks. One-dimensional pressure in MPa (1 MPa = 9.81 atm) [Hur70].

Rock	k [W m^{-1} K^{-1}]																							
	increasing pressure [MPa]													decreasing pressure [MPa]										
	4	100	200	420	840	1250	1670	2090	2510	2530	2930	3340	4180	2930	2510	2090	1670	840	620	420	200	100	4	0
Anhydrite																								
no. 16	3.58		4.19	4.27	4.35	4.40	4.46			4.48		4.52	4.53				4.47			4.38	4.32	4.23	4.19	4.10
no. 85	3.65	3.91	4.13	4.28	4.32	4.42	4.40			4.43		4.46	4.45		4.43			4.33		4.30	4.21	4.16	4.07	4.05
no. 98	3.94		4.42	4.55	4.72	4.82	4.84	4.86			4.89													
no. 100	3.14	3.40	3.68	3.91	3.62		4.20			4.31		4.35	4.40		4.36			4.28		4.24	4.11		3.90	
no. 259a	3.40		4.04	4.19	4.30	4.31	4.35			4.48		4.51	4.53		4.53					4.31				
no. 259b	3.54	3.82	4.03	4.15	4.26		4.35			4.48		4.44	4.48				4.41	4.40			4.28		4.03	3.28
no. 366	3.42	3.51	3.70	3.86	3.99	4.02	4.05			4.08		4.10	4.14			4.07					3.82	3.75	3.48	
Sandstone																								
no. 172	2.86		3.11	3.08	3.13		3.17		3.20			3.21	3.24			3.22		3.19		3.14	3.16		3.08	
no. 224	2.31		2.45	2.48	2.51		2.60		2.68			2.67	2.73		2.71			2.67		2.63	2.55			
no. 234	2.80		2.94	3.01	3.08		3.14		3.18		3.26	3.22	3.29					3.13		3.11	3.08		2.60	
no. 286	3.14		3.40	3.55	3.61		3.73		3.77			3.84	3.86		3.77			3.78		3.67	3.60		3.24	
no. 313	3.95		4.37	4.40	4.51	4.31	4.71				4.83		4.85					4.57		4.52	4.38		4.15	
no. 343	3.56		3.96	4.07					4.38				4.48	4.38							4.02		3.60	
Dolomite																								
no. 103	3.19			3.87	3.90	3.97			4.08			4.12	4.10	4.04					3.80			3.74		
no. 365	2.59		2.77	2.85	2.90	2.94			3.01			3.06	3.08		3.16					2.95			3.51	2.50
Limestone																								
no. 19	2.12		2.30		2.36		2.39		2.43			2.44	2.42			2.43		2.36			2.33		2.21	
no. 34	1.77		1.76		1.81		1.81		1.87			1.89	1.88			1.84		1.83			1.79		1.76	
no. 102	3.21		3.42	3.43	3.44	3.44					3.49		3.55					3.47			3.40		3.08	
no. 260	1.81			2.01	2.01	2.02						2.03	2.03			2.05		2.06		2.02	2.00		1.98	
no. 270	1.48		1.57	1.62	1.65		1.65		1.66				1.63								1.57		1.43	
Porphyry																								
no. 244	1.98		2.01	2.06					2.06				2.10				2.04		1.99		1.99		1.93	
no. 245	1.75		1.79	1.90	1.95		1.98		2.01			2.03	2.01				1.99	1.96			1.76		1.50	
Clastogene																								
no. 378	1.73		1.82	1.85	1.87		1.87		1.87		1.89		1.89							1.87	1.85		1.85	

Table 14. Combined temperature and pressure effect on the conductivity, k, of crystalline quartz, olivine, dunite, and sodium chloride, after [Bec78]. (Only several selected values given here, more information provided in the original paper).

(A) Crystalline quartz, heat flowing perpendicular to the optical axis

P [GPa]	k [W m^{-1} K^{-1}]				
	T [°C] 0	93	165	251	347
0.00	6.46	4.96	4.19	3.62	3.33
2.50	6.67	5.19	4.41	3.86	3.55
3.40	6.98	5.36	4.57	3.99	3.66
4.60	7.03	5.44	4.64	4.07	3.71
5.30	7.05	5.48	4.70	4.12	3.75

(B) Crystalline quartz, heat flowing parallel to the optical axis

P [GPa]	k [W m^{-1} K^{-1}]			
	T [°C] 0	106	169	244
0.00	11.39	8.15	6.99	6.19
2.50	15.85	10.54	8.52	7.32
3.40	16.76	10.96	8.79	7.47
4.60	18.27	11.92	9.39	7.94
5.30	22.08	13.60	10.32	8.37

(C) Single crystal olivine, sample no. 1

P [GPa]	k [W m^{-1} K^{-1}]			
	T [°C] 0	105	171	242
0.00	5.66	4.60	4.05	3.69
2.50	6.40	5.14	4.57	4.10
3.40	6.82	5.42	4.84	4.28
4.20	7.09	5.62	5.03	4.43
4.95	7.15	5.61	5.07	4.46

(D) Single crystal olivine, sample no. 2

P [GPa]	k [W m^{-1} K^{-1}]			
	T [°C] 0	108	194	265
0.00	9.50	7.99	6.94	6.67
2.50	11.36	9.15	7.91	7.40
3.40	12.31	10.03	8.50	7.91
4.20	12.94	10.37	8.85	8.17
4.95	12.96	10.36	8.83	8.16

(E) Carolina dunite (run 1)

P [GPa]	k [W m^{-1} K^{-1}]			
	T [°C] 0	108	190	330
0.00	3.75	3.13	2.66	2.42
2.50	4.39	3.49	2.99	2.66
3.40	4.47	3.52	3.11	2.76
4.20	4.57	3.68	3.20	2.83
4.95	4.84	3.80	3.29	2.92

(F) Twin-Sister dunite (sample no. 2)

P [GPa]	k [W m^{-1} K^{-1}]			
	T [°C] 0	95	224	298
0.00	8.65	6.92	5.39	4.99
2.50	9.30	7.37	5.76	5.44
3.40	9.90	7.69	5.79	5.39
4.20	10.08	7.88	6.02	5.70
4.95	9.33	7.50	6.19	5.68

(F) Muskox dunite

P [GPa]	k [W m^{-1} K^{-1}]			
	T [°C] 0	97	172	271
0.00	4.95	3.91	3.71	3.72
2.50	5.59	4.52	4.06	3.84
3.40	5.80	4.71	4.13	3.91
4.20	5.73	4.73	4.16	3.97
5.60	6.68	5.70	4.50	3.94

(G) Sodium chloride (large natural single crystal)

P [GPa]	k [W m^{-1} K^{-1}]			
	T [°C] 0	96	181	236
0.00	5.49	3.00	1.94	1.64
2.50	8.01	5.06	3.65	3.19
3.40	8.16	5.70	4.31	3.75
4.20	9.63	6.30	4.70	4.25
4.95	9.71	6.95	5.47	4.78

4.1.3 References for 4.1 — Literatur zu 4.1

Ano52	Anonymous: Landolt-Börnstein, Zahlenwerte und Funktionen. Bd. III, Astronomie und Geophysik, Springer, Berlin, Wien, Heidelberg, **1952**.
Ano67	Anonymous: Landolt-Börnstein, Zahlenwerte und Funktionen, Bd. IV, Technik, Teil 4, Wärmetechnik, Springer, Berlin, Wien, Heidelberg, **1967**.
Ano76	Anonymous: Proceedings of the 2nd UN Symp. on Development and Use of Geothermal Resources. US Govern. Printing Office, Washington, D.C. **2** (1976).
Bec65	Beck, J.M., Beck, A.E.: J. Geophys. Res. **70** (1965) 5227.
Bec68	Becher, D., Meincke, W.: Z. Angew. Geol. **14** (1968) 291.
Bec78	Beck, A.E., Darbha, D.M., Schloessin, H.H.: Phys. Earth Planet. Interiors **17** (1978) 35.
Ben39	Benfield, A.F.: Proc. Roy. Soc. (London) **A 173** (1939) 428.
Ben47	Benfield, A.F.: Am. J. Sci. **245** (1947) 1.
Bir40	Birch, F., Clark, H.: Am. J. Sci. **238** (1940) 529 and 613.
Bir42	Birch, F., cited in: [Bir42a], p. 245.
Bir42a	Birch, F.: Handbook of Physical Constants. Geol. Soc. America Spec. Paper no. 36, 325 p, 1942.
Bir50	Birch, F.: Geol. Soc. Am. Bull. **61** (1950) 567.
Bir54	Birch, F.: Am. J. Sci. **252** (1954) 1.
Bir66	Birch, F.: cited in: [Cla66], p. 461.
Bol63	Boldizsár, T., Gózon, J.: Acta Techn. Acad. Sci. Hung. **43** (1963) 467.
Bol64	Boldizsár, T.: Acta Techn. Acad. Sci. Hung. **47** (1964) 293.
Bol66	Boldizsár, T.: Pure Appl. Geophys. **64** (1966) 121.
Bri24	Bridgmann, P.W.: Amer. J. Sci. **7** (1924) 81.
Bul39	Bullard, E.C.: Proc. Roy. Soc. (London) **A 173** (1939) 474.
Bul51	Bullard, E.C., Niblett, E.R.: Mon. Not. Roy. Astr. Soc., Geophys. Suppl. **6** (1951) 222.
Car55	Carte, A.E.: Am. J. Sci. **253** (1955) 482.
Cer67	Čermák, V.: Chemie Erde **26** (1967) 271.
Cer67a	Čermák, V., Krčmář, B.: Věst. Ústř. úst. geol. **42** (1967) 445.
Cer68	Čermák, V., Krčmář, B.: Věst. Ústř. úst. geol. **43** (1968) 415.
Cer79	Čermák V., Rybach L. (Eds.): Terrestrial Heat Flow in Europe. Berlin-Heidelberg-New York: Springer, **1979**.
Che72	Cheremensky, G.A.: Geotermiya, Izd. Nedra, Leningrad, **1972**.
Cla56	Clark, S.P., Niblett, E.R.: Mon. Not. Roy. Astr. Soc., Geophys. Suppl. **7** (1956) 176.
Cla57	Clark, S.P.: Trans. Am. Geophys. Union **38** (1957) 239.
Cla57a	Clark, S.P.: Am. Mineralogist **42** (1957) 732.
Cla61	Clark, S.P.: Geophys. J. **6** (1961) 54.
Cla66	Clark, S.P.: Thermal Conductivity. Handbook of Physical Constants, p. 459. Geol. Soc. Am. Memoir 97. Washington, D.C., **1966.**
Col52	Colosky, B.P.: Bull. Am. Ceram. Soc. **31** (1952) 465.
Cos47	Coster, H.P.: Mon. Not. Roy. Astr. Soc., Geophys. Suppl. **5** (1947) 131.
Cre64	Creutzburg, H.: Kali Steinsalz **4** (1964) 73.
Cre65	Creutzburg, H.: Kali Steinsalz **4** (1965) 170.
D'An49	D'Ans, J., Lax, E.: Taschenbuch für Chemiker und Physiker. Berlin-Göttingen-Heidelberg: Springer, **1949**.
Dim64	Diment, W.H., Werre, R.W.: J. Geophys. Res. **69** (1964) 2143.
Dim65	Diment, W.H., Marine, I.W., Neiheisel, J., et al.: J. Geophys. Res. **70** (1965) 5635.
Dja69	Djamalova, A.S.: Glubinniy teplovoy potok na territorii Dagestana. Moskva: Izd. Nauka, **1969**.
Dre74	Dreyer, W.: Materialverhalten anisotroper Festkörper (Thermische und elektrische Eigenschaften). Wien-New York: Springer Verlag, 345 p. **1974**.
Duc74	Duchkov, A.D., Sokolova, L.S.: Geotermicheskiye issledovaniya v. Sibiri. Izd. Nauka, Sibir. otd., Novosibirsk, **1974**.
Eng78	England, P.C.: Tectonophysics **46** (1978) 21.
Eri79	Eriksson, K.G., Malmqvist, D., cited in: [Cer79], p. 267.
Euc11	Eucken, A.: Ann. Physik **34** (1911) 185.
Fly69	Flynn, D.R., Watson, T.W., in: C.Y. Ho, R.E. Taylor (Eds.): Thermal Conductivity, p. 913. New York: Plenum Press **1969**.

Gol77	Golubev, V.A., Goldyrev, G.S., Duchkov, A.D., Lysak, S.V., Kazancev, S.A.: Geologiya i geofizika, Novosibirsk, No. 8 (1977) 103.
Gor42	Goranson, J., cited in: [Bir42a], p. 228.
Gup70	Gupta, M.L., Verma, R.K., Hamza, V.M., Venkateshwar Rao, G., Rao, R.U.M.: Tectonophysics **10** (1970) 147.
Hae71	Haenel, R.: Z. Geophys. **37** (1971) 119.
Hae73	Haenel, R., Zoth, G.: Z. Geophys. **39** (1973) 425.
Hae76	Haenel, R., in: P. Giese, C. Prodehl, A. Stein (Eds.): Explosion Seismology in Central Europe, p. 32. Berlin-Heidelberg-New York: Springer, **1976**.
Hae79	Haenel, R., cited in: [Cer79], p. 49.
Her56	Herrin, E., Clark, S.P.: Geophysics **21** (1956) 1087.
Hor60	Horai, K., Uyeda, S.: Bull. Earthq. Res. Inst. **38** (1960) 199.
Hor71	Horai, K.: J. Geophys. Res. **76** (1971) 1278.
Hor72	Horai, K., Baldridge, S.: Phys. Earth Planet. Interiors **5** (1972) 151.
Hüc66	Hückel, B., Kappelmeyer, O.: Z. Deut. Geol. Ges., **117** (1966) 280.
Hur65	Hurtig, E.: Pure Appl. Geophys. **60** (1965) 85.
Hur70	Hurtig, E., Brugger, H.: Tectonophysics **10** (1970) 67.
Hut68	Hutt, J.R., Berg, J.W.: Geophysics **33** (1968) 489.
Hyn66	Hyndman, R.D., Sass, J.H.: J. Geophys. Res. **71** (1966) 587.
Hyn71	Hyndman, R.D., Jessop, A.M.: Can. J. Earth Sci. **8** (1971) 391.
Jär79	Järvimäki, P., Puranen, M., cited in: [Cer79], p. 172.
Kap74	Kappelmeyer, O., Haenel, R.: Geothermics with Special Reference to Application. Geopublication Associates, Berlin-Stuttgart: Gebrüder Borträger, **1974**.
Kaw64	Kawada, K.: Bull. Earthq. Res. Inst. **42** (1964) 631.
Kaw66	Kawada, K.: Bull. Earthq. Res. Inst. **44** (1966) 1071.
Kan68	Kanamori, H., Fujii, N., Mizutani, H.: J. Geophys. Res. **73** (1968) 595.
Kay26	Kaye, G.W.C., Higgins, W.F.: Proc. Roy. Soc. (London) **A 113** (1926) 335.
Kie76	Kieffer, S., Getting, I.C., Kennedy, G.C.: J. Geophys. Res. **81** (1976) 3018.
Kut70	Kutas, R.I., Gordiyenko, V.V.: Geofiz. Sb., Kiev, **34** (1970) 29.
Leb69	Lebedev, T.S., Kutas, R.I., Gordiyenko, V.V.: Bull. Volcanol. **33** (1969) 191.
Len66	Leney, G.W., Wilson, J.T., cited in: [Cer66], p. 464.
Leo67	Leonidov, Ya.: Geochem. Internat. **4** (1967) 400.
Lew69	Lewis, T.J.: Can. J. Earth Sci. **6** (1969) 1191.
Lew77	Lewis, T.J., Beck, A.E.: Tectonophysics **41** (1977) 41.
Lov63	Lovering, J.F., Morgan, J.W.: Nature **199** (1963) 479.
Lub64	Lubimova, E.A., Lyusova, L.N., Firsov, F.V.: Izv. Akad. Nauk SSSR, Ser. Geofiz. **11** (1964) 1622.
Lub64a	Lubimova, E.A., Lyusova, L.N., Firsov, F.V., in: E.A. Lubimova (Ed.): Geotermicheskiye issledovaniya, p. 5. Moskva: Izd Nauka, **1964**.
Lub64b	Lubimova, E.A., Starikova, G.N., Shuspanov, A.P., in: E.A. Lubimova (Ed.): Geotermicheskiye issledovaniya, p. 115. Moskva: Izd. Nauka **1964**.
Lub66	Lubimova, E.A., Starikova, G.N., in: F.A. Makarenko (Ed.): Geotermicheskiye issledovaniya i ispolzovaniye tepla Zemli, p. 135. Moskva: Izd. Nauka **1966**.
Lub72	Lubimova, E.A., Karus, E.V., Firsov, F.A., Starikova, G.N., Vlasov, V.K., Lyusova, L.N., Koperbach, E.B.: Geothermics **1** (1972) 81.
Lub75	Lubimova, E.A., Lysak, S.V., Firsov, F.V., Starikova, G.N., Efimov, A.V., Ignatev, B.I., in: N.A. Floresnsov (Ed.): Baikalskii Rift, p. 94. Izd. Nauka, Sibir. otd., Novosibirsk, **1975**.
Lub76	Lubimova, E.A., Nikitina, V.N., Tomara, G.A.: Teploviye polya vnutrennykh i okrainnykh morey SSSR, 224 pp. Moskva: Izd. Nauka **1976**.
Lyu73	Lyusova, L.N., Kutasov, I.M.: Verkhnaya mantiya, No. 12, p. 58. Moskva: Izd. Nauka **1973**.
Mei67	Meincke, W., Hurtig, E., Weiner, J.: Geophys. Geol. **11** (1967) 40.
Mis51	Misener, A.D., Thompson, L.G.D., Uffen, R.J.: Trans. Am. Geophys. Union **32** (1951) 729.
Mis55	Misener, A.D.: Trans. Am. Geophys. Union **36** (1955) 1055.
Miz70	Mizutani, H., Baba, K., Kobayashi, N., Chang, C.C., Lee, C.H., Kamg, Y.S.: Tectonophysics **10** (1970) 183.
Moi66	Moiseenko, U.I., Solovyeva, Z.A., Kutolin, V.A.: Dokl. Akad. Nauk SSSR **173** (1966) 163.
Moi68	Moiseenko, U.I.: Freiberger Forschh. **C 238** (1968) 89.
Mor75	Morgan, P.: Earth Planet. Sci. Lett. **26** (1975) 253.

Mos51	Mossop, S.C., Gafner, G.: J. Chem. Met. Miner. Soc., S. Afr. **52** (1951) 61.
Müc62	Mücke, G.: Disertation, Aachen **1962**.
Ner67	Nernst, W., in: [Ano67], p. 904.
Ple66	Plewa, S.: Regionalny obraz parametrów geotermicznych obszaru Polski. Wyd. Geofizyka i geologia naftowa, Kraków, **1966**.
Rao70	Rao, R.U.M., Verma, R.K., Venkateshwar Rao, G., Hamza, V.M., Panda, P.K., Gupta, M.L.: Tectonophysics **10** (1970) 165.
Rat59	Ratcliffe, E.H.: Brit. J. Appl. Phys. **10** (1959) 22.
Raz69	Ražnjević, K.: Tepelné tabulky a diagramy. Nakl. Alfa, Bratislava **1969**.
Ric76	Richardson, S.W., Powell, R.: Scott. J. Geol. **12** (1976) 237.
Sak73	Sakvarelidze, E.A.: Verkhnaya mantiya, No. 12, p. 125. Moskva: Izd. Nauka **1973**.
Sas63	Sass, J.H., Le Marne, A.E.: Geophys. J. **7** (1963) 477.
Sas64	Sass, J.H.: J. Geophys. Res. **69** (1964) 299.
Sas71	Sass, J.H., Lachenbruch, A.H., Munroe, R.J.: J. Geophys. Res. **76** (1971) 3391.
Sas72	Sass, J.H., Nielsen, B.L., Wollenberg, H.A., Munroe, R.J.: J. Geophys. Res. **77** (1972) 6435.
See28	Seemann, H.E.: Phys. Rev. **31** (1928) 119.
Smi79	Smirnova, E.V., in: M.P. Volarovich (Ed.): Eksperimentalnoye i teoreticheskoye izucheniye teplovykh potokov, p. 91. Moskva: Izd. Nauka, **1979**.
Smi79a	Smirnova, E.V., Lyusova, L.N., in: M.P. Volarovich (Ed.): Eksperimentalnoye i teoreticheskoye izucheniye teplovykh potokov, p. 34. Moskva: Izd. Nauka, **1979**.
Sta73	Starikova, G.N., Lubimova, E.A.: Verkhnaya mantiya, No. 12, p. 112. Moskva: Izd. Nauka **1973**.
Swa74	Swanberg, Ch.A., Chessman, M.D., Simmons, G., Smithson, S.B., Grønlie, G., Heier, K.S.: Tectonophysics **23** (1974) 31.
Tim70	Timareva, S.B., Smirnov, Ya.B., Polyak, B.G., in: F.A. Makarenko, B.G. Polyak (Eds.): Teplovoy rezhim nedr SSSR, p. 45. Moskva: Izd. Nauka, **1970**.
Ver55	Verhoogen, J.: Trans. Am. Geophys. Union **36** (1955) 866.
Ver66	Verma, R.K., Rao, R.U.M., Gupta, M.L.: J. Geophys. Res. **71** (1966) 4943.
Wat76	Watts, G.P., Adams, W.M., cited in: [Ano76], p. 1247.
Web67	Weber, R., cited in: [Ano67], p. 904.
Wei66	Weiner, J.: Diplomarbeit, Leipzig **1966**.
Wen69	Wenk, H.R., Wenk, E.: Schweiz. Min. Petr. Mitt. **49** (1969) 343.
Win52	Winkler, H.G., cited in: [Ano52], p. 324.
Woo61	Woodside, W., Messmer, J.H.: J. Appl. Phys. **32** (1961) 1699.
Yuk74	Yukutake, H.: J. Phys. Earth **22** (1974) 299.
Zie56	Zierfuss, H., Vliet, G. van der: Bull. Am. Assoc. Petrol. Geol. **40** (1956) 2475.

4.2 Thermal conductivity of soil — Wärmeleitfähigkeit des Bodens

4.2.1 Introduction — Einleitung

The measuring or the thermal conductivity of soil is generally complicated [Gem50]. A standardized method has been developed by [ERA55], and has been introduced into practice (e.g. [Kap74]). The following values have been determined by this method.

The thermal conductivity of soil is essentially influenced by moisture. Temperature, density, pressure, and anisotropy affect insignificantly the thermal conductivity of the soil in contrast to that of the rocks of the crust and the upper layer. The thermal conductivity of dry soils is small ($K = 0.2 \cdots 0.8$ W m^{-1} K^{-1}), reaches a maximum by 20 to 30 wt% of water content ($K \approx 2$ or 3 W m^{-1} K^{-1}), decreases for higher contents of water, e.g. wet bog [Sch80], and draws near the value of thermal conductivity of water ($K = 0.6$ W m^{-1} K^{-1}).

Die Messung der Wärmeleitfähigkeit des Bodens ist schwierig [Gem50]. Eine mittlerweile standardisierte Methode ist von [ERA55] ausgearbeitet und beschrieben worden und hat Eingang in die Praxis gefunden (z.B. [Kap74]). Die mitgeteilten Werte wurden nach dieser Methode bestimmt.

Temperatur, Dichte, Druck und Anisotropie beeinträchtigen die Wärmeleitfähigkeit der Böden im Gegensatz zu der der Gesteine der Kruste und des oberen Mantels unbedeutend. Entscheidend hingegen ist die Bodenfeuchte. In trockenen Böden ist die Wärmeleitfähigkeit klein ($K = 0.2 \cdots 0.8$ W m^{-1} K^{-1}). Sie erreicht ein Maximum bei einem Anteil des Wassers zwischen 20 und 30 Gew-% ($K \approx 2$ oder 3 W m^{-1} K^{-1}), nimmt bei höheren Anteilen, wie z.B. im nassen Torf [Sch80], kleinere Werte an und nähert sich der Größe der Wärmeleitfähigkeit des Wassers ($K = 0.6$ W m^{-1} K^{-1}).

4.2.2 Data — Daten

Thermal conductivity K [W K^{-1} m^{-1}] and moisture content w_g*) (% by weight of the soil) of some Central European soils

	K W m^{-1} K^{-1}	w_g*) wt%
Raised-bog peat — Hochmoortorf	0.6 $\cdots$ 0.7	391 $\cdots$ 913
Low-bog peat — Niedermoortorf	0.7 $\cdots$ 0.8	411 $\cdots$ 663
Alluvium — Auelehm	1.0 $\cdots$ 2.5	22.9 $\cdots$ 28.1
Loess clay of stagnant moisture — Staunasser Lößlehm	1.4 $\cdots$ 1.7	22.7 $\cdots$ 31.5
Amaltheen clay — Almatheen Ton	2.0	27.7
Tertiary sand — Sand aus dem Tertiär	1.8 $\cdots$ 2.2	4.6
Loess clay — Lößlehm	1.6 $\cdots$ 2.1	21.9 $\cdots$ 22.3
Horticultural soil — Gartenerde	3.3	

*) $w_g = [(G_f - G_t) 100] G_t^{-1}$, G_f = weight of moist soil sample [in g], G_t = weight of dry soil sample (105 °C) [in g].

4.2.3 References for 4.2 — Literatur zu 4.2

Gem50	Gemant, A.: J. Appl. Phys. **21** (1950) 750.
ERA55	E.R.A. (The British electrical and allied industries research association): Technical Report F/T 181, **1955**.
Kap74	Kappelmeyer, O., Haenel, R.: Geoexploration Monographs, Ser. 1 No. 4, 1–238, Hannover **1974**.
Sch80	Schuch, M.: Physik des Torfes und der Moorböden. – In: K. Göttlich, Moor- und Torfkunde, Stuttgart **1980**.

4.3 Melting temperature of rocks — Schmelztemperaturen der Gesteine
4.3.1 Introduction — Einführung

The melting process of a polymineralic rock extends from the first appearance of a melt at the temperature of the *solidus* to the complete melting up of the rock at the temperature of the *liquidus*. The crystallization process of a magma, in contrast, begins with the crystallization of the first crystalline phase at the temperature of the liquidus and terminates with the complete crystallization of the melt at the temperature of the solidus. The range between solidus and liquidus is called "melting interval". Within this interval partial melts coexist with crystalline phases. In the following, solidus and liquidus temperatures and curves will comprehensively be called "melting temperatures" and "melting curves".

Melting temperatures depend on the total pressure acting on the system. Both the solidus and liquidus of a rock are therefore represented by a curve on a P, T-diagram. Melting temperatures are also dependent on the chemical composition of the system (including the volatiles such as H_2O, CO_2, HF, HCl, H_2S in the vapour phase). The solidus and liquidus of a group of rocks with varying composition therefore cover areas rather than curves in the P, T-diagram. With increasing pressure the compositional effect of the volatiles on the melting temperatures increases and may even exceed the compositional effect of the non-volatiles, because the solubility of volatiles in silicate melts rapidly increases with pressure. This is especially true for the volatile H_2O (Fig. 2). Melting temperatures may be up to 600 °C lower if H_2O is present in the vapour phase at high pressures (e.g. curves no. 1 in Fig. 6 a, b). In the P, T-diagrams of Fig. 5 and 6 solidus and liquidus areas of the different rock groups are therefore separately plotted for the two extreme conditions:

- $P_{H_2O} = P_{total}$ (so called "wet solidus", Fig. 5a, and "wet liquidus", Fig. 6a). This condition is fulfilled, if H_2O is present in excess or just in the correct amount to saturate the melt.

- $P_{H_2O} = 0$ (so called "dry solidus", Fig. 5b, and "dry liquidus", Fig. 6b).

If insufficient water is available to saturate the liquid, the solidus and liquidus are located between the corresponding curves for wet and dry conditions, leaving the wet curve at exactly the point where the

melt incorporates just all the available H_2O for saturation (Fig. 1). If H_2O is not the only volatile in the system and if not its concentration but its molar ratio to the other volatiles is fixed, the solidus and liquidus are located between the corresponding curves for wet and dry conditions as well, but leaving these curves at zero pressure.

Shifts of the melting intervals to lower temperatures due to CO_2 are much smaller than shifts due to H_2O, at least in carbonate free rocks at pressures below 20 kbars, because of the lower solubility of this volatile. Therefore, and because the influence of other volatiles has not yet been studied thoroughly, only H_2O has been considered in this compilation of melting temperatures. The solubility of H_2O is lower in ultramafic and mafic liquids than in granitoid liquids (Fig. 2). Data on the solubility of CO_2 in silicate melts and on melting temperatures of rocks with CO_2 or $H_2O + CO_2$ in the vapour phase are available from [Bre75, Bre76, Bur62, Egg73a, Egg76, Egg78, Gre73b, Ham64, Khi73, Mys75, Mys75a, Mys76a, Wyl77a, Wyl78, Wyl79].

As supplementary information to Figs. 5 and 6, the solidus, liquidus, and near liquidus phase relationships which may be encountered in rhyolite, andesite, and basalt are given in Fig. 1. For corresponding phase relationships in the range between the wet solidus and liquidus, see [Ste75].

Short compilations on melting temperatures of various rock groups can be found in [Gre76, Kes76, LeB77, Win76, Wyl77, Yod76], further references: [Lam72, Lam74, Ste75].

Bedingungen an jenem Punkt verlassen, an welchem die Schmelze gerade alles Wasser zur Sättigung aufbraucht (Fig. 1). Wenn H_2O nicht die einzige Gasspezies im System ist und wenn nicht sein gewichtsprozentiger Anteil sondern sein molares Verhältnis zu den anderen Gasspezies fixiert ist, liegen Solidus und Liquidus ebenfalls zwischen den entsprechenden Kurven für nasse und trockene Bedingungen, gehen hier aber vom Schnittpunkt dieser Kurven bei $P=0$ aus.

Im Vergleich zu H_2O ist die durch CO_2 verursachte Schmelzpunktserniedrigung viel kleiner, zumindest in karbonatfreien Gesteinen bei Drucken unterhalb 20 kbar, weil sich CO_2 weniger gut in silikatischen Schmelzen löst als H_2O. Deshalb, und weil der Einfluß anderer Gasspezies noch nicht systematisch untersucht wurde, wird in den zusammenfassenden Darstellungen der Schmelztemperaturen (Fig. 5 und 6) nur H_2O berücksichtigt. Die Löslichkeit von H_2O ist in ultramafischen und mafischen Schmelzen deutlich niedriger als in granitoiden Schmelzen (Fig. 2). Angaben über die Löslichkeit von CO_2 in silikatischen Schmelzen sowie über die Schmelztemperaturen von Gesteinen in Gegenwart von CO_2 oder $H_2O + CO_2$ können [Bre75, Bre76, Bur62, Egg73a, Egg76, Egg78, Gre73b, Ham64, Khi73, Mys75, Mys75a, Mys76a, Wyl77a, Wyl78, Wyl79] entnommen werden.

Ergänzend zu den Fig. 5 und 6 informiert die Fig. 1 über die Solidus- und Liquidus-Phasenbeziehungen in je einer Rhyolit-, Andesit- und Basalt-Probe. Die entsprechenden Phasenbeziehungen für den Bereich zwischen dem nassen Solidus und Liquidus finden sich in [Ste75].

Kurze zusammenfassende Arbeiten über Schmelztemperaturen verschiedener Gesteinsgruppen findet man in [Gre76, Kes76, LeB77, Win76, Wyl77, Yod76], weitere Literatur in [Lam72, Lam74, Ste75].

4.3.2 Melting temperatures of individual rock groups — Schmelztemperaturen einzelner Gesteinsgruppen

4.3.2.1 General comments to Figs. 5 and 6 — Allgemeine Bemerkungen zu Fig. 5 und 6

Figs. 5 and 6 have been compiled using experimental data on melting temperatures of natural and synthetic rocks, representing, however, only a selection of all the data that have been determined in the last 30 years. The ranges of the wet and dry solidi and liquidi which have been given in these figures for the different rock groups should therefore be considered as approximations to the real melting ranges. Furthermore, for the sake of easier reading of the figures, overlapping of the melting ranges of adjacent rock groups is in most cases not shown. The curves separating the ranges have been drawn through the middle of the areas of overlap and thus do not represent strict boundaries.

Fig. 5 und 6 entstanden aus der Zusammenfassung experimentell ermittelter Schmelztemperaturen natürlicher und synthetischer Gesteine. Es handelt sich dabei nur um eine begrenzte Auswahl all jener Schmelzdaten, welche in den letzten 30 Jahren bestimmt wurden. Die Solidus- und Liquidusbereiche, die in diesen Figuren für die einzelnen Gesteinsgruppen ausgeschieden wurden, sollten deshalb nur als Annäherungen an die wahren Schmelzbereiche dieser Gesteinsgruppen betrachtet werden, umsomehr als zugunsten einer besseren Lesbarkeit der Figuren Überlappungen benachbarter Bereiche, wie sie normalerweise auftreten, meist nicht dargestellt sind. Die Grenzlinien wurden in solchen Fällen einfach durch die Mitte des Überlappungsbereichs gezogen.

All shaded areas in Figs. 5 and 6 are documented by experiments in the quoted literature, each type of shading referring to one and the same main rock group abbreviated in these figures and in Fig. 4 by means of the following capital letters:

G	rhyolitoids + dacites/granitoids — Rhyolitoide + Dacite/Granitoide	[Hua75, Lut64, Piw73, Ste75]
B	basalts + andesites/gabbroids + dioritoids — Basalte + Andesite/Gabbroide + Dioritoide	[All75, Coh67, Egg72, Egg73, Gre67, Gre67a, Gre72, Gre67b, Gre68, Gre72a, Ito68, Ito71, Ito74, Lam72, Lam74, Til64, Yod62] and as further ref.: [Ste75]
B_a	= andesites/dioritoids — Andesite/Dioritoide	
B_b	= basalts — Basalte	
F_a	phonolitoids + tephritoids/foid-syenitoids + foid-dioritoids and foid-gabbroids — Phonolitoide + Tephritoide/Foid-Syenitoide + Foid-Dioritoide und -Gabbroide	[All73, Bar78, Bul71, Edg76, Gre73, Hay71, Mer75, Mil74]
F_b	foiditoids (mainly nephelinites)/foidolites — Foiditoide (hauptsächlich Nephelinite)/Foidolite	
PB	picritic basalts (and leucitites) — pikritische Basalte (und Leucitite)	
U	ultramafic rocks — ultramafische Gesteine	[Arn76, Bic77, Boe73, Boy64, Dav64, Gre67c, Gre73a, Gre75, Haw75, Ito67, Kus68, Mac64, Mys76, Nis70, Rea64]
(M	modal content of mafic minerals in the rock — modaler Anteil mafischer Gemengeteile im Gestein)	

The criteria for the assignment of the experimentally investigated samples to one of these rock groups was for plutonic rocks their location in the modal QAPF (quartz-alkalifeldspar-plagioclase-feldspathoid) triangles (Fig. 4) according to [Str74], and for the effusive rocks two chemical norm ratios according to [Str79] using the terminology suggested in [Str78].

Trachytoids/syenitoids (projecting into the unshaded field between A and B in Fig. 4) in most cases exhibit melting relationships similar to G but may also range in the fields B (rarely F_a) of Figs. 5 and 6. Not included in the compilation are carbonatites and melilitites. Information concerning melting temperatures of these rocks are available from [Bre75, Bre76, LeB77, Wyl77a].

The melting temperatures of phyllites, mica schists and quartz-feldspar gneisses match those of rock group G, and the melting ranges of amphibolites and eclogites cover those of rock group B.

Die experimentell untersuchten Proben wurden auf Grund folgender Kriterien einer dieser Gesteinsgruppen zugeordnet: Die plutonischen Gesteine auf Grund ihrer Lage in den QAPF (Quarz-Alkalifeldspat-Plagioklas-Foid)-Dreiecken (Fig. 4) entsprechend [Str74], die effusiven Gesteine auf Grund zweier chemischer Normparameter nach [Str79] unter Verwendung der in [Str78] vorgeschlagenen Terminologie.

Trachytoide/Syenitoide (welche das unschraffierte Feld zwischen A und B in Fig. 4 besetzen) zeigen ein ähnliches Schmelzverhalten wie die Gesteinsgruppe G, seltener B (oder F_a). Nicht behandelt ist hier das Schmelzverhalten der Karbonatite und Melilitite. Nähere Angaben hierzu finden sich in [Bre75, Bre76, LeB77, Wyl77a].

Die Schmelztemperaturen von Phylliten, Glimmerschiefern und Quarz-Feldspat-Gneisen stimmen weitgehend mit denjenigen der Gesteinsgruppe G überein, und die Schmelzbereiche der Amphibolite und Eklogite decken sich ungefähr mit denjenigen der Gesteinsgruppe B.

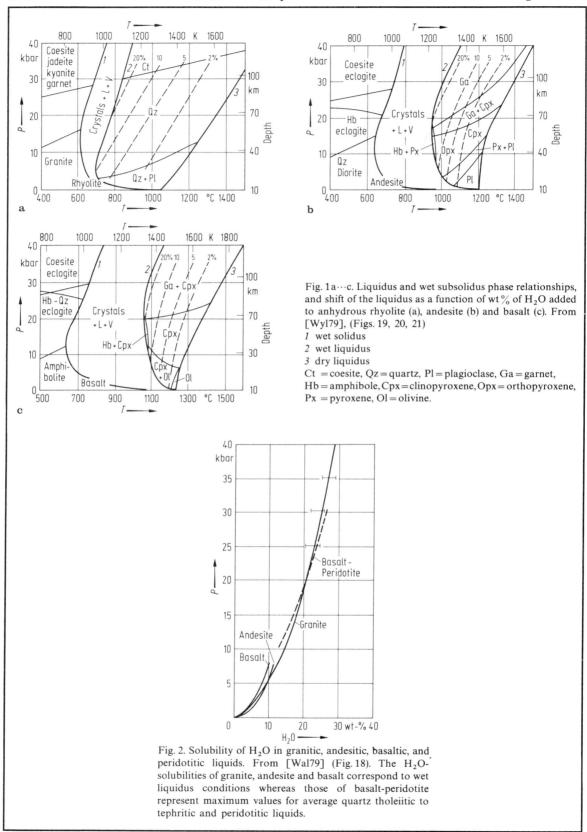

Fig. 1a···c. Liquidus and wet subsolidus phase relationships, and shift of the liquidus as a function of wt% of H_2O added to anhydrous rhyolite (a), andesite (b) and basalt (c). From [Wyl79], (Figs. 19, 20, 21)
1 wet solidus
2 wet liquidus
3 dry liquidus
Ct = coesite, Qz = quartz, Pl = plagioclase, Ga = garnet, Hb = amphibole, Cpx = clinopyroxene, Opx = orthopyroxene, Px = pyroxene, Ol = olivine.

Fig. 2. Solubility of H_2O in granitic, andesitic, basaltic, and peridotitic liquids. From [Wal79] (Fig. 18). The H_2O-solubilities of granite, andesite and basalt correspond to wet liquidus conditions whereas those of basalt-peridotite represent maximum values for average quartz tholeiitic to tephritic and peridotitic liquids.

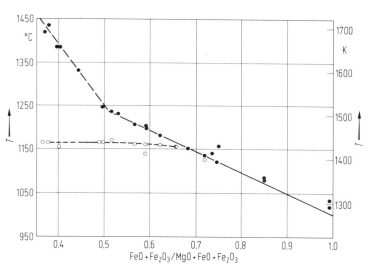

Fig. 3. Correlation of the liquidus temperature with iron enrichment in natural mafic rocks at 1 bar. Solid dots represent liquidus temperatures, open circles mark the temperature where all major phases begin to precipitate together. From [Yod76], (Fig. 2-1). The ratios for the abscissa are calculated using weight percentages of the oxides.

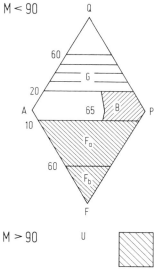

Fig. 4. Subdivision of effusive and plutonic rocks in the QAPF triangles according to [Str74, Str78, Str79]. Explanations of G, B, F_a, F_b, U and M in the text.

4.3.2.2 Comments on the solidi (Fig. 5) — Bemerkungen zur Solidusdarstellung (Fig. 5)

- Field G in Fig. 5a is confined to the left by the "granite minimum" which may be shifted to lower temperatures if other volatiles such as F, Cl (e.g. in pegmatites) coexist with H_2O in the vapour phase.

- Field B_b is subdivided into B_b', mainly representing tholeiitic and Al-rich basalts, and B_b'', mainly representing alkali basalts and some olivine rich basalts.

- The dry solidus part of field U represents, besides pyroxenites and peridotites, also picritic basalts. This field theoretically extends to the congruent melting of forsterite Mg_2SiO_4 (line 4 in Fig. 5b).

- The shaded area representing the wet solidus of rock group U does not include the experimental results of [Mys75a]. According to [Gre76a] the temperatures for the wet solidus determined by [Mys75a] may be too low.

- The absence of data prevented a reliable plotting of the solidus for rock groups F_a and F_b. The solidus presumably covers parts of B_b'' and an area to the right of it.

- Within the individual rock groups the solidus temperature generally increases with increasing $Mg/(Mg+Fe^{2+}+Fe^{3+})$ and Ca/Al.

- Der Bereich des nassen Solidus für G ist links durch das „Granitminimum" begrenzt. Falls noch andere Gasspezies wie z.B. F und Cl (in Pegmatiten) neben H_2O auftreten würden, kann dieses Minimum zu tieferen Temperaturen verschoben sein.

- Das Feld B_b ist in zwei weitere Felder unterteilt, nämlich B_b' und B_b''. In B_b' fallen hauptsächlich tholeiitische und Al-reiche Basalte, in B_b'' vorwiegend Alkali- und einige Olivin-reiche Basalte.

- Der Bereich für den trockenen Solidus von U beinhaltet, neben Pyroxeniten und Peridotiten, auch pikritische Basalte. Dieses Feld erstreckt sich theoretisch bis zum kongruenten Aufschmelzen von Forsterit Mg_2SiO_4 (Linie 4 in Fig. 5b).

- Die experimentellen Resultate von [Mys75a] fallen außerhalb des in Fig. 5 dargestellten Bereichs des nassen Solidus der Gesteinsgruppe U. Nach [Gre76a] dürften die in [Mys75a] veröffentlichten Solidustemperaturen zu niedrig sein.

- Mangels Daten ist der Solidus der Gesteinsgruppen F_a und F_b nicht eingezeichnet worden. Er dürfte sich teilweise mit B_b'' decken, sich aber auch noch weiter nach rechts ausdehnen.

- Innerhalb der einzelnen Gesteinsgruppen nimmt die Solidustemperatur im allgemeinen mit steigenden Quotienten $Mg/(Mg+Fe^{2+}+Fe^{3+})$ und Ca/Al zu.

4.3.2.3 Comments on the liquidi (Fig. 6) — Bemerkungen zur Liquidusdarstellung (Fig. 6)

Only one characteristic example (curves no. 1 in Fig. 6a, b) is given for the liquidus of granitoids covering, under wet conditions, a wide area to the left of field B_a. Attention should be drawn to the fact that at elevated pressures the dry liquidus for granites may be located at higher temperatures than that of andesites or even basalts. Apart from this the liquidus temperature fairly regularly changes with changing bulk composition, e.g. it increases at decreasing normative quartz contents, increasing normative olivine contents (ol) and increasing $X_{Mg}=Mg/(Mg+Fe^{2+}+Fe^{3+})$. The experimentally investigated samples exhibit approximately the following chemical changes: From rock group G to rock group B the normative quartz content decreases and may disappear, and X_{Mg} increases to 0.5. Within B ol increases from none to 20% and X_{Mg} increases to 0.6. The adjoining fields F_a and F_b incorporate also olivine rich basalts with ol up to 30% and X_{Mg} up to 0.7. Field PB represents picritic basalts with ol up to 50% and X_{Mg} up to 0.8. In the adjoining field, U, still higher values are reached. Fig. 3 clearly demonstrates the dependence between Mg enrichment and liquidus temperature at 1 bar for mafic rocks.

Für den Liquidus granitoider Gesteine ist nur ein einziges charakteristisches Beispiel (Kurve 1 in Fig. 6a, b) gegeben. Der Liquidus dieser Gesteine dürfte ein weites Feld links von B_a bedecken. Zu beachten ist allerdings, daß bei hohem Druck der trockene Liquidus von Graniten bei höheren Temperaturen liegen kann als derjenige von Andesiten oder sogar von Basalten. Abgesehen davon hängt aber die Liquidustemperatur recht deutlich vom Chemismus der Gesteine ab. So nimmt sie z.B. mit abnehmendem normativem Quarzgehalt, steigendem normativem Olivingehalt (ol) und steigendem $X_{Mg}=Mg/(Mg+Fe^{2+}+Fe^{3+})$ zu. Diese chemischen Parameter variieren in den Gesteinsproben, deren Schmelzdaten der Fig. 6 zugrunde liegen, etwa wie folgt: Von der Gesteinsgruppe G zur Gesteinsgruppe B kann der normative Quarzgehalt bis gegen Null abnehmen, wogegen X_{Mg} auf etwa 0.5 ansteigt. Innerhalb B steigt ol von Null auf etwa 20% und X_{Mg} auf 0.6. Die anschließenden Felder F_a und F_b enthalten auch Olivin-reiche Basalte mit ol bis 30% und X_{Mg} bis 0.7. Das Feld PB stellt pikritische Basalte dar mit ol bis 50% und X_{Mg} bis 0.8. Im anschließenden Feld U werden dann noch höhere Werte für ol und X_{Mg} erreicht. Aus der Fig. 3 ist die Abhängigkeit zwischen Mg/Fe-Verhältnis und Liquidustemperatur in mafischen Gesteinen bei Atmosphärendruck klar ersichtlich.

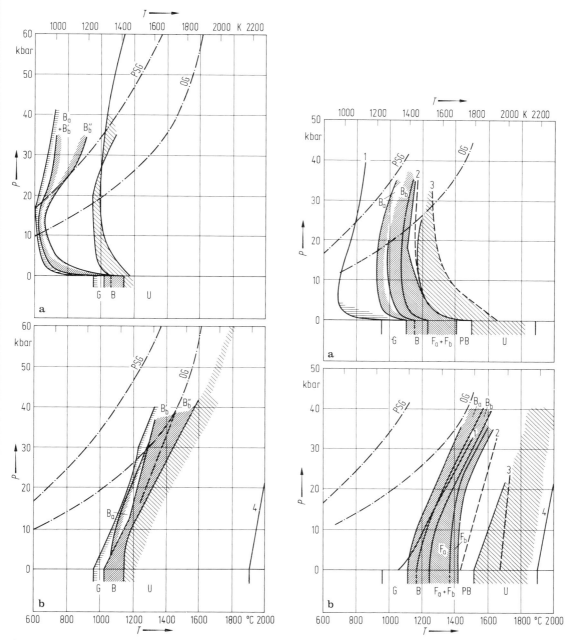

Fig. 5. Wet (a) and dry (b) solidus ranges for rock groups G, B (B_a, B'_b, B''_b) and U (for explanation, see text).
PSG = precambrian shield geotherm [Rin64]
OG = oceanic geotherm [Rin64]
4 = curve for congruent melting of forsterite Mg_2SiO_4 [Dav64].

Fig. 6. Wet (a) and dry (b) liquidus ranges for rock groups G, B (B_a, B_b), $F_a + F_b$, PB and U (for explanation, see text). Liquidus curves for individual rock specimens and curve for congruent melting of forsterite:
1 granite [Ste75]
2 olivine leucitite [Edg76]
3 garnet websterite [Mys76]
4 forsterite Mg_2SiO_4 [Dav64]
PSG = precambrian shield geotherm [Rin64]
OG = oceanic geotherm [Rin64].

4.3.3 References for 4.3.1 and 4.3.2 — Literatur zu 4.3.1 und 4.3.2

All73	Allen, J.C., Boettcher, A.L.: EOS Am. Geophys. Union Trans. **54** (1973) 481.
All75	Allen, J.C., Boettcher, A.L., Marland, G.: Am. Mineral. **60** (1975) 1069.
Arn76	Arndt, N.T.: Carnegie Inst. Washington Yearb. **75** (1976) 555.
Bar78	Barton, M., Hamilton, D.L.: Contr. Miner. Petrol. **66** (1978) 41.
Bic77	Bickle, M.J., Ford, C.E., Nisbet, E.G.: Earth Planet. Sci. Lett. **37** (1977) 97.
Boe73	Boettcher, A.L.: Tectonophysics **17** (1973) 223.
Boy64	Boyd, F.R., England, J.L., Davis, B.T.C.: J. Geophys. Res. **69** (1964) 2101.
Bre75	Brey, G., Green, D.H.: Contr. Miner. Petrol. **49** (1975) 93.
Bre76	Brey, G., Green, D.H.: Contr. Miner. Petrol. **55** (1976) 217.
Bul71	Bultitude, R.J., Green, D.H.: J. Petrol. **12** (1971) 121.
Bur62	Burnham, C.W., Jahns, R.H.: Am. J. Sci. **260** (1962) 721.
Coh67	Cohen, L.H., Ito, K., Kennedy, G.C.: Am. J. Sci. **5** (1967) 475.
Dav64	Davis, B.T.C., England, J.L.: J. Geophys. Res. **69** (1964) 1113.
Edg76	Edgar, A.D., Green, D.H., Hibberson, W.O.: J. Petrol. **17** (1976) 339.
Egg72	Eggler, D.H.: Contr. Miner. Petrol. **34** (1972) 261.
Egg73	Eggler, D.H., Burnham, C.W.: Geol. Soc. Am. Bull. **84** (1973) 2517.
Egg73a	Eggler, D.H.: Carnegie Inst. Washington Yearb. **72** (1973) 457.
Egg76	Eggler, D.H., Mysen, B.O.: Contr. Miner. Petrol. **55** (1976) 231.
Egg78	Eggler, D.H.: Am. J. Sci. **278** (1978) 305.
Gre67	Green, D.H., Ringwood, A.E.: Contr. Miner. Petrol. **15** (1967) 103.
Gre67a	Green, D.H., Ringwood, A.E.: Geochim. Cosmochim. Acta **31** (1967) 767.
Gre67b	Green, T.H.: Contr. Miner. Petrol. **16** (1967) 84.
Gre67c	Green, D.H., Ringwood, A.E.: Earth Planet. Sci. Lett. **3** (1967) 151.
Gre68	Green, T.H., Ringwood, A.E.: Contr. Miner. Petrol. **18** (1968) 105.
Gre72	Green, D.H., Ringwood, A.E.: J. Geol. **80** (1972) 277.
Gre72a	Green, T.H.: Contr. Miner. Petrol. **34** (1972) 150.
Gre73	Green, D.H.: Earth Planet. Sci. Lett. **17** (1973) 456.
Gre73a	Green, D.H.: Earth Planet. Sci. Lett. **19** (1973) 37.
Gre73b	Green, D.H.: Tectonophysics **17** (1973) 285.
Gre75	Green, D.H., Nicholls, I.A., Viljoen, M., Viljoen, R.: Geology **3** (1975) 11.
Gre76	Green, D.H.: Earth Sci. Reviews **12** (1976) 99.
Gre76a	Green, D.H.: Can. Mineral. **14** (1976) 255.
Ham64	Hamilton, D.L., Burnham, C.W., Osborn, E.F.: J. Petrol. **5** (1964) 21.
Hay71	Haygood, C., Allen, J.C., Boettcher, A.L.: Geol. Soc. Am. Ann. Meeting Abstr. **3** (1971) 594.
How75	Howells, S., Begg, C., O'Hara, M.J., in: Ahrens, L.H.: Phys. Chem. Earth, Oxford: Pergamon **9** (1975) 895.
Hua75	Huang, W.-L., Wyllie, P.J.: J. Geol. **83** (1975) 737.
Ito67	Ito, K., Kennedy, G.C.: Am. J. Sci. **265** (1967) 519.
Ito68	Ito, K., Kennedy, G.C.: Contr. Miner. Petrol. **19** (1968) 177.
Ito71	Ito, K., Kennedy, G.C., in: Heacock, J.G.: The structure and physical properties of the earth's crust, Washington D.C., Am. Geophys. Union **1971**.
Ito74	Ito, K., Kennedy, G.C.: J. Geol. **82** (1974) 383.
Kes76	Kesson, S.E., Lindsley, D.H.: Rev. Geophys. and Space Phys. **14** (1976) 361.
Khi73	Khitarov, N.I., Kadik, A.A.: Contr. Miner. Petrol. **41** (1973) 205.
Kus68	Kushiro, I., Syono, Y., Akimoto, S.: J. Geophys. Res. **73** (1968) 6023.
Lam72	Lambert, I.B., Wyllie, P.J.: J. Geol. **80** (1972) 693.
Lam74	Lambert, I.B., Wyllie, P.J.: J. Geol. **82** (1974) 88.
LeB77	Le Bas, M.J.: Carbonatite-nephelinite volcanism; an African case history, London: Wiley **1977**.
Lut64	Luth, W.C., Jahns, R.H., Tuttle, O.F.: J. Geophys. Res. **69** (1964) 759.
Mac64	MacGregor, I.D.: Carnegie Inst. Washington Yearb. **63** (1964) 156.
Mer75	Merrill, R.B., Wyllie, P.J.: Bull. Geol. Soc. Am. **86** (1975) 555.
Mil74	Milhollen, G.L., Wyllie, P.J.: J. Geol. **82** (1974) 589.
Mys75	Mysen, B.O., Arculus, R.J., Eggler, D.H.: Contr. Miner. Petrol. **53** (1975) 227.
Mys75a	Mysen, B.O., Boettcher, A.L.: J. Petrol. **16** (1975) 520.
Mys76	Mysen, B.O., Boettcher, A.L.: J. Petrol. **17** (1976) 1.
Mys76a	Mysen, B.O., Eggler, D.H., Seitz, M.G., Holloway, J.R.: Am. J. Sci. **276** (1976) 455.

Nis70	Nishikawa, M., Kono, S., Aramaki, S.: Phys. Earth Planet. Interiors **4** (1970) 138.
Piw73	Piwinskii, A.J.: Tschermaks Miner. Petr. Mitt. **20** (1973) 107.
Rea64	Reay, A., Harris, P.G.: Bull. Volcanol. **27** (1964) 115.
Rin64	Ringwood, A.E., MacGregor, I.D., Boyd, F.R.: Carnegie Inst. Washington Yearb. **63** (1964) 147.
Ste75	Stern, Ch.R., Huang, W.L., Wyllie, P.J.: Earth Planet. Sci. Lett. **28** (1975) 189.
Str74	Streckeisen, A.: Geol. Rundschau **63** (1974) 773.
Str78	Streckeisen, A.: N. Jb. Miner. Abh. **134** (1978) 1.
Str79	Streckeisen, A., Le Maître, R.W.: N. Jb. Miner. Abh. **136** (1979) 169.
Til64	Tilley, C.E., Yoder, H.S., Jr., Schairer, J.F.: Carnegie Inst. Washington Yearb. **63** (1964) 92.
Win76	Winkler, H.G.F.: Petrogenesis of Metamorphic Rocks, Heidelberg: Springer **1976**.
Wyl77	Wyllie, P.J.: Tectonophysics **43** (1977) 41.
Wyl77a	Wyllie, P.J.: Nature **266** (1977) 45.
Wyl78	Wyllie, P.J.: J. Geol. **86** (1978) 687.
Wyl79	Wyllie, P.J.: Am. Mineral. **64** (1979) 469.
Yod62	Yoder, H.S., Jr., Tilley, C.E.: J. Petrol. **3** (1962) 342.
Yod76	Yoder, H.S., Jr.: Generation of Basaltic Magma, Washington D.C., Natl., Acad. Sci. **1976**.

4.4 Radioactive heat generation in rocks — Radioaktive Wärmeproduktion in Gesteinen

4.4.1 Introduction — Einleitung

4.4.1.1 General remarks — Allgemeine Bemerkungen

Radioactive heat generation is a scalar petrophysical property independent of *in situ* temperature and pressure. It is usually expressed in terms of heat generated per unit volume and time (e.g. in $\mu W\,m^{-3}$). The heat generated by the decay of naturally radioactive elements in the earth's crust contributes a substantial portion to terrestrial heat flow. On the average, heat flow at the earth's surface amounts to about $65\,mW\,m^{-2}$ and the heat flow from the mantle – in continental areas – is around $20\,mW\,m^{-2}$; the difference is due to radioactive heat generation in crustal rocks.

Die radioaktive Wärmeproduktion ist eine skalare gesteinsphysikalische Größe, welche von *in situ*-Druck und -Temperatur unabhängig ist. Die Wärmeproduktion wird meist in Einheiten der pro Zeit- und Volumeneinheit erzeugten Wärme angegeben (z.B. in $\mu W\,m^{-3}$). Wesentliche Anteile des terrestrischen Wärmeflusses stammen von der Zerfallswärme der natürlichen Radioisotope. Der Wärmefluß an der Erdoberfläche beträgt im Durchschnitt etwa $65\,mW\,m^{-2}$, der Wärmefluß aus dem Mantel – in kontinentalen Gebieten – rund $20\,mW\,m^{-2}$; die Differenz rührt von der radioaktiven Wärmeproduktion der Krustengesteine her.

4.4.1.2 Heat generation by radioactive decay — Wärmeproduktion durch natürliche Radioaktivität

Radioactive decay converts mass into energy. The energy is released at first as the kinetic energy of the particles and nuclei involved in the decay process (emitted α- and β-particles, recoil nuclei), further as the energy of γ- and/or X-rays and of the neutrino. It has been demonstrated by [Hur53] that all energy, except the amount carried away by the neutrino, is converted to heat in the immediate vicinity of the decaying nucleus.

All naturally radioactive isotopes generate heat to a certain degree. It can be shown, however, that the only significant contributions arise from the decay series of ^{238}U, ^{235}U and ^{232}Th, and from the isotope ^{40}K.

Beim radioaktiven Zerfall wird Masse in Energie umgewandelt. Die freigesetzte Zerfallsenergie erscheint zunächst als die kinetische Energie der beteiligten Teilchen und Kerne (α- und β-Teilchen, Rückstoßkerne), ferner als die Strahlungsenergie der begleitenden γ- und/oder Röntgen-Quanten, sowie des Neutrinos. Es steht fest, daß mit Ausnahme des Neutrino-Anteils die gesamte Energie in Wärme umgewandelt wird [Hur53].

Als mögliche Wärmeproduzenten kommen zunächst alle natürlichen radioaktiven Isotope in Frage. Es läßt sich jedoch zeigen, daß nur die Beiträge der ^{238}U-, ^{235}U- und ^{232}Th-Zerfallsreihe sowie des Isotops ^{40}K signifikant sind.

The radioactive heat generation of a given rock A (in $\mu W\,m^{-3}$) can be calculated by taking into account the heat generation constants (amount of heat released per gram U, Th and K per unit time) and from the uranium, thorium and potassium concentrations c_U, c_{Th} and c_K present in the rock:

$$A = 10^{-5}\varrho(9.52\,c_U + 2.56\,c_{Th} + 3.48\,c_K)$$

where ϱ is the density of the rock (in $kg\,m^{-3}$); c_U and c_{Th} are in weight ppm, c_K in weight %. The revised constants are from [Ryb76].

Die radioaktive Wärmeproduktion A (in $\mu W\,m^{-3}$) eines gegebenen Gesteins läßt sich aus den Uran-, Thorium- und Kalium-Gehalten (c_U, c_{Th}, c_K) berechnen, unter Berücksichtigung der Wärmetönungen für U, Th und K (die pro Gramm radioaktive Substanz pro Zeiteinheit freigesetzte Wärme):

wobei ϱ die Gesteinsdichte (in $kg\,m^{-3}$) bedeutet; c_U und c_{Th} sind in Gewichts-ppm, c_K in Gewichts-%. Die revidierten Zahlenwerte sind in [Ryb76] zu finden.

4.4.1.3 Geochemical control of heat generation — Geochemische Gesetzmäßigkeiten

It is evident from Eq. (1) that heat generation in a given rock is governed by the amounts of uranium, thorium and potassium present. These amounts vary greatly with rock type, but exhibit certain regularities due to the similar geochemical behaviour of U, Th and K in the processes which determine the distribution of the naturally radioactive elements in rocks (magmatic differentiation, sedimentation, metamorphism).

Wie im vorhergehenden Abschnitt dargelegt, wird die Wärmeproduktion der Gesteine durch die Gehalte der natürlichen radioaktiven Elemente U, Th, K bestimmt. Die Gehalte variieren von Gesteinstyp zu Gesteinstyp z.T. sehr stark, unterliegen aber bestimmten Gesetzmäßigkeiten. Dies hängt mit dem geochemischen Verhalten des U und Th bzw. des K während der Vorgänge, welche die Verteilung der radioaktiven Elemente bestimmen (magnetische Differentiation, Sedimentation, Metamorphose), zusammen.

4.4.1.3.1 Igneous rocks — Magmatische Gesteine

During magmatic differentiation U and Th as well as K to some extent have a tendency towards enrichment in progressively more acidic phases. Geochemically similar behaviour of U and Th is due to their close correspondence in ion radius, high valence, electronegativity, and coordination number with respect to oxygen.

Whereas the abundances vary over several orders of magnitude according to the degree of differentiation in igneous rocks (see Table 1), the mean Th/U and K/U ratios remain fairly constant (about 4 and $1\cdot10^4$, respectively). Consequently, because of the different heat generation constants of U, Th and K (cf. numerical factors in Eq. (1)), U and Th contribute in most igneous rocks a comparable amount, whereas K contributes always a substantially smaller amount to the total heat generation, in proportions of about 40%:45%:15%.

In rocks with comparable bulk chemistry, heat generation is higher in the "wet" facies (containing biotite and hornblende) than in the "dry" phase (containing mainly pyroxene) as was clearly demonstrated by [Smi71] and [Smi73].

Während der magnetischen Differentiation haben U, Th und z.T. auch K die Tendenz, sich in den zunehmend sauren Phasen anzureichern. Das geochemisch ähnliche Verhalten von U und Th hat seinen Grund darin, daß die beiden Elemente bezüglich Valenz, Sauerstoff-Koordinationszahl und Ionenradius weitgehend übereinstimmen.

Während die Gehalte je nach Maß der Differentiation über mehrere Größenordnungen variieren (Tabelle 1), sind die mittleren Th/U und K/U-Verhältnisse erstaunlich konstant (um 4 bzw. $1\cdot10^4$). Dies hat zur Folge, daß wegen der verschiedenen Wärmeproduktionskonstanten (numerische Faktoren in Gl. (1)) U und Th meist den ungefähr gleichen, K jedoch immer einen wesentlich kleineren Beitrag liefert, etwa im Verhältnis 40%:45%:15%.

Von Gesteinen mit vergleichbarem Pauschalchemismus hat die wasserhaltige (Biotit und Hornblende enthaltende) Varietät stets die höhere Wärmeproduktion, als die „dehydrierte" (hauptsächlich Pyroxen enthaltende) Phase [Smi71], [Smi73].

4.4.1.3.2 Sedimentary rocks — Sedimentgesteine

U, Th, and K abundances as well as heat generation values for sedimentary rocks are given in Table 2. Uranium and thorium do not exhibit similar behaviour as they do in igneous rocks, mainly because U is more readily oxidized ($U^{4+} \to U^{6+}$) in aqueous solutions.

Wärmeproduktionswerte für Sedimentgesteine sind in Tabelle 2 gegeben. Der gleichartige Verlauf von U und Th ist nicht mehr so ausgeprägt wie bei der magmatischen Differentiation, dies vor allem wegen der höheren Oxidierbarkeit des U (von U^{4+} zu U^{6+}) in

Consequently, Th/U ratios show considerable variation which reflects pH $-E_h$ conditions during sedimentation: Therefore, conclusions can be drawn concerning the conditions for sedimentation: high ratios (Th/U > 6) in continental formations deposited in an oxidizing medium but low ratios (Th/U < 2) in sediments of a reducing marine environment.

wäßrigen Lösungen. Die Th/U-Verhältnisse sind entsprechend variabel und spiegeln die pH $-E_h$-Verhältnisse während der Sedimentation wider. Deshalb kann man aufgrund der Th/U-Verhältnisse auf die Sedimentationsbedingungen schließen: Th/U > 6 bei kontinentalem, oxydierenden Milieu; Th/U < 2 bei marinen, reduzierenden Bedingungen.

4.4.1.3.3 Metamorphic rocks — Metamorphe Gesteine

Metamorphic rocks are formed primarily from igneous and/or sedimentary materials. Their U, Th and K content is absorbed and redistributed according to the degree of the metamorphic transformation. As a general rule, a decrease of the content of the elements is observed with increasing degree of metamorphosis (Table 3).

The depletion of U and Th, caused by progressive metamorphism, is most markedly evident in rocks of the granulte facies. U^{4+} and Th^{4+} ions, with large ion radii can only exist in 8-fold coordination with oxygen and, thus, can not be accomodated in the closely packed structures newly formed under the higher *PT* conditions. Therefore, U and Th have a tendency towards upward migration in the earth's crust because of reactions due to dehydration (middle level of crust) or because of partial melting near the base of the crust (migmatites). Since metamorphic mineral transformations (e.g. biotite → orthoclase + hyperstene + water) cause loss of volatiles, upward migration of U and Th is accompanied by an H_2O and/or CO_2 phase. Potassium seems to be more or less unaffected by these processes.

Metamorphe Gesteine entstehen primär aus magmatischen und/oder sedimentären Ausgangsmaterialien. Ihre U-, Th- und K-Gehalte werden übernommen und entsprechend dem Metamorphosegrad neu verteilt. Allgemein läßt sich eine Abnahme der Gehalte mit zunehmendem Metamorphosegrad feststellen (Tabelle 3).

Diese Abnahme der U- und Th-Gehalte, von der progressiven Metamorphose herrührend, ist am stärksten ausgeprägt in Gesteinen der Granulit-Facies. Die U^{4+} und Th^{4+}-Ionen mit großem Ionenradius können mit Sauerstoff nur in Achter-Koordination vorkommen und können deshalb in den bei höheren *P*/*T*-Bedingungen neugebildeten, dichter gepackten Mineralphasen nicht eingebaut werden. U und Th haben deshalb die Tendenz, in der Erdkruste nach oben zu migrieren, ausgelöst durch Dehydratationsreaktionen (in mittleren Krustenniveaus) bzw. durch partielle Aufschmelzung (Migmatitbildung; nahe der Basis der Kruste). Da metamorphe Mineralreaktionen zur Bildung von leichtflüchtigen Bestandteilen führen (z.B. Biotit → Kalifeldspat + Hypersthen + Wasser), erfolgt die Aufwärtsmigration von U und Th in Begleitung einer H_2O und/oder CO_2-Phase. K scheint von diesem Vorgang nicht erfaßt zu werden.

4.4.1.4 Arrangement of data for heat generation — Anordnung der Wärmeproduktionsdaten

The data for heat generation in Tables 1⋯3 were recalculated from the references cited, by applying Eq. (1). If no densities were reported, mean densities as given in [Ryb76] were used.

Figs. 1⋯3 display the range of variation of heat generation in igneous, sedimentary and metamorphic rock types, along with the mean values (dots), in case of different data for the same rock type the weighted average was calculated, by taking into account the corresponding number (n) of samples.

Die Wärmeproduktionswerte in Tabelle 1⋯3 wurden aufgrund der Literaturangaben über die U-, Th- und K-Gehalte mit Hilfe von Gl. (1) neu berechnet. Bei fehlenden Dichteangaben wurden mittlere Dichtewerte aus [Ryb76] eingesetzt.

Die Variationsbreiten der Wärmeproduktion in den verschiedenen magmatischen, sedimentären und metamorphen Gesteinen sind in Fig. 1⋯3 dargestellt, samt den Mittelwerten (Punkte; bei mehreren Literaturangaben für den gleichen Gesteinstyp wurden nach der Proben-Anzahl (n) gewichtete Mittelwerte berechnet).

Acknowledgements

Part of the data material reported here was acquired during a research project supported by the Swiss National Science Foundation (grant no. 2.108-0.78). Special thanks are due to Mr. G. Guzzi (Zurich) for his help with the compilation.

Verdankung

Die Kompilation enthält viele Originaldaten, welche im Rahmen eines vom Schweizerischen Nationalfonds (Projekt-Nr. 2.108-0.78) unterstützten Forschungsprojekts gewonnen wurden. Herrn G. Guzzi (Zürich) gebührt besonderer Dank für seine große Hilfe.

4.4.2 Data — Daten

4.4.2.1 Igneous rocks — Magmatische Gesteine

Table 1. Content of U, Th, and K (c_U, c_{Th}, c_K, respectively) and radioactive heat generation A of igneous rocks (n = number of samples).

Rock	Locality	Ref.	n	c_U wt ppm range	mean	c_{Th} wt ppm range	mean	c_K wt % range	mean	A µW m^{-3} range	mean
Plutonic rocks											
Granite	N.E. Minnesota, USA	Rye78	38	0.39···6.86	2.54	1.08···44.72	17.94	2.88···5.42	4.48	0.5···4.5	2.28
	Norway	Swa74	104		4.58		26.98		4.04		3.49
	Indian shield	Rao76	12							0.59···1.81	1.15
	Ivrea Zone, Italy	Höh75	22	2.0···7.4	4.32	1.6···28.9	17.75		4.20	0.87···3.56	2.72
	Sierra Nevada, California	Wol68	3		5.3		31.4		3.79		3.9
	Sierra Nevada, California	Wol68	5		2.8		15.8		2.37		2.1
	Sierra Nevada, California	Wol68	6		6.3		15.3		3.61		3.1
	Sierra Nevada, California	Wol68	17		3.8		15.2		3.54		2.4
	South Norway	Kil75a	52		5.7		45.7		4.70		5.2
	South Norway	Kil75a	134		9.9		50.2		4.52		6.6
	South Norway	Kil75a	5		5.5		63.5		5.14		6.4
	South Norway	Kil75a	109		4.7		27.1		4.26		3.54
	South Norway	Kil75a	22		4.6		21.1		3.86		3.08
	South West England	Tam74	10		15.9		27.1		4.6		6.53
	South West England	Tam74	10		13.8		16.9		5.0		5.28
	N.E. Sudan, Africa	Eva74	8		1.0		3.4		1.9		0.67
	New Hampshire, USA	Sto65	145		7.6		23.4		4.0		3.9
	Mont Blanc, France	Ryb66	8	3.71···25.86	12.37	19.0···58.0	36.2	3.34···4.15	3.79	2.55···10.14	5.96
	Kamchatka, USSR	Puz77	90		2.28		6.52		2.19		3.03
	Czechoslovakia	Cer77	524							0.8···3.5	1.95
	Swiss Alps	Ryb73	8	1.0···12.0	6.3	1.9···38.8	17.98	0.40···4.90	3.40	0.06···6.05	3.1
	Swiss Alps	Ryb81	9	3.84···8.92	5.69	10.33···30.29	19.36	2.78···4.29	3.73	2.12···4.64	3.11
	Schwarzwald, Germany	Ryb81	7	1.71···7.85	4.67	0.73···26.81	15.08	3.47···4.97	4.03	0.82···3.94	2.59
	New South Wales, Australia	Bun75	7	2.12···5.84	4.25	13.24···31.15	20.09	2.82···4.14	3.46	1.83···3.99	2.77

(continued)

4.4 Radioactive heat generation in rocks

Rock	Locality	Ref.	n	c_U wt ppm range	c_U wt ppm mean	c_{Th} wt ppm range	c_{Th} wt ppm mean	c_K wt % range	c_K wt % mean	A µW m^{-3} range	A µW m^{-3} mean
Granite (continued)	Snowy Mountains, Australia	Bun75	86	1.79···12.82	5.18	5.91···25.85	15.17	0.47···3.93	2.77	1.38···5.07	2.61
Granite (from drillholes)	Western Australia	Bun75	13	2.18···2.81	2.43	5.99···7.90	6.50	1.70···2.20	1.97	1.14···1.26	1.45
(from drillholes)	Western Australia	Bun75	6	1.22···3.31	2.33	14.21···14.98	14.74	2.55···2.97	2.78	1.59···2.11	1.86
(from drillholes)	Western Australia	Bun75	4	1.22···1.27	1.24	6.95···8.45	7.89	2.39···3.10	2.80	1.01···1.16	1.11
Granite	South Australia	Bun75	4	15.23···15.45	15.35	47.93···49.73	48.86	4.33···4.64	4.50	7.56···7.73	7.65
(from drillholes)	South Australia	Bun75	20	2.73···4.76	3.61	15.75···24.69	19.65	4.28···4.55	4.45	2.31···3.11	2.67
(from drillholes)	South Australia	Bun75	17	3.05···6.76	5.37	20.04···62.19	45···53	4.17···5.13	4.75	2.62···6.26	4.91
(from drillholes)	Eyre Peninsula, South Australia	Bun75	35	1.12···19.50	6.32	4.14···91.05	41.83	1.87···5.44	4.30	0.60···11.38	4.86
	N. Territory, Australia	Bun75	7	5.04···8.35	6.12	24.08···27.80	26.48	3.07···4.65	4.25	3.39···4.22	3.76
	N. Territory, Australia	Hei66	26	2.8···22.1	11.4	19.0···73.0	47.0	2.8···6.6	4.6	2.7···12.9	6.5
	N. Territory, Australia	Hei66	14	6.2···13.0	10.5	14.0···49.0	31.0	2.8···4.7	4.1	1.8···12.2	5.2
Young Granite	Swiss Alps	Kis78	161	2.3···15.9	8.2	6.8···43.8	26.4	0.5···5.2	4.0	1.5···7.7	4.45
Rotondo-Granite	Idaho, USA	Swa72	6		5.8		29.4		3.7		4.0
Medium-grained Granite	South Africa	Kol66	9	4.3···20.0	12.0	17.0···62.0	36.9	4.0···5.0	4.36		6.0
Fine-Grained Granite	South Africa	Kol66	8	5.9···19.1	11.6	4.8···35.0	23.5	3.76···4.76	4.17		4.9
Porphyritic Biotite-Granite	South Africa	Kol66	17	2.3···20.0	6.5	14.0···59.0	21.6	3.1···4.7	4.15		3.5
Biotite-Granite	Snowy Mountains, Australia	Bun75	29	1.80···5.39	3.99	5.91···18.63	14.87	0.81···3.77	2.58	1.04···2.68	2.27
Mica-poor Granite	Snowy Mountains, Australia	Bun75	27	2.61···12.37	6.87	4.61···12.79	8.73	0.23···3.89	2.87	1.16···3.92	2.61
Muscovite Granite	N.E. Minnesota, USA	Rye78	4	0.87···2.99	1.89	1.35···10.19	6.28	3.66···4.51	4.06	0.65···1.86	1.29
Biotite/Muscovite Granite	N.E. Minnesota, USA	Rye78	2	1.43···8.10	4.77	3.49···11.44	7.47	3.72···3.89	3.81	0.95···3.20	2.07
Hornblende Granite	Brit. Columbia, Canada	Lew76	3		8.33		31.3		4.91		4.8
Alkaligranite	Swiss Alps	Ryb81	1		3.84		20.99		4.29		2.81
Granodiorite (felsic)	Southern California	Til69	15	1.0···3.3	1.9	3.1···11.3	7.9	1.9···3.8	2.4	0.8···2.0	1.3
(felsic)	Montana, USA	Til69	7	0.8···2.7	1.5	5.4···9.3	7.1	2.1···3.6	2.5	0.9···1.6	1.1
(mafic)	Montana, USA	Til69	4	1.7···3.9	2.8	8.8···10.8	9.5	2.0···3.4	2.7	1.3···1.8	1.6
(continued)	Minnesota, USA	Rye78	2	1.30···1.85	1.58	7.61···10.26	8.94	1.64···2.84	2.24	1.00···1.43	1.22

(continued)

Table 1 (continued)

Rock	Locality	Ref.	n	c_U wt ppm range	mean	c_{Th} wt ppm range	mean	c_K wt % range	mean	A µW m⁻³ range	mean
Granodiorite (mafic) (continued)	S. Nevada, California	Wol68	2		5.8		12.8		2.17		2.6
	S. Nevada, California	Wol68	8		2.8		11.2		2.62		1.8
	S. Nevada, California	Wol68	14		4.9		21.5		4.02		3.2
	S. Nevada, California	Wol68	42		3.4		11.2		1.86		1.8
	S. Nevada, California	Wol68	16		6.3		21.1		2.53		3.4
	S. Nevada, California	Wol68	44		6.8		21.5		2.78		3.5
	Coast Range, Western Canada	Lew76	19		1.9		4.1		2.0		1.0
	S. Nevada, California	Wol64	9	1.72···6.45	3.95	4.9···19.9	12.8	1.40···2.64	2.05		2.1
	S. Nevada, California	Wol64	7	4.71···9.48	7.61	16.6···29.6	23.3	2.46···3.05	2.83		3.9
	Bergell, Italy	Reu80	85	2.6···14.1	5.95	9.5···45.4	21.2	2.45···4.81	3.57	1.57···6.23	3.29
	Bergell, Italy	Reu80	10	3.4···16.0	10.26	10.4···28.0	20.18	3.30···5.57	4.10	2.3···7.4	4.4
	Bergell, Italy (Comoer Molasse)	Reu80	12	4.6···17.3	10.86	16.3···31.9	22.48	3.39···5.81	3.83	3.43···6.51	4.65
	Kamchatka, USSR	Puz77	12		1.4		3.6		2.0		1.94
	Czechoslovakia	Cer77	13							1.1···3.6	2.49
	Western Australia	Sas76										1.92
	Swiss Alps/Schwarzwald, Germany	Ryb81	12	1.18···14.58	5.77	3.16···31.54	15.88	0.47···4.50	2.83	0.57···5.88	2.84
	New South Wales, Australia	Bun75	12	2.18···3.44	2.58	5.35···17.21	8.49	1.66···1.86	1.77	1.11···2.03	1.40
(from drillholes)	South Australia	Bun75	12	7.31···10.01	8.23	15.31···23.96	20.56	1.46···2.87	2.36	3.02···4.30	3.71
Hornblende-Biotite Granodiorite	N.E. Minnesota, USA	Rye78	7	0.37···1.80	0.94	2.45···5.84	3.86	2.01···2.65	2.32	0.46···0.95	0.72
Hornblende-Granodiorite	New South Wales, Australia	Bun75	1		7.83		19.03		2.98		3.56
Ferro-Granodiorite	N.E. Minnesota, USA	Hei63	3	0.94···2.00	1.48	2.6···6.9	4.93	0.79···1.50	1.22	0.50···1.13	0.84
Tonalite	Bergell, Italy	Höh75	5	2.7···4.4	3.30	11.9···22.6	17.1	3.0···4.0	3.3	1.87···3.21	2.25
	Bergell, Italy	Reu80	8	1.2···7.9	3.74	4.3···22.8	14.3	1.55···2.96	2.39	0.74···3.65	2.15
	Bergell, Italy	Ryb81	8	0.22···5.55	2.48	3.15···12.41	7.88	1.48···3.65	2.04	0.85···2.36	1.41
Quartz-Tonalite	N.E. Minnesota, USA	Rye78	7	0.17···0.72	0.47	0.46···1.58	0.95	0.76···1.42	1.11	0.17···0.41	0.29
Quartz-Monzonite	Swiss Alps	Ryb81	1		6.54		27.24		3.36		3.75
Quartz-Diorite and Diorite	S.Nevada, California	Wol68	9		0.7		1.8		0.8		0.4

Rock	Locality	Ref.	n	c_U wt ppm range	c_U wt ppm mean	c_{Th} wt ppm range	c_{Th} wt ppm mean	c_K wt % range	c_K wt % mean	A µW m^{-3} range	A µW m^{-3} mean
Hornblende-Quartz-Diorite	British Columbia, Western Canada	Lew76	3		2.7		8.3		2.0		1.5
Quartz-Diorite and Diorite	Coast Range, Western Canada	Lew76	20		0.9		2.0		1.0		0.5
Pyrite-Quartz-Diorite	S. Nevada, California	Wol64	2		0.65		1.10		0.45		0.3
Quartz-Diorite and Tonalite	Idaho	Swa72	6		0.4		2.6		1.3		0.4
Syenite	Norway	Smi74	70		2.72		15.32		4.50		2.23
	Czechoslovakia	Cer77	49							1.7···7.8	5.91
Hornblende-Syenite	Wyoming, USA	Smi73	20		3.2		24.8		5.0		3.0
Pyrite-Syenite	Wyoming, USA	Smi73	23		1.4		3.9		5.1		1.1
Monzonite	Norway	Orm78	101		0.4		1.1		5.1		0.6
	Norway	Orm78	29		1.1		2.6		4.8		0.8
	Norway, Oslo region	Orm78	266		4.2		15.1		4.2		2.5
	Swiss-Alps	Fah67	5		5.3		17.2		3.56		2.9
Hornblende-Monzonite	Montana, USA	Til69	1		0.8		4.1		2.2		0.69
Diorite	Kamchatka, USSR	Puz77	32		1.26		2.86		1.50		1.61
	Czechoslovakia	Cer77	36							0.1···2.4	1.0
	Northern Territory, Australia	Hei66	2	2.8···4.5	3.7	8.5···22.0	15.3	2.1···2.2	2.2	1.6···3.1	2.4
Diorite and Quartz-Diorite	Swiss Alps	Föh67	2		0.35		0.4		0.3		0.2
	S. Nevada, California	Wol68	9		0.7		1.8		0.8		0.4
Hornblende-Diorite	Swiss Alps/Schwarzwald, Germany	Ryb81	7	0.87···3.13	1.87	1.97···7.13	3.31	0.73···2.09	1.35	0.46···1.35	0.9
Hornblende-Pyrite-Diorite	N.E. Minnesota, USA	Rye78	2	0.37···0.56	0.47	1.02···5.99	3.51	1.72···2.66	2.19	0.41···0.71	0.56
Gabbro	N.E. Minnesota, USA	Rye78	1		0.40		1.18		0.48		0.23
	Ivrea Zone, Italy	Höh75	4	0.11···0.17	0.14	0.09···0.37	0.285	0.13···0.35	0.23	0.07···0.10	0.09
	Kamchatka, USSR	Puz77	8		0.5		1.3		0.65		0.68
	Czechoslovakia	Cer77	19							0.04···0.5	0.36
(continued)	Baltic shield, USSR	Ars72, Ars73	3	0.2···0.5	0.32	0.2···0.6	0.39	0.2···0.71	0.38	0.1···0.3	0.17

(continued)

Rybach/Čermák

Table 1 (continued)

Rock	Locality	Ref.	n	c_U wt ppm range	mean	c_{Th} wt ppm range	mean	c_K wt % range	mean	A µW m^{-3} range	mean
Gabbro (continued)	Swiss Alps	Ryb81	5	0.03···0.57	0.27	0.0···1.16	0.49	0.04···0.85	0.21	0.014···0.33	0.14
	Minnesota, USA	Hei63	3	0.14···0.27	0.26	0.74···0.96	0.82	0.31···0.55	0.42	0.15···0.21	0.18
Hornblende-Gabbro	Southern California	Hei63	3	0.31···1.1	0.64	0.94···3.1	1.85	0.38···0.52	0.46	0.20···0.59	0.37
	Idaho, USA	Swa72	4		0.2		1.1		0.6		0.2
Alkaligabbro	Ivrea Zone, Italy	Höh75	2	0.78···0.98	0.88	0.98···1.17	1.08	4.2···4.7	4.45	0.74···0.81	0.77
	Baltic shield, USSR	Ars72, Ars73	7	0.2···1.6	0.62	0.2···1.9	0.74	0.26···0.76	0.48	0.1···0.7	0.30
Syenogabbro	Montana, USA	Til69	2	1.7···2.3	2.0	5.6···5.9	5.7	2.3···2.6	2.5	1.1···1.3	1.2
Syenodiorite	Montana, USA	Til69	1		0.4		2.7		3.1		0.57
Hornblende-Syeno-diorite	N.E. Minnesota, USA	Rye78	5	0.23···0.76	0.43	1.18···4.57	3.40	2.19···3.08	2.66	0.50···0.62	0.59
Anorthosite	Norway	Swa74	5		0.11		0.30		0.46		0.092
	Minnesota, USA	Hei63	1		2.1		7.1		0.43		1.1
Anorthositic Gabbro	Minnesota, USA	Hei63	3	1.9···2.5	2.4	6.1···8.8	7.9	1.17···1.90	1.56	1.1···1.62	1.4
Norite	South Africa	Hei63	2	0.10···0.17	0.14	0.47···0.60	0.54	0.25···0.40	0.33	0.090···0.14	0.11
Alaskite	Montana, USA	Til69	5	4.3···20.0	9.2	26.0···42.0	36.3	3.9···4.9	4.5	3.3···8.3	5.2
	S. Nevada, California	Wol68	4		4.7		17.0		3.62		2.8
Trondjemite	N.E. Minnesota, USA	Rye78	1		0.34		0.55		1.12		0.23
Sparagmite	Norway	Swa74	5		2.87		6.28		1.34		1.32
Adamellite	N.E. Minnesota, USA	Til69	5	1.58···6.89	3.16	7.22···22.31	15.08	3.75···3.96	3.85	1.28···2.30	2.19
Silexite	Swiss Alps	Ryb81	1		6.54		27.24		3.36		3.75

Ultramafic Rocks

Rock	Locality	Ref.	n	c_U wt ppm range	mean	c_{Th} wt ppm range	mean	c_K wt % range	mean	A µW m^{-3} range	mean
Peridotite	Ivrea Zone, Italy	Höh75	3	0.0···0.07	0.025	0.0···0.07	0.023	0.0···0.19	0.057	0.015···0.040	0.028
	Swiss Alps	Ryb81	6							0.003···0.027	0.016
Olivine-Peridotite	Ivrea Zone, Italy	Höh75	4							0.006···0.018	0.014
Olivine-Pyrite-Peridotite	Ivrea Zone, Italy	Höh75	2							0.032···0.092	0.062
Pyroxenite	Baltic shield, USSR	Ars72, Ars73	2	0.1···0.3	0.19	0.1···0.3	0.22		1.25	0.1···0.3	0.21
	Swiss Alps	Ryb81	2	0.2···0.22	0.21	0.53···0.54	0.54	0.03···0.06	0.05		0.11
	Japan	Hei63	1		0.70		2.5		0.76		0.50
Hornblendite	Swiss Alps	Ryb81	2	0.02···0.75	0.39	0.18···0.44	0.31	0.31···1.93	1.12	0.05···0.44	0.25
Lherzolite Nodules	Victoria, S.E. Australia	Gre68	6	0.030···0.114	0.0343	0.139···0.457	0.148	0.0012···0.011	0.0055	0.0022···0.736	0.0233

Rock	Locality	Ref.	n	c_U wt ppm range	mean	c_{Th} wt ppm range	mean	c_K wt % range	mean	A µW m^{-3} range	mean
Lherzolite (oceanic)	Hawaii/Pacific, Ocean	Wak67	10	0.76···4.02	2.09	2.37···12.4	6.07	0.0017···0.0093	0.0057	0.428···2.25	1.14
Lherzolite (continental)	Eifel, Germany	Wak67	1		3.78		3.60		0.00788		1.46
Garnet-free Lherzolite	Oga Peninsula, Japan	Wak67	2	1.28···3.79	2.54	1.07···1.68	1.38		0.00867	0.531···1.250	0.891
	Oga Peninsula, Japan	Wak67	2	4.12···4.74	4.43	9.58···11.1	10.34		0.0516	2.18···2.24	2.21
Lherzolite (continental)	New Mexico, USA	Wak67	1		4.12		13.5		0.0162		2.38
	Central Massiv, France	Wak67	1		4.8		77.0		0.0362		7.81
	Brit. Columbia, Canada	Wak67	1		0.58		0.41		0.00337		0.212
	New Zealand	Wak67	1		14.0		4.75		0.00357		4.68
Dunite	Ivrea Zone, Italy	Höh75	3		0.0032		0.0101		0.0011	0.012···0.031	0.019
		Kap74		0.0031···0.0050	0.0040	0.01···0.011	0.0103	0.088···0.094	0.091		0.0019
		Kap74			0.0160		0.0500		0.00031		0.012
		Kap74									0.009

Volcanic Rocks

Rock	Locality	Ref.	n	c_U wt ppm range	mean	c_{Th} wt ppm range	mean	c_K wt % range	mean	A µW m^{-3} range	mean
Andesite	Montana, USA	Til69	7	0.9···2.6	2.0	2.7···8.8	5.5	1.1···5.0	2.5	0.5···1.5	1.1
	Kamchatka, USSR	Puz77	32		1.34		1.9		1.61		0.62
	Czechoslovakia	Cer77	358							0.7···1.7	1.18
Basalt	Kamchatka, USSR	Puz77	95		0.49		1.28		1.62		0.40
	Czechoslovakia	Cer77	79								0.95
	Japan	Hei63	4	0.44···1.4	1.01	2.0···8.8	5.13	0.92···1.57	1.38	0.37···1.21	0.81
	Montana, USA	Til69	1		0.71		2.5		1.7		0.56
Pyrite-Basalt	Oregon, USA	Hei63	2	0.22···0.5	0.36	0.94···1.1	1.02	0.34···0.53	0.44	0.17···0.28	0.22
Nepheline-Basalt	Hawaii	Hei64	1		1.29		5.58		0.65		0.85
Olivine-Basalt	Hawaii	Hei64	1		0.32		2.46		0.95		0.37
Tholeitic Basalt	Hawaii	Hei64	6	0.10···0.24	0.18	0.53···0.87	0.69	0.12···0.36	0.26	0.095···0.17	0.13
Alkali-Basalt	Japan	Hei63	2	0.13···0.28	0.21	0.45···1.1	0.78	0.34···0.62	0.48	0.11···0.23	0.17
Alkali-Olivine-Basalt	Hawaii	Hei64	1		0.68		2.12		1.05		0.46
	Japan	Hei63	2	0.48···0.57	0.53	3.6···4.2	3.9	0.81···1.00	0.91		0.54
Rhyodacite	Swiss Alps	Ryb81	2	2.5···3.84	3.17	18.36···19.26	18.81	4.22···4.41	4.32	2.28···2.70	2.49

(continued)

Table 1 (continued)

Rock	Locality	Ref.	n	c_U wt ppm range	mean	c_{Th} wt ppm range	mean	c_K wt % range	mean	A µW m^{-3} range	mean
Rhyolite	Czechoslovakia	Cer77	34		5.6		30.9		3.4	1.9···4.0	3.58
	Montana, USA	Til69	1								3.6
Dacite	Montana, USA	Til69	4	0.8···2.4	1.8	3.1···8.9	5.7	1.7···4.0	2.7	0.7···1.3	1.1
	Swiss Alps	Rub81	3	4.78···7.74	5.78	13.99···20.58	15.73	3.22···4.34	3.6	2.38···3.21	2.88
	Kamchatka, USSR	Puz77	25		1.7		2.8		2.26		0.8
	Czechoslovakia	Cer77	19							1.3···1.8	1.52
Trachyte	Montana, USA	Til69	3	1.8···2.9	2.3	4.9···8.9	7.5	2.6···3.6	3.1	1.3···1.6	1.4
	Hawaii	Hei64	1		2.42		7.55		2.75		1.39
	Kamchatka, USSR	Puz77	8		4.0		6.3		3.53		1.7
Trachybasalt	Montana, USA	Til69	1		0.4		2.1		2.6		0.51
Latite	Swiss Alps	Ryb81	1		2.71		10.99		2.85		1.69
	Montana, USA	Til69	2	4.0···6.0	5.0	9.3···12	10.7		4.3		2.2
Quartz-Latite	Montana, USA	Til69	9	1.9···3.6	2.5	6.3···9.0	7.8	2.5···6.0	2.9	1.0···1.7	1.4
Basanite	Victoria, Australia and Tasmania	Gre68	6	1.32···2.41	1.93	4.25···8.00	6.26	1.9···3.7 1.38	1.76	0.91···1.66	1.30
Phonolite	Czechoslovakia	Cer77	15							3.7···4.0	3.80
Essexite	Japan	Hei63	1		3.1		10.4		2.41		1.93
Spilite	Swiss Alps	Ryb81	2	0.29···0.38	0.34	0.09···0.28	0.19	0.26···0.37	0.32		0.13

Anchimetamorphic volcanic rocks

Rock	Locality	Ref.	n	c_U wt ppm range	mean	c_{Th} wt ppm range	mean	c_K wt % range	mean	A µW m^{-3} range	mean
Quartz-Porphyry	Ivrea Zone, Italy	Höh75	4	3.6···10.0	5.6	15.4···17.8	16.8	0.8···4.9	3.4	2.11···4.1	2.9
	Czechoslovakia	Cer77	9							1.1···1.8	1.26
	Swiss Alps	Ryb81	1		5.00		21.5		4.68		3.03
Porphyrite	Swiss Alps	Ryb81	1		2.0		7.20		2.53		1.23
	Kamchatka, USSR	Puz77	21		1.46		2.22		1.65		0.68
	Czechoslovakia	Cer77	12							0.5···1.5	1.03
Pyrite-Porphyrite	Swiss Alps	Föh67	4		4.1		6.8		1.57		1.7
Porphyry	Upper York Peninsula, Australia	Bun75	3	15.62···43.02	25.18	8.06···133.35	91.21	0.90···5.45	3.86	11.56···13.80	12.98
Porphyroide	Czechoslovakia	Cer77	65							0.92···2.22	1.69

Hypabyssal rocks

Rock	Locality	Ref.	n	c_U wt ppm range	mean	c_{Th} wt ppm range	mean	c_K wt % range	mean	A µW m^{-3} range	mean
Granophyre	Swiss Alps	Ryb81	3	3.63···5.16	4.32	22.30···22.74	22.50	4.2···4.97	4.48	2.72···3.14	2.90
	Minnesota, USA	Hei63	8	2.3···4.6	3.26	10.0···21.3	12.73	1.76···3.52	2.44	1.58···2.78	1.82
Melaphyre	Czechoslovakia	Cer77	18							0.7···1.6	0.86

Rock	Locality	Ref.	n	c_U wt ppm range	c_U wt ppm mean	c_{Th} wt ppm range	c_{Th} wt ppm mean	c_K wt % range	c_K wt % mean	A µW m^{-3} range	A µW m^{-3} mean
Aplite	Kamchatka, USSR	Puz77	33		3.11		2.44		2.71		1.2
	Czechoslovakia	Cer77	13							1.26···2.93	1.69
	Swiss Alps	Ryb81	1		0.28		0.14		4.10		0.45
	Swiss Alps	Kis78	6	11.0···29.3	21.1	15.9···33.9	22.4	2.1···5.1	3.7	5.8···9.5	7.5
	Swiss Alps	Ryb73	3	19.0···22.0	20.3	0.5···1.4	0.9	0.16···1.93	1.18	4.9···5.9	5.3
Pegmatite	Czechoslovakia	Cer77	34							0.80···1.84	1.09
Lamprophyre	Swiss Alps	Kis78	2	0.9···1.4	1.2	4.1···6.9	5.5	4.0···4.2	4.1	1.0···1.3	1.2
Kersantite	Swiss Alps	Ryb81	4	3.53···7.97	5.22	10.88···31.34	25.79	3.72···5.86	4.56	2.57···4.39	3.42
Diabase	Hawaii	Hei64	3	0.56···0.69	0.61	2.26···2.42	2.34	0.49···0.51	0.50	0.38···0.43	0.40
	Minnesota, USA	Hei63	1		0.41		1.6		0.83		0.33
	Kamchatka, USSR	Puz77	9		0.06		0.12		0.14		0.040
	Czechoslovakia	Cer77	19							0.4···1.5	0.87
	Baltic shield, USSR	Ars72, Ars73	4	0.07···0.10	0.08	0.09···0.12	0.11	0.22···0.45	0.36	0.05···0.09	0.074
Pyrite-Diabase	Oregon, USA	Hei63	5	0.23···0.48	0.35	1.5···2.4	1.84	0.36···0.78	0.74	0.22···0.46	0.32
Gabbro-Diabase	Baltic shield, USSR	Ars72, Ars73	8	0.02···0.28	0.16	0.02···0.34	0.19	0.12···0.63	0.31	0.02···0.16	0.10
Dolerite	South Africa	Hei63	1		0.42		1.6		0.67		0.31
Tholeitic Quartz-Dolerite	Hawaii	Hei63	1		0.06		0.41		0.36		0.085

4.2.2 Sedimentary rocks — Sedimentgesteine

Table 2. Content of U, Th, and K (c_U, c_{Th}, c_K, respectively) and radioactive heat generation A of sedimentary rocks (n = number of samples).

Rock	Locality	Ref.	n	c_U wt ppm range	mean	c_{Th} wt ppm range	mean	c_K wt % range	mean	A µW m^{-3} range	mean
Deep Sea clays	North Atlantic, Caribbean Sea	Kap74	18		2.1		11.0		2.5		1.5
Beach sands		Kap74	83		2.97		6.42		0.33		1.2
Beach sands	Southern USA Galveston Island	Mur58	10	0.31···0.64	0.5	1.5···2.2	1.9	0.77···1.1	0.9	0.29···0.39	0.34
Sand	Czechoslovakia	Cer77	82							0.38···1.21	0.75
Gravel	Czechoslovakia	Cer77	10							0.59···1.09	0.95
Loam	Czechoslovakia	Cer77	35							1.26···1.72	1.47
Conglomerate	Czechoslovakia	Cer77	107							0.33···1.55	0.84
Sandstone	Czechoslovakia	Cer77	911							0.54···1.84	0.94
Sandstone	Kamchatka, USSR	Puz77	111		1.40		4.92		1.33		0.79
Sandstone	Scotland	Ric76	10	0.82···2.45	1.51	4.6···11.9	7.98	0.48···3.51	2.35	0.66···1.63	1.62
Sandstone (metamorphic)	Coast Range, California	Wol67	13	1.23···2.54	1.80	3.39···8.67	5.33	0.82···1.64	1.19	0.7···1.4	0.99
Quartz Sandstone	Czechoslovakia	Cer77	92							0.29···1.42	0.56
Arkose	Czechoslovakia	Cer77	29							0.88···2.60	1.57
Graywacke	Czechoslovakia	Cer77	225							1.05···1.80	1.47
	Wyoming, USA	Rog69	18	0.9···2.6	1.6	4.3···16.8	9.1	1.3···3.0	1.9	0.7···2.0	1.2
	Coast Range, California	Wol67	24	1.06···3.15	2.05	3.44···12.1	7.19	0.53···1.88	1.30	0.6···1.8	1.2
	Coast Range, California	Wol67	29	1.47···2.53	2.02	4.52···11.4	7.68	0.97···2.26	1.61	0.8···1.6	1.2
	Coast Range, California	Wol67	20	0.12···2.30	1.06	1.67···9.63	3.49	0.19···2.56	0.94	0.1···1.5	0.6
Tuff	Czechoslovakia	Cer77	196							0.5···4.2	1.31
Tuffite	Czechoslovakia	Cer77	20							0.63···1.13	0.73
Shale	Kamchatka, USSR	Puz77	18		2.4		7.0		2.45		1.28
Shale	Czechoslovakia	Cer77	352							0.75···2.18	1.44
Mancos Shale	Utah, Colorado, New Mexico, Arizona	Pli62	135	0.9···12.3	3.7	5.3···23.0	10.2	0.9···3.3	1.9	0.6···3.1	1.53
Shale		Kap74	75	1.0···13.0	3.7	2.0···4.7	12.0		2.7		2.0

Rock	Locality	Ref.	n	c_U wt ppm range	mean	c_{Th} wt ppm range	mean	c_K wt % range	mean	A µW m^{-3} range	mean
Clay	Czechoslovakia	Cer77	190							1.05···2.85	1.43
Argillite	Kamchatka, USSR	Puz77	76		2.15		6.54		2.27		1.17
	Czechoslovakia	Cer77	205							0.46···3.27	2.04
Argillaceous Quartzwackes	Minnesota, USA	Rog69	2	1.7···2.0	1.9	8.0···9.4	8.7	0.6···1.9	1.3	1.1···1.2	1.2
Argillaceous Sandstone	Minnesota, USA	Rog69	5	0.6···0.9	0.7	3.6···4.1	3.8	1.1···4.0	2.1	0.5···0.9	0.6
Lithic Quartzwackes	Minnesota, USA	Rog69	11	0.3···0.5	0.4	1.2···2.0	1.5	0.6···3.6	1.7	0.2···0.6	0.4
Sandy + Silty shales	Minnesota, USA	Rog69	8	1.5···2.1	1.7	7.4···9.0	8.2	1.6···2.9	2.3	1.1···1.3	1.2
Siltstone	Kamchatka, USSR	Puz77	138		1.96		5.03		0.99		0.91
Limestone	Czechoslovakia	Cer77	387							0.04···0.54	0.28
		Kap74	10	0.5···6.0	2.0	0.2···4.0	1.5				0.6
Carbonatic rocks	Swiss Alps	Ryb73	12	0.0···2.7	0.87	0.01···4.4	0.88	0.01···1.21	0.28	0.2···0.9	0.3
Dolomite	Czechoslovakia	Cer77	111							0.13···0.54	0.28
Travertine	Czechoslovakia	Cer77	5							0.08···0.25	0.10
Marlstone	Czechoslovakia	Cer77	107							0.54···1.47	0.93

4.4.2.3 Metamorphic rocks — Metamorphe Gesteine

Table 3. Content of U, Th, and K (c_U, c_{Th}, c_K, respectively) and radioactive heat generation A of metamorphic rocks (n = number of samples).

Rock	Locality	Ref.	n	c_U wt ppm range	mean	c_{Th} wt ppm range	mean	c_K wt % range	mean	A µW m^{-3} range	mean
A. Monomict metamophites											
Marble	Czechoslovakia	Cer77	103							0.1···0.5	0.34
	Swiss Alps	Ryb73	2	0.2···0.6	0.4	2.1···2.3	2.2	0.95···1.16	1.06	0.3···0.4	0.35
Quartzite	Czechoslovakia	Cer77	77							0.4···2.2	0.83
Muscovite-Quartzite	Swiss Alps	Ryb81	1		3.43		13.07		3.22		2.09
Serpentinite	Swiss Alps	Ryb81	6	0.0···0.13	0.03	0.0···0.23	0.067	0.0···0.02	0.005	0.0···0.0054	0.013
	Czechoslovakia	Cer77	21							0.0···0.25	0.09
	Swiss Alps	Ryb73	1		0.0128		0.038		0.0021		0.00682

(continued)

Table 3 (continued)

Rock	Locality	Ref.	n	c_U wt ppm range	mean	c_{Th} wt ppm range	mean	c_K wt % range	mean	A µW m^{-3} range	mean
B. Fels-types											
Hornblende-Pyrite-Fels	Ivrea Zone, Italy	Höh75	1		0.22		0.28		0.42		0.14
Hornblende-Pyrite-Fels	Ivrea Zone, Italy	Höh75	3							0.015···0.039	0.026
Garnet-Olivine-Fels	Swiss Alps	Ryb73	1		0.0136		0.041		0.0175		0.009
Quartzitic Hornfels	Snowy Mountains, Australia	Bun75	1		4.35		22.21		2.15		2.82
Hornfels	Czechoslovakia	Cer77	16							0.1···0.8	0.49
C. Polymict metamorphites											
Gneiss	Indian shield	Rao76	21							1.31···4.93	2.80
	Indian shield	Rao76	10							2.22···3.62	2.93
	Swiss Alps	Ryb73	55	0.9···24.9	4.95	1.2···25.7	13.1	0.32···4.71	3.11	0.69···8.20	2.44
	Western Australia	Sas76	84								2.13
Potassium-feldspar-gneiss	Czechoslovakia	Cer77	411							1.1···2.3	1.7
	Baltic shield, USSR	Ars72, Ars73	8	0.26···1.63	0.62	1.43···9.16	3.49	1.13···6.11	2.90	0.32···1.32	0.70
Plagioclase-gneiss	Kamchatka, USSR	Puz77	122		1.71		5.59		1.95		1.0
	Czechoslovakia	Cer77	79							0.96···2.09	1.45
	Baltic shield	Ars72, Ars73	12	0.12···0.75	0.41	0.67···4.23	2.36	1.43···2.51	1.94	0.27···0.71	0.47
Amphibolite gneiss	N.E. Minnesota, USA	Rye78	4	0.43···1.31	0.71	0.82···5.75	3.75	0.64···1.09	0.85	0.27···0.78	0.51
Hornblende-Pyrite-Garnet-Gneiss	Ivrea Zone, Italy	Höh75	3							0.016···0.046	0.026
Hornblende-Pyrite Epidote-Gneiss	Ivrea Zone, Italy	Höh75	1		0.58		1.7		0.44		0.36
Hornblende-Epidote-Gneiss	Bergell, Italy	Höh75	10	3.4···4.5	4.1	6.4···18.8	12.5	2.1···2.4	2.2	1.8···2.9	2.3

Rock	Locality	Ref.	n	c_U wt ppm range	mean	c_{Th} wt ppm range	mean	c_K wt % range	mean	A µW m^{-3} range	mean
Biotite-Gneiss	Ivrea Zone, Italy	Höh75	2		3.9		13.3		3.05		2.3
Granitic Gneiss	N.E. Minnesota	Rye78	4	0.56···0.78	0.72	0.84···3.16	2.23	0.85···1.25	1.07	0.33···0.50	0.43
	Norway	Swa74	9		4.33		25.49		4.05		3.31
	Indian shield	Rao76	29							0.67···2.40	1.50
	South Norway	Kil75	15		4.5		20.1		3.09		2.9
	Northern Territory, Australia	Hei66	2	3.2···7.4	5.3	16···30	23.0	3.5···4.4	4.0	2.2···4.3	3.3
Granitic Gneiss	Western Australia	Sas76									8.90
	Snowy Mountains, Australia	Bun75	21	2.35···7.90	4.32	8.15···35.35	18.89	1.13···5.07	2.54	1.12···3.50	2.66
(from drillholes)	South Australia	Bun75	11	2.67···5.06	3.33	19.33···34.39	27.41	2.36···3.93	2.96	2.31···3.82	2.99
Granodiorite Gneiss (Amphibolite facies)	Canadian shield	Fah67	15	0.6···2.3	1.1	4.8···13.0	9.4	1.6···2.9	2.3	0.8···1.6	1.3
Granodiorite Gneiss (Hornblende-Granulite-Facies)	Canadian shield	Fah67	7	0.4···1.3	0.8	4.1···10.7	7.1	1.4···2.9	2.1	0.6···1.4	0.95
Migmatized Gneiss	Czechoslovakia	Cer77	18							1.34···2.09	1.55
Mica-Schist Gneiss	Czechoslovakia	Cer77	49							1.09···2.34	1.65
Schist	Norway	Swa74	14		1.20		2.75		1.21		0.62
	Indian shield	Rao76	41							2.37···3.27	2.93
	Swiss Alps	Ryb73	18	0.4···3.7	2.14	1.6···17.2	9.73	0.39···4.44	2.23	0.25···2.28	1.41
Biotite-Schist	N.E. Minnesota	Rye78	2		1.69		4.96		1.99		0.95
Mica-Schist	Kamchatka, USSR	Puz77	57		2.2		7.3		2.38		1.4
	Czechoslovakia	Cer77	31							1.17···1.84	1.46
	Baltic shield, USSR	Ars73	3	0.07···0.73	0.34	0.09···0.88	0.41	0.12···0.43	0.22	0.04···0.34	0.16
	Wyoming, USA	Smi74	7		4.0		14.3		2.7		2.3
Two-Mica-Schist	Ivrea Zone, Italy	Höh75	3	2.9···6.0	3.97	6.7···9.8	8.50	4.3···4.8	4.5	1.61···2.69	2.04
Quartz-Mica-Schist	Norway	Swa74	7		2.58		8.50		1.19		1.33
Quartz-Sericite-Schist	New South Wales, Australia	Bun75	3	3.46···3.92	3.63	13.02···14.25	13.58	6.02···6.74	6.39	2.34···2.59	2.44
Hornblende-Schist	Indian shield	Rao76	15							0.063···0.48	0.25
Chloritic-Schist	Czechoslovakia	Cer77	16							0.33···0.92	0.39
Antigorite-Schist	Swiss Alps	Ryb73	2		0.0215		0.065		0.0065		0.011

(continued)

Table 3 (continued)

Rock	Locality	Ref.	n	c_U wt ppm range	mean	c_{Th} wt ppm range	mean	c_K wt % range	mean	A µW m⁻³ range	mean
Slate	Kamchatka, USSR	Puz77	16		2.0		6.4		1.83		1.1
Phyllite	Indian shield	Rao76	11							1.66···3.47	2.19
+Argillite	Indian shield	Rao76	20							1.64···3.08	2.61
Phyllite	Kamchatka, USSR	Puz77	42		1.96		5.85		1.80		1.18
	Czechoslovakia	Cer77	192							0.67···2.43	1.60
Chlorite-Phyllite	Scotland	Ric76	4	0.89···2.54	1.55	4.30···19.90	12.50	1.00···5.87	3.54	0.65···2.56	1.65
Garnet-Pelite	Scotland	Ric76	18	0.95···3.50	2.19	4.93···21.60	12.64	0.48···8.32	3.05	0.61···2.77	1.78
Migmatite	Czechoslovakia	Cer77	147							0.75···2.72	1.61
	Wyoming, USA	Smi74	30		2.7		18.5		4.0		2.4
Biotite-Migmatite	N.E. Minnesota, USA	Rye78	11	0.51···9.10	3.16	1.30···28.98	12.16	2.73···4.59	3.33	0.48···3.87	1.94
Amphibolite-Migmatite	N.E. Minnesota, USA	Rye78	1		1.45		7.78		3.68		1.24
	N.E. Minnesota, USA	Rye78	2		0.82		2.38		0.76		0.44
Schistic Migmatite	N.E. Minnesota, USA	Rye78	2		1.57		5.10		2.01		0.93
Amphibolite	Norway	Swa74	8		0.86		2.62		1.23		0.53
	Norway	Swa74	2		0.12		0.20		0.33		0.079
	Ivrea Zone, Italy	Höh75	4	0.23···0.85	0.53	0.03···3.3	1.57	0.78···1.54	1.03	0.18···0.61	0.40
	Wyoming, USA	Smi74	5		1.4		5.6		1.3		0.88
	Swiss Alps	Ryb73	8	0.00···7.8	1.65	0.01···13.7	3.00	0.11···2.22	1.23	0.01···3.5	0.82
	Swiss Alps	Ryb81	4	0.18···0.71	0.46	0.15···2.02	1.05	0.0···1.11	0.63	0.15···0.40	0.27
	Kamchatka, USSR	Puz77	38		0.7		1.2		0.6		0.4
	Czechoslovakia	Cer77	112							0.13···0.96	0.32
Granulite	Kamchatka, USSR	Puz77	9		1.0		4.4		1.76		0.9
	Czechoslovakia	Cer77	58							0.54···1.06	0.77
Pyrite-Granulite	Western Australia	Sas76	60								0.54
Kinzingite	Ivrea Zone, Italy	Höh75	8	2.9···5.2	4.14	13.1···16.1	14.44	2.2···3.6	3.13	2.04···2.78	2.52
Prasinite	Swiss Alps	Ryb81	3	0.11···0.27	0.22	0.26···0.55	0.40	0.11···0.30	0.18	0.08···0.13	0.11
Mylonite	Mont Blanc, France	Ryb66	1		3.46		13.2		2.47		2.01
Greenstone	Norway	Swa74	10		0.12		0.20		0.22		0.059
	Norway	Swa74	6		0.25		0.44		0.18		0.112
	Norway	Swa74	3		0.47		0.45		0.21		0.18

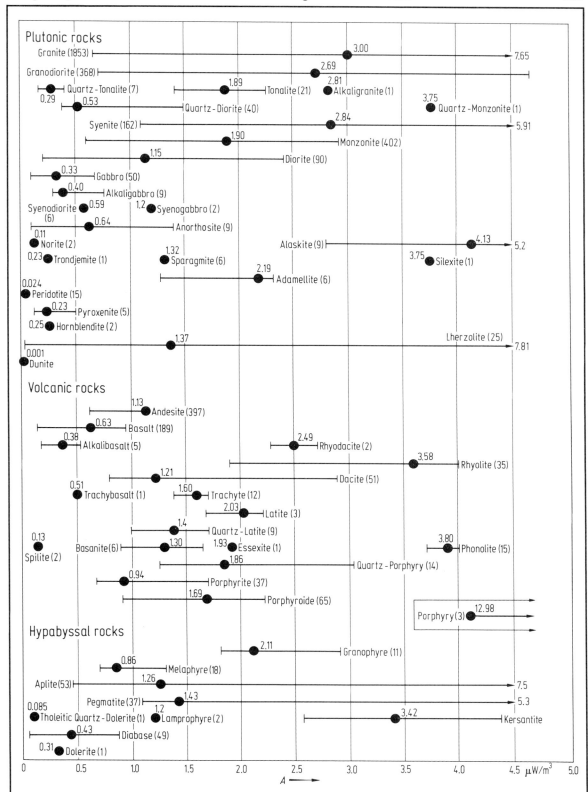

Fig. 1. Radioactive heat generation A of magmatic rocks. Figures in parentheses: number of measurements.

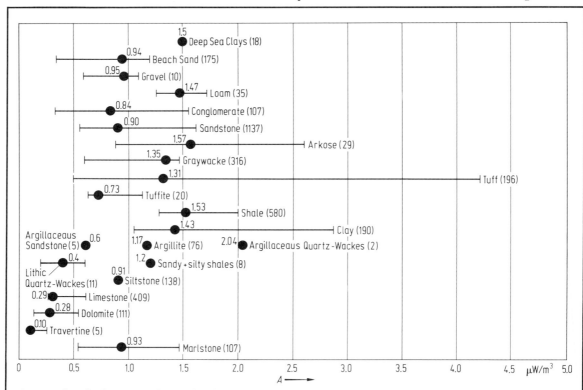

Fig. 2. Radioactive heat generation A of sedimentary rocks. Figures in parentheses: number of measurements.

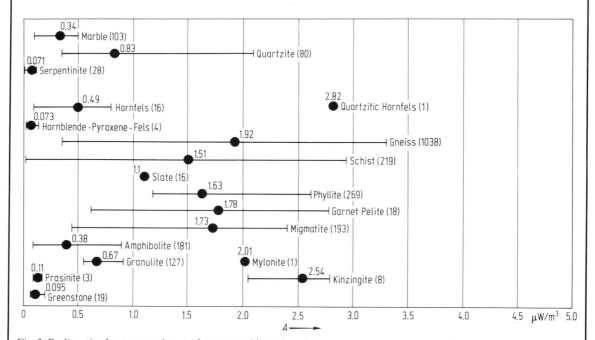

Fig. 3. Radioactive heat generation A of metamorphic rocks. Figures in parentheses: number of measurements.

4.4.3 References for 4.4.1 and 4.4.2 — Literatur zu 4.4.1 und 4.4.2

Ars72	Arshavskaya, N.I., Berzina, I.G., Lubimova, E.A.: Geothermics **1** (1972) 25–30.
Ars73	Arshavskaya, N.I., in: Vlodavec, V.I., Lubimova, E.A. (Eds.): Teploviye potoki iz kori i verckhney mantiyi Zemli. Verckhnaya mantiya No. 12. Izd. Nauka, Moskva, **1973**, p. 26–31.
Bun75	Bunker, C.M., Bush, C.A., Munroe, R.J., Sass, J.H.: Open-File Report 75-393. U.S. Dep. of the Int. Geol. Survey, **1975**.
Cer77	Čermák, V., Vaňková, V., Matolín, M., Bartošek, J.: Stud. Geophys. Geod. **21** (1977) 70–80.
Eva74	Evans, T.R., Tammemagi, H.Y.: Earth Planet. Sci. Lett. **23** (1974) 349–356.
Fah67	Fahrig, W.F., Eade, K.E., Adams, J.A.S.: Nature **214** (1967) 1002–1003.
Föh67	Föhn, P., Rybach, L.: Schweiz. Min. Petr. Mitt. **47** (1967) 581–598.
Gre68	Green, D.H., Morgan, J.W., Heier, K.S.: Earth Planet. Sci. Lett. **4** (1968) 155–166.
Hei63	Heier, K.S., Rogers, J.J.W.: Geochim. Cosmochim. Acta **27** (1963) 137–154.
Hei64	Heier, K.S., Mc Dougall, I., Adams, J.A.S.: Nature **201** (1964) 54.
Hei66	Heier, K.S., Rhodes, J.: Econ. Geology **61** (1966) 563–571.
Höh75	Höhndorf, A.: Schweiz. Min. Petr. Mitt. **55** (1975) 89–102.
Hur53	Hurley, P.M., Fairbairn, H.: Bull. Geol. Soc. Amer. **64** (1953) 659–673.
Kap74	Kappelmeyer, O., Haenel, R.: Geothermics with Special References to Application. Bornträger, Berlin/Stuttgart, **1974**.
Kil75	Killeen, P.G., Heier, K.S.: Contr. Miner. Petrol. **48** (1975) 171–177.
Kil75a	Killeen, P.G., Heier, K.S.: Geochim. Cosmochim. Acta **39** (1975) 1515–1524.
Kil75b	Killeen, P.G., Heier, K.S.: Oet Norske Videnskaps-Akademi, 1. Mat.-Naturv. Klasse, Skrifter Ny Serie, No. 35, Universitetsforlaget, Oslo, **1975**.
Kis78	Kissling, E., Labhart, T.P., Rybach, L.: Schweiz. Min. Petr. Mitt. **58** (1978) 357–388.
Kol66	Kolbe, P., Taylor, S.R.: Contr. Miner. Petrol. **12** (1966) 202–222.
Lew76	Lewis, T.J.: Can. J. Earth Sci. **13** (1976) 1634–1642.
Mur58	Murray, E.G., Adams, J.A.S.: Geochim. Cosmochim. Acta **13** (1958) 260–269.
Orm78	Ormaasen, D.E., Raade, G.: Earth Planet. Sci. Lett. **39** (1978) 145–150.
Pli62	Pliler, R., Adams, J.A.S.: Geochim. Cosmochim. Acta **26** (1962) 1115–1135.
Puz77	Puzankov, Y.M., Bobrov, V.A., Duchkov, A.D.: Radioaktivniye elementy i teplovoy potok Zemnoy kori poluostrova Kamchatki. Izd. Nauka, Sibir. otd., Novosibirsk, **1977**, 126pp.
Rao76	Rao, R.U.M., Rao, G.V., Narain, H.: Earth Planet. Sci. Lett. **30** (1976) 57–64.
Reu80	Reusser, E.: Radiometrische Untersuchungen am Bergeller Granodiorit, Diplomarbeit am Institut für Geophysik, ETH Zürich, **1980**.
Ric76	Richardson, S.W., Powell, R.: Scott. J. Geol. **12**(3) (1976) 237–268.
Rog69	Rogers, J.J.W., Condie, K.C., Mahan, S.: Chem. Geology **5** (1969/70) 207–213.
Ryb66	Rybach, L., Von Raumer, J., Adams, J.A.S.: Pure Applied Geophys. **63** (1966) 153.
Ryb73	Rybach, L.: Beiträge zur Geologie der Schweiz, Geotechn. Serie **51**, Mitt. 82, Institut Geophysik ETH Zürich, **1973**.
Ryb76	Rybach, L., in: R.G.J. Strens (Ed.): The Physics and Chemistry of Minerals and Rocks, Wiley & Sons, London, **1976**, p. 309–318.
Ryb81	Rybach, L., Buntebarth, G.: Earth Planet. Sci. Lett. (1981) (in press).
Rye78	Rye, R.M., Roy, R.F.: Am. J. Sci. **278** (1978) 354–378.
Sas76	Sass, J.H., Jaeger, J.C., Munroe, R.J.: Open-File Report 76-250. U.S. Dep. pf the Int. Geol. Survey, **1976**.
Smi71	Smithson, S.B., Heier, K.S.: Earth Plan. Sci. Lett. **12** (1971) 325–326.
Smi73	Smithson, S.B., Decker, E.R.: Earth Planet. Sci. Lett. **19** (1973) 131–134.
Smi74	Smithson, S.B., Decker, E.R.: Earth Planet. Sci. Lett. **22** (1974) 215–225.
Sto65	Stow, S.H., Adams, J.A.S.: Transaction Am. Geoph. Union (Abstr.), **46**/3 (1965) 548.
Swa72	Swanberg, C.A.: J. Geoph. Res. **77** (1972) 2508–2513.
Swa74	Swanberg, C.A., Chessman, M.D., Simmons, G., Gronlie, G., Heier, K.S.: Tectonophys. **23** (1974) 31–48.
Tam74	Tammemagi, H.Y., Wheildon, J.: Geophys. J.R. Astron. Soc. **38** (1974) 83–94.
Til69	Tilling, R.I., Gottfried, D.: U.S. Geol. Survey Prof. Paper 614 E, 29 p. (1969).
Wak67	Wakita, H., Nagasawa, H., Uyeda, S., Kuno, S.: Geochem. J. **1** (1967) 183.
Wol64	Wollenberg, H.A., Smith, A.R.: J. Geophys. Res. **69** (1964) 3471.
Wol67	Wollenberg, H.A., Smith, A.R., Bailey, E.H.: J. Geophys. Res. **72** (1967) 4139–4150.
Wol68	Wollenberg, H.A., Smith, A.R.: J. Geophys. Res. **73** (1968) 1481–1495.

Two-dimensional survey of contents

The numbers in the survey correspond to the number of the respective chapter, section or subsection where the specified information is to be found. Both subvolumes are considered (chapters 0, 1, 2, 4 in subvolume V/1a, chapters 3, 5···9 in subvolume V/1b).

Physical properties of rocks Physikalische Eigenschaften der Gesteine		Chemical elements, special materials Chemische Elemente, spezielle Materialien	Minerals Minerale
Nomenclature, chemical and mineralogical components of rocks Nomenklatur, chemische und mineralogische Komponenten der Gesteine		0.1, 0.2	0.2
Density, Dichte	normal conditions	1.2.6	1.1, 3.1
	extreme conditions	1.3.3.1, 1.3.3.2, 1.3.3.4	1.3.3.2
Mean atomic weight, Mittleres Atom-Gewicht			3.1.3.1.7
Porosity, Porosität	normal conditions		
	extreme conditions		
Permeability, internal surfaces, capillarity Permeabilität, interne Oberflächen, Kapillarität			
Wave velocity and constants of elasticity Wellen-Geschwindigkeiten und Elastizitäts-Konstanten			
	normal conditions	3.1.2.4	
	extreme conditions		3.1.4.4.1.1
Inelasticity, Inelastizität			
Thermal conductivity, specific heat Wärme-Leitfähigkeit, spezifische Wärme			4.1.2.1
Thermal expansion, Thermische Ausdehnung			
Parameter of melting processes Parameter der Schmelzprozesse			
Radiogenic heat, Radiogene Wärme			
Electrical conductivity Elektrische Leitfähigkeit	normal conditions	5.1.1, 5.3.1	5.1.1
	extreme conditions		5.4
Dielectric constant, Dielektrizitäts-Konstante		5.2	5.2
Magnetic properties, Magnetische Eigenschaften		6.1	6.1
Optical properties, Optische Eigenschaften			
Radioactivity, Radioaktivität		7.1	7.1
Absolute ages, Absolute Alter			

Zwei-dimensionale Inhaltsübersicht

Die Ziffern in der Übersicht entsprechen der Nummer des jeweiligen Kapitels, Abschnitts oder Unterabschnitts, in dem die spezielle Information zu finden ist. Beide Teilbände sind berücksichtigt (Kapitel 0, 1, 2, 4 in Teilband V/1a, Kapitel 3, 5···9 in Teilband V/1b).

Igneous rocks Magmatische Gesteine	Metamorphic rocks Metamorphe Gesteine	Sediments Sedimente	Ice Eis	Moon Mond
0.2, 1.3.3.3.1, 3.1.4.3	0.3, 1.3.3.3.2, 3.1.4.3	0.4, 1.3.3.3.3, 3.1.4.3		1.3.3.3
1.2.1, 1.2.2, 1.2.3, 3.1	1.2.4, 3.1	1.2.5, 2.1, 3.1	8.1	9.1
1.3.3.3.1	1.3.3.3.2	1.3.3.3.3		
3.1.4.3	3.1.4.3			
	2.1	2.1, 3.1.4.3		9.1
		2.1.4		
		2.2, 2.3		
3.1.2, 3.1.3	3.1.2, 3.1.3	3.1.2, 3.1.3	8.4	9.2, 9.3
3.1.4	3.1.4	3.1.4		9.2, 9.3
3.2	3.2	3.2	8.4	
4.1.2.2	4.1.2.2	4.1.2.2, 4.2	8.1	9.4
3.1.4.3	3.1.4.3		8.1	9.4
0.2.5, 4.3, 3.1.4.4.3			8.1	
4.4, 7.1	4.4, 7.1	4.4, 7.1		
5.1.2	5.1.2	5.1.2, 5.3.2	8.2	9.5
5.4				
5.2	5.2	5.2	8.2	9.5
6.2	6.2	6.2		9.6
			8.3	
7.1	7.1	7.1		
7.2	7.2	7.2		